D1101141
WITHDRAWN

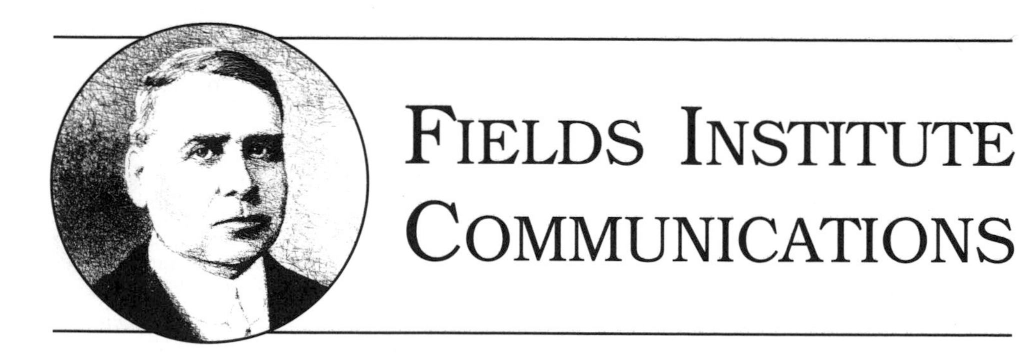

THE FIELDS INSTITUTE FOR RESEARCH IN MATHEMATICAL SCIENCES

Nonlinear Dynamics and Evolution Equations

Hermann Brunner
Xiao-Qiang Zhao
Xingfu Zou
Editors

American Mathematical Society
Providence, Rhode Island

The Fields Institute for Research in Mathematical Sciences

The Fields Institute is a center for mathematical research, located in Toronto, Canada. Our mission is to provide a supportive and stimulating environment for mathematics research, innovation and education. The Institute is supported by the Ontario Ministry of Training, Colleges and Universities, the Natural Sciences and Engineering Research Council of Canada, and seven Ontario universities (Carleton, McMaster, Ottawa, Toronto, Waterloo, Western Ontario, and York). In addition there are several affiliated universities and corporate sponsors in both Canada and the United States.

2000 *Mathematics Subject Classification.* Primary 34C20, 34K05, 35B40, 35J55, 37C05, 37C65, 37K55, 53D17, 92D25, 92D30.

Library of Congress Cataloging-in-Publication Data

Nonlinear dynamics and evolution equations / Hermann Brunner, Xiao-Qiang Zhao, Xingfu Zou, editors.

p. cm. — (Fields Institute communications, ISSN 1069-5265 ; 48)

Includes bibliographical references (alk. paper)

ISBN 0-8218-3721-4

1. Evolution equations, Nonlinear–Congresses. 2. Dynamics–Research–Congresses. 3. Differential equations–Congresses. 4. Differential dynamical systems–Congresses. 5. Differential equations, Partial–Congresses. 6. Mathematical models–Congresses. I. Brunner, H. (Hermann), 1941– II. Zhao, Xiao-Qiang. III. Zou, Xingfu, 1958– IV. Fields Institute communications ; 48.

QA377.I56 2004 2006042831

515′.353–dc22 CIP

∞ The paper used in this book is acid-free and falls within the guidelines
established to ensure permanence and durability.
This publication was prepared by the Fields Institute.
`http://www.fields.utoronto.ca`
Visit the AMS home page at `http://www.ams.org/`

10 9 8 7 6 5 4 3 2 1 11 10 09 08 07 06

Contents

Preface

The papers in this volume of the *Fields Institute Communications* reflect a broad spectrum of current research activities on the theory and applications of nonlinear dynamics and evolution equations. They are based on lectures given during the International Conference on Nonlinear Dynamics and Evolution Equations at Memorial University of Newfoundland, St. John's, NL, Canada, July 6–10, 2004. The aim of that conference was to bring together leading experts, researchers, and graduate students for five days of high-level lectures and informal research interactions; this was made possible by generous financial support from the Centre de recherches mathématiques (CRM), Montréal, the Fields Institute for Research in Mathematical Sciences, Toronto, the Pacific Institute for the Mathematical Sciences (PIMS), Vancouver, and the Atlantic Association for Research in the Mathematical Sciences (AARMS).

This volume contains thirteen invited (and refereed) papers. Nine of these are *survey papers*, introducing the reader to, and describing the current state of the art in, major areas of dynamical systems, ordinary, functional and partial differential equations, and applications of such equations in the mathematical modelling of various biological and physical phenomena. These papers are complemented by four *research papers* that examine particular problems in the theory of dynamical systems (asymptotic properties of systems that are comparable to quasi-monotone systems; smoothness of center manifolds for state-dependent delay differential equations; dynamics of the asymmetric generalized May-Leonard model of three species competition; exact Poisson structures on manifolds).

Four of the survey papers deal with various aspects of the *theory* of partial differential equations and dynamical systems. Motivated by the facts that many of the celebrated (nonlinear) PDEs of mathematical physics can be viewed as Hamiltonian systems and that solutions of their linearized versions exhibit periodic or quasi-periodic solutions, Walter Craig (McMaster University, Canada) presents an overview of some of the techniques and results of KAM-like methods for analyzing analogous phenomena in solutions to nonlinear PDEs. Norman Dancer (University of Sydney, Australia) surveys a number of open questions for PDEs with small diffusion, $-\varepsilon^2 \Delta u = f(u)$, in various geometries: of interest are results on the number of solutions and their asymptotic shapes. The contribution by Christiane Rousseau (Université de Montréal, Canada), on normal forms for germs of analytic families of planar vector fields unfolding a generic saddle-node or resonant saddle, is concerned with the case in which the normal forms are not polynomial but analytic and for which the formal change of coordinates to normal form generically diverges.

Finally, Radu Saghin (University of Toronto, Canada) and Zhihong Xia (Northwestern University, USA) study generic properties of two particular classes of dynamical systems, namely symplectic diffeomorphisms and Hamiltonian systems.

Applications in biological sciences and in materials science are surveyed in five contributions. Two of these papers discuss biological invasions and disease spread: Julien Arino (University of Manitoba, Canada) and Pauline van den Driessche (University of Victoria, Canada) look at extensions of continuous time and discrete space metapopulation models addressing cross infection between several species and keeping track of the patches in which the species reside, while Stephen Gourley (University of Surrey, UK) and Jianhong Wu (York University, Canada) study nonlocal diffusive equations that arise in the modelling and analysis of long-term behaviors of biological and epidemiological systems where individuals move randomly and the feedback nonlinearity involves time lags.

Predator-prey and competition models are the subjects of the survey by Yihong Du (University of New England, Australia) and Junping Shi (College of William and Mary, USA) and the research paper by Gail Wolkowicz (McMaster University, Canada), respectively. The former presents recent results on diffusive predator-prey models in spatially heterogeneous environments and then examines the influence of a protection zone in a particular diffusive model. The paper by Wolkowicz studies the global dynamics of a Lotka-Volterra system of three species competition by using a model of a food web in a chemostat involving three species competing for a single (non-reproducing) nutrient, with one of the competitors also predating on one of the other two competitors.

Brian Sleeman (University of Leeds, UK) presents an illuminating description of how mathematical models may be formulated on the basis of the complex biochemical processes involved in the modelling of tumour angiogenesis (the formation of new blood vessels), and he describes various properties of solutions to such models (such as local existence and uniqueness, and the development of spikes).

A different field of applications (materials science) is reviewed in the survey by Peter Bates (Michigan State University, USA). In his discussion of nonlinear evolution equations arising in materials science, he describes various properties such as well-posedness, asymptotics, travelling waves or pulses. These systems represent lattice or nonlocal versions of the Allen-Cahn, Cahn-Hilliard, phase-field, or Klein-Gordon equations.

The other invited research papers are by Jifa Jiang (Tongji University, China) who analyzes the asymptotic behavior for systems that are comparable to quasi-monotone systems; by Tibor Krisztin (University of Szeged, Hungary), on C^1-smoothness of center manifolds for differential equations with state-dependent delay; and by Yingfei Yi (Georgia Institute of Technology, USA) and Xiang Zhang (Shanghai Jiaotong University, China) who present, among other things, a characterization of exact Poisson structures which are invariant under the flow of a class of completely integrable systems.

It has been the aim of the editors to create a proceedings volume that may serve as an important resource, both for new researchers and experts in this promising area, for many years to come. The editors wish to thank the Fields Institute for Research in Mathematical Science and its Editorial Board (chaired by Carl Riehm) for agreeing to include this set of papers in the Fields Institute Communications series.

Hermann Brunner (Memorial University of Newfoundland)
Xiao-Qiang Zhao (Memorial University of Newfoundland)
Xingfu Zou (University of Western Ontario)

Fields Institute Communications
Volume **48**, 2006

Disease Spread in Metapopulations

Julien Arino
Department of Mathematics
University of Manitoba
Winnipeg, MB, Canada R3T 2N2
arinoj@cc.umanitoba.ca

P. van den Driessche
Department of Mathematics and Statistics
University of Victoria
Victoria, BC, Canada V8W 3P4
pvdd@math.uvic.ca

Abstract. Some continuous time, discrete space, metapopulation models that have been formulated for disease spread are presented. Motivation for such a formulation with travel between discrete patches is presented. A system of $4p$ ordinary differential equations describes disease spread in an environment divided into p patches. The basic reproduction number $\mathcal{R}_0$ is calculated, with the disease dying out in each patch if $\mathcal{R}_0 < 1$. If travel is assumed to be independent of disease status, then numerical results are cited that indicate that for $\mathcal{R}_0 > 1$ solutions tend to an endemic equilibrium with the disease present in each patch. The system is extended to include cross infection between several species. A second extension involves keeping track of both the current patch and the patch in which an individual usually resides. Travel can change disease spread in a complicated way; it may help the disease to persist or may aid disease extinction. Complexity that can be built into metapopulation models is illustrated by three case-study examples from the literature.

1 Motivation for spatial epidemic models

Classical deterministic epidemic models implicitly assume that space is homogeneous, and so do not include spatial variation. There are, however, many reasons why epidemic models should include spatial variation. Firstly, initial conditions of disease are often heterogeneous, with disease spreading geographically with time. For example, plague (black death) spread east to west and south to north along the trade routes of Europe between 1347 and 1350, and fox rabies spread west from the

---

2000 *Mathematics Subject Classification.* Primary 92D30; Secondary 34D23.
Research partially supported by NSERC and MITACS.

Russian-Polish border in 1940 to reach France by 1968. More recently, West Nile virus arrived in New York in 1999 and spread to the west coast of North America by 2004. Secondly, the environment is heterogeneous both in a geographical sense and in a human sense with birth rates, death rates and health care facilities varying with location. Thirdly, different species have different travel rates, a factor that is especially important for diseases involving many species (for example, the foot-and-mouth disease outbreak in the UK in 2001) and for vector transmitted diseases. For bubonic plague, the vectors, which are fleas, travel quickly over short distances; whereas the reservoir mammals, which are rodents, travel more slowly but over longer distances. Mosquitoes and birds, the vectors and reservoirs for West Nile virus, respectively, have different flight patterns that also depend on season, topography, and geographic conditions. In the case of rabies, foxes have different travel patterns when infective. Fourthly, for human diseases, social groupings and mixing patterns vary with geography and age. This can be illustrated by comparing humans in a hospital setting with those in isolated communities in Canada's North and with children in schools. Currently most humans live in cities and travel along defined routes. These influence the spatial spread of disease, for example, HIV spread along highways in the USA during the latter part of the twentieth century, and SARS in 2003 was spread by air travellers.

Continuous spatial models with continuous time yield partial differential equations of reaction-diffusion type. For example, such models have been formulated and analyzed for rabies by Murray and coauthors, see [21], and for West Nile virus in [18]. Discrete spatial models with continuous time yield systems of ordinary differential equations, which are *metapopulation* models involving movement of individuals between discrete spatial patches. This movement is captured by a digraph (or a multi-digraph) with the patches as vertices. Such compartmental models have been discussed for influenza spread due to air travel between cities by Hyman and LaForce [15] using a model with structure similar to that developed by Rvachev and Longini [23]. Such models have also been formulated for measles and influenza by Sattenspiel and coauthors [25, 26], further analyzed by Arino and coauthors [2, 3, 4, 5], and Wang and coauthors [16, 22, 30, 31, 32]. Here we review some of these models and survey other metapopulation models in the literature. We focus in particular on the basic reproduction number, $\mathcal{R}_0$, which is the average number of secondary cases produced by a single infected introduced into a totally susceptible population. This parameter $\mathcal{R}_0$ is a key concept in the study of infectious diseases and can aid in guiding measures to control disease. If $\mathcal{R}_0 < 1$, then the disease should die out if introduced at a low level, whereas if $\mathcal{R}_0 > 1$, then the disease is able to invade the population. To calculate $\mathcal{R}_0$, the next generation matrix method is used, details are given in [10, 29].

We remark that other types of spatial models have also been formulated in the literature; see, for example, [19, Part 2]. Included there are papers by Cliff [8] on geographic mapping methods to trace spatial disease spread, Metz and van den Bosch [20] on velocities of epidemic spread and Durrett [11] on disease spread on a lattice. Epidemics among a population partitioned into households are considered by Ball and Lyne [6], disease dynamics in discrete-time patchy environments are formulated by Castillo-Chavez and Yakubu [9], the rate of spread of endemic infections using integrodifference equations is investigated by Allen and Ernest [1], and urban social networks using a bipartite graph are explored by Eubank et al [12].

Due in part to increasing capacities of computers and to advances in mathematical analysis, there has been a recent surge of interest in metapopulation models. We hope that this review, although personal and not exhaustive, will encourage readers to delve further into the literature and to formulate new metapopulation models for disease spread.

2 Metapopulation model on p patches

We begin with the formulation of a general metapopulation SEIRS epidemic model. The structure of our model is based on that of Arino *et al.* [4], in which a multi-species epidemic model is constructed with the assumption that travel rates are independent of disease status. However, our model here is for disease transmission in one species, but allows for travel rates to depend on disease status. To formulate the deterministic model, assume that the environment under consideration is divided into p patches, which may be cities, geographic regions or communities. Within each patch conditions are assumed to be homogeneous. The population in patch i, is divided into compartments of susceptible, exposed (latent), infective and recovered individuals with the number in each compartment denoted by $S_i(t)$, $E_i(t)$, $I_i(t)$ and $R_i(t)$, respectively, for $i = 1, \ldots, p$. The total number of individuals in patch i is $N_i(t) = S_i(t) + E_i(t) + I_i(t) + R_i(t)$. The rates of travel of individuals between patches are assumed to depend on disease status, and individuals do not change disease status during travel. Let m_{ij}^S, m_{ij}^E, m_{ij}^I, m_{ij}^R denote the rate of travel from patch j to patch i of susceptible, exposed, infective, recovered individuals, respectively, where $m_{ii}^S = m_{ii}^E = m_{ii}^I = m_{ii}^R = 0$. This structure defines a multi-digraph with patches as vertices and arcs given by the travel rates, which can be represented by the nonnegative matrices $M^S = [m_{ij}^S]$, $M^E = [m_{ij}^E]$, $M^I = [m_{ij}^I]$ and $M^R = [m_{ij}^R]$. It is assumed that these matrices are irreducible.

Birth (or input) in patch i is assumed to be into the susceptible class at a rate $A_i(N_i) > 0$ individuals per unit time, and natural death is assumed to be independent of disease status with rate constant $d_i > 0$. The disease is assumed to be transmitted by horizontal incidence $\beta_i(N_i) S_i I_i$, thus an average individual makes $\beta_i(N_i) N_i$ contacts per unit time. It is reasonable to take $\beta_i(N_i)$ as a nonnegative nonincreasing function of N_i. Once infected, a susceptible individual harbors an agent of disease and moves to the exposed compartment, then into the infective compartment as the individual becomes able to transmit the disease. On recovering from the disease, an individual moves to the recovered compartment, and then back to the susceptible compartment as disease immunity fades. The period in the exposed, infective and recovered compartment is taken to be exponentially distributed with rate constant α_i, γ_i, δ_i, respectively. Thus $1/\alpha_i$, $1/\gamma_i$, $1/\delta_i$ is the average period (without accounting for death) of latency, infection, immunity, respectively. For a disease that causes mortality, the death rate constant for infectives is denoted by ε_i. The epidemic parameters are assumed to be nonnegative, with limiting cases giving simpler models. For example, if a disease confers permanent immunity, then $\delta_i = 0$ and an SEIR model results. If a disease has a very short latent period that can be ignored, then $\alpha_i \to \infty$ (an SIRS model); and if in addition the period of immunity is so short that it can be ignored, then $\delta_i \to \infty$ and an SIS model results. Such a model is appropriate for gonorrhea.

The above assumptions lead to a system of $4p$ ordinary differential equations (ODEs) describing the disease dynamics. For $i = 1, \ldots, p$ these equations are

$$\frac{dS_i}{dt} = A_i(N_i) - \beta_i(N_i) S_i I_i - d_i S_i + \delta_i R_i + \sum_{j=1}^{p} m_{ij}^S S_j - \sum_{j=1}^{p} m_{ji}^S S_i \quad (2.1)$$

$$\frac{dE_i}{dt} = \beta_i(N_i) S_i I_i - (\alpha_i + d_i) E_i + \sum_{j=1}^{p} m_{ij}^E E_j - \sum_{j=1}^{p} m_{ji}^E E_i \quad (2.2)$$

$$\frac{dI_i}{dt} = \alpha_i E_i - (\varepsilon_i + \gamma_i + d_i) I_i + \sum_{j=1}^{p} m_{ij}^I I_j - \sum_{j=1}^{p} m_{ji}^I I_i \quad (2.3)$$

$$\frac{dR_i}{dt} = \gamma_i I_i - (d_i + \delta_i) R_i + \sum_{j=1}^{p} m_{ij}^R R_j - \sum_{j=1}^{p} m_{ji}^R R_i \quad (2.4)$$

with initial conditions $S_i(0) > 0, E_i(0), I_i(0), R_i(0) \geq 0, \sum_{i=1}^{p} E_i(0) + I_i(0) > 0$.
The population of patch i, namely N_i, evolves according to the sum of equations (2.1)-(2.4). Solutions of (2.1)-(2.4) remain nonnegative with N_i positive for all $t \geq 0$. The total population in all patches $N = N_1 + N_2 + \ldots + N_p$ satisfies

$$\frac{dN}{dt} = \sum_{i=1}^{p} (A_i(N_i) - \varepsilon_i I_i - d_i N_i) \quad (2.5)$$

The metapopulation model is at equilibrium if the time derivatives in (2.1)-(2.4) are zero. Patch i is at a disease free equilibrium (DFE) if $E_i = I_i = 0$, and the p-patch model is at a DFE if $E_i = I_i = 0$ for all $i = 1, \ldots, p$. Thus at a DFE, for all $i = 1, \ldots, p$, $S_i = N_i$ and satisfies

$$A_i(N_i) - d_i N_i + \sum_{j=1}^{p} m_{ij}^S N_j - \sum_{j=1}^{p} m_{ji}^S N_i = 0 \quad (2.6)$$

Assume that (2.6) has a solution that gives the DFE $S_i^* = N_i^*$, which is unique. This is certainly true if $A_i(N_i) = d_i N_i$ (i.e., birth rate equal to the death rate) and $\varepsilon_i = 0$ (i.e., no disease related death) giving a constant total population from (2.5). Arino *et al* [4] make these assumptions for a multi-species epidemic model. It is also true if $A_i(N_i) = A_i$ as assumed in [24].

Linear stability of the disease free equilibrium can be investigated by using the next generation matrix [10, 29]. Using the notation of [29], and ordering the infected variables as $E_1, \ldots, E_p, I_1, \ldots, I_p$ the matrix of new infections F and the matrix of transfer between compartments V are given in partitioned form by

$$F = \begin{bmatrix} 0 & F_{12} \\ 0 & 0 \end{bmatrix} \text{ and } V = \begin{bmatrix} V_{11} & 0 \\ -V_{21} & V_{22} \end{bmatrix} \quad (2.7)$$

Here $F_{12} = \operatorname{diag}(\beta_i(N_i^*) N_i^*)$, $V_{11} = -M^E + \operatorname{diag}\left(\alpha_i + d_i + \sum_{j=1}^{p} m_{ji}^E\right)$,
$V_{21} = \operatorname{diag}(\alpha_i)$, $V_{22} = -M^I + \operatorname{diag}\left(\varepsilon_i + \gamma_i + d_i + \sum_{j=1}^{p} m_{ji}^I\right)$.

Matrices V_{11} and V_{22} are $p \times p$ irreducible M-matrices [7] and thus have positive inverses. The next generation matrix

$$FV^{-1} = \begin{bmatrix} F_{12}V_{22}^{-1}V_{21}V_{11}^{-1} & F_{12}V_{22}^{-1} \\ 0 & 0 \end{bmatrix}$$

has spectral radius, denoted by ρ, given by $\rho\left(FV^{-1}\right) = \rho\left(F_{12}V_{22}^{-1}V_{21}V_{11}^{-1}\right)$. As shown in [29], the Jacobian matrix of the infected compartments at the DFE, which is given by $F - V$, has all eigenvalues with negative real parts if and only if $\rho\left(FV^{-1}\right) < 1$. The number $\rho\left(FV^{-1}\right)$ is the basic reproduction number $\mathcal{R}_0$ for the disease transmission model, thus

$$\mathcal{R}_0 = \rho\left(F_{12}V_{22}^{-1}V_{21}V_{11}^{-1}\right), \tag{2.8}$$

and the DFE is linearly stable if $\mathcal{R}_0 < 1$, but unstable if $\mathcal{R}_0 > 1$. If $A_i\left(N_i\right) = A_i$ and $\beta_i\left(N_i\right) = \beta_i/N_i$ (standard incidence), then a comparison theorem argument can be used to show that if $\mathcal{R}_0 < 1$, then the DFE is globally asymptotically stable [24]. This extends the results for 2 patches given by [30, Theorem 2.1] for a constant population. Wang and Zhao [31] assume mass action incidence and show that population travel in an SIS model can either intensify or reduce the spread of disease in a metapopulation. Moreover, for this SIS model, the disease persists for $\mathcal{R}_0 > 1$, and if susceptible and infective individuals have the same travel rates, then there exists a unique, globally attracting endemic equilibrium [16, Theorem 3.1].

3 Travel rates independent of disease status

For mild diseases it may be reasonable to simplify the model of the previous section by assuming that individuals do not die from disease ($\varepsilon_i = 0$) and travel rates are independent of disease status, thus $M^S = M^E = M^I = M^R = M = [m_{ij}]$ (irreducible). Travel rates are thus specified on a digraph. These assumptions are made in the multi-species model formulated by Arino *et al.* [4], in which it is also assumed that $A_i(N_i) = d_i N_i$ and $\beta_i(N_i) = \beta_i/N_i$ (i.e., standard incidence). With these assumptions the one species model given by equations (2.1)-(2.4) becomes for $i = 1, \ldots, p$

$$\frac{dS_i}{dt} = d_i\left(N_i - S_i\right) - \beta_i \frac{S_i I_i}{N_i} + \delta_i R_i + \sum_{j=1}^{p} m_{ij} S_j - \sum_{j=1}^{p} m_{ji} S_i \tag{3.1}$$

$$\frac{dE_i}{dt} = \beta_i \frac{S_i I_i}{N_i} - \left(\alpha_i + d_i\right) E_i + \sum_{j=1}^{p} m_{ij} E_j - \sum_{j=1}^{p} m_{ji} E_i \tag{3.2}$$

$$\frac{dI_i}{dt} = \alpha_i E_i - \left(\gamma_i + d_i\right) I_i + \sum_{j=1}^{p} m_{ij} I_j - \sum_{j=1}^{p} m_{ji} I_i \tag{3.3}$$

$$\frac{dR_i}{dt} = \gamma_i I_i - \left(d_i + \delta_i\right) R_i + \sum_{j=1}^{p} m_{ij} R_j - \sum_{j=1}^{p} m_{ji} R_i \tag{3.4}$$

Summing (3.1)-(3.4) gives

$$\frac{dN_i}{dt} = \sum_{i=1}^{p} m_{ij} N_j - \sum_{j=1}^{p} m_{ji} N_i \tag{3.5}$$

Thus the N_i equation uncouples from the epidemic variables. This linear system of equations has coefficient matrix $M - \text{diag}\left(\sum_{j=1}^{p} m_{ji}\right)$ which is the negative of a singular M-matrix (since each column sum is zero). From (3.5), see also (2.5), the total population N is constant. Subject to this constraint, it can be shown that (3.5) has a unique positive equilibrium $N_i = N_i^*$ that is asymptotically stable [4, Theorem 3.3]. Thus the disease free equilibrium of (3.1)-(3.4) is given by $(S_i, E_i, I_i, R_i) = (N_i^*, 0, 0, 0)$ and is unique.

The basic reproduction number $\mathcal{R}_0$ is calculated as in Section 2 with $F_{12} = \text{diag}\,(\beta_i)$ and $M^E = M^I = M$. The linear stability result for $\mathcal{R}_0 < 1$ can be strengthened to a global result as follows. Since $S_i \leq N_i$, equation (3.2) gives the inequality

$$\frac{dE_i}{dt} \leq \beta_i I_i - (\alpha_i + d_i) E_i + \sum_{j=1}^{p} m_{ij} E_j - \sum_{j=1}^{p} m_{ji} E_i \tag{3.6}$$

For comparison, define a linear system given by (3.6) with equality, namely

$$\frac{dE_i}{dt} = \beta_i I_i - (\alpha_i + d_i) E_i + \sum_{j=1}^{p} m_{ij} E_j - \sum_{j=1}^{p} m_j E_i$$

and by equation (3.3). This system has coefficient matrix $F - V$, and so by the argument in Section 2, satisfies $\lim_{t\to\infty} E_i = 0$ and $\lim_{t\to\infty} I_i = 0$ for $\mathcal{R}_0 = \rho(FV^{-1}) < 1$. Using a comparison theorem [17, Theorem 1.5.4], [28, Theorem B.1] and noting (3.6), it follows that these limits also hold for the nonlinear system (3.2) and (3.3). That $\lim_{t\to\infty} R_i = 0$ and $\lim_{t\to\infty} S_i = N_i^*$ follow from (3.4) and (3.1). Thus for $\mathcal{R}_0 < 1$, the disease free equilibrium is globally asymptotically stable and the disease dies out.

The existence and stability of endemic equilibria if $\mathcal{R}_0 > 1$ are open analytical questions. As in many high dimensional epidemic models, these are hard problems. It is sometimes possible to prove that the disease is globally uniformly persistent by appealing to the techniques of persistence theory; see [33].

Arino *et al* [4, Section 4] state that numerical simulations of (3.1)-(3.4) indicate that solutions of their metapopulation model specialized to one species on p patches with $\mathcal{R}_0 > 1$ tend to a unique endemic equilibrium with disease present in each patch. They display [4, Figure 1] solutions in the case of $p = 2$ patches with parameter values compatible with influenza that give $\mathcal{R}_0^{(1)} = 1.015$ and $\mathcal{R}_0^{(2)} = 0.952$. With no travel between patches, disease is endemic in patch 1, but dies out in patch 2. With small travel rates $m_{12} = m_{21} = 0.001$, equation (2.8) gives $\mathcal{R}_0 \approx 1.0095 > 1$, and the system approaches an epidemic equilibrium in both patches. However if travel rates are increased to $m_{12} = m_{21} = 0.05$, then $\mathcal{R}_0 \approx 0.985 < 1$ and the system approaches the DFE in both patches. Thus small travel rates help the disease to persist, whereas slightly higher travel rates stabilize the DFE. For a single species, the effect of quarantine where the patches are arranged in a ring is numerically investigated in [5].

4 Multi-species model

Spatial spread is all the more important for diseases that involve several species, for example, bubonic plague and West Nile virus. In [4] an SEIR epidemic model for a population consisting of s species and occupying p spatial patches is considered. This is extended in [5] to allow for temporary immunity, giving an SEIRS model. Here, we also allow rates of travel between patches to depend on disease status. With assumptions as in Section 2 and using standard incidence, the dynamics for species $j = 1, \ldots, s$ in patch $i = 1, \ldots, p$ is given by the following system of $4sp$ equations

$$
\begin{aligned}
\frac{dS_{ji}}{dt} &= A_{ji}(N_{ji}) - \sum_{k=1}^{s} \beta_{jki} S_{ji} \frac{I_{ki}}{N_{ki}} - d_{ji} S_{ji} + \delta_{ji} R_{ji} + \\
&\quad \sum_{q=1}^{p} m_{jiq}^{S} S_{jq} - \sum_{q=1}^{p} m_{jqi}^{S} S_{ji} && (4.1) \\
\frac{dE_{ji}}{dt} &= \sum_{k=1}^{s} \beta_{jki} S_{ji} \frac{I_{ki}}{N_{ki}} - (\alpha_{ji} + d_{ji}) E_{ji} + \sum_{q=1}^{p} m_{jiq}^{E} E_{jq} - \sum_{q=1}^{p} m_{jqi}^{E} E_{ji} && (4.2) \\
\frac{dI_{ji}}{dt} &= \alpha_{ji} E_{ji} - (\varepsilon_{ji} + \gamma_{ji} + d_{ji}) I_{ji} + \sum_{q=1}^{p} m_{jiq}^{I} I_{jq} - \sum_{q=1}^{p} m_{jqi}^{I} I_{ji} && (4.3) \\
\frac{dR_{ji}}{dt} &= \gamma_{ji} I_{ji} - (d_{ji} + \delta_{ji}) R_{ji} + \sum_{q=1}^{p} m_{jiq}^{R} R_{jq} - \sum_{q=1}^{p} m_{jqi}^{R} R_{ji} && (4.4)
\end{aligned}
$$

where the total population of species j in patch i is denoted by $N_{ji} = S_{ji} + E_{ji} + I_{ji} + R_{ji}$. The parameters are defined similarly to those in Section 2, but now the first subscript denotes the species, for example, $1/\gamma_{ji}$ is the average period of infection for species j in patch i and β_{jki} is the rate of disease transfer from species k to species j in patch i. Each species has its own travel matrices, for example $M_j^I = [m_{jiq}^I]$ where m_{jiq}^I denotes the rate of travel of an infective individual of species j from patch q to patch i. With nonnegative initial conditions having $N_{ji}(0) > 0$ the solutions remain nonnegative with $N_{ji}(t) > 0$ for all $t \geq 0$.

For a simplified version of (4.1)-(4.4) in which the birth (input) term $A_{ji}(N_{ji}) = d_{ji} N_{ji}$, recovered individuals have permanent immunity ($\delta_{ji} = 0$) and travel is independent of disease status, there is a unique DFE; see [4, Theorem 3.3]. Assume this is true for (4.1)-(4.4) with $S_{ji}^* = N_{ji}^*$ at the DFE. Then the basic reproduction number, $\mathcal{R}_0$, can be calculated by the method used in Section 2. This is illustrated for the case of two species on three patches. The infected variables are ordered as $E_{11}, E_{21}, E_{12}, E_{22}, E_{13}, E_{23}, I_{11}, I_{21}, I_{12}, I_{22}, I_{13}, I_{23}$. The nonnegative matrix F has the form given by (2.7) with $F_{12} = G_1 \oplus G_2 \oplus G_3$ where for $r = 1, 2, 3$,

$$
G_r = \begin{bmatrix} \beta_{11r} & \beta_{12r} \frac{N_{1r}^*}{N_{2r}^*} \\ \beta_{21r} \frac{N_{2r}^*}{N_{1r}^*} & \beta_{22r} \end{bmatrix}
$$

Matrices V_{11}, V_{21} and V_{22} in V given by (2.7) are now block matrices with each block being a 2×2 diagonal matrix. Writing

$$V_{11} = \begin{bmatrix} A_{11} & A_{12} & A_{13} \\ A_{21} & A_{22} & A_{23} \\ A_{31} & A_{32} & A_{33} \end{bmatrix}$$

the (i,i) entry of A_{kk} is $\alpha_{ik} + d_{ik} + \sum_{q=1}^{3} m^E_{iqk}$, and the (i,i) entry of A_{jk} for $j \neq k$ is $-m^E_{ijk}$. Similarly writing V_{22} as the block matrix B_{jk} , the (i,i) entry of B_{kk} is $\varepsilon_{ik} + \gamma_{ik} + d_{ik} + \sum_{q=1}^{3} m^I_{iqk}$ and for $j \neq k$, the (i,i) entry of B_{jk} is $-m^I_{ijk}$. The matrix $V_{21} = C_1 \oplus C_2 \oplus C_3$ with C_r having (i,i) entry equal to α_{ir}.

In terms of the above matrices, the basic reproduction number $\mathcal{R}_0$ is given by (2.8). The block structure enables $\mathcal{R}_0$ to be easily calculated for a given set of disease parameters. Note that $\mathcal{R}_0$ depends explicitly on the travel rates of exposed and infective individuals, and implicitly (through N^*_{jr}) on the travel rates of susceptible individuals. In the case in which travel is independent of disease status and there is no disease death, a comparison theorem argument can be used as in Section 3 to show that if $\mathcal{R}_0 < 1$, then the DFE is globally asymptotically stable; whereas if $\mathcal{R}_0 > 1$, then the DFE is unstable. A ring of patches with one-way travel is used to model low pathogenecity avian influenza in birds and humans [5].

5 Model including residency patch

Sattenspiel and Dietz [26] introduced a single species, multi-patch model that describes the travel of individuals, and keeps track of the patch where an individual is born and usually resides as well as the patch where an individual is at a given time. This model has subsequently been studied numerically in various contexts, including the effects of quarantining [27]. We studied this model [2, 3] giving some analytical results and calculating the basic reproduction number. In this model, if a resident from patch i travels to patch j then they are assumed to return home to patch i before traveling to another patch k, i.e., such an individual does not travel directly from patch j to patch k (for j, $k \neq i$). We now extend the SIR model in [26] by removing this restriction and allowing for such an individual to travel between two patches that are not their residency patch. The assumption of [26] may be appropriate for travel between isolated communities; see [25] and references therein for situations linked to the spread of influenza in the Canadian subarctic. However, our formulation allows for a wider range of travel patterns.

To formulate our model, let $N_{ij}(t)$ be the number of residents of patch i who are present in patch j at time t, with $S_{ij}(t)$, $I_{ij}(t)$ and $R_{ij}(t)$ being the number that are susceptible, infective and recovered, respectively. Matrix $M^S_i = [m^S_{ijk}]$ gives the travel rates of susceptible individuals resident in patch i from patch k to patch j. Similarly $M^I_i = [m^I_{ijk}]$ and $M^R_i = [m^R_{ijk}]$ give these rates for infective and recovered individuals. Taking standard incidence as in previous models, β_{ikj} denotes the proportion of adequate contacts in patch j between a susceptible from patch i and an infective from patch k that results in disease transmission and κ_j denotes the average number of such contacts in patch j. For all patches, the recovery rate of infectives is denoted by γ, the loss of immunity rate by δ, birth is assumed to occur in the residency patch at rate d and natural death to occur (in all disease states) at the same rate d. For p patches, the model takes the following form for $i, j = 1, \ldots, p$.

$$\begin{aligned}
\frac{dS_{ii}}{dt} &= d\sum_{k=1}^{p} N_{ik} - \sum_{k=1}^{p} \kappa_i \beta_{iki} S_{ii} \frac{I_{ki}}{\mathbf{N}_i} - dS_{ii} + \delta R_{ji} + \sum_{k=1}^{p} m_{iik}^S S_{ik} - \sum_{k=1}^{p} m_{iki}^S S_{ii} \\
\frac{dI_{ii}}{dt} &= \sum_{k=1}^{p} \kappa_i \beta_{iki} S_{ii} \frac{I_{ki}}{\mathbf{N}_i} - (d+\gamma) I_{ii} + \sum_{k=1}^{p} m_{iik}^I I_{ik} - \sum_{k=1}^{p} m_{iki}^I I_{ii} \\
\frac{dR_{ii}}{dt} &= \gamma I_{ii} - (d+\delta) R_{ii} + \sum_{k=1}^{p} m_{iik}^R R_{ik} - \sum_{k=1}^{p} m_{iki}^R R_{ii}
\end{aligned}$$

and for $i \neq j$

$$\begin{aligned}
\frac{dS_{ij}}{dt} &= -\sum_{k=1}^{p} \kappa_j \beta_{ikj} S_{ij} \frac{I_{kj}}{\mathbf{N}_i} - dS_{ij} + \delta R_{ij} + \sum_{k=1}^{p} m_{ijk}^S S_{ik} - \sum_{k=1}^{p} m_{ikj}^S S_{ij} \\
\frac{dI_{ij}}{dt} &= \sum_{k=1}^{p} \kappa_j \beta_{ikj} S_{ij} \frac{I_{kj}}{\mathbf{N}_j} - (d+\gamma) I_{ij} + \sum_{k=1}^{p} m_{ijk}^I I_{ik} - \sum_{k=1}^{p} m_{ijk}^I I_{ij} \\
\frac{dR_{ij}}{dt} &= \gamma I_{ij} - (d+\delta) R_{ij} + \sum_{k=1}^{p} m_{ijk}^R R_{ik} - \sum_{k=1}^{p} m_{ijk}^R R_{ij}
\end{aligned}$$

where $\mathbf{N}_i = \sum_{j=1}^{p} N_{ji}$, the number present in patch i. Properties of this model remain to be explored.

For the simpler case formulated in [26] in which travel is independent of disease status and individuals return to their residency patch after traveling to another patch, some analysis is given in [2, 3] for corresponding SIS and SEIRS models. These models have a unique DFE and the basic reproduction number is calculated by the method used in Section 2 with F and V being block matrices. Numerical simulations show that a change in travel rates can lead to a bifurcation at $\mathcal{R}_0 = 1$; thus travel can stabilize or destabilize the disease free equilibrium.

6 Other discrete spatial models

We end this survey with a brief description of three other metapopulation models from the recent literature and we emphasize their novel features. The first is for a human disease, whereas the last two model specific animal diseases. Together they illustrate the possible complexity that can be built into patch models. We hope that these descriptions encourage readers to consult the original papers as well as to formulate and analyze other metapopulation models that are applicable to disease spread.

6.1 Spread of influenza. Hyman and LaForce [15] formulate a multi-city transmission model for the spread of influenza between cities (patches) with the assumption that people continue to travel when they are infectious and there is no death due to influenza. Because influenza is more likely to spread in the winter than in the summer, they assume that the infection rate has a periodic component. In addition, they introduce a new disease state P in which people have partial immunity to the current strain of influenza. Thus they have an SIRPS model in which both susceptible and partially immune individuals can be infected, but this is more likely for susceptibles. A symmetric travel matrix $M = [m_{ij}]$ with $m_{ij} = m_{ji}$ is assumed, thus the population of each city remains constant. Their model for p cities is formulated as a $4p$ system of non autonomous ODEs.

The authors take epidemic parameters appropriate for influenza virus, in particular for strains of H3N2 in the 1996-2001 influenza seasons with an infectious period of $1/\alpha = 4.1$ days in all cities. Parameters modeling the number of adequate contacts per person per day and the seasonal change of infectivity are estimated by a least squares fit to data. The populations of the largest 33 cities in the US are taken from 2000 census data, and migration between cities is approximated by airline flight data. A sensitivity analysis reveals that the parameter α is the single most important parameter. From numerical simulations on the network of 33 cities, the authors find that the peak of the epidemic lags behind the seasonal peak in infectivity. A comparison of model results with data is given for several cities, and the model is seen to capture the essential features of the yearly influenza epidemics.

6.2 Tuberculosis in possums. The spread of bovine tuberculosis amongst the common brushtail possum in New Zealand, is modeled by Fulford *et al* [14]. Since only maturing possums (1 to 2 year old males) travel large distances, the authors formulate a two-age class metapopulation model with juvenile and adult possums. As this disease is fatal, an SEI model is appropriate. In addition to horizontal transmission between both age-classes, pseudo-vertical transmission is included since juveniles may become infected by their mothers. Susceptible and exposed juveniles (but not infective juveniles) travel between patches as they mature. For p patches, the authors formulate a system of $6p$ ODEs to describe the disease dynamics. Using the next generation matrix method [10, 29], the authors explicitly calculate $\mathcal{R}_0$ for $p = 1$ and for $p = 2$, and give the structures of the next generation matrices for $p = 4$ and three spatial topologies, namely a spider, chain and loop.

Fulford *et al* [14] give numerical results and compute $\mathcal{R}_0$ with appropriate parameters [14, Table 1] for $p = 2$ and for $p = 4$ with the above topologies. The design of control strategies (culling) based on these three spatial topologies is considered. The critical culling rates are calculated and the spatial aspects are shown to be important.

6.3 Feline leukemia virus. Fromont *et al* [13] derive a model appropriate for Feline Leukemia Virus among a population of domestic cats. There are p patches called farms or villages depending on the magnitude of the patch carrying capacity. Dispersal (which depends on disease state) can take place between any pair of patches or into/out of non-specified populations surrounding the patches (representing transient feral males). Infected cats become either infectious or immune and remain so for life, thus the model is of SIR type, but a proportion of cats go directly from the susceptible to the immune state. A density dependent mortality function is assumed, as well as different incidence functions depending on the population density (mass action for cats on farms, but standard incidence for cats in villages).

The model consists of $3p$ ODEs and is analyzed for the case $p = 2$. Fromont *et al* [13] take data appropriate for the virus with one patch being a village and one patch being a farm, or both patches being farms. For a set of parameters such that in isolation the virus develops in the village but goes extinct on the farm, travel between the patches of either susceptible and immune cats or of infective cats can result in the virus persisting in both patches. Thus results show that, in general for this model, spatial heterogeneity promotes disease persistence.

References

[1] Allen, L.J.S. and Ernest, R.K.: The impact of long-range dispersal on the rate of spread in population and epidemic models. In: Castillo-Chavez, C. with Blower, S., van den Driessche, P., Kirschner, D. and Yakubu A-A. (eds) Mathematical Approaches for Emerging and Reemerging Infectious Diseases. An Introduction: IMA Volumes in Math. and Appln. **125**, 183-197 (2002)

[2] Arino, J. and van den Driessche, P.: A multi-city epidemic model. Mathematical Population Studies. **10**, 175-193 (2003)

[3] Arino, J. and van den Driessche, P.: The basic reproduction number in a multi-city compartment model. LNCIS. **294**, 135-142 (2003)

[4] Arino, J., Davis, J.R., Hartley, D., Jordan, R., Miller, J.M. and van den Driessche, P.: A multi-species epidemic model with spatial dynamics. Math. Medicine and Biology, **22**, 129-142 (2005)

[5] Arino, J., Jordan, R. and van den Driessche, P.: Quarantine in a multi-species epidemic model with spatial dynamics. Math. Bios., to appear.

[6] Ball, F.G. and Lyne, O.D.: Epidemics among a population of households. In: Castillo-Chavez, C. with Blower, S., van den Driessche, P., Kirschner, D. and Yakubu A-A. (eds) Mathematical Approaches for Emerging and Reemerging Infectious Diseases. Models, Methods and Theory: IMA Volumes in Math. and Appln. **126**, 115-142 (2002)

[7] Berman, A. and Plemmons, R.J.: Nonnegative Matrices in the Mathematical Sciences, Academic Press (1979)

[8] Cliff, A.: Incorporating spatial components into models of epidemic spread. In: Mollison, D. (ed) Epidemic Models: Their Structure and Relation to Data, Cambridge University Press 119-149 (1995)

[9] Castillo-Chavez, C. and Yakubu A-A.: Intraspecific competition, dispersal and disease dynamics in discrete-time patchy invironments. In: Castillo-Chavez, C. with Blower, S., van den Driessche, P., Kirschner, D. and Yakubu A-A. (eds) Mathematical Approaches for Emerging and Reemerging Infectious Diseases. An Introduction: IMA Volumes in Math. and Appln. **125**, 165-181 (2002)

[10] Diekmann, O. and Heesterbeek, J.A.P.: Mathematical Epidemiology of Infectious Diseases, Model Building, Analysis and Interpretation. Wiley, New York (2000)

[11] Durrett, R.: Spatial epidemic model. In: Mollison, D. (ed) Epidemic Models: Their Structure and Relation to Data, Cambridge University Press 187-201 (1995)

[12] Eubank, S., Guciu, H., Kumar, V.S.A., Marathe, M.V., Srinivasan, A., Toroczkai, Z. and Wang, N., Modeling disease outbreaks in realistic urban social networks. Nature **429**, 180-184 (2004)

[13] Fromont, E., Pontier, D. and Langlais, M.: Disease propagation in connected host population with density-dependent dynamics: the case of the Feline Leukemia Virus. J. Theoretical Biol. **223**, 465-475 (2003)

[14] Fulford, G.R., Roberts, M.G. and Heesterbeek, J.A.P.: The metapopulation dynamics of an infectious disease: tuberculosis in possums. Theor. Pop Biol. **61**, 15-29 (2002)

[15] Hyman, J.M. and LaForce, T., Modeling the spread of influenza among cities. In: Banks, H. and Castillo-Chávez, C. (eds) Bioterrorism: Mathematical Modeling Applications in Homeland Security. SIAM 211-236 (2003)

[16] Jin, Y. and Wang, W.: The effect of population dispersal on the spread of a disease, J. Math. Anal. Appl., **308**, 343-364 (2005)

[17] Lakshmikantham, V., Leela, S. and Martynyuk, M. Stability Analysis of Non-linear Systems. Dekker (1989)

[18] Lewis, M.A., Rencławowicz, J. and van den Driessche, P.: Traveling waves and spread rates for a West Nile virus model. Bull. Math. Biology, to appear.

[19] Mollison, D. (ed) Epidemic Models: Their Structure and Relation to Data, Cambridge University Press (1995)

[20] Metz, H. and van den Bosch, F.: Velocities of epidemic spread. In: Mollison, D. (ed) Epidemic Models: Their Structure and Relation to Data, Cambridge University Press 150-186 (1995)

[21] Murray, J.D.: Mathematical Biology II: Spatial Models and Biomedical Applications, Springer-Verlag (2003)

[22] Ruan, S., Wang, W. and Levin, S.A.: The effect of global travel on the spread of SARS. Math. Bios. Eng. **3** (2006).

[23] Rvachev, L.A. and Longini, I.M.: A mathematical model for the global spread of influenza. Math. Bios. **75**, 3-22 (1985)

[24] Salmani, M. and van den Driessche, P.: A model for disease transmission in a patchy environment. Discrete Contin. Dynam. Systems Ser.B, to appear.

[25] Sattenspiel, L.: Infectious diseases in the historical archives: a modeling approach. In: Herring, D.A. and Swedlund, A.C. (eds) Human Biologists in the Archives. Cambridge University Press. 234-265 (2003)

[26] Sattenspiel, L. and Dietz, K.: A structured epidemic model incorporating geographic mobility among regions. Math. Bios. **128**, 71-91 (1995)

[27] Sattenspiel, L. and Herring, D.A.: Simulating the effect of quarantine on the spread of the 1918-19 flu in central Canada. Bull. Math. Biol. **65**, 1-26 (2003)

[28] Smith, H.L. and Waltman, P.: The Theory of the Chemostat, Cambridge University Press (1995)

[29] van den Driessche, P. and Watmough, J.: Reproduction numbers and sub-threshold endemic equilibria for compartmental models of disease transmission. Math. Bios. **180**, 29-48 (2002)

[30] Wang, W. and Mulone, G.: Threshold of disease transmission in a patch environment. J. Math. Anal. Appl. **285**, 321-335 (2003)

[31] Wang, W. and Zhao, X.-Q.: An epidemic model in a patchy environment. Math. Biosci. **190**, 97-112 (2004)

[32] Wang, W. and Zhao, X.-Q.: An age-structured epidemic model in a patchy environment, SIAM J. Appl. Math., **65**, 1597-1614 (2005)

[33] Zhao, X.-Q. Dynamical Systems in Population Biology, Springer-Verlag (2003)

Fields Institute Communications
Volume **48**, 2006

On Some Nonlocal Evolution Equations Arising in Materials Science

Peter W. Bates
Department of Mathematics
Michigan State University
East Lansing, MI 48824 USA
bates@math.msu.edu

Abstract. Equations for a material that can exist stably in one of two homogeneous states are derived from a microscopic or lattice viewpoint with the assumption that the evolution follows a gradient flow of the free energy with respect to some metric. Alternatively, Newtonian dynamics can be considered. The resulting lattice dynamical systems are analyzed, as are equations on the continuum where the lattice interaction energy is viewed as an approximation to a Riemann integral. These equations are lattice or nonlocal versions of the Allen-Cahn, Cahn-Hilliard, Phase-Field, or Klein-Gordon equations. Some results presented here provide for the well-posedness of the equations, while others give asymptotics or qualitative behavior of special solutions, such as traveling waves or pulses. This summarizes results previously reported in papers with co-authors Xinfu Chen, Adam Chmaj, Jianlong Han, Chunlei Zhang, and Guangyu Zhao.

1 Introduction

We view a material sample as a collection of 'atoms' occupying an n-dimensional lattice Λ.

These atoms will be assigned 'spin' A or B but we view this as an order parameter that could represent many different things, such as true spin, local concentration or degree of solidification, etc. (when each 'atom' is really a small block of material itself). Allowing for fluctuations, we take the occupancy of type A at site $r \in \Lambda$ in the statistical sense to be a fraction denoted by $a(r) \in [0, 1]$. The lattice is filled; what is not type A is type B, and so the occupancy of B is given by $1 - a(r)$.

The *state* of the material is the function $a : \Lambda \to [0, 1]$, which also evolves in time. To postulate an evolution mechanism we first follow van der Waals [76] who, in 1893, wrote

---

2000 *Mathematics Subject Classification.* Primary 35K55, 37L60; Secondary 82B20, 82B26, 34K30, 34K60, 35B30, 35B40, 35R10, 37L25.

Supported by NSF DMS 9974340 and NSF DMS 0200961.

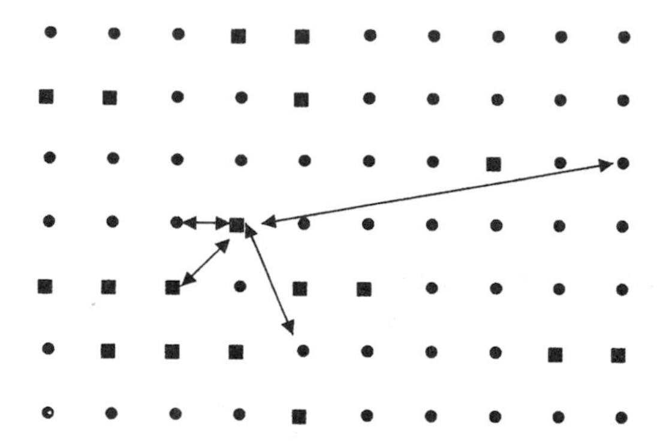

Figure 1 Lattice with long-range interactions

"We shall obtain a complete solution of the problem ... if we can express the free energy at each point as a function of the density at that point and of the differences of density in the neighboring phases, out to a distance limited by the range over which the molecular forces act."

It is possible that some molecular forces act, albeit with very small strength, at great distances and we adopt that point of view, choosing to include all pairwise interactions. The following reasoning was described in more detail in [11] and [12] but we include a brief description here for completeness.

The *Helmholtz free energy* of a state is given by

$$E = H - TS,$$

where H = interaction energy, T = absolute temperature, and S = total entropy. We include all pairwise interaction, allowing for the possibility that pairs of type A interact differently than pairs of type B and both differently from the interaction of mixed pairs. Thus,

$$\begin{aligned} H(a) &\equiv -\frac{1}{2}\sum_{r,r'\in\Lambda}\Big[J^{AA}(r-r')a(r)a(r') + J^{BB}(r-r')(1-a(r))(1-a(r')) \\ &\quad + J^{AB}(r-r')(a(r)(1-a(r')) + a(r')(1-a(r)))\Big]. \end{aligned}$$

We expect the interaction, through the $J's$, to be symmetric and translation-invariant, but possibly anisotropic.

Rearranging:

$$\begin{aligned} H &= \frac{1}{4}\sum_{r,r'\in\Lambda} J(r-r')(a(r)-a(r'))^2 \\ &\quad - D\frac{1}{2}\sum_{r\in\Lambda}(a(r)^2 - a(r)) + d\sum_{r\in\Lambda} a(r) + \text{const.} \end{aligned}$$

where $J(r) = J^{AA}(r) + J^{BB}(r) - 2J^{AB}(r)$, $D = \sum J(r)$, and $d = \sum(J^{BB}(r) - J^{AA}(r))/2$.

At site r the entropy $s(a(r))$ for aN particles in N identical sites is given by

$$e^{Ns/K} = \frac{N!}{(aN)!(N-aN)!}$$

where K is Boltzman's constant.

Hence,

$$s(a) \simeq -K[a \ln a + (1-a)\ln(1-a)].$$

The total entropy, $S(a) = \sum_{r \in \Lambda} s(a(r))$ and so

$$\begin{aligned} E(a) = H - TS &= \frac{1}{4} \sum_{r,r' \in \Lambda} J(r-r')(a(r)-a(r'))^2 \\ &+ \sum_{r \in \Lambda} \big[KT\{a(r)\ln a(r) + (1-a(r))\ln(1-a(r))\} \\ &-D(a(r)^2 - a(r)) + da(r)\big]. \end{aligned}$$

There is a critical temperature T_c such that for $T \geq T_c$ the term $[\ \cdots]$ is strictly convex and so there is a unique homogeneous state which minimizes $E(a)$, while for $T < T_c$, this term has two local minima and so two distinct a-states (say $\alpha < \beta$) give spatially homogeneous local minimizers of E.

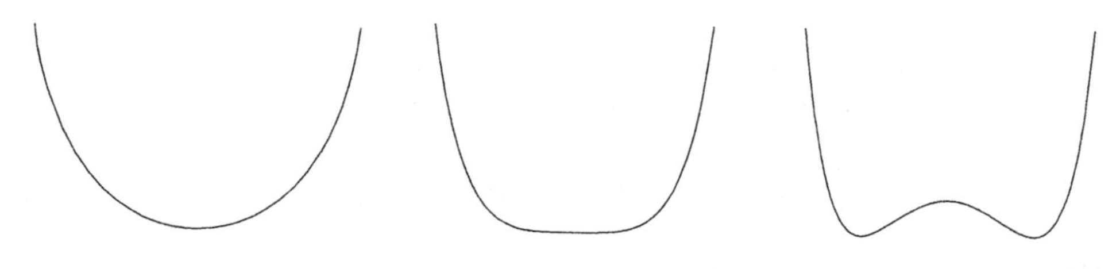

Figure 2a. $T < T_c$, **Figure 2b.** $T = T_c$, **Figure 2c.** $T > T_c$

This is the origin of phase transition in spin systems (e.g. ferromagnets.) We will fix $T < T_c$. If we were to take the continuum limit by using a scaling so that the summation could be viewed as an approximation to a Riemann integral, then we would obtain a free energy in the isothermal case of the form

$$E(u) = \frac{1}{4} \iint J(x-y)(u(x)-u(y))^2 dxdy + \int F(u)dx,$$

where F is a double well function, having minima at ± 1 (after changing variables, using u instead of a to represent the order parameter), and J is assumed to be integrable with positive integral and with $J(-x) = J(x)$.

It is interesting to compare with the Ginzburg-Landau functional:

$$\int (\frac{\varepsilon^2}{2}|\nabla u|^2 + F(u))dx.$$

This is easily obtained from the above nonlocal energy by assuming the atomic interaction is short ranged so that for each state u, one could be justified by approximating $(u(x) - u(y)) \simeq (x-y) \cdot \nabla u(x)$. With that, the coefficient ε^2 in the energy is a second moment of J in the isotropic case.

In fact van der Waals took this approach. The resulting Euler-Lagrange equation,

$$\varepsilon^2 \Delta u - F'(u) = 0,$$

has been well studied and provides some insight into phase transitions. We do not make this short-range approximation however, believing that while it may be good for a single smooth function u, it is not a good approximation in the operator sense. It is worth noting here that for several results we do not require that J be nonnegative, although it is assumed to have positive integral (or sum) and we sometimes will assume that it has a positive second moment.

Away from equilibrium we take as a fundamental principle the postulate that a material structure evolves in such a way that its free energy decreases as quickly as possible. That is, the spatial function u will evolve in such a way that $E(u)$ decreases, and does so optimally in some sense as u evolves in a function space, X. This suggests the evolution law

$$\frac{\partial u}{\partial t} = -\mathrm{grad} E(u), \tag{1.1}$$

where grad $E(u) \in X^*$, the dual of X, is defined by

$$< \mathrm{grad}\ E(u), v >= \frac{d}{dh} E(u + hv)|_{h=0}.$$

If $X = L^2$ then (1.1) becomes what we call the *nonlocal Allen-Cahn equation*,

$$\frac{\partial u}{\partial t} = J * u - Du - F'(u), \tag{1.2}$$

where $*$ is convolution and $D = \int J$ is assumed positive.

The above equation is for the case when the domain $\Omega = \mathbb{R}^n$ but for general Ω we have

$$\frac{\partial u}{\partial t} = J * u - u \int_\Omega J(x-y)dy - F'(u), \tag{1.3}$$

where $J * u(x) \equiv \int_\Omega J(x-y)u(y)dy$.

On the infinite lattice, Λ (a special case being $\mathbb{Z}$) the equation is

$$\dot{u}_r = \sum_{s \in \Lambda, s \neq r} J(r-s)u_s - Du_r - F'(u_r), \;\; r \in \Lambda,\, t \geq 0, \tag{1.4}$$

where $D = \sum_{s \in \Lambda, s \neq 0} J(s)$.

If we had made the Ginzburg-Landau approximation, the resulting gradient flow would be the Allen-Cahn equation [5],

$$\frac{\partial u}{\partial t} = \varepsilon^2 \Delta u - F'(u).$$

Note that the operator

$$J * u - u \int_\Omega J(x-y)dy$$

may be thought of as an approximation to the Laplacian, especially in the case $J \geq 0$, since then it is a nonpositive selfadjoint operator which has a maximum principle. However, unlike the Laplacian, it is bounded and so (1.3) does not smooth in forward time and has solutions that exist locally backwards in time.

For both the continuous and lattice versions, there is now a large body of work giving qualitative behavior of solutions, traveling waves, propagation failure, stability and pattern formation (see, e.g., [58], [78], [42], [55], [41], [43], [15], [66],

[12], [11], [27], [36], [37], [35], [56], [57], [28], [29], [30], [31], [38], [7], [9], [46], [10], and the references therein). There is other recent work on nonlocal equations in applications (see [70], [71], and [50]) but the earliest theoretical results are perhaps those due to Weinberger [77].

In the case that u represents local concentration of one species in a binary alloy, then, with the idea of conserving species, we take $X = H_0^{-1}$ (the dual of H^1 with zero mean). Then (1.1) becomes what we call the *nonlocal Cahn-Hilliard equation*

$$\frac{\partial u}{\partial t} = -\Delta(J * u - u \int_\Omega J(x-y)dy - F'(u)). \tag{1.5}$$

Of course, the original Cahn-Hilliard equation, introduced in [24], has undergone intensive study (see [68], [26], [39], [74], [67], [40], [2], [3], [13], [14], [16], and the references therein) but little has been written on the nonlocal version. To the best of our knowledge, the first was Giacomin and Liebowitz [53], [54], but more recently other results have appeared (see [52] and [32]). Here, we extend some of those results, summarizing the findings in [17] and [18] on the well posedness of (1.5) and long term behavior of solutions.

When temperature is allowed to evolve and latent heat of fusion is included in the model then the free energy, E, is often taken to be

$$\frac{1}{4}\int\int J(x-y)(u(x)-u(y))^2 dxdy + \int (F(u(x)) + \frac{1}{2}\theta^2)dx, \tag{1.6}$$

where u represents degree of solidification, θ is absolute temperature, and l is a latent heat coefficient. The internal energy density is given by $e = \theta + lu$ and in order to conserve the total internal energy, $I \equiv \int e$, the simplest gradient flow is with respect to $(u, e) \in L^2 \times H_0^{-1}$. This leads to the *nonlocal phase field system*:

$$u_t = J * u - u \textstyle\int_\Omega J(x-y)dy - F'(u) + l\theta, \tag{1.7}$$

$$(\theta + lu)_t = \Delta\theta. \tag{1.8}$$

The local phase-field system, introduced by Fix [47], Langer [65], and Caginalp [22], has also undergone much analysis and generalization (see, e.g., Caginalp and Fife [23], Penrose and Fife [69], Kenmochi and Kubo and Ito [59], [63], Colli and Laurencot [33], Colli and Sprekels, [34], etc.) and still is finding many new and important applications. With hysteresis and nonlocal effects there is also the work of Krejči, Sprekels, and S. Zheng, [60], [61], [73] and some previous results in [8], all of which influenced this work on well-posedness and long term behavior of solutions for the nonlocal version (1.7), (1.8). We would also like to point out recent interesting results in [45] giving stabilization in the case of analytic nonlinearity.

Finally, we are interested in Newtonian dynamics with a force derived from the isothermal energy according to

$$\frac{\partial^2 u}{\partial t^2} = -\operatorname{grad} E(u).$$

In L^2 this leads to a *nonlocal wave equation*

$$\frac{\partial^2 u}{\partial t^2} = J * u - u \int_\Omega J(x-y)dy - F'(u). \tag{1.9}$$

For this equation we will establish the existence of traveling pulses with $\Omega = \mathbb{R}$ and J replaced by a large amplitude, short range kernel, namely, $\frac{1}{\varepsilon^2} j_\varepsilon$ where $j_\varepsilon(x) = \frac{1}{\varepsilon} J(\frac{x}{\varepsilon})$. Thus, we will consider

$$\frac{\partial^2 u}{\partial t^2} = \frac{1}{\varepsilon^2}(j_\varepsilon * u - u) - F'(u). \tag{1.10}$$

In the discrete case, the corresponding equation takes the form

$$\ddot{u}_n = \frac{1}{\varepsilon^2} \sum_{k=-\infty}^{\infty} \alpha_k u_{n-k} - F'(u_n), \qquad n \in \mathbb{Z} \tag{1.11}$$

where $\varepsilon > 0$ and the coefficients α_k satisfy $\sum_{k=-\infty}^{\infty} \alpha_k = 0, \quad \alpha_0 < 0, \quad \alpha_k = \alpha_{-k}, \quad \sum_{k \geq 1} \alpha_k k^2 = d > 0$. This may be viewed as a generalized lattice Klein-Gordon equation. Several studies exist for versions of lattice Klein-Gordon equations (e.g., see [25], [44], [49], [62], [72], and [75], etc.), but to the best of our knowledge, there are no prior results for (1.10).

In the following three sections we outline results for the nonlocal Cahn-Hilliard, phase-field, and Klein-Gordon equations, respectively. In Section 4. we also prove existence of traveling pulses for the Klein Gordon lattice system. We give details of proofs in many cases, and where details are lacking, we indicate the route. Our hope is that the reader will gain an appreciation for the variety of techniques that may be brought to bear on these nonlocal evolution equations. Missing here are variational methods but the reader may turn to [10], for instance, to see that those methods may also be applied in some cases.

2 Nonlocal Cahn-Hilliard equation

The first issue to address is whether or not (1.5) is well-posed with suitable boundary conditions. These results are to be found in more detail in papers with Jianlog Han, [17] and [18]. Since the equation is second order in space (while the usual Cahn-Hilliard equation is fourth order), only one boundary condition is expected to be necessary and sufficient for existence and uniqueness of the solution. We will therefore consider both the Dirichlet and no-flux boundary condition, the latter being more natural in the sense that species should then be conserved. Thus, we consider

$$\frac{\partial u}{\partial t} = -\Delta(J * u - u \int_\Omega J(x-y)dy - f(u)) \tag{2.1}$$

with either

$$u = 0 \quad \text{on} \quad \partial\Omega \tag{2.2}$$

or

$$\frac{\partial}{\partial n}(\int_\Omega J(x-y)dy u(x) - \int_\Omega J(x-y)u(y)dy + f(u)) = 0 \quad \text{on} \quad \partial\Omega, \tag{2.3}$$

where $f = F'$ is of bistable type (e.g., $f(u) = u - u^3$). This second condition of Neumann type (2.3) may look peculiar but simply states that the chemical potential has no flux across the boundary. We append the initial condition

$$u(x,0) = u_0(x), \quad \text{for} \quad x \in \bar{\Omega}.$$

We treat the Neumann problem first, discussing the main points of [17]. In order to prove the existence of a classical solution to (2.1)–(2.3) we need the initial data to satisfy the boundary condition. So we assume $u_0(x) \in C^{2+\beta, \frac{2+\beta}{2}}(\bar{\Omega})$ for some $\beta > 0$, and $u_0(x)$ satisfies the compatibility condition:

$$\frac{\partial(\int_\Omega J(x-y)dy u_0(x) - \int_\Omega J(x-y)u_0(y)dy + f(u_0))}{\partial n} = 0 \quad \text{on} \quad \partial\Omega. \tag{2.4}$$

Rewrite the initial-boundary value problem as

$$\begin{cases} \frac{\partial u}{\partial t} = a(x,u)\triangle u + b(x,u,\nabla u) & \text{in} \quad \Omega,\, t>0, \\ a(x,u)\frac{\partial u}{\partial n} + \frac{\partial a(x)}{\partial n}u(x) - \int_\Omega \frac{\partial J(x-y)}{\partial n}u(y)dy = 0 & \text{on} \quad \partial\Omega,\, t>0, \\ u(x,0) = u_0(x), \end{cases} \tag{2.5}$$

where

$$\begin{aligned} a(x,u) &= a(x) + f'(u), \\ a(x) &= \int_\Omega J(x-y)dy, \\ b(x,u,\nabla u) &= 2\nabla a \cdot \nabla u + f''(u)|\nabla u|^2 + u\triangle a - (\triangle J) * u. \end{aligned}$$

We assume the following conditions:

(A_1) $a(x) \in C^{2+\beta}(\bar{\Omega})$, $f \in C^{2+\beta}(\mathbb{R})$.

(A_2) There exist $c_1 > 0$, $c_2 > 0$, and $r > 0$ such that

$$a(x,u) = a(x) + f'(u) \geq c_1 + c_2|u|^{2r}.$$

(A_3) $\partial\Omega$ is of class $C^{2+\beta}$.

With regard to (A_2), note that if $a(x) + f'(u(x,t)) < 0$, for some (x,t) then there is no solution beyond that point in general, since the equation is essentially a backward heat equation. Note also that (A_2) implies

$$F(u) = \int_0^u f(s)ds \geq c_3|u|^{2r+2} - c_4 \tag{2.6}$$

for some positive constants c_3, and c_4.

For any $T > 0$, denote $Q_T = \Omega \times (0,T)$. We first establish an *a priori* bound for solutions of (2.1)–(2.3).

Theorem 2.1 *If $u(x,t) \in C^{2,1}(\bar{Q}_T)$ is a solution of equation (2.1)-(2.3), then*

$$\max_{Q_T} |u(x,t)| \leq \bar{C}(u_0) \tag{2.7}$$

for some constant $\bar{C}(u_0)$.

In order to prove the theorem, we need the following lemma.

Lemma 2.2 *If $u(x,t) \in C^{2,1}(\bar{Q}_T)$ is a solution of equation (2.1)–(2.3), then there is a constant $C(u_0)$ such that*

$$\sup_{0\leq t\leq T} ||u(\cdot,t)||_q \leq C(u_0) \tag{2.8}$$

for any $q \leq 2r+2$.

Proof Let

$$E(u) = \frac{1}{4}\int\int J(x-y)(u(x)-u(y))^2 dxdy + \int F(u(x))dx. \tag{2.9}$$

Since we have a gradient flow

$$\frac{dE(u)}{dt} \le 0.$$

Therefore $E(u) \le E(u_0)$, i.e.,

$$\begin{aligned}&\frac{1}{4}\int\int J(x-y)(u(x)-u(y))^2 dxdy + \int F(u(x))dx \\ &\quad \le \frac{1}{4}\int\int J(x-y)(u_0(x)-u_0(y))^2 dxdy + \int F(u_0(x))dx.\end{aligned}$$

From condition (A_1), (2.6), and Young's inequality, we obtain

$$\int_\Omega |u|^{2r+2} dx \le C(u_0).$$

Since this is true for any $t > 0$, we have

$$\sup_{0\le t\le T}\int_\Omega u^{2r+2} dx \le C(u_0),$$

where $C(u_0)$ does not depend on T.

Since Ω is bounded, it follows that

$$\sup_{0\le t\le T} ||u||_q \le C(u_0)$$

for any $q \le 2r+2$. □

We will prove the theorem with an iteration argument, similar to that found in [1]

Proof For $p > 1$, multiply equation (2.1) by $u|u|^{p-1}$ and integrate over Ω, to obtain

$$\begin{aligned}\int u|u|^{p-1}u_t dx = &-\int a\,(x,u)\nabla u\cdot\nabla(u|u|^{p-1}(x))dx \\ &-\int\int \nabla J(x-y)u(x)\nabla(u|u|^{p-1}(x))dydx \\ &+\int\int \nabla J(x-y)u(y)\nabla(u|u|^{p-1}(x))dydx.\end{aligned} \tag{2.10}$$

Since

$$\int_\Omega a(x,u)\nabla u\cdot\nabla(u|u|^{p-1})dx = p\int_\Omega a\,(x,u)|u|^{p-1}|\nabla u|^2 dx \tag{2.11}$$

and

$$|\nabla|u|^{\frac{p+1}{2}}|^2 = \frac{(p+1)^2}{4}|u|^{p-1}|\nabla u|^2, \tag{2.12}$$

with condition (A_2), we have

$$\begin{aligned}\int_\Omega a\,(x,u)\nabla u\cdot\nabla(u|u|^{p-1})dx \ge &\frac{4pc_1}{(p+1)^2}\int_\Omega |\nabla|u|^{\frac{p+1}{2}}|^2 dx \\ &+\frac{4pc_2}{(p+2r+1)^2}\int_\Omega |\nabla|u|^{\frac{p+2r+1}{2}}|^2 dx.\end{aligned} \tag{2.13}$$

This yields

$$\begin{aligned}\frac{1}{p+1}\frac{d}{dt}\int_\Omega |u|^{p+1}dx + \frac{4pc_1}{(p+1)^2}\int_\Omega |\nabla|u|^{\frac{p+1}{2}}|^2 dx \\ \leq -\int\int \nabla J(x-y)u(x)\nabla(u|u|^{p-1}(x))dydx \\ +\int\int \nabla J(x-y)u(y)\nabla(u|u|^{p-1}(x))dydx.\end{aligned} \tag{2.14}$$

From Cauchy-Schwartz and Young's inequalities we have

$$\begin{aligned}&-\int\int \nabla J(x-y)u(x)\nabla(u|u|^{p-1}(x))dydx \\ &\leq M_1 p\int_\Omega ||u|^p\nabla u(x)| \\ &\leq \frac{c_1 p}{(p+1)^2}\int_\Omega |\nabla|u|^{\frac{p+1}{2}}|^2dx + M_2 p\int_\Omega |u|^{p+1}dx,\end{aligned} \tag{2.15}$$

for some positive constant M_2 which does not depend on p, and $M_1 = sup\int J(x-y)dy$. Also we have

$$\begin{aligned}&\int\int |\nabla J(x-y)||u(y)|\nabla(u|u|^{p-1}(x))dydx \\ &= p\int\int |\nabla J(x-y)||u(y)||u(x)|^{p-1}|\nabla u(x)|dxdy \\ &\leq \frac{c_1 p}{(p+1)^2}\int_\Omega |\nabla|u|^{\frac{p+1}{2}}|^2dx + M_3 p\int_\Omega |u|^{p+1}dx\end{aligned} \tag{2.16}$$

for some constant M_3 which does not depend on p. Inequalities (2.14)-(2.16) imply

$$\frac{d}{dt}\int_\Omega |u|^{p+1}dx + \frac{2pc_1}{(p+1)}\int_\Omega |\nabla|u|^{\frac{p+1}{2}}|^2dx \leq C\cdot(p+1)^2\int_\Omega |u|^{p+1}dx. \tag{2.17}$$

Now we need the following Nirenberg-Gagliado inequality,

$$||D^j v||_{L^s} \leq C_1||D^m v||_{L^r}^a||v||_{L^q}^{1-a} + C_2||v||_{L^q}, \tag{2.18}$$

where

$$\frac{j}{m} \leq a \leq 1,\ \frac{1}{s} - \frac{j}{n} + a(\frac{1}{r} - \frac{m}{n}) + (1-a)\frac{1}{q}. \tag{2.19}$$

In (2.18), set $s=2$, $j=0$, $r=2$, $m=1$, to get

$$||v||_2^2 \leq C_1||Dv||_2^{2a}||v||_q^{2(1-a)} + C_2||v||_q^2. \tag{2.20}$$

Let $v = |u|^{\frac{\mu_k+1}{2}}$, $\mu_k = 2^k$, $q = \frac{2(\mu_{k-1}+1)}{\mu_k+1}$, and

$$a = \frac{n(2-q)}{n(2-q)+2q} = \frac{n}{n+2+2^{2-k}}. \tag{2.21}$$

Using Young's inequality this yields

$$\int_\Omega |u|^{\mu_k+1}dx \leq \epsilon\int_\Omega |\nabla|u|^{\frac{\mu_k+1}{2}}|^2dx + c\epsilon^{-\frac{a}{1-a}}(\int_\Omega |u|^{\mu_{k-1}+1}dx)^{\frac{\mu_k+1}{\mu_{k-1}+1}}. \tag{2.22}$$

If we set $p = \mu_k$ in (2.17) and plug (2.22) into (2.17), we obtain

$$\begin{aligned}&\frac{d}{dt}\int_\Omega |u|^{\mu_k+1}dx + \frac{2c_1\mu_k}{\mu_k+1}\int_\Omega |\nabla |u|^{\frac{\mu_k+1}{2}}|^2 dx \\ &\leq C(\mu_k+1)^2(\epsilon\int_\Omega |\nabla|u|^{\frac{\mu_k+1}{2}}|^2dx + c\epsilon^{-\frac{a}{1-a}}(\int_\Omega |u|^{\mu_{k-1}+1}dx)^{\frac{\mu_k+1}{\mu_{k-1}+1}}).\end{aligned} \tag{2.23}$$

Choosing $\epsilon = \frac{1}{C(\mu_k+1)^2}\cdot\frac{c_1\mu_k}{\mu_k+1}$, we have

$$\frac{d}{dt}\int_\Omega |u|^{\mu_k+1}dx + C_1(k)\int_\Omega |\nabla|u|^{\frac{\mu_k+1}{2}}|^2dx \leq C_2(k)(\int_\Omega |u|^{\mu_{k-1}+1}dx)^{\frac{\mu_k+1}{\mu_{k-1}+1}}, \tag{2.24}$$

where $C_1(k) = \frac{c_1\mu_k}{\mu_k+1}$, $C_2(k) = C^{\frac{1}{1-a}}\cdot c\cdot(\frac{c_1\mu_k}{\mu_k+1})^{-\frac{a}{1-a}}\cdot(\mu_k+1)^{\frac{2}{1-a}}$.

Choosing $\epsilon = 1$ in (2.22), this and (2.24) also imply

$$\frac{d}{dt}\int_\Omega |u|^{\mu_k+1}dx + C_1(k)\int_\Omega |u|^{\mu_k+1}dx \leq C_4(k)(\int_\Omega |u|^{\mu_{k-1}+1}dx)^{\frac{\mu_k+1}{\mu_{k-1}+1}}$$

where $C_4(k) = C_2(k) + c$.

By Gronwall's inequality, we have

$$\begin{aligned}\int_\Omega |u|^{\mu_k+1}dx &\leq \int_\Omega |u_0|^{\mu_k+1}dx + \frac{C_4(k)}{C_1(k)}(\sup_{t\geq 0}\int_\Omega |u|^{\mu_{k-1}+1}dx)^{\frac{\mu_k+1}{\mu_{k-1}+1}} \\ &\leq \delta(k)max\{M_0^{\mu_k+1}|\Omega|, (\sup_{t\geq 0}\int_\Omega |u|^{\mu_{k-1}+1}dx)^{\frac{\mu_k+1}{\mu_{k-1}+1}}\},\end{aligned} \tag{2.25}$$

where $\delta(k) = c(1+\mu_k)^\alpha$, $\alpha = \frac{2}{1-a}$, and $M_0 = \sup_{x\in\Omega}|u_0|$. This implies

$$\begin{aligned}\int_\Omega |u|^{\mu_k+1}dx &\leq \delta(k)max\{M_0^{\mu_k+1}|\Omega|, (\sup_{t\geq 0}\int_\Omega |u|^{\mu_{k-1}+1}dx)^{\frac{\mu_k+1}{\mu_{k-1}+1}}\} \\ &\leq \prod_{i=0}^{k}(|\Omega|\delta(k-i))^{\frac{\mu_k+1}{\mu_{k-i}+1}}max\{M_0^{\mu_k+1}, (\sup_{t\geq 0}\int_\Omega |u|^2dx)^{\frac{\mu_k+1}{2}}\}.\end{aligned} \tag{2.26}$$

Since $\frac{\mu_k+1}{\mu_{k-i}+1} < 2^i$, we have

$$\begin{aligned}&\delta(k)\delta(k-1)^{\frac{\mu_k+1}{\mu_{k-1}+1}}\delta(k-2)^{\frac{\mu_k+1}{\mu_{k-2}+1}}\cdots\delta(1)^{\frac{\mu_k+1}{2}} \\ &\leq c^{2^k-1}(2^\alpha)^{-k+2^{k+1}-2}\end{aligned} \tag{2.27}$$

and

$$|\Omega|\cdot|\Omega|^{\frac{\mu_k+1}{\mu_{k-1}+1}}\cdots|\Omega|^{\frac{\mu_k+1}{2}} \leq |\Omega|^{2^k+1}. \tag{2.28}$$

Estimates (2.26)-(2.28) and Lemma 2.2 imply

$$(\int_\Omega |u|^{\mu_k+1}dx)^{\frac{1}{\mu_k+1}} \leq C|\Omega|2^{2\alpha}max\{M_0, \sup_{t\geq 0}(\int_\Omega |u|^2dx)^{\frac{1}{2}}\} \leq \bar{C}(u_0) \tag{2.29}$$

where $\bar{C}(u_0)$ does not depend on k. Since this is true for any k, letting $k\to\infty$ in (2.29), we have

$$||u||_\infty \leq \bar{C}(u_0),$$

and therefore,

$$\sup_{0\leq t\leq T}||u||_\infty \leq \bar{C}(u_0). \tag{2.30}$$

Since $u \in C(\bar{Q}_T)$, it follows that

$$\max_{Q_T} |u(x,t)| \leq \bar{C}(u_0)$$

□

Remark 2.3 In (2.30), since $\bar{C}(u_0)$ does not depend on T, we also obtain a global bound for u whenever there is global existence of a classical solution.

Since $\max_{Q_T} |u| \leq M$, after a slight modification of the proof of Theorem 7.2 in Chapter V in [64], using the equivalent form (2.5) we have

Theorem 2.4 *For any solution $u \in C^{2,1}(\bar{Q}_T)$ of equation (2.1)-(2.3) having $\max_{Q_T} |u| \leq C$, one has the estimates*

$$\max_{Q_T} |\nabla u| \leq K_1, \ |u|_{Q_T}^{(1+\delta)} \leq K_2, \tag{2.31}$$

where constants K_1, K_2, and δ depend only on C, $||u_0||_{C^2(\bar{\Omega})}$ and Ω, $|\cdot|_{Q_T}^{(1+\delta)}$ is a Hölder norm in [64].

We will now set the stage for using the Leray-Schauder Fixed Point Theorem, first formulating the problem with homogeneous initial as well as boundary conditions, and then embedding our problem in a continuous family of initial-boundary value problems.

In (2.5), setting $v(x,t) = u(x,t) - u_0(x)$, we obtain the equivalent form

$$\begin{cases} \frac{\partial v}{\partial t} = \tilde{a}(x,v,u_0)\triangle v + \tilde{b}(x,v,\nabla v,u_0) & \text{in} \quad \Omega,\ t>0, \\ \tilde{a}(x,v,u_0)\frac{\partial v}{\partial n} + \tilde{\psi}(x,v,u_0) = 0 & \text{on} \quad \partial\Omega,\ t>0, \\ v(x,0) = 0, \end{cases} \tag{2.32}$$

where

$$\tilde{a}(x,v,u_0) = a(x,v+u_0),$$
$$\tilde{b}(x,v,\nabla v,u_0) = a(x,v+u_0)\triangle u_0 + b(x,v+u_0,\nabla(v+u_0)),$$

and

$$\begin{aligned} \tilde{\psi}(x,v,u_0) &= \frac{\partial a(x)}{\partial n}(v(x,t)+u_0(x)) + \tilde{a}(x,v,u_0)\frac{\partial u_0}{\partial n} \\ &\quad - \int_\Omega \frac{\partial J(x-y)}{\partial n}(v(y,t)+u_0(y))dy. \end{aligned}$$

Since (2.4) implies $\tilde{\psi}(x,0,u_0) = 0$, the compatibility condition for (2.32) is also satisfied.

Denote

$$Lv = \frac{\partial v}{\partial t} - \tilde{a}(x,v,u_0)\triangle v - \tilde{b}(x,v,\nabla v,u_0),$$

and

$$L_0 v = \frac{\partial v}{\partial t} - c_1 \triangle v,$$

where c_1 is the constant in condition (A_2).

Consider the following family of problems:

$$\begin{cases} \lambda Lv + (1-\lambda)L_0 v = 0 \qquad in\ Q_T, \\ \lambda(\tilde{a}(x,v,u_0)\frac{\partial v}{\partial n} + \tilde{\psi}(x,v,u_0)) + (1-\lambda)(c_1(\frac{\partial v}{\partial n})) = 0 \text{on}\, \partial\Omega \times [0,T], \\ v(x,0) = 0. \end{cases} \tag{2.33}$$

Lemma 2.5 *If $v(x,t,\lambda) \in C^{2,1}(\bar{Q}_T)$ is a solution of (2.33), then*

$$\max_{Q_T} |v(x,t,\lambda)| \le K, \tag{2.34}$$

where K does not depend on λ.

Proof Since $\lambda\tilde{a}(x,v,u_0) + (1-\lambda)c_1 \ge \lambda c_1 + (1-\lambda)c_1 = c_1 > 0$, the terms in (2.33) also satisfy $(A_1)-(A_2)$ and so (2.34) follows from Theorem 2.1.

□

Consequently one may also conclude from Lemma 2.5 and Theorem 2.4 that:

Lemma 2.6 *If $v(x,t,\lambda) \in C^{2,1}(\bar{Q}_T)$ is a solution of equation (2.33), then*

$$\max_{Q_T} |\nabla v(x,t,\lambda)| \le K_1, \quad |v(x,t,\lambda)|_{Q_T}^{(1+\delta)} \le K_2, \tag{2.35}$$

where constants K_1, K_2, and δ do not depend on λ.

Define a Banach space

$$X = \{v(x,t) \in C^{1+\beta,\frac{1+\beta}{2}}(\bar{Q}_T) : v(x,0) = 0\}$$

with the usual Hölder norm.

For any function $w \in X$ satisfying conditions $max_{Q_T}|w| \le M$ and $max_{Q_T}|w_x| \le M_1$, we consider the following linear problem

$$\begin{cases} v_t - (\lambda\tilde{a}(x,w,u_0) + (1-\lambda)c_1)\triangle v + \lambda\tilde{b}(x,w,\nabla w,u_0) = 0 \quad in\ Q_T, \\ \lambda(\tilde{a}(x,w,u_0)\frac{\partial v}{\partial n} + \tilde{\psi}(x,w,u_0)) + (1-\lambda)c_1\frac{\partial v}{\partial n} = 0 \quad \text{on} \quad \partial\Omega \times [0,T], \\ v(x,0) = 0. \end{cases} \tag{2.36}$$

It is clear that there exists a unique solution $v(x,t,\lambda) \in C^{2+\beta,\frac{2+\beta}{2}}(\bar{Q}_T)$ of (2.36).

Define $T(w,\lambda)$ by

$$v(x,t,\lambda) = T(w,\lambda).$$

It is fairly straightforward to show that for w being in a bounded set of X, $T(w,\lambda)$ is uniformly continuous in λ, and that for any fixed λ, $T(x,\lambda)$ is continuous in x.

Since $C^{2+\beta,\frac{2+\beta}{2}}(\bar{Q}_T) \hookrightarrow C^{1+\beta,\frac{1+\beta}{2}}(\bar{Q}_T)$ is compact, we have that for any fixed λ, $T(w,\lambda)$ is a compact transformation.

The Leray-Schauder Fixed Point Theorem (see, e.g., [64]) gives the existence of a solution $v(x,t)$ of (2.32), and therefore:

Theorem 2.7 *For $u_0 \in C^{2+\beta}(\bar{\Omega})$ for some $\beta > 0$ satisfying the boundary condition (2.4), there exists a solution u to (2.1)–(2.3) with $u \in C^{2+\beta,\frac{2+\beta}{2}}(\bar{Q}_T)$.*

To continue with the well-posedness question, we have

Theorem 2.8 *(Uniqueness and continuous dependence on initial data)*
If $u_1(x,t)$ and $u_2(x,t)$ are two solutions corresponding initial data $u_{10}(x)$ and $u_{20}(x)$ of equation (2.1)-(2.3), then for some C depending only on T,

$$\sup_{0\le t\le T}\int_\Omega |u_1-u_2|dx \le C\int_\Omega |u_{10}-u_{20}|dx. \tag{2.37}$$

Proof For any $\theta\in C^{2,1}(\bar{Q}_T)$ with $\frac{\partial\theta}{\partial n}=0$ *on* $\partial\Omega$, we have

$$\begin{aligned}\int_\Omega u_i(x,\tau)\theta(x,\tau)dx &= \int_\Omega u_i(x,0)\theta(x,0)dx+\int_0^\tau\int_\Omega (u_i\theta_t+B(x,u_i)\triangle\theta)dxdt\\ &+\int_0^\tau\int_\Omega \theta\triangle J*u_idxdt+\int_0^\tau\int_{\partial\Omega}\theta\frac{\partial J}{\partial n}*u_idxdt,\end{aligned} \tag{2.38}$$

where $B(x,u)=a(x)u+f(u)$. Hence,

$$\begin{aligned}&\int_\Omega (u_1-u_2)\theta(x,\tau)dx=\int_\Omega (u_{10}-u_{20})\theta(x,0)dx\\ &+\int_0^\tau\int_\Omega (u_1-u_2)(\theta_t+H\triangle\theta)dxdt+\int_0^\tau\int_\Omega \theta\triangle J*(u_1-u_2)dxdt\\ &+\int_0^\tau\int_{\partial\Omega}\theta\frac{\partial J}{\partial n}*(u_1-u_2)dxdt,\end{aligned} \tag{2.39}$$

where

$$H(x,t)=\begin{cases}\frac{B(x,u_1)-B(x,u_2)}{u_1-u_2} & \text{for}\quad u_1\ne u_2\\ \frac{\partial B(x,u_1)}{\partial u} & \text{for}\quad u_1=u_2.\end{cases}$$

Let θ be the solution to the final value problem

$$\begin{cases}\frac{\partial\theta}{\partial t}=-H(x,t)\triangle\theta+\beta\theta & \text{in}\quad \Omega,\ 0\le t\le\tau,\\ \frac{\partial\theta}{\partial n}=0 & \text{on}\quad \partial\Omega,\\ \theta(x,\tau)=h(x), \end{cases} \tag{2.40}$$

where $h(x)\in C_0^\infty(\Omega)$, $0\le h\le 1$ and $\beta>0$ is a constant.

By the comparison theorem, we have

$$0\le\theta\le e^{\beta(t-\tau)}.$$

Therefore, from (2.39) we have

$$\begin{aligned}&\int_\Omega (u_1-u_2)hdx\\ &=\int_\Omega (u_{10}-u_{20})\theta(x,0)dx+\int_0^\tau\int_\Omega (u_1-u_2)\beta\theta dxdt\\ &+\int_0^\tau\int_\Omega \theta\triangle J*(u_1-u_2)dxdt+\int_0^\tau\int_{\partial\Omega}\theta\frac{\partial J}{\partial n}*(u_1-u_2)dxdt.\end{aligned} \tag{2.41}$$

Hence,

$$\begin{aligned}&\int_\Omega (u_1 - u_2) h dx \\ &\le \int_\Omega |u_{10} - u_{20}| e^{-\beta\tau} dx + \int_0^\tau \int_\Omega |u_1 - u_2| \beta e^{\beta(t-\tau)} dx dt \\ &+ C_1 \int_0^\tau \int_\Omega |u_1 - u_2| e^{\beta(t-\tau)} dx dt + C_2 \int_0^\tau \int_\Omega |u_1 - u_2| e^{\beta(t-\tau)} dx dt.\end{aligned} \tag{2.42}$$

Letting $\beta \to 0$ and $h \to sign(u_1 - u_2)^+$ in (2.42), we have

$$\int_\Omega (u_1 - u_2)^+ dx \le \int_\Omega |u_{10} - u_{20}| dx + C_3 \int_0^\tau \int_\Omega |u_1 - u_2| dx dt. \tag{2.43}$$

Interchanging u_1 and u_2 gives

$$\int_\Omega |u_1 - u_2| dx \le \int_\Omega |u_{10} - u_{20}| dx + C_3 \int_0^\tau \int_\Omega |u_1 - u_2| dx dt. \tag{2.44}$$

By Gronwall's inequality, (2.44) yields

$$\int_\Omega |u_1 - u_2| dx \le C(T) \int_\Omega |u_{10} - u_{20}| dx. \tag{2.45}$$

□

We have been discussing classical solutions where the initial data is smooth and satisfies a compatibility condition.

If $u_0(x) \in L^\infty(\Omega)$, we can consider weak solutions as follows:

Define

$$X = \{f(x) \in C_0^\infty(\Omega) : \ g(x) \equiv \int_\Omega J(x-y) f(y) dy = 0 \text{ on } \partial\Omega\}$$

and

$$B = \text{Closure of X in the L}^2 \text{ norm.}$$

Definition 2.9 A weak solution of (2.1)–(2.3) is a function $u \in C([0,T], L^2(\Omega)) \cap L^\infty(Q_T) \cap L^2([0,T], H^1(\Omega))$, $u_t \in L^2([0,T], H^{-1}(\Omega))$, $\nabla h(x,u) \in L^2((0,T), L^2(\Omega))$ such that

$$\begin{aligned}< u_t(x,t), \psi(x) > &+ \int_\Omega \nabla h(x,u) \cdot \nabla \psi(x) dx \\ &- \int_\Omega (\nabla J * u(\cdot, s)) \cdot \nabla \psi(x) dx = 0\end{aligned} \tag{2.46}$$

for all $\psi \in H^1(\Omega)$ and a.e. time $0 \le t \le T$, where $h(x,u) = a(x)u + f(u)$, $a(x) = \int_\Omega J(x-y) dy$, and

$$u(x,0) = u_0(x). \tag{2.47}$$

Theorem 2.10 *If $(A_1), (A_2)$, and (A_3) are satisfied and $u_0 \in L^\infty(\Omega) \cap B$, then there exists a unique weak solution u of (2.1)–(2.3)*

The proof of existence proceeds as follows: Since $u_0 \in L^\infty(\Omega) \cap B$, there exists a sequence $u_0^{(k)} \in X$ such that

$$\begin{aligned}&||u_0^{(k)} - u_0||_{L^2} \to 0, \\ &||u_0^{(k)}||_\infty < C,\end{aligned} \tag{2.48}$$

where C does not depend on k. Consider equation (2.1)–(2.3) with initial data $u_0^{(k)}$. There exists a unique classical solution $u^{(k)}$. By the energy estimate and other *a priori* bounds, one can find a subsequence and a weak limit u such that

$$\begin{aligned} u^{(k)} &\rightharpoonup u \ \ in \ \ L^2((0,T),H^1(\Omega)),\\ h(x,u^{(k)}) &\rightharpoonup h(x,u) \ \ in \ \ L^2((0,T),H^1(\Omega)),\\ u_t^{(k)} &\rightharpoonup u_t \ \ in \ \ L^2((0,T),H^{-1}(\Omega)), \end{aligned} \tag{2.49}$$

and u satisfies equation (2.46).

Now we turn to discussing the long-term behavior of solutions in the L^p norm.

We establish a nonlinear version of the *Poincaré* inequality, that could be useful in other settings.

Proposition 2.11 *Let $\Omega \subset \mathbb{R}^n$ be smooth and bounded. For $p \geq 1$, there is a constant $C(\Omega)$ such that for all $u \in W^{1,2p}(\Omega)$ with $\int_\Omega u = 0$*

$$\int_\Omega |u|^{2p}dx \leq C(\Omega)\int_\Omega |\nabla |u|^p|^2 dx. \tag{2.50}$$

Proof If (2.50) is not true, there exists a sequence $\{u_k\} \subset W^{1,2p}(\Omega)$ such that

$$\int_\Omega u_k = 0, \ \int_\Omega |u_k|^{2p}dx > k\int_\Omega |\nabla |u_k|^p|^2 dx. \tag{2.51}$$

If $w_k = \frac{u_k}{||u_k||_{2p}}$, then it follows that

$$\int_\Omega w_k = 0, \ \int_\Omega |w_k|^{2p}dx = 1, \ \int_\Omega |\nabla |w_k|^p|^2 dx < \frac{1}{k}. \tag{2.52}$$

Therefore, there exists a subsequence (still denoted by $\{|w_k|^p\}$) and $w \in H^1(\Omega)$ such that

$$|w_k|^p \rightharpoonup w \ in \ H^1 \ and \ |w_k|^p \to w \ \ in \ \ L^2. \tag{2.53}$$

Since $\int_\Omega |\nabla |w_k|^p|^2 dx \leq \frac{1}{k}$, for any $\varphi \in C_0^\infty(\Omega)$, we have

$$\int_\Omega \frac{\partial |w_k|^p}{\partial x_i}\varphi dx \to 0 \tag{2.54}$$

for i=1,$\cdots$,n. Therefore,

$$\int_\Omega \frac{\partial w}{\partial x_i}\varphi dx = 0 \tag{2.55}$$

for i=1,$\cdots$,n and $\varphi \in C_0^\infty(\Omega)$. So $\nabla w = 0$ a.e in Ω, and w is constant in Ω. By taking a subsequence, (2.52) and (2.53) yield

$$w = (\frac{1}{|\Omega|})^{\frac{1}{2}}, \ and \ |w_k|^p \to (\frac{1}{|\Omega|})^{\frac{1}{2}} \ \ a.e \ \ in \ \Omega. \tag{2.56}$$

So, we have

$$|w_k| \to (\frac{1}{|\Omega|})^{\frac{1}{2p}} \ \ a.e \ \ in \ \Omega. \tag{2.57}$$

Since $\int w_k = 0$, there exists a unique solution φ_k to

$$\begin{cases} -\Delta\varphi = w_k & in \ \Omega,\\ \frac{\partial\varphi}{\partial n} = 0 & on \ \partial\Omega,\\ \int_\Omega \varphi dx = 0. \end{cases} \tag{2.58}$$

From (2.58), we obtain

$$\int |\nabla \varphi_k|^2 = \int w_k \varphi_k \le ||w_k||_{L^2}||\varphi_k||_{L^2}. \tag{2.59}$$

Since $\int_\Omega \varphi_k dx = 0$, by *Poincaré*'s inequality, $||\varphi_k||_{L^2} \le c||\nabla \varphi_k||_{L^2}$, therefore (2.58) and (2.59) imply

$$||\nabla \varphi_k||_{L^2} \le c||w_k||_{L^2} \tag{2.60}$$

and

$$\int_\Omega \nabla(|w_k|^{p-1} w_k) \nabla \varphi_k dx = \int_\Omega |w_k|^{p+1} dx. \tag{2.61}$$

Since

$$\nabla(|w_k|^{p-1} w_k) = p|w_k|^{p-1} \nabla w_k, \tag{2.62}$$

we have

$$|\nabla(|w_k|^{p-1} w_k)| = p|w_k|^{p-1}|\nabla w_k| = |\nabla |w_k|^p|. \tag{2.63}$$

Hence, from (2.61), we have

$$\begin{aligned} \int_\Omega |w_k|^{p+1} dx &= \int_\Omega \nabla(|w_k|^{p-1} w_k) \nabla \varphi_k dx \\ &\le \int_\Omega |\nabla |w_k|^p||\nabla \varphi_k| dx \\ &\le ||\nabla |w_k|^p||_{L^2}||\nabla \varphi_k||_{L^2} \\ &\to 0 \end{aligned} \tag{2.64}$$

as $k \to \infty$, by (2.52) and (2.60).

Hence, along a subsequence,

$$|w_k|^{p+1} \to 0 \ a.e \ in \ \Omega,$$

i.e,

$$|w_k| \to 0 \ a.e \ in \ \Omega. \tag{2.65}$$

This contradicts (2.57).

□

Remark 2.12 After this was complete, we became aware of a similar result by Alikakos and Rostamian in [4] but we include our result for completeness.

The next step is to establish the existence of an absorbing set in L^q for all $q > 2$. This is done by first writing the equation in terms of $v = u - u_0$, multiplying that equation by $v|v|^{q-2}$, and integrating. Then one uses Proposition 2.11 and a uniform Gronwall inequality found in [74] to obtain the following:

Proposition 2.13 *Let* $\alpha_0 < (\frac{c_1}{c_2})^{\frac{1}{2r}}$, *where* c_1 *and* c_2 *are the constants in assumption* (A_2), *and let* $\bar{u}_0 = \frac{1}{|\Omega|}\int_\Omega u_0 dx$. *If* $u(x)$ *is a solution of (2.1)-(2.3), and* $|\bar{u}_0| \le \alpha_0$, *then for any* $q > 2$, *we have*

$$\int_\Omega |u - \bar{u}_0|^q dx < C_1 + (\frac{C_2 rt}{q})^{-\frac{q}{2r}} \tag{2.66}$$

where C_1 depends on α_0 and q, and C_2 depends on q. Consequently, one has for any solution of (2.1)-(2.3) with $\frac{1}{|\Omega|}|\int u_0 dx| = |\bar{u}_0| \leq \alpha_0$, there exists a time $t_0(\alpha_0, q) \geq 0$ such that

$$||u||_q < \mu, \text{ for all } t > t_0(\alpha_0, q), \tag{2.67}$$

where

$$\mu > (\frac{C_3(q)}{C_4(q, \alpha_0)})^{\frac{1}{q+2r}} + \alpha_0 |\Omega|^{\frac{1}{q}}.$$

We have shown that in L^p there exists a "local absorbing set" in the sense that if $|\int u_0|$ is not too large, the solution enters a fixed bounded set in the affine space $\bar{u}_0 + L^p$ in finite time (note that $\bar{u}_0 = \frac{1}{|\Omega|}\int_\Omega u_0$ is conserved by the evolution). Now we consider the long term behavior of the solution in the H^1 norm. In this case, we do not need any restriction on $|\int u_0|$.

Note that (A_2) implies

$f(u)u \geq c_5|u|^{2r+2} - c_6$ for some constants c_5 and c_6.

We make additional assumptions on the nonlinearity,

(A_4) $|f(u)| \leq c_7|u|^{2r+1} + c_8$,

(A_5) $F(u) = \int_0^u f(s)ds \leq c_9|u|^{2r+2} + c_{10}$, and $c_5 > c_9$.

Remark 2.14 (A_2), (A_4), and (A_5) hold for $f(u) = c|u|^{2r}u+$ lower terms.

Denote $\bar{\psi} = \frac{1}{|\Omega|}\int_\Omega \psi dx$, write $\varphi = \psi - \bar{\psi}$.

For $\varphi \in L^2(\Omega)$, satisfying $\bar{\varphi} = 0$, we consider the following equation:

$$\begin{cases} -\Delta\theta = \varphi \\ \frac{\partial\theta}{\partial n}|_{\partial\Omega} = 0 \\ \int_\Omega \theta = 0 \end{cases} \tag{2.68}$$

The equation (2.68) has a unique solution $\theta := (-\Delta_0)^{-1}(\varphi)$. Denote $||\varphi||_{-1} = (\int_\Omega(-\Delta_0)^{-1}(\varphi)\varphi dx)^{\frac{1}{2}}$. This is a continuous norm on $L^2(\Omega)$.

Since $\bar{u} = \bar{u}_0$ is constant, we may write the equation as

$$\frac{\partial(u - \bar{u})}{\partial t} = \Delta K(u), \tag{2.69}$$

where $K(u) = \int_\Omega J(x-y)dyu(x) - \int_\Omega J(x-y)u(y)dy + f(u)$. Applying the operator $(-\Delta_0)^{-1}$ to both sides of equation (2.69), we obtain

$$\frac{d(-\Delta_0)^{-1}(u - \bar{u})}{dt} + K(u) = 0. \tag{2.70}$$

Taking the scalar product with $u - \bar{u}$ in $L^2(\Omega)$, we have

$$\frac{1}{2}\frac{d}{dt}||u - \bar{u}||_{-1}^2 + (K(u), u - \bar{u}) = 0. \tag{2.71}$$

From condition (A_2)–(A_5), we have

$$\begin{aligned} &(K(u), u - \bar{u}) \\ &\geq \frac{1}{2}\int\int J(x-y)(u(x) - u(y))^2 dxdy + c_5\int |u|^{2r+2}dx \\ &- \epsilon\int |u|^{2r+2}dx - c(\bar{u}, \epsilon) \end{aligned} \tag{2.72}$$

for any $\epsilon > 0$. Choosing $\epsilon = c_5 - c_9$, we have

$$\begin{aligned}
&(K(u), u - \bar{u}) \\
&\geq \frac{1}{2}\int\int J(x-y)(u(x)-u(y))^2 dxdy + c_9 \int |u|^{2r+2}dx - c(\bar{u}) \\
&\geq E(u) - c(\bar{u}) \\
&= E(u) - c(\bar{u}_0).
\end{aligned} \tag{2.73}$$

Also from (2.73), we have

$$(K(u), u - \bar{u}) \geq c\int |u|^{2r+2}dx - c(\bar{u}_0) \tag{2.74}$$

for some positive constants c and $c(\bar{u}_0)$.

Since $||.||_{-1}$ is a continuous norm on $L^2(\Omega)$, we have

$$||u - \bar{u}||_{-1} \leq C||u - \bar{u}||_2. \tag{2.75}$$

Therefore,

$$\begin{aligned}
||u - \bar{u}_0||_{-1} &\leq C||u - \bar{u}_0||_2 \\
&\leq C||u||_{2r+2} + C(\bar{u}_0)
\end{aligned} \tag{2.76}$$

for some positive constants C and $C(\bar{u}_0)$. From (2.71), (2.74), and (2.76), it follows that

$$\frac{d}{dt}||u - \bar{u}_0||_{-1}^2 + C||u - \bar{u}_0||_{-1}^{2r+2} \leq C(\bar{u}_0). \tag{2.77}$$

By the uniform Gronwall inequality mentioned above, we obtain

$$||u - \bar{u}_0||_{-1}^2 \leq (\frac{C(\bar{u}_0)}{C})^{\frac{1}{r+1}} + (C(r)t)^{\frac{-1}{r}}. \tag{2.78}$$

Thus, we have proved:

Theorem 2.15 *There exists $M(\bar{u}_0)$ such that for any $\rho > M(\bar{u}_0)^{\frac{1}{2r+2}}$, there exists a time t_0 such that*

$$||u - \bar{u}_0||_{-1} \leq \rho, \ \ \forall\, t \geq t_0. \tag{2.79}$$

From (2.71) and (2.72), we also obtain

$$\frac{1}{2}\frac{d}{dt}||u - \bar{u}_0||_{-1}^2 + E(u) \leq c(\bar{u}_0). \tag{2.80}$$

Integrating from t to $t+1$, then (2.79) implies

$$\int_t^{t+1} E(u(s))ds \leq c^*(\bar{u}_0) \equiv c(\bar{u}_0) + \frac{\rho^2}{2} \tag{2.81}$$

for $t \geq t_0$. Since $E(u(t))$ is decreasing, (2.81) implies

$$E(u(t)) \leq c^*(\bar{u}_0) \tag{2.82}$$

for $t \geq t_0 + 1$.

Since, from (2.6),

$$\begin{aligned}
E(u(t)) &\geq \frac{1}{4}\int\int J(x-y)(u(x)-u(y))^2 dxdy + \int F(u)dx \\
&\geq c_3 \int |u|^{2r+2} - c_4,
\end{aligned} \tag{2.83}$$

inequalities (2.82) and (2.83) yield

$$\int |u|^{2r+2} \leq c_*(\bar{u}_0) \tag{2.84}$$

for $t \geq t_0$.

Corollary 2.16 *There exists $c_*(\bar{u}_0) > M(\bar{u}_0)^{\frac{1}{2r+2}}$ such that for any $\rho > c_*(\bar{u}_0)$, there exists a time t_0^* such that*

$$\int |u|^{r+1} \leq c_*(\bar{u}_0) \ for \ t \geq t_0^*. \tag{2.85}$$

Next we estimate $||\nabla u||_2$.

Letting $h(x,u) = a(x)u + f(u)$, multiplying (2.1) by $h(x,u)$ and integrating over Ω, we have

$$\int h(x,u)u_t + \int |\nabla h(x,u)|^2 = \int \nabla J * u \cdot \nabla h(x,u). \tag{2.86}$$

Since

$$h(x,u)u_t = (a(x)u + f(u))u_t = \frac{\partial}{\partial t}[\frac{1}{2}a(x)u^2 + F(u)], \tag{2.87}$$

and

$$\int \nabla J * u \cdot \nabla h(x,u) \leq c||u||_2^2 + \frac{1}{2}||\nabla h(x,u)||_2^2, \tag{2.88}$$

equation (2.86) yields

$$\frac{d}{dt}\int [\frac{1}{2}a(x)u^2 + F(u)] + \frac{1}{2}\int |\nabla h(x,u)|^2 \leq c||u||_2^2. \tag{2.89}$$

Integrate (2.89) from t to $t+1$, and use assumption (A_2) and Corollary 2.16, to obtain

$$\int_t^{t+1} \int |\nabla h(x,u)|^2 \leq c \tag{2.90}$$

for some constant c and all $t \geq t_0^*$.

Multiply (2.1) by $h(x,u)_t$ and integrate on Ω, to obtain

$$\int h(x,u)_t u_t + \int \nabla h(x,u) \cdot \nabla h(x,u)_t = \int \nabla J * u \cdot \nabla h(x,u)_t. \tag{2.91}$$

Since

$$\begin{aligned} &h(x,u)_t u_t = a(x)u_t^2 + f'(u)u_t^2 \geq c_1 u_t^2, \\ &\int \nabla h(x,u) \cdot \nabla h(x,u)_t = \frac{1}{2}\frac{d}{dt}\int |\nabla h(x,u)|^2, \\ \text{and} \\ &\int \nabla J * u \cdot \nabla h(x,u)_t = \frac{d}{dt}\int \nabla J * u \cdot \nabla h(x,u) - \int \nabla J * u_t \cdot \nabla h(x,u), \end{aligned} \tag{2.92}$$

we have

$$\begin{aligned} c_1 \int |u_t|^2 &+ \frac{1}{2}\frac{d}{dt}\int |\nabla h(x,u)|^2 \\ &\leq \frac{d}{dt}\int \nabla J * u \cdot \nabla h(x,u) - \int \nabla J * u_t \cdot \nabla h(x,u). \end{aligned} \tag{2.93}$$

Estimate (2.93), with the Cauchy-Schwartz and Young's inequalities, implies

$$\frac{d}{dt}\int |\nabla h(x,u)|^2 \leq \frac{d}{dt}\int 2\nabla J * u\cdot \nabla h(x,u) + \gamma\int |\nabla h(x,u)|^2 \tag{2.94}$$

for some constant $\gamma > 0$.

For $t < s < t+1$, multiplying (2.94) by $e^{\gamma(t-s)}$, we have

$$\frac{d}{ds}[e^{\gamma(t-s)}\int |\nabla h(x,u)|^2] \leq e^{\gamma(t-s)}\frac{d}{ds}\int 2\nabla J * u\cdot \nabla h(x,u). \tag{2.95}$$

Integrating (2.95) between s and $t+1$, we obtain

$$\begin{aligned} &e^{-\gamma}\int_\Omega |\nabla h(x,u(x,t+1))|^2 - e^{\gamma(t-s)}\int |\nabla h(x,u(x,s))|^2 \\ &\leq \int_s^{t+1} e^{\gamma(t-\mu)}\frac{d}{d\mu}\int_\Omega 2\nabla J * u(\cdot,\mu)\cdot\nabla h(x,u(x,\mu))dxd\mu. \end{aligned} \tag{2.96}$$

Since

$$\begin{aligned} &\int_s^{t+1} e^{\gamma(t-\mu)}\frac{d}{d\mu}\int_\Omega 2\nabla J * u(\cdot,\mu)\cdot\nabla h(x,u(x,\mu))dxd\mu \\ &= e^{\gamma(t-\mu)}\int_\Omega 2\nabla J * u(\cdot,\mu)\cdot\nabla h(x,u(x,\mu))dx|_s^{t+1} \\ &\quad - \int_s^{t+1}(-\gamma)e^{\gamma(t-\mu)}\int_\Omega 2\nabla J * u(\cdot,\mu)\cdot\nabla h(x,u(x,\mu))dxd\mu \\ &= I_1 + I_2. \end{aligned} \tag{2.97}$$

These may be individually estimated yielding

$$\begin{aligned} &e^{-\gamma}\int_\Omega |\nabla h(x,u(x,t+1))|^2 - e^{\gamma(t-s)}\int_\Omega |\nabla h(x,u(x,s))|^2 \leq \\ &\frac{e^{-\gamma}}{2}\int_\Omega |\nabla h(x,u(x,t+1))|^2 + C\int_\Omega |u(x,t+1)|^2 + \int |\nabla h(x,u(x,s))|^2 \\ &+ C\int_\Omega |u(x,s)|^2 + C\int_s^{t+1}[\int_\Omega |\nabla h(x,u(x,\mu))|^2 + \int_\Omega |u(x,\mu)|^2]d\mu. \end{aligned} \tag{2.98}$$

Therefore,

$$\begin{aligned} &\frac{e^{-\gamma}}{2}\int_\Omega |\nabla h(x,u(x,t+1))|^2 \\ &\leq e^{\gamma(t-s)}\int_\Omega |\nabla h(x,u(x,s))|^2 + C\int_\Omega |u(x,t+1)|^2 + \int_\Omega |\nabla h(x,u(x,s))|^2 \\ &+ C\int_\Omega |u(x,s)|^2 + C\int_s^{t+1}[\int_\Omega |\nabla h(x,u(x,\mu))|^2 + \int_\Omega |u(x,\mu)|^2]d\mu. \end{aligned} \tag{2.99}$$

Integrating (2.99) from t to $t+1$ with respect to s, we have

$$\begin{aligned}
&\frac{e^{-\gamma}}{2}\int_{\Omega}|\nabla h(x,u(x,t+1))|^2dx \\
&\leq \int_t^{t+1}\int_{\Omega}|\nabla h(x,u(x,s))|^2dxds + C\int_{\Omega}|u(x,t+1)|^2dx \\
&+\int_t^{t+1}\int_{\Omega}|\nabla h(x,u(x,s))|^2dxds + C\int_t^{t+1}\int_{\Omega}|u(x,s)|^2dxds \\
&+C\int_t^{t+1}(\mu-t)[\int_{\Omega}|\nabla h(x,u(x,\mu))|^2dx+\int_{\Omega}|u(x,\mu)|^2dx]d\mu.
\end{aligned} \tag{2.100}$$

By (2.85) and (2.90), estimate (2.100) yields

$$\int_{\Omega}|\nabla h(x,u(x,t+1))|^2dx \leq C(\bar{u}_0) \tag{2.101}$$

for $t \geq t_0(\bar{u}_0)$ and some $C(\bar{u}_0) > 0$.

Since

$$\nabla h(x,u(x,t+1)) = (a(x)+f'(u(t+1)))\nabla u(x,t+1) - u(x,t+1)\nabla a(x), \tag{2.102}$$

we have

$$\begin{aligned}
\int_{\Omega}|\nabla h(x,u(x,t+1))|^2 &\geq \frac{1}{2}\int_{\Omega}|(a(x)+f'(u(t+1)))|^2|\nabla u(x,t+1)|^2 \\
&-\int_{\Omega}|u(x,t+1)\nabla a(x)|^2 \\
&\geq \int_{\Omega}\frac{1}{2}c_1^2|\nabla u(x,t+1)|^2 - D(\bar{u}_0)
\end{aligned} \tag{2.103}$$

for $t \geq t_0(\bar{u}_0)$ and some constant $D(\bar{u}_0)$.

Estimates (2.101) and (2.103) imply

$$\int_{\Omega}|\nabla u(x,t+1)|^2 \leq G(\bar{u}_0), \tag{2.104}$$

for $t \geq t_0^*(\bar{u}_0)$ and $G(\bar{u}_0) > 0$. Thus, we have

Theorem 2.17 *There exists a time $t_0^*(\bar{u}_0)$ such that*

$$||u||_{H^1} \leq c(\bar{u}_0) \ for \ t \geq t_0^*(\bar{u}_0). \tag{2.105}$$

Remark 2.18 [74] gives a similar result for the Cahn-Hilliard equation.

This boundedness gives weak convergence of subsequences as $t_n \to \infty$ but more is true, as can be demonstrated by calculations similar to the foregoing:

Theorem 2.19 *If u is a solution of (2.1)-(2.3), and $Q(u) = (\int_{\Omega} J(x-y)dy)u(x) - J * u(x) + f(u(x))$, then there exist a sequence $\{t_k\}$ and u^* such that*

$$\begin{aligned}
u(t_k) &\rightharpoonup u^* \ weakly \ in \ H^1 \\
Q(u(t_k)) &\rightharpoonup Q(u^*) \ weakly \ in \ H^1
\end{aligned} \tag{2.106}$$

and $Q(u^*)$ is a constant, i.e. u^* is a steady state solution of (2.1)-(2.3).

We may also use the above techniques for the following integrodifferential equation that may be derived from interacting particle systems with Kawasaki dynamics

$$\begin{cases} \frac{\partial u}{\partial t} = \triangle(u - tanh\beta J * u) & \text{in} \quad \Omega \\ \\ \frac{\partial(u - tanh\beta J*u)}{\partial n} = 0 & \text{on} \quad \partial\Omega \\ \\ u(x,0) = u_0(x) \end{cases} \tag{2.107}$$

where β is a constant, J is a smooth function.

Wellposedness and regularity of solutions is established along the lines used for (2.1)-(2.3) with the usual smoothness assumptions on J, f, Ω, and the initial data. Note that the average of u is constant in time and one can show that there is an absorbing set in every constant mass affine subspace of H^1.

Returning to the nonlocal Cahn-Hilliard equation (2.1), we may also append the homogeneous Dirichlet boundary condition

$$u(x) = 0 \quad \text{for} \quad x \in \partial\Omega. \tag{2.108}$$

While we no longer have conservation of the integral, this boundary condition is strongly dissipative and so we expect results similar to those above. In particular with condition

$(A_2)_D$ There exists $c_1 > 0$ such that $a(x,u) \equiv a(x) + f'(u) \geq c_1$,

one can prove

Proposition 2.20 *Assume* $(A_1), (A_2)_D$, *and* (A_3). *If* $u(x,t) \in C(\bar{Q}_T) \cap C^{2,1}(Q_T)$ *is a solution of (2.1),(2.108), with initial data* u_0 *then*

$$\max_{\bar{Q}_T} |u| \leq C(\Omega, T, u_0) \tag{2.109}$$

for some positive constant $C(\Omega, T, u_0)$.

This is proved by letting $u(x,t) = v(x,t)e^{\sigma t}$ for appropriate choice of σ, multiplying the v-equation by v to obtain an equation for v^2, and applying a maximum principle. A result similar to Theorem 2.4 gives gradient and Hölder-$(1+\alpha)$ bounds on the solution for small $\alpha > 0$. Then it is straightforward to apply the Schauder Fixed Point Theorem to establish the existence of a classical solution.

Under the assumption $(A_2)_D$, equation (2.1) is a nondegenerate parabolic equation. We may also consider the degenerate case. Consider the following equation with $u_0 \in L^\infty(\Omega)$

$$\begin{cases} \frac{\partial u}{\partial t} = \triangle(h(x,u)) - \int_\Omega u(y) \triangle J(x-y)dy & \text{in} \quad Q_T \\ u = 0 & \text{on} \quad S_T \\ u(x,0) = u_0(x), \end{cases} \tag{2.110}$$

where

$$h(x,u) = a(x)u(x) + f(u).$$

Instead of nondegeneracy condition $(A_2)_D$, we assume:

(B_1) For every fixed x, $h(x,0) = 0$, and $\frac{\partial h(x,u)}{\partial u} \geq d_1|u|^{r_1}$ for some positive constants r_1 and d_1.

Definition 2.21 A generalized solution of (2.110) is a function $u \in C([0,T] : L^1(\Omega)) \cap L^\infty(Q_T)$ such that

$$\int_\Omega u(x,t)\psi(x,t)dx - \int\int_{Q_t} u(x,t)\psi_s(x,s)dxds = \int\int_{Q_t} h(x,u)\triangle\psi(x,s)dxds - \int\int_{Q_t} (\triangle J * u(\cdot,s))\psi(x,s)dxds + \int_\Omega u(x,0)\psi(x,0)dx \tag{2.111}$$

for all $\psi \in C^{2,1}(\bar{Q}_T)$ such that $\psi(x,t) = 0$ for $x \in \partial\Omega$ and $0 \leq t \leq T$, and

$$u(x,0) = u_0(x). \tag{2.112}$$

We first prove the uniqueness.

Proposition 2.22 *Let u_1, u_2 be two solutions of equation (2.110) with initial data u_{10}, $u_{20} \in L^\infty(\Omega)$, then*

$$||u_1(\tau) - u_2(\tau)||_{L^1(\Omega)} \leq C(T)||u_{10} - u_{20}||_{L^1(\Omega)}$$

for each $\tau \in (0,T)$, and some constant $C(T)$.

Proof For any $\tau \in (0,T)$, and $\psi \in C^{2,1}(\bar{Q}_\tau)$ with $\psi|_{\partial\Omega} = 0$ for $0 < t < \tau$, after multiplying (2.110) by ψ and integrating over $\Omega \times (0,\tau)$, we have

$$\int_\Omega u_i(x,\tau)\psi(x,\tau)dx = \int_\Omega u_i(x,0)\psi(x,0)dx + \int_0^\tau\int_\Omega (u_i\psi_t + h(x,u_i)\triangle\psi)dxdt + \int_0^\tau\int_\Omega (\triangle J * u_i)\psi dxdt. \tag{2.113}$$

Setting $z = u_1 - u_2$ and $z_0 = u_{10} - u_{20}$, equation (2.113) gives

$$\int_\Omega z(x,\tau)\psi(x,\tau)dx = \int_\Omega z_0(x)\psi(x,0)dx + \int_0^\tau\int_\Omega z(\psi_t + b(x,t)\triangle\psi)dxdt + \int_0^\tau\int_\Omega (\triangle J * z)\psi dxdt, \tag{2.114}$$

where

$$b(x,t) = \begin{cases} \frac{h(x,u_1)-h(x,u_2)}{u_1-u_2} & \text{for} \quad u_1 \neq u_2, \\ h_u(x,u_1) & \text{for} \quad u_1 = u_2. \end{cases}$$

Following the ideas in [6], we consider problem:

$$\begin{cases} \frac{\partial\psi}{\partial t} = -b\triangle\psi + \nu\psi & \text{in} \quad \Omega, \ 0 < t < \tau, \\ \psi = 0 & \text{on} \quad \partial\Omega, \ 0 < t < \tau, \\ \psi(x,\tau) = g(x), \end{cases} \tag{2.115}$$

where $g(x) \in C_0^\infty(\Omega)$, $0 \leq g \leq 1$, and $\nu > 0$ is constant.

Since b just belongs to $L^\infty(Q_T)$ and may be equal to zero, we perturb to get a nondegenerate equation, by setting $b_n = \rho_n * b + \frac{1}{n}$, where ρ_n is a mollifier in $\mathbb{R}^n$, and $\int_0^\tau \int_\Omega (\rho_n * b - b)^2 dxdt \le \frac{1}{n^2}$. Consider

$$\begin{cases} \frac{\partial \psi}{\partial t} = -b_n \triangle \psi + \nu \psi & \text{in} \quad \Omega, \ 0 < t < \tau, \\ \psi = 0 & \text{on} \quad \partial\Omega, \ 0 < t < \tau, \\ \psi(x, \tau) = g(x). \end{cases} \tag{2.116}$$

Since $b_n \ge \frac{1}{n}$, the equation is a nondegenerate parabolic equation, and so there exists a solution $\psi_n \in C^{2,1}(\bar{Q}_\tau)$.

The following, whose proof we omit, is easily established.

Lemma 2.23 *The solution of (2.116) has the following properties*

$$(i) \ \ 0 \le \psi_n \le e^{\nu(t-\tau)},$$

$$(ii) \ \ \int_0^\tau \int_\Omega b_n |\triangle(\psi_n)|^2 dxdt \le C,$$

$$(iii) \ \ \sup_{0 \le t \le \tau} \int_\Omega |\nabla \psi_n|^2 dx \le C,$$

where the constant C depends only on g.

Replacing ψ by ψ_n in (2.114), and using (2.116) we obtain

$$\begin{aligned} \int_\Omega z(x,\tau) g(x) dx - \int_0^\tau \int_\Omega z(b - b_n) \triangle \psi_n dxdt \\ = \int_\Omega z(x,0)\psi_n(0) dx + \int\int_{Q_\tau} (\triangle J * z + \nu z)\psi_n dxdt. \end{aligned} \tag{2.117}$$

Since

$$\begin{aligned} &\int_0^\tau \int_\Omega z(b - b_n) \triangle \psi_n dxdt \\ &\le C\left(\int_0^\tau \int_\Omega \frac{(b-b_n)^2}{b_n} dxdt\right)^{\frac{1}{2}} \left(\int_0^\tau \int_\Omega b_n |\triangle \psi_n|^2 dxdt\right)^{\frac{1}{2}} \\ &\le \frac{C}{\sqrt{n}} \to 0, \end{aligned}$$

equation (2.117) implies

$$\begin{aligned} &\int_\Omega z(x,\tau) g(x) dx \\ &\le \int_\Omega |z(x,0)| e^{\nu(t-\tau)} dx + \int\int_{Q_\tau} |\triangle J * z + \nu z| e^{\nu(t-\tau)} dxdt. \end{aligned} \tag{2.118}$$

Letting $\nu \to 0$ and $g(x) \to signz^+(x,\tau)$ in (2.118), we have

$$\int_\Omega (u_1 - u_2)^+ dx \le \int_\Omega |u_{10} - u_{20}| dx + \int\int_{Q_\tau} |\triangle J * z| dxdt. \tag{2.119}$$

Interchanging u_1 and u_2 yields

$$\int_\Omega |u_2 - u_1| dx \le \int_\Omega |u_{20} - u_{10}| dx + C \int\int_{Q_\tau} |u_2 - u_1| dxdt. \tag{2.120}$$

(2.120) and Gronwall's inequality imply the conclusion. □

Remark 2.24 Since every classical solution is also a weak solution, this also proves the uniqueness and continuous dependence on initial values for classical solutions.

To prove the existence of a solution to (2.111), we consider the regularized problem and take $u_0 \in C^{2+\alpha}(\bar{\Omega})$ for some $\alpha > 0$, with $u_0|_{\partial\Omega} = 0$:

$$\begin{cases} \frac{\partial u}{\partial t} = \triangle(h^\epsilon(x,u)) - \int_\Omega \triangle J(x-y)u(y)dy & \text{in} \quad Q_T, \\ u = 0 & \text{on} \quad S_T, \\ u(x,0) = u_0(x), \end{cases} \tag{2.121}$$

where

$$h^\epsilon(x,u) = a(x)u(x) + f(u) + \epsilon u.$$

We have shown that there exists a classical solution $u_\epsilon(x,t) \in C^{2+\alpha, \frac{2+\alpha}{2}}(\bar{Q}_T)$. It is easy to show that these solutions are uniformly bounded on $\bar{Q}_T$.

Using the growth conditions and Arzela-Ascoli's lemma, one can then prove

Theorem 2.25 *For any $T > 0$ and $u_0 \in L^\infty(\Omega)$, if conditions (A_1), (B_1), and (A_3) are satisfied, then there exists a unique function $u \in C([0,T], L^1(\Omega)) \cap L^\infty(Q_T)$ which satisfies equation (2.111).*

Results concerning the long-term behavior of solutions to the Dirichlet problem follow from similar ideas introduced for the case of no-flux boundary conditions but this time a nonlinear version of the *Poincaré* inequality is not used.

In order to prove the existence of an absorbing set, instead of $(A_2)_D$, we assume the original (A_2) and

(A_4) There exist positive constants c_3 and c_4 such that $a(x,u) \leq c_3|u|^r + c_4$.

First one establishes L^p bounds for solutions by using a Gronwall inequality after multiplying the equation by a power of u and integrating, performing the usual tedious manipulations.

Then one proves

Proposition 2.26 *If $u_0 \in L^\infty(\Omega)$, then*

$$\sup_{t \geq 0} ||u||_\infty \leq C(u_0). \tag{2.122}$$

Also, gradient bounds may be obtained using the above and further calculations:

Theorem 2.27 *Assume that u is a solution of (2.1), (2.108) and conditions (A_1) - (A_4) are satisfied. There exists $t_0 > 0$ such that if $t \geq t_0$ then*

$$\sup_{t \geq t_0} ||\nabla u||_2 < C, \tag{2.123}$$

where constant C does not depend on initial data.

If we restrict our attention to one space dimension where better embedding theorems are in force, one can then prove:

Theorem 2.28 *For $n = 1$, if conditions (A_1) - (A_4) are satisfied, then the semigroup associated with (2.1) with Dirichlet boundary conditions possesses an attractor $\mathcal{A} \subset H^1(\Omega) \cap X$ which is maximal and compact.*

3 Nonlocal phase-field system

We now turn to the system where the temperature evolves and the order parameter represents local solidification, partially driven by temperature and phase change in turn producing or absorbing heat energy, thus driving temperature. The following presents some results reported in [19]. As outlined above, this system has the form:

$$u_t = J * u - u \int_\Omega J(x-y)dy - f(u) + l\theta, \tag{3.1}$$

$$(\theta + lu)_t = \Delta\theta, \tag{3.2}$$

which is complemented by the initial and boundary conditions

$$u(0,x) = u_0(x),\ \theta(0,x) = \theta_0(x), \tag{3.3}$$

$$\frac{\partial\theta}{\partial n}|_{\partial\Omega} = 0, \tag{3.4}$$

where $T > 0$ and $\Omega \subset \mathbb{R}^n$ is a bounded domain. We are interested in the well posedness of this initial and boundary value problem.

In order to prove the existence, we make the following assumptions

(P_1) $M \equiv \sup \int_\Omega |J(x-y)|dy < \infty$ and $f \in C(\mathbb{R})$.
(P_2) There exist $c_1 > 0$, $c_2 > 0$, $c_3 > 0$, $c_4 > 0$ and $r > 2$ such that $f(u)u \geq c_1|u|^r - c_2|u|$, and $|f(u)| \leq c_3|u|^{r-1} + c_4$.

Note that (P_2) implies

$$F(u) = \int_0^u f(s)ds \geq c_5|u|^r - c_6|u| \tag{3.5}$$

for some positive constants c_5 and c_6.

We prove the existence of a solution to (3.1)-(3.4) by the method of successive approximation.

Define $\theta^{(0)}(t,x) := \theta_0(x)$ and for $k \geq 1$ define $(u^{(k)}, \theta^{(k)})$ iteratively to be solutions to the system

$$u_t^{(k)} = \int_\Omega J(x-y)u^{(k)}(y)dy - \int_\Omega J(x-y)dyu^{(k)}(x) - f(u^{(k)}) + l\theta^{(k-1)}, \tag{3.6}$$

$$\theta_t^{(k)} - \triangle\theta^{(k)} + \theta^{(k)} = -lu_t^{(k)} + \theta^{(k-1)} \tag{3.7}$$

in $(0,T)\times\Omega$, with initial and boundary conditions

$$u^{(k)}(0,x) = u_0(x),\ \theta^{(k)}(0,x) = \theta_0(x), \tag{3.8}$$

$$\frac{\partial\theta^{(k)}}{\partial n}|_{\partial\Omega} = 0. \tag{3.9}$$

Lemma 3.1 *With $k = 1$, for any $T > 0$, if $u_0 \in L^\infty(\Omega)$, and $\theta_0 \in H^1 \cap L^\infty(\Omega)$, then there exists a unique solution (u,θ) to system (3.6) -(3.9). Furthermore, $u^{(1)}$, $u_t^{(1)} \in L^\infty((0,T), L^\infty(\Omega))$ and $\theta^{(1)} \in L^\infty((0,T), L^\infty(\Omega)) \cap L^2((0,T), H^2(\Omega))$.*

Proof Since the right hand side of equation (3.1) is locally Lipschitz continuous in $L^\infty((0,T), L^\infty(\Omega))$, local existence follows from standard ODE theory. In order to prove the global existence, we prove global boundedness of the solutions. For any

$p > 1$, multiplying equation (3.1) by $|u^{(1)}|^{p-1}u$ and integrating over Ω, we obtain

$$\begin{aligned}&\frac{1}{p+1}\frac{d}{dt}\int |u^{(1)}|^{p+1}dx + \int f(u^{(1)})|u^{(1)}|^{p-1}udx \\ &= \int\int J(x-y)u^{(1)}(y)|u^{(1)}|^{p-1}u^{(1)}dxdy \\ &- \int\int J(x-y)u^{(1)}(x)|u^{(1)}|^{p-1}u^{(1)}dxdy + l\int \theta^{(0)}|u^{(1)}|^{p-1}udx.\end{aligned} \tag{3.10}$$

Using *Hölder*'s and Young's inequalities and conditions (P_1) and (P_2), we have

$$\begin{aligned}&\frac{1}{p+1}\frac{d}{dt}\int |u^{(1)}|^{p+1}dx + C\int |u^{(1)}|^{p+r-1}udx \\ &\leq C(p){C_1}^{p+1},\end{aligned} \tag{3.11}$$

where C_1 is a constant independent of p and $lim_{p\to\infty}C(p)^{\frac{1}{p+1}} \leq C_2$ with C_2 independent of p.

Using the uniform Gronwall inequality and (3.11), we have

$$||u^{(1)}||_{p+1}^{p+1} \leq (C(p)C_1^{p+1})^{\frac{p+1}{p+r-1}} + (C(r-2)t)^{\frac{-(p+1)}{r-2}}. \tag{3.12}$$

Therefore,

$$||u^{(1)}||_{p+1} \leq C(p)^{\frac{1}{p+1}}(C_1)^{\frac{p+1}{p+r-1}} + (C(r-2)t)^{\frac{-1}{r-2}}. \tag{3.13}$$

Letting $p \to \infty$, we have

$$||u^{(1)}||_\infty \leq C. \tag{3.14}$$

for some constant C.

Also from condition (P_2) and equation (3.6), we have

$$||u_t^{(1)}||_\infty \leq C. \tag{3.15}$$

Since equation (3.7) is a linear parabolic equation, by inequality (3.15) and standard parabolic theory, we have $\theta^{(1)} \in L^\infty((0,T), L^\infty(\Omega)) \cap L^2((0,T), H^2(\Omega))$. □

By induction, there exists a unique solution $(u^{(k)}, \theta^{(k)})$ of system (3.6)-(3.8). Furthermore, $u^{(k)}$, $u_t^{(k)} \in L^\infty((0,T), L^\infty(\Omega))$ and

$$\theta^{(k)} \in L^\infty((0,T), L^\infty(\Omega)) \cap L^2((0,T), H^2(\Omega))$$

for every k. Now we prove that there exists a uniform bound for $u^{(k)}$, $u_t^{(k)}$ and $\theta^{(k)}$.

Multiplying equation (3.7) by $|\theta^{(k)}|^{p-1}\theta^{(k)}(x)$ for $p > \frac{n}{2}$, and integrating over Ω, we have

$$\begin{aligned}&\int |\theta^{(k)}|^{p-1}\theta^{(k)}\theta_t^{(k)}dx + \int \nabla(|\theta^{(k)}|^{p-1}\theta^{(k)})\cdot\nabla\theta^{(k)}dx + \int |\theta^{(k)}|^{p+1}dx \\ &= -l\int\int J(x-y)u^{(k)}(y)|\theta^{(k)}|^{p-1}\theta^{(k)}dydx + l\int f(u^{(k)})|\theta^{(k)}|^{p-1}\theta^{(k)}dx \\ &+ l\int\int J(x-y)u^{(k)}(x)|\theta^{(k)}|^{p-1}\theta^{(k)}dydx + (1-l^2)\int |\theta^{(k)}|^{p-1}\theta^{(k)}\theta^{(k-1)}dx.\end{aligned} \tag{3.16}$$

Since

$$|\nabla|\theta|^{\frac{p+1}{2}}|^2 = \frac{(p+1)^2}{4}|\theta|^{p-1}|\nabla\theta|^2 = \frac{(p+1)^2}{4p}\nabla(|\theta|^{p-1}\theta)\cdot\nabla\theta, \tag{3.17}$$

using *Hölder*'s and Young's inequalities, we obtain

$$\begin{aligned}&\frac{1}{p+1}\frac{d}{dt}\int_{\Omega}|\theta^{(k)}|^{p+1}dx+\frac{4p}{(p+1)^2}\int|\nabla|\theta^{(k)}|^{\frac{p+1}{2}}|^2dx+\frac{1}{2}\int|\theta^{(k)}|^{p+1}dx\\&\leq c_1(l,p)\int|u^{(k)}|^{p+1}dx+c_2(l,p)\int|\theta^{(k-1)}|^{p+1}dx+\int|f(u^{(k)})|^{p+1}dx\end{aligned}\tag{3.18}$$

for some positive constants $c_1(l,p)$ and $c_2(l,p)$, depending only on p and l.

Multiplying equation (3.6) by $|u^{(k)}|^{(r-1)p-1}u^{(k)}$, and integrating over Ω, we obtain

$$\begin{aligned}&\frac{1}{(r-1)p+1}\frac{d}{dt}\int|u^{(k)}|^{(r-1)p+1}dx+\int f(u^{(k)})|u^{(k)}|^{(r-1)p-1}u^{(k)}dx\\&=\int\int J(x-y)u^{(k)}(y)|u^{(k)}|^{(r-1)p-1}u^{(k)}dxdy+l\int\theta^{(k-1)}|u^{(k)}|^{(r-1)p-1}u^{(k)}dx\\&\quad-\int\int J(x-y)u^{(k)}(x)|u^{(k)}|^{(r-1)p-1}u^{(k)}dxdy.\end{aligned}\tag{3.19}$$

Condition (P_2) implies

$$f(u)|u|^{(r-1)p-1}u\geq c_1|u|^{(r-1)(p+1)}-c_2|u|^{(r-1)p}\tag{3.20}$$

and

$$|f(u)|^{p+1}\leq c_7|u|^{(r-1)(p+1)}+c_8\tag{3.21}$$

for some positive constants c_7 and c_8. ¿From equation (3.19), inequality (3.20), *Hölder*'s and Young's inequalities, we have

$$\begin{aligned}&\frac{1}{(r-1)p+1}\frac{d}{dt}\int|u^{(k)}|^{(r-1)p+1}dx+\frac{c_1}{2}\int|u^{(k)}|^{(r-1)(p+1)}dx\\&\qquad\leq c(r,p)+c_1(r,p,l)\int|\theta^{(k-1)}|^{p+1}dx\end{aligned}\tag{3.22}$$

for some positive constants $c(r,p)$ and $c_1(r,p,l)$.

Integrating (3.22) from 0 to t, we obtain

$$\begin{aligned}&\frac{1}{(r-1)p+1}\int|u^{(k)}|^{(r-1)p+1}dx+\frac{c_1}{2}\int_0^t\int|u^{(k)}|^{(r-1)(p+1)}dx\\&\qquad\leq c(r,p)t+c_1(r,p,l)\int_0^t\int\theta^{(k-1)}|^{p+1}dx+\int|u_0|^{(r-1)p+1}\\&\qquad\leq c(u_0,T,r,p)+c_1(r,p,l)\int_0^t\int\theta^{(k-1)}|^{p+1}dx\end{aligned}\tag{3.23}$$

for some positive constants $c(u_0,T,r,p)$ and $c_1(r,p,l)$.

Integrating inequality (3.18) from 0 to t, using (3.21) and (3.23), we have

$$\int_{\Omega}|\theta^{(k)}|^{p+1}dx\leq c(u_0,\theta_0,p,r,l,T)(1+\int_0^t\int|\theta^{(k-1)}|^{p+1}dxds)\tag{3.24}$$

for some positive constant $c(u_0,\theta_0,p,r,l,T)$ which does not depend on k.

By induction, we have

$$\int_{\Omega}|\theta^{(k)}|^{p+1}dx\leq ce^{ct}\tag{3.25}$$

for some positive constant c which does not depend on k.

Similarly from inequalities (3.22) and (3.25), we also have

$$\int_{\Omega} |u^{(k)}|^{p+1} dx \le C, \tag{3.26}$$

and

$$\int_{\Omega} |f(u^{(k)})|^{p+1} dx \le C \tag{3.27}$$

for some positive constant C which does not depend on k.

Equation (3.6), inequalities (3.25)-(3.27), and Young's inequality imply

$$\int_{\Omega} |u_t^{(k)}|^{p+1} dx \le C \tag{3.28}$$

for some positive constant C which does not depend on k.

This implies $-lu_t^{(k)} + \theta^{(k-1)} \in L^{p+1}((0,T), L^{p+1}(\Omega))$ and

$$|| - lu_t^{(k)} + \theta^{(k-1)}||_{p+1} \le C \tag{3.29}$$

for some positive constant C which does not depend on k.

Applying standard parabolic estimates to equation (3.7), and using inequality (3.28), we have

$$||\theta^{(k)}||_\infty \le C. \tag{3.30}$$

Multiplying equation (3.7) by θ_t^k, and integrating equation (3.7) over Ω, using *Hölder* and Young's inequalities and (3.30), we have

$$\int_0^T \int_{\Omega} |\theta_t^{(k)}|^2 dxdt \le C \tag{3.31}$$

for some constant C which does not depend on k.

Equation (3.7), inequalities (3.28), (3.30), and (3.31) yield

$$\int_0^T \int_{\Omega} |\triangle\theta^{(k)}|^2 dxdt \le C \tag{3.32}$$

for some constant C which does not depend on k.

Since $||\theta^{(k)}||_\infty \le C$, using a similar argument to that in the proof of Lemma 3.1, we have

$$||u^{(k)}||_\infty \le C, \tag{3.33}$$

and

$$||u_t^{(k)}||_\infty \le C \tag{3.34}$$

for some constant C which does not depend on k.

Next we prove the convergence of $\{\theta^{(k)}\}$ in $C([0,T], L^2(\Omega))$. From equation (3.7), we have

$$\begin{aligned}(\theta^{(k+1)} - \theta^{(k)})_t - \triangle(\theta^{(k+1)} - \theta^{(k)}) + (\theta^{(k+1)} - \theta^{(k)}) \\ = -l(u^{(k+1)} - u^{(k)})_t + (\theta^{(k)} - \theta^{(k-1)}).\end{aligned} \tag{3.35}$$

Multiplying equation (3.35) by $(\theta^{(k+1)}-\theta^{(k)})$, and integrating over Ω, using *Hölder*'s and Young's inequalities, we have

$$\begin{aligned}\frac{1}{2}\frac{d}{dt}\int|\theta^{(k+1)}-\theta^{(k)}|^2dx+\int|\nabla(\theta^{(k+1)}-\theta^{(k)})|^2dx+\frac{1}{2}\int(\theta^{(k+1)}-\theta^{(k)})^2\\ \leq l^2\int|u_t^{(k+1)}-u_t^{(k)}|^2dx+\int|\theta^{(k)}-\theta^{(k-1)}|^2dx.\end{aligned}\tag{3.36}$$

Since $||u^{(k)}||_\infty\leq C$, from equation (3.6), and condition (P_2), we have

$$\int|u^{(k+1)}-u^{(k)}|^2dx\leq C(T)\int|\theta^{(k)}-\theta^{(k-1)}|^2dx,\tag{3.37}$$

and

$$\begin{aligned}\int|f(u^{(k+1)})-f(u^{(k)})|^2dx &= \int|f'(\lambda u^{(k+1)}+(1-\lambda)u^{(k)})(u^{(k+1)}-u^{(k)})|^2dx\\ &\leq C(T)\int|u^{(k+1)}-u^{(k)}|^2dx.\end{aligned}\tag{3.38}$$

Therefore, equation (3.6), and inequalities (3.37)-(3.38) imply

$$\begin{aligned}&\int|u_t^{(k+1)}-u_t^{(k)}|^2dx\\ &\leq 4\int|\int J(x-y)(u^{(k+1)}-u^{(k)})dy|^2dx\\ &+4\int(\int J(x-y)dy)^2(u^{(k+1)}-u^{(k)})^2dx\\ &+4\int(f(u^{(k+1)})-f(u^{(k)})^2dx+4\int(\theta^{(k)}-\theta^{(k-1)})^2dx\\ &\leq C_1(T)\int|\theta^{(k)}-\theta^{(k-1)}|^2dx\end{aligned}\tag{3.39}$$

for some positive constant $C_1(T)$ which does not depend on k.

Inequalities (3.36)-(3.39) yield

$$\frac{d}{dt}\int|\theta^{(k+1)}-\theta^{(k)}|^2dx\leq C(T)\int|\theta^{(k)}-\theta^{(k-1)}|^2dx\tag{3.40}$$

for some positive constant $C(T)$ which does not depend on k.

By induction, this implies

$$\int|\theta^{(k+1)}-\theta^{(k)}|^2dx\leq\frac{(ct)^{(k-1)}}{(k-1)!}\int_0^t\int|\theta^1-\theta^0|dxds.\tag{3.41}$$

So $\theta^{(k)}$ is a Cauchy sequence in $C([0,T],L^2(\Omega))$. Therefore, there exists $\theta\in C([0,T],L^2(\Omega))$ such that $\theta^{(k)}\to\theta$ in $C([0,T],L^2(\Omega))$. From (3.30)-(3.32), we have

$$||\theta||_\infty\leq C,\tag{3.42}$$

$$\int_0^T\int_\Omega|\triangle\theta|^2dxdt\leq C,\tag{3.43}$$

$$\int_0^T\int_\Omega|\theta_t|^2dxdt\leq C.\tag{3.44}$$

Also from (3.33), (3.37)-(3.39), we have

$$u^{(k)} \to u \; in \; C([0,T], L^2(\Omega)), \tag{3.45}$$

$$u_t^{(k)} \to u_t \; in \; C([0,T], L^2(\Omega)), \tag{3.46}$$

$$f(u^{(k)}) \to f(u) \; in \; C([0,T], L^2(\Omega)). \tag{3.47}$$

Therefore, letting $k \to \infty$ in equation (3.6), we have

$$u_t = \int_\Omega J(x-y)u(y)dy - \int_\Omega J(x-y)dyu(x) - f(u) + l\theta \tag{3.48}$$

for $t > 0$ and a.e. $x \in \Omega$.

Since $u_t^{(k)} \rightharpoonup u_t$, $\theta_t^{(k)} \rightharpoonup \theta_t$, $\triangle\theta^{(k)} \rightharpoonup \triangle\theta$ in $L^2((0,T), L^2(\Omega))$, letting $k \to \infty$ in the weak form of equation (3.7), we have

$$\int_0^T \int_\Omega (lu_t + \theta_t)\xi(t,x)dxdt = \int_0^T \int_\Omega \triangle\theta\xi(t,x)dxdt \tag{3.49}$$

for $\xi(t,x) \in L^2((0,T), L^2(\Omega))$.

Since it is true of $\theta^{(k)}$, we also have

$$\int_0^T \int_\Omega \eta(t)(\triangle\theta\varphi + \nabla\theta \cdot \nabla\varphi)dxdt = 0 \tag{3.50}$$

for any $\varphi \in W^{1,2}(\Omega)$ and $\eta \in L^2(0,T)$. This implies $\frac{\partial\theta}{\partial n} = 0$ a.e on $(0,T) \times \partial\Omega$. Also we have

$$\int_\Omega |\theta(0,x) - \theta_0|^2 dx \le 3(\int_\Omega |\theta(0,x) - \theta(t,x)|^2 dx + \int_\Omega |\theta(t,x) - \theta^{(k)}(t,x)|^2 dx + \int_\Omega |\theta^{(k)}(t,x) - \theta_0|^2 dx). \tag{3.51}$$

Since $\theta^{(k)}(t,x) \to \theta$ in $C([0,T], L^2(\Omega))$, and since $\theta^{(k)}(t,x)$ and $\theta(t,x)$ are continuous with respect to t in $L^2(\Omega)$, by taking k arbitrarily large we can see that $\theta(0,x) = \theta_0$ a.e. in Ω. Similarly, $u(0,x) = u_0$ a.e. in Ω.

Equations (3.48)-(3.51) imply that u and θ are solutions of system (3.1)-(3.4) in a weak sense.

To prove uniqueness and continuous dependence on initial data, let $\theta_{i0} \in L^\infty(\Omega) \cap W^{1,2}(\Omega)$, $u_{i0} \in L^\infty(\Omega)$, and for $R > 0$, $||\theta_{i0}||_{L^\infty} < R$, $||u_{i0}||_{L^\infty} < R$, where $i = 1, 2$.

Let u_i and θ_i be solutions corresponding to initial data u_{i0} and θ_{i0}, then we have $||\theta_i||_{L^\infty} \le C(T,R)$, and $||u_i||_{L^\infty} \le C(T,R)$.

Denote $v = u_1 - u_2$, $w = \theta_1 - \theta_2$. We have

$$v_t = \int_\Omega J(x-y)v(y)dy - \int_\Omega J(x-y)dyv(x) - f'(\lambda u_1 + (1-\lambda)u_2)v + lw, \tag{3.52}$$

$$(w + lv)_t = \triangle w \tag{3.53}$$

in $(0,T) \times \Omega$, for some $\lambda(x,t) \in [0,1]$. We also have initial and boundary conditions

$$v(0,x) = v_0(x), \; w(0,x) = w_0(x), \tag{3.54}$$

$$\frac{\partial w}{\partial n}|_{\partial\Omega} = 0. \tag{3.55}$$

Multiplying equation (3.52) by v_t, integrating over Ω, multiplying equation (3.52) by v, integrating over Ω, multiplying equation (3.53) by w, integrating over Ω, we have

$$\begin{aligned} \int |v_t|^2 = \int\int J(x-y)v(y)dyv_t dx - \int J(x-y)dyv(x)v_t \\ - \int (f'(\lambda u_1 + (1-\lambda)u_2)vv_t + lwv_t)dx, \end{aligned} \tag{3.56}$$

$$\begin{aligned} \int v_t v = \int\int_\Omega J(x-y)v(y)dyvdx - \int_\Omega J(x-y)dyv^2 \\ - \int (f'(\lambda u_1 + (1-\lambda)u_2)v^2 + lwv)dx, \end{aligned} \tag{3.57}$$

$$\int (w_t w + l v_t w) = -\int |\nabla w|^2 dx, \tag{3.58}$$

Adding equations (3.56)-(3.58) together, using *Hölder*'s and Young's inequalities, we have

$$\frac{d}{dt}\int [w^2+v^2]dx \le C_2(T,R)\int [w^2+v^2]dx \tag{3.59}$$

for some positive constant $C_2(T,R)$.

Inequality (3.59) and Gronwall's inequality imply the uniqueness and continuous dependence on initial data of the solution of (3.6)-(3.7).

Denote $Q_T = (0,T)\times\Omega$, we have the following theorem:

Theorem 3.2 *If assumptions $(P_1),(P_2)$ are satisfied, $u_0 \in L^\infty(\Omega)$ and $\theta_0 \in L^\infty \cap H^1(\Omega)$, then there exists a unique solution $(u,\theta) \in C([0,T], L^\infty(\Omega))$ to the system (3.6)-(3.9) such that $u_t \in L^\infty(Q_T)$, and u_{tt}, θ_t, $\triangle\theta \in L^2(Q_T)$.*

Results concerning the asymptotic behavior of solutions follow along similar, though somewhat more complicated, lines as for the nonlocal Cahn-Hilliard equation above. Here the results are summarized without proof.

Recall that $I_0 \equiv \int(\theta_0 + lu_0)$ is conserved.

Theorem 3.3 *There exists a constant $C(I_0)$ otherwise independent of initial data such that*

$$\begin{aligned} ||u||_r \le C(I_0), \\ ||\theta||_{H^1} \le C(I_0) \end{aligned}$$

for $t \ge t_0(I_0)$.

For the following results let

$$X = \{\phi : A\phi \equiv \phi - \triangle\phi \in L^p, \partial_\nu\phi = 0\}$$

and let X^α be the space $D(A^\alpha)$ endowed with the graph norm $||.||_\alpha$ of A^α for $\frac{n}{2p} < \alpha < 1$.

Theorem 3.4 *Suppose that conditions (P_1) and (P_2) are satisfied and $(u_0,\theta_0) \in L^\infty \times X^\alpha$. Then the solution $(u,\theta) \in L^\infty \times X^\alpha$ satisfies*

$$\sup_{0\le t<\infty} ||\theta(t)||_{X^\alpha} \le C_1(||\theta_0||_{X^\alpha}, ||u_0||_\infty),$$

$$\sup_{0\le t<\infty} ||u(t)||_\infty \le C_2(||\theta_0||_{X^\alpha}, ||u_0||_\infty),$$

$$\lim_{t\to\infty} ||\theta - \bar{\theta}||_{W^{1,q}} = \lim_{t\to\infty} ||u_t||_2 = 0,$$

for some $q > n$.

If $u_0(x) \in W^{1,\sigma}(\Omega)$, $\theta_0(x) \in W^{2,\sigma}(\Omega)$ for $\sigma > n$, and $f' + \int_\Omega J(x-y)dy \ge C > 0$, then there exists a subsequence t_k such that

$$(u(t_k), \theta(t_k)) \to (u^*, c) \text{ in } C^\gamma(\Omega),$$

where (u^*, c) is a steady state solution.

4 Pulses for the nonlocal wave equation

The results here are with Chunlei Zhang and are given in full in [20]. We consider the nonlocal wave equation for $u(x,t)$:

$$u_{tt} - \frac{1}{\varepsilon^2}(j_\varepsilon * u - u) + f(u) = 0, \quad \text{for} \quad t > 0 \quad \text{and} \quad x \in \mathbb{R}, \tag{4.1}$$

where ε is a positive parameter and the kernel j_ε of the convolution is defined by

$$j_\varepsilon(x) = \frac{1}{\varepsilon} j(\frac{x}{\varepsilon}),$$

where $j(\cdot)$ is an even function with unit integral. We assume that f is a C^2 function, satisfying $f(0) = 0$, $f'(0) < 0$, and the first positive zero of

$$F(\zeta) = \int_0^\zeta f(s)ds$$

is nondegenerate. Typical examples include the quadratic function $f(u) = u(u-a)$ or the cubic $f(u) = u(u-b)(u+c)$ with $a, b, c > 0$.

We also consider a lattice version

$$\ddot{u}_n - \frac{1}{\varepsilon^2} \sum_{k=-\infty}^{\infty} \alpha_k u_{n-k} + f(u_n) = 0, \qquad n \in \mathbb{Z}. \tag{4.2}$$

Note that, as $\varepsilon \to 0$, $\frac{1}{\varepsilon^2}(j_\varepsilon * u - u) \to du_{xx}$, formally and in some weak sense described in [10], where d is a constant determined by j. So we can also regard (4.1) as a nonlocal version of the standard nonlinear wave equation

$$u_{tt} - du_{xx} + f(u) = 0. \tag{4.3}$$

In this paper we will study homoclinic traveling wave solutions of (4.1), i.e., solutions of the form $u(x,t) = u(x-ct)$ which decay at infinity.

It is worth mentioning here that the parabolic versions

$$u_t - \frac{1}{\varepsilon^2}(j_\varepsilon * u - u) + f(u) = 0, \quad \text{for} \quad t > 0 \quad \text{and} \quad x \in \mathbb{R}, \tag{4.4}$$

and

$$\dot{u}_n - \frac{1}{\varepsilon^2} \sum_{k=-\infty}^{\infty} \alpha_k u_{n-k} + f(u_n) = 0, \qquad n \in \mathbb{Z}, \tag{4.5}$$

where f is bistable (e.g., the cubic above) were treated in [10] and [9], respectively, where traveling or stationary waves were shown to exist, connecting the stable zeros of f. Certain assumptions are needed upon j and the α_k's but we note that they are not required to be non-negative, i.e., they may change sign. When the wave has nonzero velocity, then the results in those papers are perturbative and rely upon spectral theory that we develop for the linearized operators for $\varepsilon > 0$ sufficiently small. When the wave is stationary (the potential has wells of equal depth), then under the conditions imposed on the coefficients, solutions exist for all $\varepsilon > 0$.

In this paper, we make slightly different assumptions on j and the α_k's than in [10] and [9], and the proofs are very different, but in some sense spectral analysis is still involved. To be more precise, we assume

(W_1) $f \in C^2(\mathbb{R}), f(0) = 0, f'(0) = -a < 0;\quad f(\zeta_0) > 0$, where

$$\zeta_0 \equiv \inf\{\zeta > 0 : F(\zeta) = 0\} \text{ and } F(\zeta) = \int_0^\zeta f(s)ds.$$

(W_2) $j(x) \in L^1(R)$ is even, has unit integral,

$$\lim_{z\to 0} \frac{\widehat{j}(z) - 1}{z^2} = -d \qquad \text{and} \qquad \widehat{j}(z) \geq 1 - d_1 z^2,$$

where $0 \leq d \leq d_1$ are constants and the Fourier transform is given by $\widehat{j}(z) \equiv \int_{-\infty}^{\infty} e^{-izx} j(x)dx$.

Remark 4.1 If $\widehat{j} \in C^2$ then $d = \frac{1}{2}\int_{-\infty}^{\infty} j(x)x^2 dx$.

In (4.1), let $u(x,t) = u(x - ct) = u(\eta)$, so that $u(\eta)$ satisfies the equation

$$c^2 u'' - \frac{1}{\varepsilon^2}(j_\varepsilon * u - u) = -g(u) + au, \tag{4.6}$$

where $g(u) = f(u) + au$. Applying the Fourier transform, equation (4.6) becomes

$$-c^2\xi^2\hat{u} - \frac{1}{\varepsilon^2}(\widehat{j_\varepsilon} \cdot \hat{u} - \hat{u}) = -\widehat{g(u)} + a\hat{u}$$

or

$$(c^2\xi^2 + l_\varepsilon(\xi) + a)\hat{u} = \widehat{g(u)},$$

where $l_\varepsilon = \frac{1}{\varepsilon^2}(\widehat{j_\varepsilon} - 1)$. Thus, an equivalent formulation is

$$\hat{u} = p_\varepsilon(\xi)\widehat{g(u)},$$

where

$$p_\varepsilon(\xi) = \frac{1}{c^2\xi^2 + l_\varepsilon(\xi) + a}. \tag{4.7}$$

The inverse Fourier transform gives

$$u = \check{p}_\varepsilon * g(u), \tag{4.8}$$

where $\check{p}_\varepsilon$ is the inverse transform of p_ε.

Define the operator

$$P_\varepsilon(u) \equiv \check{p}_\varepsilon * g(u),$$

and write (4.8) as

$$u = P_\varepsilon(u). \tag{4.9}$$

Note that, due to (W_2),

$$l_\varepsilon(\xi) = \frac{1}{\varepsilon^2}(\widehat{j_\varepsilon}(\xi) - 1) = \frac{\widehat{j}(\varepsilon\xi) - 1}{(\varepsilon\xi)^2} \cdot \xi^2 \to -d\xi^2$$

as $\varepsilon \to 0$, so $p_\varepsilon \approx \dfrac{1}{(c^2 - d)\xi^2 + a}$ for ε small. Thus, when $\varepsilon \to 0$, (4.9) formally becomes

$$u = P_0(u), \tag{4.10}$$

where $P_0(u) \equiv \check{p}_0 * g(u)$ and

$$p_0(\xi) = \frac{1}{(c^2 - d)\xi^2 + a}.$$

Clearly, (4.10) is equivalent to

$$u = ((d - c^2)\partial^2 + a)^{-1}(g(u)),$$

that is,

$$(c^2 - d)u'' = au - g(u),$$

or

$$(c^2 - d)u'' + f(u) = 0. \tag{4.11}$$

By the results in [21], under the assumption (W_1), (4.11) has a unique even, positive homoclinic solution for each $c^2 > d$, which we denote by u_0. Thus, u_0 is a fixed point of operator P_0. We can write equation (4.6) in the form

$$u = P_0(u) + (P_\varepsilon - P_0)(u),$$

and look for a fixed point near u_0.

In the case of the lattice Klein-Gordon equation (4.2), we assume

(W_3)

$$\sum_{k=-\infty}^{\infty} \alpha_k = 0, \quad \alpha_0 < 0, \quad \alpha_k = \alpha_{-k}, \quad \sum_{k \geq 1} \alpha_k k^2 = d > 0$$
$$\text{and} \quad \sum_{k \geq 1} |\alpha_k| k^2 = \bar{d} < \infty.$$

With the ansatz $u_n(t) = u(\varepsilon n - ct) = u(\eta)$, we get the following differential equation with infinitely many advanced and delayed terms:

$$c^2 \frac{d^2 u}{d\eta^2} - \frac{1}{\varepsilon^2} \sum_{k=-\infty}^{\infty} \alpha_k u(\eta - kc) + f(u) = 0. \tag{4.12}$$

Applying the Fourier transform, equation (4.12) becomes

$$-c^2\xi^2\hat{u} - \frac{1}{\varepsilon^2} \sum_{k=-\infty}^{\infty} \alpha_k e^{i\varepsilon k\xi}\hat{u} + \widehat{f(u)} = 0.$$

Using (W_3) we may write

$$\begin{aligned} \frac{1}{\varepsilon^2} \sum_{k=-\infty}^{\infty} \alpha_k e^{i\varepsilon k\xi}\hat{u} &= \frac{1}{\varepsilon^2} \sum_{k \geq 1} \alpha_k (e^{i\varepsilon k\xi} - 2 + e^{-i\varepsilon k\xi})\hat{u} \\ &= \frac{4}{\varepsilon^2} \sum_{k \geq 1} \alpha_k \sin^2(\frac{\varepsilon k\xi}{2})\hat{u}. \end{aligned}$$

Therefore, we may write our equation as

$$[(c^2 - \sum_{k\geq 1} \alpha_k k^2 \mathrm{sinc}^2(\frac{\varepsilon k \xi}{2}))\xi^2 + a]\hat{u} = \widehat{g(u)},$$

where $\mathrm{sinc}(z) = \dfrac{\sin z}{z}$.

We define

$$D_1 =: \sup_z \sum_{k\geq 1} \alpha_k k^2 \mathrm{sinc}^2(kz)$$

and note that $D_1 \geq d$.

Let

$$q_\varepsilon(\xi) = \frac{1}{(c^2 - \sum_{k\geq 1} \alpha_k k^2 \mathrm{sinc}^2(\frac{\varepsilon k \xi}{2}))\xi^2 + a},$$

then we can write the equation in the form

$$u = Q_\varepsilon(u),$$

where Q_ε is the operator defined by $Q_\varepsilon(u) =: \check{q}_\varepsilon * g(u)$. Since

$$\mathrm{sinc}^2(\frac{\varepsilon k \xi}{2}) = 1 - \frac{1}{8}\varepsilon^2\xi^2 + o(\varepsilon^4\xi^4),$$

$q_\varepsilon(\xi)$ has the limit

$$p_0(\xi) = \frac{1}{(c^2 - d)\xi^2 + a}$$

as $\varepsilon \to 0$. Therefore, formally when $\varepsilon \to 0$, we have the limit equation

$$u = P_0(u),$$

where, as before, $P_0(u) =: \check{p}_0 * g(u)$, and the integral equation is equivalent to (4.11), which has the nondegenerate homoclinic solution, u_0, discussed previously.

This time we have the fixed point problem

$$u = P_0(u) + (Q_\varepsilon - P_0)(u),$$

and the idea is to show that $Q_\varepsilon - P_0$ (or $P_\varepsilon - P_0$ in the previous case) is sufficiently small in some neighborhood of u_0 that a fixed point exists. An abstract lemma in [51], from which we borrowed this plan of attack, gives conditions under which this is the case. Basically, the lemma is a variant of the Implicit Function Theorem. The conditions are essentially that P_0 and P_ε (or Q_ε) are close as C^1 mappings on a ball about u_0 in a Banach space, with $I - DP(u_0)$ invertible with small norm.

Armed with this lemma we are able to prove the following theorem by showing that the various hypotheses hold.

Theorem 4.2 *Under the assumptions* (W_1) *and* (W_2) *(or* (W_3)*), there exits an* $\varepsilon_0 > 0$ *such that for any* $\varepsilon \in (0, \varepsilon_0)$ *and speed* c *satisfying* $c^2 > d_1$*, (or* $c^2 > D_1$*) equation* (4.6) *(or* (4.12)*) has a unique nonzero solution* u_ε *in the set*

$$\{u \in H^1(\mathbb{R}) : \ u \text{ is even}, \ \|u - u_0\|_{H^1} < \delta\},$$

where u_0 *is the even, positive homoclinic solution of* (4.11) *and* $\delta > 0$ *depends on* j *and* f *and satisfies* $\delta < \|u_0\|_{H^1}$.

We remark that even though these traveling pulses are H^1-close to u_0 and are even, we do not know if they are monotone decreasing away from the midpoint. In fact, this may not be the case in general, since the interaction coefficients (J or the α_k's) can change sign. In the case of heteroclinic wave for the parabolic version considered in [10], an example showed that monotonicity of the wave did not necessarily hold.

Another point worth mentioning is the fact that for any $c > 0$ there is a traveling pulse for ε sufficiently large, depending upon c. This is due to the boundedness of the convolution operator, making that term a regular perturbation from the case where $\varepsilon = \infty$. It is still an open question what happens as ε is decreased: Do high speed pulses exist for ε large, disappear as ε decreases below a certain threshold, only to appear again as ε decreases below a second threshold? If so, how do they appear and disappear?

References

[1] N. D. Alikakos, L^p bounds of solutions of reaction-diffusion equations, *Comm. P. D. E.*, **4**(8) (1979), 827-868.

[2] N. D. Alikakos, P. W. Bates and X. Chen, Convergence of the Cahn-Hilliard equation to the Hele-Shaw model, *Arch. Rat. Mech. Anal.* **128** (1994), 165-205.

[3] N. D. Alikakos, P. W. Bates and G. Fusco, Slow motion for the Cahn-Hilliard equation in one space dimension, *J. Diff. Eq.* **90** (1990), 81-135.

[4] N. D. Alikakos and R. Rostamian, Large time behavior of solutions of Neumann boundary value problem for the porous medium equation, *Indiana U. Math. J.* **30** (1981), 749–785.

[5] S. Allen and J.W. Cahn, A microscopic theory for antiphase boundary motion and its application to antiphase domain coarsening, *Acta Metall.* **27** (1979), 1084–1095

[6] D. Aronson, M. G. Crandall and L. A. Peletier, Stabilization of solutions of a degenerate nonlinear diffusion problem, Nonlinear Analysis, T.M.A. (1982), 1001-1022.

[7] P. W. Bates and F. Chen, Spectral analysis and multidimensional stability of traveling waves for nonlocal Allen-Cahn equation, *J. Math. Anal. Appl.* **273** (2002), no. 1, 45–57.

[8] P. W. Bates, F. Chen, and J. Wang, Global existence and uniqueness of solutions to a nonlocal phase-field system, in US-Chinese Conference on Differential Equations and Applications, P. W. Bates, S-N. Chow, K. Lu, X. Pan, Eds., *International Press*, Cambridge, MA, 1997, 14-21.

[9] P. W. Bates, X. Chen, and A. J. J. Chmaj, Traveling waves of bistable dynamics on a lattice. *SIAM J. Math. Anal.*, **35** (2003) no.2, 520-546.

[10] P. W. Bates, X. Chen, and A. J. J. Chmaj, Waves in the van der Waals model of phase transition. *Calc Var PDE*, available online September, 2005.

[11] P. W. Bates and A. J. J. Chmaj, A discrete convolution model for phase transitions, *Arch. Rat. Mech. Anal.* **150** (1999), 281-305.

[12] P. W. Bates and A. J. J. Chmaj, An integrodifferential model for phase transitions: stationary solutions in higher space dimensions. *J. Statist. Phys.*, **95** (1999), no. 5-6, 1119–1139.

[13] P. W. Bates and P. C. Fife, Spectral comparison principles for the Cahn-Hilliard and phase-field equations, and the scales for coarsening, *Phys. D* **43** (1990), 335-348.

[14] P. W. Bates and P. C. Fife, The dynamics of nucleation for the Cahn-Hilliard equation, *SIAM J. Appl. Math.* **53** (1993), 990-1008.

[15] P. W. Bates, P. C. Fife, X. Ren, and X. Wang, Traveling waves in a convolution model for phase transitions. *Arch. Rational Mech. Anal.*, **138** (1997), no. 2, 105–136.

[16] P. W. Bates and G. Fusco, Equilibria with many nuclei for the Cahn-Hilliard equation, *J. Diff. Eq.* **160** (2000), 283-356.

[17] P. W. Bates and J. Han, Neumann boundary problem for the nonlocal Cahn-Hilliard equation, *J. Diff. Eq.* **212** (2005), no. 2, 235–277.

[18] P. W. Bates and J. Han, The Dirichlet boundary problem for a nonlocal Cahn-Hilliard equation, *J. Math. Anal. Appl.* **311** (2005), 289-312.

[19] P. W. Bates, J. Han and G. Zhao, On a Nonlocal Phase-Field System, to appear *J. Nonlin. Anal.*

[20] P. W. Bates and C. Zhang, Traveling Pulses for the Klein-Gordon Equation on a Lattice or Continuum with Long-range Interaction, to appear, *J. Discrete Cont. Dyn. Syst.*

[21] H. Berestycki and P. -L. Lions, Nonlinear scalar field equations. I. Existence of a ground state. *Arch. Rational Mech. Anal.*, **82** (1983), no. 4, 313–345.

[22] G. Caginalp, Analysis of a phase field model of a free boundary, *Arch. Rat. Mech. Anal.* **92** (1986) 205-245.

[23] G. Caginalp and P. C. Fife Dynamics of layered interfaces arising form phase boundaries, *SIAM J. Appl. Math.* **48** (1988) 506-518.

[24] J. W. Cahn and J. E. Hilliard, Free energy of a nonuniform system I. Interfacial free energy, *J. Chem. Phys.* **28**, 258-267.

[25] A. Carpio and L. L. Bonilla, Oscillatory wave fronts in chains of coupled nonlinear oscillators. *Phys. Rev. E*, **67** (2003), no. 5, 056621.

[26] J. G. Carr, M. E. Gurtin, M. Slemrod, Structured phase transitions on a finite interval, *Arch. Rat. Mech. Anal.* **86** (1984) 317-351.

[27] X. Chen, Existence, uniqueness, and asymptotic stability of traveling waves in nonlocal evolution equations. *Adv. Differential Equations*, **2** (1997), no. 1, 125–160.

[28] A. Chmaj and X. Ren, Homoclinic solutions of an integral equation: existence and stability, *J. Diff. Eqs.* **155** (1999), 17-43.

[29] A. J. J. Chmaj and X. Ren, Multiple solutions of the nonlocal bistable equation, *Phys. D* **147** (2000), 135-154.

[30] A. J. J. Chmaj and X. Ren, Pattern formation in the nonlocal bistable equation, *Methods Appl. Anal.* **8** (2001), 369–386.

[31] A. J. J. Chmaj and X. Ren, The nonlocal bistable equation: stationary solutions on a bounded interval, *Electronic J. Diff. Eqs.* **2002** (2002) 1-12

[32] P. Colli, P. Krejci, E. Rocca, and J. Sprekels, Nonlinear evolutions arising from phase change models, preprint.

[33] P. Colli and Ph. Laurencot, Weak solutions to the Penrose-Fife phase field model for a class of admissible heat flux laws, *Physica D.* **111** (1998) 311-334.

[34] P. Colli and J. Sprekels, Stefan problems and the Penrose-Fife phase field model, *Adv. Math. Sci. Appl.* **7** (1997) 911-934.

[35] A. de Masi, T. Gobron and E. Presutti, Traveling fronts in non-local evolution equations, *Arch. Rational Mech. Anal.* **132** (1995), 143-205.

[36] A. de Masi, E. Orlandi, E. Presutti, L. Triolo, Stability of the interface in a model of phase separation, *Proc. Roy. Soc. Edin.* **124A** (1994), 1013-1022.

[37] A. de Masi, E. Orlandi, E. Presutti, L. Triolo, Uniqueness of the instanton profile and global stability in nonlocal evolution equations, *Rend. Math.* **14** (1994), 693-723.

[38] D.Duncan, M. Grinfeld, and I. Stoleriu, Coarsening in an integro-differential model of phase transitions, *Euro. J. Appl. Math.* **11** (2000) 511-572.

[39] C. M. Elliott and S. Zheng, On the Cahn-Hilliard equation, *Arch. Rat. Mech. Anal.,* **96** (1986), 339-357.

[40] C. M. Elliott and S. Zheng, Global existence and stability of solutions to the phase field equations, in Free Boundary Problems, K. H. Hoffmann and J. Sprekels(eds.), *Internat. Ser. Numerical Math.*, **95**, Birkhauser Verlag, Basel, (1990), 46-58.

[41] C.E. Elmer and E.S. Van Vleck, Computation of traveling waves for spatially discrete bistable reaction-diffusion equations, *Appl. Numer. Math.* **20** (1996), 157–169.

[42] T. Erneux and G. Nicolis, Propagating waves in discrete bistable reaction diffusion systems, *Physica D* **67** (1993), 237–244.

[43] G. Fath, Propagation failure of traveling waves in a discrete bistable medium, *Phys. D* **116** (1998), 176–190.

[44] M. Feckan, Blue sky catastrophes in weakly coupled chains of reversible oscillators. *Discrete Contin. Dyn. Syst. Ser. B* , **3** (2003), no. 2, 193–200.

[45] E. Feireisl, F. I. Roch, and H. Petzeltova, A non-smooth version of the Lojasiewicz-Simon theorem with applications to non-local phase-field systems, *J. Diff. Eq.* **199** (2004), 1-21.

[46] P. C. Fife, Well-posedness issues for models of phase transitions with weak interaction, *Nonlinearity* **14** (2001), 221-238.

[47] G. Fix, Phase field methods for free boundary problems, in *Free Boundary Problems*, A Fasano and M. Primicerio, eds., pp. 580-589, Pitman, London, 1983.

[48] S. Flach, C. R. Willis, Discrete breathers. *Phys. Rep.***295** (1998), no. 5, 181–264.

[49] S. Flach, Y. Zolotaryuk, and K. Kladko, Moving lattice kinks and pulses: An inverse method. *Physical Review E*, **59** (1999), 61056115 .

[50] R. L. Fosdick and D. E. Mason, Single phase energy minimizers for materials with nonlocal spatial dependence, *Quart. Appl. Math.* **54** (1996), 161-195.

[51] G. Friesecke and R. L. Pego, Solitary waves on FPU lattices. I. Qualitative properties, renormalization and continuum limit. *Nonlinearity*, **12** (1999), no. 6, 1601–1627.

[52] H. Gajewski and K. Zacharias, On a nonlocal phase separation model, *J. Math. Anal. Appl.*, **286** (2003), 11-31.

[53] G. Giacomin and J. L. Lebowitz, Phase segregation dynamics in particle systems with long range interactions II: Interface motion, *SIAM J. Appl. Math.* **58** (1998), 1707-1729.

[54] G. Giacomin and J.L. Lebowitz, Exact macroscopic description of phase segregation in model alloys with long range interactions, *Phys. Rev. Lett.* **76** (7) (1996), 1094-1097.

[55] C. Grant and E. Van Vleck, Slowly migrating transition layers for the discrete Allen-Cahn and Cahn-Hilliard Equations, *Nonlinearity* **8** (1995), 861-876.

[56] M. Katsoulakis and P.E. Souganidis, Interacting particle systems and generalized mean curvature evolution, *Arch. Rat. Mech. Anal.* **127** (1994).

[57] M. Katsoulakis and P.E. Souganidis, Generalized motion by mean curvature as a macroscopic limit of stochastic Ising models with long range interactions and Glauber dynamics, *Comm. Math. Phys.* **169** (1995), 61-97.

[58] J.P. Keener, Propagation and its failure in coupled systems of discrete excitable fibers, *SIAM J. Appl. Math.* **47** (1987), 56-572.

[59] N. Kenmochi and M. Kubo, Weak solutions of nonlinear systems for non-isothermal phase transitions, *Adv. in Math. Sci. and Appl.* **9** (1999), 499-521.

[60] P. Krejči and J. Sprekels, Phase-field models with hysteresis, J. Math. Anal. Appl. 252 (2000), 198-219.

[61] P. Krejči and J. Sprekels, A hysteresis approach to phase-field models, Nonlinear. Anal. 39 (2000), 569-586.

[62] O. Kresse and L. Truskinovsky, Mobility of lattice defects: discrete and continuum approaches. *J. Mech. Phys. Solids*, **51** (2003), no. 7, 1305–1332.

[63] M. Kubo, A. Ito, and N. Kenmochi, Non-isothermal phase separation models: Weak well-posedness and global estimates, N. Kenmochi (Ed.) Free boundary problems: Theory and applications II (Chiba, 1999), *Gakuto Int. Ser. Math. Sci. Appl.* **14**, Gakkotosho, Tokyo, 2000, 311-323.

[64] O. A. Ladyzenskaja, V. A. Solonnikov, and N. N. Uralceva, Linear and quasilinear equations of parabolic type, Volume 23, Translations of Mathematical Monographs, AMS, Providence, 1968.

[65] J. Langer, Models of pattern formation in first-order phase transitions, in *Directions in Condensed Matter Physics*, pp. 164-186, World Science Publ., 1986

[66] J. Mallet-Paret, The global structure of traveling waves in spatially discrete dynamical systems, *J. Dyn. Diff. Eq.* **11** (1999), 49-127.

[67] B. Nicolaenko, B. Scheurer and R. Temam, Some global dynamical properties of a class of pattern formation equations, *Comm. P. D. E.* **14** (1989), 245-297.

[68] A. Novick-Cohen and L. A. Segel, Nonlinear aspects of the Cahn-Hilliard equation, *Phys. D* **10** (1984), 277-298.

[69] O. Penrose and P. C. Fife, Thermodynamically consistent models of phase field type from the kinetics of phase transitions, *Physica D* **43** (1990), 44-62.

[70] R.C. Rogers, A nonlocal model for the exchange energy in ferromagnetic materials, *J. Integral Eqs. Appl.*, **3** (1991), 85-127.

[71] R.C. Rogers, Some remarks on nonlocal interactions and hysteresis in phase transitions, *Continuum Mech. Thermodynamics*, **8** (1996), 65-73.

[72] A. V. Savin, Y. Zolotaryuk and J. C. Eilbeck, Moving kinks and nanopterons in the nonlinear Klein-Gordon lattice. *Phys. D*, **138** (2000), no. 3-4, 267–281.

[73] J. Sprekels and S. Zheng, Global existence and asymptotic behavior for a nonlocal phase-field model for non-isothermal phase transitions, *J. Math. Anal. Appl.* **279** (2003), 97-110.

[74] R. Temam, Infinite dimensional dynamical systems in mechanics and physics, Springer-Verlag, New York 1988.

[75] L. Truskinovsky and A. Vainchtein, Kinetics of martensitic phase transitions: Lattice model. To appear in *SIAM J. Appl. Math.*

[76] J. D. van der Waals, The thermodynamic theory of capillarity under the hypothesis of a continuous variation of density, (in Dutch) *Verhandel Konink. Akad. Weten. Amsterdam* **8** (1893). Translation by J. S. Rowlinson, *J. Statist. Phys.* **20** (1979), 197-244.

[77] H.F. Weinberger, Long-time behavior of a class of biological models, *SIAM J. Math. Anal.* **13** (1982), 353-396.

[78] B. Zinner, Existence of traveling wavefront solutions for the discrete Nagumo equation, *J. Diff. Eqs.* **96** (1992), 1-27.

Fields Institute Communications
Volume **48**, 2006

Invariant Tori for Hamiltonian PDE

Walter Craig
Department of Mathematics and Statistics
McMaster University
Hamilton, Ontario L8S 4K1, Canada
craig@math.mcmaster.ca

Abstract. Hamiltonian partial differential equations, dynamical systems Many of the central equations of mathematical physics, the nonlinear wave equation, the nonlinear Schrödinger equation, the Euler equations for free surfaces, can be posed as Hamiltonian systems with infinitely many degrees of freedom. In a neighborhood of an equilibrium, the linearized equations are those of a harmonic oscillator and thus solutions exhibit periodic and quasi-periodic motion. To construct solutions of the same nature for the nonlinear partial differential equations (PDE)s is a small divisor problem in general. This article gives an overview of some of the techniques and results of KAM-like methods for PDE, which have been developed to address the analysis of this problem.

1 Introduction

A Hamiltonian partial differential equation is an evolution equation whose initial value problem takes the form

$$\partial_t v = J \mathrm{grad}_v H(v) \ , \quad v(x,0) = v_0(x) \ , \tag{1.1}$$

where $v \in \mathcal{H}$ which is a Hilbert space playing the rôle of the phase space, which has a symplectic form given by

$$\omega(X,Y) = \langle X, J^{-1}Y \rangle_{\mathcal{H}} \ , \quad J^T = -J \ . \tag{1.2}$$

The flow of this dynamical system, if it exists on a neighborhood of $v_0(x)$, is denoted by

$$v(x,t) = \varphi_t(v_0(x)) \ ,$$

tracing a curve in $\mathcal{H}$ through v_0. Our aim is to describe some analytic results constructing special compact invariant sets for this flow in the phase space $\mathcal{H}$.

2000 *Mathematics Subject Classification.* Primary 35Q55, 37K55; Secondary 76B15.

This research has been supported in part by the Canada Research Chairs Program, the NSERC under operating grant #238452, and the NSF under grant #DMS-0070218.

The outline of this article are as follows:

2 Principal examples

2.1 Nonlinear wave equations. This is a model for a class of nonlinear wave equations for which the function $u(x,t)$ represents a scalar field. We seek solutions of the equation

$$\partial_t^2 u - \Delta u + g(x,u) = 0 \ , \tag{2.1}$$

satisfying a boundary condition on a domain $\Omega \subseteq \mathbb{R}^d$, which could be Dirichlet or Neumann conditions. More conveniently one often poses periodic boundary conditions on a torus, which is to say that $\Omega = \mathbb{T}^d = \mathbb{R}^d/\Gamma$ for a given period lattice Γ. This is what we will do for the most part of this article. The Hamiltonian functional for problem (2.1) is given by

$$H(u,p) = \int_{\mathbb{T}^d} \tfrac{1}{2}p^2 + \tfrac{1}{2}|\nabla u|^2 + G(x,u)\, dx \ , \tag{2.2}$$

with respect to which the equations (2.1) can be rewritten as

$$\begin{aligned} \partial_t u &= p = \mathrm{grad}_p H(u,p) \\ \partial_t p &= \Delta u - \partial_u G(x,u) = -\mathrm{grad}_u H(u,p) \ , \end{aligned} \tag{2.3}$$

where $g(x,\cdot) = \partial_u G(x,\cdot)$. This is to say that the system has the form

$$\partial_t \begin{pmatrix} u \\ p \end{pmatrix} = \begin{pmatrix} 0 & I \\ -I & 0 \end{pmatrix} \begin{pmatrix} \mathrm{grad}_u H(u,p) \\ \mathrm{grad}_p H(u,p) \end{pmatrix} \ , \tag{2.4}$$

and the symplectic form is given by

$$\omega(X,Y) = \int_{\mathbb{T}^d} X_2(x)Y_1(x) - X_1(x)Y_2(x)\, dx$$

for two vector fields $X = (X_1,X_2), Y = (Y_1,Y_2) \in \mathcal{H}$. Recognizing the canonical form for

$$J = \begin{pmatrix} 0 & I \\ -I & 0 \end{pmatrix} \ . \tag{2.5}$$

we say that this system is posed in *Darboux coordinates.* Assume that the nonlinear term $G(x,u)$ is sufficiently smooth, and expand in its Taylor series around $(u,p) = 0$;

$$G(x,u) = \frac{1}{2}g_1(x)u^2 + \frac{1}{3}g_2(x)u^3 + \dots$$

then the point $(u,p) = 0$ is a stationary point for the dynamical system (2.3), and an expression for the Hamiltonian reflecting this expansion about this point is given by

$$H = H^{(2)} + H^{(3)} + \cdots$$

An elegant way to linearize this problem about zero is to truncate the Hamiltonian at quadratic order. In doing so we obtain

$$\begin{aligned} H^{(2)} &= \int_{\mathbb{T}^d} \tfrac{1}{2}p^2 + \tfrac{1}{2}|\nabla u|^2 + \frac{1}{2}g_1(x)u^2\, dx \\ &= \sum_{k\in\Gamma'} \tfrac{1}{2}|p(k)|^2 + \tfrac{1}{2}\omega(k)^2|u(k)|^2 \ , \end{aligned} \tag{2.6}$$

where the coefficients $(u(k), p(k))$ are the generalized Fourier coefficients of an eigenfunction expansion

$$\begin{pmatrix} u(x) \\ p(x) \end{pmatrix} = \sum_{k \in \Gamma'} \begin{pmatrix} u(k) \\ p(k) \end{pmatrix} \psi_k(x)$$

and where the eigenfunction, eigenvalue pairs $(\psi_k(x), \omega^2(k))$ satisfy the problem

$$L(g_1)\psi_k = (-\Delta + g_1(x))\psi_k = \omega(k)^2 \psi_k \ . \tag{2.7}$$

This is evidently a harmonic oscillator, with frequencies $\{\omega(k)\}$, for $k \in \Gamma'$.

All solutions of the linearized equations

$$\partial_t \begin{pmatrix} u \\ p \end{pmatrix} = \begin{pmatrix} 0 & I \\ -I & 0 \end{pmatrix} \begin{pmatrix} \mathrm{grad}_u H^{(2)}(u,p) \\ \mathrm{grad}_p H^{(2)}(u,p) \end{pmatrix} , \tag{2.8}$$

are given by this eigenfunction expansion for the linear flow (for reasons of simplicity of the discussion we have not pursued cases in which some eigenvalues of (2.7) may be negative, $\omega^2(k) < 0$);

$$\begin{aligned} \begin{pmatrix} u(x,t) \\ p(x,t) \end{pmatrix} &= \sum_{k \in \Gamma'} \psi_k(x) \begin{pmatrix} \cos(\xi(k,t)) & \frac{1}{\omega(k)} \sin(\xi(k,t)) \\ -\omega(k)\sin(\xi(k,t)) & \cos(\xi(k,t)) \end{pmatrix} \begin{pmatrix} u_0(k) \\ p_0(k) \end{pmatrix} \\ &= \Phi_t \begin{pmatrix} u_0(x) \\ p_0(x) \end{pmatrix} . \end{aligned} \tag{2.9}$$

In this expression, the phases are given by

$$\xi(k,t) = \omega(k)t \ . \tag{2.10}$$

The simple facts about the flow (2.9) are that

(1) the Hamiltonian $H^{(2)}$ is preserved by the evolution;

$$H^{(2)}(\Phi_t(v)) = H^{(2)}(v).$$

(2) Furthermore, the action functionals are preserved by the flow, where the action is defined by;

$$\begin{aligned} I_k(u,p) &= \tfrac{1}{2}\Big(\omega(k)|u(k)|^2 + \frac{1}{\omega(k)}|p(k)|^2\Big) , \\ I_k(\Phi_t(v)) &= I_k(v) \ ; \end{aligned} \tag{2.11}$$

(3) The phases evolve linearly in time, as in (2.10). Hence all solutions are either periodic in time, quasiperiodic in time, or at least almost periodic in time. Namely, the orbit $(u(x,t), p(x,t)) = \Phi_t(u_0(x), p_0(x))$ of (2.8) is *periodic* when each of the frequencies $\omega(k_j)$ for which at least one of the Fourier coefficients $(u_0(k_j), p_0(k_j))$ is nonzero (the *active frequencies*) is an integer multiple of a basic frequency ω_0;

$$\omega(k_j) = \ell_j \omega_0 , \quad \ell_j \in \mathbb{Z} \ .$$

The orbit $\Phi_t(u_0(x), p_0(x))$ is *quasiperiodic* with a m-dimensional frequency base if there is a m-dimensional frequency vector $\omega_0 = (\omega_0(1), \ldots \omega_0(m)) \in \mathbb{R}^m$ such that the active frequencies satisfy

$$\omega(k_j) = \langle \ell_j, \omega_0 \rangle , \quad \ell_j \in \mathbb{Z}^m \ .$$

That is, the linear span of the set $\{\omega(k_j)\}$ over the rationals $\mathbb{Q}$ has dimension m. A solution $(u(x,t), p(x,t))$ is *almost periodic* if no finite number of base frequencies suffices.

These facts motivate a number of *basic questions* regarding the nonlinear problem (2.3). Namely

(1) whether some solutions of the nonlinear problem have the same properties of periodicity, quasiperiodicity, or almost periodicity as the solutions of the linearized problem. This question is in the area of *KAM theory*, the study of invariant tori for such Hamiltonian systems.

(2) Whether all solutions with initial data $v_0(x) \in \mathcal{H}$ remain in the function space $\mathcal{H}$ for all time $t \in \mathbb{R}$, which is the question of (global) well posedness; whether for any $\delta > 0$ there exists a $\varepsilon > 0$ such that for $v(x) \in B_\varepsilon(0)$ then $\varphi_t(v) \in B_\delta(0)$ for all $t \in \mathbb{R}$, which is a question of stability; or whether the action variables $\{I_k\}$ vary by controlled amounts over long time intervals. More precisely,

$$|I_k(\varphi_t(v)) - I_k(v)| < \varepsilon^\alpha$$

over time intervals

$$|t| < T(\varepsilon) \sim \exp(\frac{1}{\varepsilon^\beta}) ,$$

which is known as a *Nekhoroshov stability* result.

(3) Whether there are lower bounds on the growth of the action functionals, or on higher Sobolev norms of the solutions of (2.3). In finite dimensional Hamiltonian systems this phenomenon is known as *Arnold diffusion.*

2.2 Nonlinear Schrödinger equations. The second example of a Hamltonian PDE that we address is the nonlinear Schrödinger equation, which appears in the form

$$i\partial_t u - \tfrac{1}{2}\Delta u + Q(x, u, \overline{u}) = 0 , \qquad (2.12)$$

where we take $x \in \mathbb{T}^d$ representing periodic boundary conditions. One could also take either Dirichlet or Neumann, or other self-adjoint boundary conditions on a domain $\Omega \subseteq \mathbb{R}^d$. Setting initial data $u(x, 0) = u_0(x) \in \mathcal{H}$, we suppose that we can solve the initial value problem, and we denote the flow of this dynamical system by

$$u(x, t) = \varphi_t(u_0(x)) .$$

The Hamiltonian is given by the expression

$$H(u) = \int_{\mathbb{T}^d} \tfrac{1}{2}|\nabla u|^2 + G(x, u, \overline{u})\, dx \qquad (2.13)$$

where the nonlinearity G is asked to satisfy two conditions; that $\partial_{\overline{u}} G = Q$, and the complex function $G(x, z, w)$ is real valued whenever $w = \overline{z}$. Then (2.12) can be rewritten as

$$\partial_t u = i \ \mathrm{grad}_{\overline{u}} H(u) , \qquad (2.14)$$

where $J = iI$ is the canonical form of *complex symplectic coordinates.*

2.3 Korteweg deVries equations. A third example of a class of Hamiltonian PDE is the Korteweg deVries equation, whose original application was to waves in the free surface of a fluid. In general form this is

$$\partial_t q = \tfrac{1}{6}\partial_x^3 q - \partial_x(\partial_q G(x, q)) , \qquad (2.15)$$

where one takes $x \in \mathbb{R}^1/\Gamma = \mathbb{T}^1$. The Hamiltonian is given by

$$H(q) = \int_{\mathbb{T}^1} \tfrac{1}{12}(\partial_x q)^2 + G(x, q)\, dx , \qquad (2.16)$$

and the symplectic form is a non-classical one, given by

$$\omega(X,Y) = \int_{\mathbb{T}^1} (\partial_x^{-1} X(x))Y(x)\, dx$$

which is to say that the symplectic form is given in terms of the operator $J = -\partial_x$.

2.4 Large amplitude long waves in an interface. The equations of motion of a free interface between two immiscible fluids of different densities, which are acted on by gravity, can be described as a Hamiltonian PDE. Suppose that there is a fixed flat bottom boundary at a finite depth $\{y = -h < 0\}$, a top boundary at $\{y = +h_1 > 0\}$ which is a rigid lid, and a sharp interface between the two fluids given by $\{y = \eta(x,t)\}$. Denote the acceleration of gravity by g, the density of the lower fluid by ρ, and that of the upper fluid by $\rho_1 < \rho$. In both of the upper and lower fluid regions the velocity field of the fluid is given as a potential flow, namely

$$\mathbf{u}_1 = \nabla\varphi_1 \ \text{ for } \ \eta(x,t) < y < h_1 \ , \quad \mathbf{u} = \nabla\varphi \ \text{ for } \ -h < y < \eta(x,t) \ .$$

Both φ and φ_1 are harmonic functions in their respective fluid regions, and they satisfy Neumann boundary conditions on the two rigid boundaries. Such situations occur in the laboratory, and under certain conditions these equations model the behavior of a sharp interface in the ocean between two water layers of different temperatures, such as occurs in the tropics (called a thermocline layer), or between two layers of different salinity (a pycnocline layer), such as occurs in deep fjords subject to large freshwater sources, or on the Mediterranean side of the Gibralter straits. It was discovered by Benjamin & Bridges [BB97] that this can be written as a Hamiltonian system, in terms of the canonical variables $(\eta(x), \xi(x))$ where one defines $\xi(x) = \rho\varphi(x,\eta(x)) - \rho_1\varphi_1(x,\eta(x))$. A systematic study of the long wave scaling limits of the resulting system of equations is presented in [CGK03, CGK04]. Perhaps the most interesting limiting system of equations when the slope of the interface is of order $\mathcal{O}(\varepsilon)$, without assuming that the actual interface displacement is small. Changing variables through the transformation

$$u(x) = \partial_x \xi(x)$$

the resulting equations for the interface take the form

$$\partial_t \begin{pmatrix} \eta \\ u \end{pmatrix} = \begin{pmatrix} 0 & \partial_x \\ -\partial_x & 0 \end{pmatrix} \begin{pmatrix} \mathrm{grad}_\eta H \\ \mathrm{grad}_u H \end{pmatrix} \ . \tag{2.17}$$

This is evidently a Hamiltonian system with symplectic form given by the expression

$$\omega(X,Y) = \int (\partial_x^{-1} X_1)Y_2 + (\partial_x^{-1} X_2)Y_1 \, dx \ , \tag{2.18}$$

which is the same as that occurring for the more well known Boussinesq system. In the present case the Hamiltonian functional is given by the expression

$$\begin{aligned} H(\eta,u) \quad = \quad & \int \tfrac{1}{2}R_0(\eta)u^2 + \tfrac{1}{2}g(\rho-\rho_1)\eta^2 \\ & + R_1(\eta)(\partial_x u)^2 + \tfrac{1}{4}R_2(\eta)\partial_x\eta(\partial_x u^2) + R_3(\eta)(\partial_x\eta)^2u^2 \, dx \end{aligned} \tag{2.19}$$

The nonlinear coefficients of this integrand are rational functions in η, which are precisely

$$\begin{aligned} R_0(\eta) &= \frac{(h+\eta)(h_1-\eta)}{\rho_1(h+\eta)+\rho(h_1-\eta)} \\ R_1(\eta) &= -\tfrac{1}{3}\frac{(h+\eta)^2(h_1-\eta)^2[\rho_1(h_1-\eta)+\rho(h+\eta)]}{[\rho_1(h+\eta)+\rho(h_1-\eta)]^2} \\ R_2(\eta) &= -\tfrac{1}{3}\rho\rho_1\frac{(h+h_1)(h+\eta)(h_1-\eta)[(h_1-\eta)^2-(h+\eta)^2]}{[\rho_1(h+\eta)+\rho(h_1-\eta)]^3} \\ R_3(\eta) &= -\tfrac{1}{3}\rho\rho_1\frac{(h+h_1)^3[\rho_1(h+\eta)^3+\rho(h_1-\eta)^3]}{[\rho_1(h+\eta)+\rho(h_1-\eta)]^4} \ . \end{aligned}$$

One can view the first as the nonlinear propagation velocity, and the other three as coefficients of nonlinear dispersion. This system and its derivation are discussed in [CGK04].

2.5 Free surface water waves. The water waves problem in d-dimensions concerns the dynamics of a free surface $\{y=\eta(x,t) : x \in \mathbb{T}^{d-1}\}$ bounding a fluid region in which the fluid velocity field is given by a potential flow;

$$\mathbf{u} = \nabla\varphi \ , \quad \Delta\varphi = 0 \ . \tag{2.20}$$

The potential satisfies Neumann boundary conditions on the bottom boundary of the fluid region $\{y=-h\}$, and as above the fluid has gravity as a restoring force. Given a fluid domain $S(\eta)$ fixed by $\eta(x)$, we parametrize the set of harmonic functions on $S(\eta)$ which satisfy Neumann bottom boundary conditions by their top boundary values $\xi(x)=\varphi(x,\eta(x))$. It was shown by Zakharov in a classical paper [Z68] that the full Euler's equations with a free surface are equivalent to the system

$$\partial_t \begin{pmatrix}\eta\\ \xi\end{pmatrix} = \begin{pmatrix}0 & I\\ -I & 0\end{pmatrix}\begin{pmatrix}\mathrm{grad}_\eta H(\eta,\xi)\\ \mathrm{grad}_\xi H(\eta,\xi)\end{pmatrix} , \tag{2.21}$$

where the Hamiltonian is given by

$$H(\eta,\xi) = \int \tfrac{1}{2}\xi G(\eta)\xi + \tfrac{1}{2}g\eta^2 \, dx \ . \tag{2.22}$$

This particular version of the problem is given in Craig & Sulem [CS93]. The Dirichlet-Neumann operator $G(\eta)$ is an integral operator on the free surface, which solves the following harmonic extension problem:

$$\begin{aligned} \xi(x) &\mapsto \varphi(x,y) && \text{harmonic extension} \\ &\mapsto N\cdot\nabla\varphi(x,\eta(x))\,dx && \text{normal derivative} \\ &:= G(\eta)\xi(x)\,dx \end{aligned} \tag{2.23}$$

Notice that the equations of motion (2.21) involve the gradient of the Dirichlet integral with respect to variations of the domain through an expression for the quantity $\delta_\eta G(\eta)\xi$. This question was discussed by Hadamard [H10], and was anticipated by him to be relevant to the study of free surface water waves in [H16].

3 A variational formulation

Consider the mapping of a m-dimensional torus $\mathbb{T}^m$ into our phase space $\mathcal{H}$;

$$S(\xi) \ : \ \mathbb{T}^m \mapsto \mathcal{H} \ , \tag{3.1}$$

which will be flow invariant under the evolution of a Hamiltonian PDE (1.1) in the sense that for some frequency vector $\Omega \in \mathbb{R}^m$,

$$S(\xi + t\Omega) = \varphi_t(S(\xi)) \quad , \forall \xi \in \mathbb{T}^m \ . \tag{3.2}$$

This implies that both

$$\begin{aligned} \partial_t S &= J \mathrm{grad}_v H(S) \ , \\ \partial_t S &= \Omega \cdot \partial_\xi S \ . \end{aligned}$$

We may rewrite these two relations in the form

$$J^{-1}\Omega \cdot \partial_\xi S - \mathrm{grad}_v H(S) = 0 \ , \tag{3.3}$$

and the problem of constructing embedded m-dimensional tori which are invariant under the flow of the Hamiltonian PDE is the problem of finding solutions $(S(\xi), \Omega)$ of equation (3.3). This is a bifurcation problem with m parameters, which is in general a small divisor problem.

There have been numerous papers on this subject over the last decade, with several different points of view. Without presenting an exhaustive survey at this point, it is nonetheless useful to point out the principal contributions to date. S. Kuksin [K87, K00] and E. Wayne [W90] first addressed the problem of invariant tori for PDE using extensions of the classical approach of KAM theory, namely the construction of a convergent sequence of canonical transformations. This point of view was pursued in subsequent work, including for example the papers of S. Kuksin and J. Pöschel [KP96] and J. Pöschel [P96]. An alternate and somewhat more direct technique was developed by W. Craig and E. Wayne in [CW93, CW94], which relies on different methods for analysing the linearized operators arising in the problem. This approach was adopted and extended by J. Bourgain in a series of papers [B95, B96, B98]. An overview of these results is given in [B00] and in [C00]. In the present article, we describe an approach to the problem of resonant tori based on the latter method. Resonant tori for completely integrable systems arise in parameter families, as can be seen in the unperturbed harmonic oscillator problem. Under perturbation, a family of m-dimensional resonant tori will be expected to break up, with only a few m-dimensional invariant tori surviving the perturbation. These survivors can be characterized by a variational problem, through which furthermore one can give an estimate on their number. This is the theme of the section below.

3.1 The variational problem. On a formal level, one can restate the equations (3.3) as the Euler - Lagrange equations for a variational problem for mappings of a torus into phase space. Consider the space of such mappings;

$$S \in X := \{S(\xi) \ : \ \mathbb{T}^m \mapsto \mathcal{H}\} \ . \tag{3.4}$$

At this level of discussion we will not specify a particular topology for X, realizing however that this is an important consideration in such analysis. Define the *action functionals* to be

$$I_j(S) = \tfrac{1}{2} \int_{\mathbb{T}^m} \langle S, J^{-1}\partial_{\xi_j} S \rangle_{\mathcal{H}} \, d\xi \ , \quad j = 1, \dots m \ , \tag{3.5}$$

whose variation satisfies

$$\delta_S I_j = J^{-1}\partial_{\xi_j} S \ . \tag{3.6}$$

Consider the *average Hamiltonian* to be

$$\overline{H}(S) = \int_{\mathbb{T}^m} H(S(\xi))\, d\xi \ , \tag{3.7}$$

where evidently

$$\delta_S \overline{H} = \text{grad}_v H(S) \ . \tag{3.8}$$

Consider the subvariety of the space X of torus mappings defined by fixing the values of the action integrals;

$$M_a = \{S \in X \ : \ I_1(S) = a_1, \ldots I_m(S) = a_m\} \subseteq X \ . \tag{3.9}$$

The equation (3.3) can be viewed as the Lagrange multiplier rule for critical points of $\overline{H}(S)$ restricted to the variety M_a; suffiently smooth critical points correspond to solutions of equation (3.3) with Lagrange multipliers the components of the frequency vector Ω

$$\delta_S \overline{H} - \Omega \cdot \delta_S I = 0 \ . \tag{3.10}$$

It is important to note that both the action functionals $I(S)$ and the average Hamiltonian $\overline{H}(S)$ are invariant under the group action of a torus $\mathbb{T}^m$, namely under the transformations

$$\tau_\alpha : S(\xi) \mapsto S(\xi + \alpha) \ , \quad \alpha \in \mathbb{T}^m \ , \tag{3.11}$$

both $I(\tau_\alpha S) = I(S)$ and $\overline{H}(\tau_\alpha S) = \overline{H}(S)$. Therefore the constraint variety M_a is invariant under the action of τ_α, and subsequently each critical point is in fact a member of a critical orbit by this m-torus action. Such considerations will have consequences in counting the number of geometrically distinct solutions of the equations (3.3).

Two questions come to mind at this point.

1. Do critical points exist on M_a? In fact the operators $L(\Omega)S = \Omega \cdot \partial_\xi S$ are degenerate on the space of mappings X, and their analysis depends in a sensitive way on the diophantine properties of the frequency vector Ω.

2. How to understand the multiplicity of solutions? Each critical point is in fact a member of a critical orbit, but different members of this orbit are geometrically indistinguishable. However, the space $M_a/\mathbb{T}^m$ possesses nontrivial topology, which implies certain lower bounds on the number of critical orbits of τ_α invariant functionals on M_a. This is despite the fact that the torus action is not in general a free action, and therefore the set $M_a/\mathbb{T}^m$ is not in general a smooth variety.

Answers that can be obtained in some cases are that: (1) One uses the Nash - Moser method to overcome the inherent small divisor problem that is present. Our own approach has been to use a method developed in [CW93] which employs Fröhlich - Spencer resolvant estimates for the linearized equations from (3.3). (2) there are well developed methods for counting critical points and critical orbits of functionals on infinite dimensional spaces, which go under the name of Morse - Bott theory. In cases of degeneracies, in which the group action is not necessarily a free action, this theory employs the analog of a classical construction of Borel. It is slightly unusual to require the details of this construction in the case of an m-dimensional torus action, such as we are encountering, so for the most part the particular details have had to be worked out independently.

3.2 The linearized problem. To start with, the equations of evolution (1.1) linearized around the equilibrium solution $v = (q,p) = 0$ are given in terms of the quadratic part of the Hamiltonian

$$\begin{aligned} H^{(2)}(v) &= \tfrac{1}{2}\sum_{k\in\Gamma'} \omega(k)\Big(\frac{1}{\omega(k)}p(k)^2 + \omega(k)q(k)^2\Big) \\ &= \sum_{k\in\Gamma'} \omega(k)I(k) \ , \end{aligned} \tag{3.12}$$

where we have represented v in terms of its eigenfunction expansion, as in (2.6). The linearized equations (1.1) around the trivial solution $v = 0$ are

$$\partial_t \begin{pmatrix} q(k) \\ p(k) \end{pmatrix} = \begin{pmatrix} 0 & 1 \\ -1 & 0 \end{pmatrix} \begin{pmatrix} \omega^2(k)q(k) \\ p(k) \end{pmatrix} \ , \qquad k \in \Gamma' \ .$$

Representing mappings $S : \mathbb{T}^m \mapsto \mathcal{H}$ of a torus into our phase space,

$$\begin{aligned} S &= S(x,\xi) = \sum_k S_k(\xi)\psi_k(x) \\ &= \sum_{j,k} s(j,k)\psi_k(x)e^{ij\cdot\xi} \ , \quad j \in \mathbb{Z}^m \ , \end{aligned} \tag{3.13}$$

the equations (3.3) linearized about the zero mapping $S = 0$ is written as

$$\Big(\delta_S^2\overline{H}^{(2)}(0) - \Omega\cdot\delta_S^2 I(0)\Big)S = \sum_{j,k}\begin{pmatrix} \omega(k) & i\Omega\cdot j \\ -i\Omega\cdot j & \omega(k)\end{pmatrix}\begin{pmatrix} s_1(j,k) \\ s_2(j,k)\end{pmatrix}\psi_k(x)e^{ij\cdot\xi} \ . \tag{3.14}$$

This represents a decomposition into 2×2 block diagonal problems, whose eigenvalues are

$$\mu(j,k) = \omega(k) \pm \Omega\cdot j \ , \qquad j\in\mathbb{Z}^m \ , k\in\Gamma' \ . \tag{3.15}$$

Chose the bifurcation point to be a frequency vector $\Omega^0 = (\Omega_1^0, \dots \Omega_m^0) \in \mathbb{R}^m$ which is the solution of m-many elementary resonance relations

$$\omega(k_\ell) - \Omega^0\cdot j_\ell = 0 \ , \quad \ell = 1,\dots m \ . \tag{3.16}$$

This represents a choice of null eigenspace $X_1 \subseteq X$ within the space of mappings, spanned by the set $\{\psi_k(x)e^{ij\cdot\xi} \ : \ \omega(k) \pm \Omega^0\cdot k = 0\}$. The set of indices satisfying the resonance relations

$$\omega(k) - \Omega^0\cdot j = 0$$

includes at least the set $\{(j_\ell, k_\ell) : \ell = 1\dots m\}$ given as solutions of the elementary resonance relations (3.16), but it is possible that there are many more.

Proposition 3.1 *The null space X_1 is even dimensional, with dimension $2M \geq 2m$. It is possible infinite dimensional, but in any case it is a symplectic subspace of X.*

The nonresonant case is when $\dim(X_1) = 2m$, which is the analog of a simple bifurcation of invariant tori. Otherwise $\dim(X_1) = 2M > 2m$, and we say that the case is *resonant.* The other eigenvalues (3.15) are

$$\mu(j,k) = \omega(k) \pm \Omega^0\cdot j \neq 0 \ ,$$

which typically forms a dense set in $\mathbb{R}$. Accumulating at $\mu = 0$, these are the small divisors.

3.3 Lyapunov-Schmidt decomposition. The space X of torus mappings $S : \mathbb{T}^m \mapsto \mathcal{H}$ can be decomposed into the null space of the linearized operator (3.14) and its orthogonal complement; $X = X_1 \oplus X_2$. Denote the associated projections by

$$QX = X_1 \ , \qquad PX = (I - Q)X = X_2 \ .$$

We consider the resulting decomposition of mappings;

$$S = QS + PS = S_1 + S_2.$$

The LHS of the nonlinear equation (3.3) is a transformation from the space of mappings X to a range space Y, which can also be decomposed into the range Y_2 of the linearized operator (3.14) and its co-range Y_1. By abuse of notation we will also denote orthogonal projections onto Y_1 and Y_2 respectively, by Q and P. The equation to solve (3.3) are equivalent to

$$Q\big(J^{-1}\Omega \cdot \partial_\xi S \quad - \quad \text{grad}_v H(S)\big) = 0 \ , \tag{3.17}$$

$$P\big(J^{-1}\Omega \cdot \partial_\xi S \quad - \quad \text{grad}_v H(S)\big) = 0 \ . \tag{3.18}$$

3.4 Critical point theory. In case $M < +\infty$ the Q-equation is finite dimensional, it is in any case called the bifurcation equation. The hard work in our problem at hand is in fact to solve the P-equation, for $S_2 = S_2(S_1, \Omega)$ as a function of the parameters S_1 and Ω. If this is achieved, the variational problem (3.10) is replaced by a reduced variational problem, which is the analog of the one proposed by A. Weinstein [W73] and J. Moser [M76] for periodic orbits of Hamiltonian systems in the resonant case of the Lyapunov center theorem. Define

$$\begin{aligned} I_j^1(S_1) &= I_j(S_1 + S_2) \\ \overline{H}^1(S_1) &= \overline{H}(S_1 + S_2) \\ M_a^1 &= \{S_1 \in X_1 \ : \ I_j^1(S_1) = a_j \ , \ j = 1 \dots m\} \ . \end{aligned} \tag{3.19}$$

Critical points of $\overline{H}^1$ on the subvariety $M_a^1 \subseteq X_1$ satisfy

$$\delta_{S_1} \overline{H}^1 - \Omega \cdot \delta_{S_1} I^1 = 0 \ , \tag{3.20}$$

and these correspond to solutions of the Q-equation (3.17). This fact focuses our attention on the reduced constraint variety M_a^1 itself. The solution of the P-equation is equivariant with respect to the group action τ_α, and it follows that the reduced functionals $I_j^1(S_1)$ and $\overline{H}^1(S_1)$ are invariant with respect to the $\mathbb{T}^m$ action τ_α. By its topology alone we can deduce a lower bound for the number of distinct critical $\mathbb{T}^m$ orbits of $\overline{H}^1(S_1)$ on M_a^1. This answer, which is still partially conjectural, is that in fact that, given the m-dimensional action parameter a, there exist integers $p_1, \dots p_m$ such that M_a^1 is a product of m-many odd dimensional spheres,

$$M_a^1 \simeq \otimes_{j=1}^m S^{2p_j - 1} \ . \tag{3.21}$$

The dimensions $p_j = p_j(a)$ satisfy $\sum_j p_j = m$. They are not constant, but change as a crosses certain singular homogeneous varieties. This process of transfer of dimension from some spheres to other ones is an example of symplectic surgery. For each choice of the sequence $(p_j(a))_{j=1}^m$, the $\mathbb{T}^m$ equivariant Morse-Bott inequalities give a lower bound on the number of critical points of sufficiently nondegenerate $\mathbb{T}^m$-invariant functionals on M_a^1. A simple answer which is uniform under all choices of $(p_j)_{j=1}^m$ is the following.

Theorem 3.1 *The number of distinct critical $\mathbb{T}^m$ orbits of $\overline{H}^1$ on M_a^1 is bounded below:*

$$\#\{\text{critical orbits of } \overline{H}^1\} \geq (M - m + 1) . \tag{3.22}$$

This lower bound corresponds to the simplest estimate given by equivariant cohomology on $M_a^1/\mathbb{T}^m$. In fact the estimate holds independently of the orbit-transversal nondegeneracy condition usually encountered in Morse-Bott theory, using multiplicative rather than additive techniques.

It is useful to check this inequality in the endpoint cases. Set $m = 1$ for periodic orbits, and the number of resonant frequencies $M > 1$ arbitrary. Then $M_a^1 \simeq S^{2M-1}$ and $M_a^1/\mathbb{T}^1 \simeq \mathbb{C}P(M-1)$, a (possibly weighted) complex projective space. The Morse inequalilies in this case imply that the number of distinct critical orbits of $\mathbb{T}^1$-invariant functions on M_a^1 is at least M. This corresponds to the case of the Weinstein - Moser theorem on the resonant Lyapunov center theorem.

Now set $m = M$, the case of quasiperiodic motion on a nonresonant m-torus. In this case,

$$M_a^1 \simeq \otimes_{j=1}^m S^1$$

which is a product of circles. Under the group action by $\mathbb{T}^m$, we have $M_a^1/\mathbb{T}^m \simeq *$, a point (or several points), corresponding to having one critical m-torus. This is a classical KAM torus. In finite dimensions, for Lagrangian tori, this variational principle has been posed (on a formal level) by I. Percival [P74]. In recent work (Craig & Nicholls [CN00]) on doubly periodic surface water wave patterns which are resonant, the case $m = 2$, $M \geq 2$ occurs, and the estimate on the number of geometrically distinct doubly periodic solutions is that

$$\#\{\text{critical orbits}\} \geq (M - 1) .$$

4 Estimates of the linear problem

The critical issue is to solve the P-equation (3.18) for the component of an m-torus embedding $S_2 = S_2(S_1, \Omega)$. This is typically a small divisor problem, which we address with a Nash - Moser implicit function theorem. Central to the workings of the method is an analysis of the linearized equation. The problem (3.18) is for an m-torus mapping, projected by P. With $S_1 = 0$ and then differentiating with respect to $S_2 \in P\mathcal{H}$ at the point $S_2 = 0$, the linearization of (3.18) is

$$P(\delta_{S_2}^2 H(0) - \Omega \cdot \delta_{S_2}^2 I(0)) PV = G , \tag{4.1}$$

where $PV = V$ and $PG = G$. In coordinates given by the eigenfunction expansion, equation (4.1) is expressed as

$$\begin{pmatrix} \omega(k) & i\Omega \cdot j \\ -i\Omega \cdot j & \omega(k) \end{pmatrix} \begin{pmatrix} v_1(j,k) \\ v_2(j,k) \end{pmatrix} = \begin{pmatrix} g_1 \\ g_2 \end{pmatrix} , \qquad j \in \mathbb{Z}^m , k \in \Gamma' . \tag{4.2}$$

It is seen that the spectrum of the operator on $P\mathcal{H}$ is the set $\{\mu(j,k) = \omega(k) - \Omega \cdot j \neq 0 : (j,k) \in \mathbb{Z}^m \oplus \Gamma'\}$. Typically this is a dense set of $\mathbb{R}$, the eigenvalues accumulating at zero constituting the small denominators of the problem. The Newton iteration scheme that is used in the Nash - Moser technique requires inversion of the linearized operator about an approximate torus embedding $S^0 = S_1 + S_2^0$, more than just the linearization about zero. That is, for $V = PV \in X_2$ we are drawn to the analysis of the operator

$$P\left(\delta^2_{S_2}\overline{H}(S_1+S_2^0)-\Omega\cdot\delta^2_{S_2}I(S_1+S_2^0)\right)V$$
$$=P\left(\mathrm{diag}_{2\times 2}\begin{pmatrix}\omega(k) & i\Omega\cdot j\\ -i\Omega\cdot j & \omega(k)\end{pmatrix}+W((j,k),(j',k'))\right)V=G\ . \tag{4.3}$$

The additional off-diagonal terms W come from the linearization of nonlinear terms of the equation (3.3). They represent a perturbation of a 2×2-block diagonal operator with dense spectrum. Estimates of this linearized operator are obtained by an adaptation of the method of J. Fröhlich and T. Spencer [FS83] using resolvant estimates and block decompositions of the lattice of indices of eigenmodes $\{y:=(j,k)\in\mathbb{Z}^m\oplus\Gamma'\}$. We work with a scale of Hilbert spaces of sequences on this lattice, which I will not specify precisely in this note. However we will denote the appropriate operator norm for this scale of spaces by $|\cdot|_{Op}$. A more detailed account of the analytic procedure appears in [CW93, C00] and [B95, B96, B98]. One first separates the lattice sites into the singular and the regular regions.

Definition 4.1. A lattice site $y=(j,k)\in\mathbb{Z}^m\oplus\Gamma'$ is d_0-*singular* for the frequency Ω when

$$|\omega(k)\pm\Omega\cdot j|<d_0\ . \tag{4.4}$$

Otherwise, the lattice site y is *regular.*

Let B be a subset of the lattice $\mathbb{Z}^m\oplus\Gamma'$, and let the orthogonal projection within $\mathcal{H}$ onto the subspace spanned by sequences $(s(y):y\in B)$ be denoted by P_B. The restriction of an operator H to the subset $P_B\mathcal{H}$ is denoted by $H_B=P_BHP_B$.

Proposition 4.1 *For a set $A\subseteq\mathbb{Z}^m\oplus\Gamma'$ consisting of regular lattice sites, and for $|W|_{Op}<d_0/2$, then the linearized operator* (4.3) *is invertible, and satisfies the estimate*

$$|P\left(\delta^2_{S_2}\overline{H}(S_1+S_2^0)-\Omega\cdot\delta^2_{S_2}I(S_1+S_2^0)\right)_A^{-1}|_{Op}\le\frac{4}{d_0}\ . \tag{4.5}$$

Following [FS83], resolvant estimates are used in order to add the singular sites of the lattice $\mathbb{Z}^m\oplus\Gamma'$ in the construction of the inverse operator $P(\delta^2_{S_2}\overline{H}(S_1+S_2^0)-\Omega\cdot\delta^2_{S_2}(S_1+S_2^0))^{-1}$. These will be quantified in the iteration scheme, over enlarging subsets of the lattice $\mathbb{Z}^m\oplus\Gamma'$ which grow to exhaust the lattice. Indeed, take regions $B_n=\{y=(j,k)\in\mathbb{Z}^m\oplus\Gamma':|y|\le R_n:=L_02^n\}$. Our ability to carry through the resolvant estimates depend upon two properties of the frequencies $\Omega\in\mathbb{R}^m$ and the linearized operator

$$P\left(\delta^2_{S_2}\overline{H}(S_1+S_2^0)-\Omega\cdot\delta^2_{S_2}I(S_1+S_2^0)\right):=D(\Omega)+W\ . \tag{4.6}$$

Property 1. Nonresonance. Suppose that $y=(j,k)$ and $y'=(j',k')$ in $\mathbb{Z}^m\oplus\Gamma'$ are two singular sites within the region $B_{n+1}\backslash B_n$. Then both

$$\begin{aligned} d_n &< |\omega(k)-\Omega\cdot j|<d_0\\ d_n &< |\omega(k')-\Omega\cdot j'|<d_0\ . \end{aligned} \tag{4.7}$$

The lower bound is adapted in the iteration scheme, $d_n=d_0\exp(-\beta n)$, and the frequency parameters Ω for which (4.7) is violated are excised. This is at the origin of the Cantor set over which the resulting existence theorem will hold.

Property 2. Separation. Suppose that y,y' are two singular sites within $B_{n+1}\backslash B_n$. We ask that either

(i) $\operatorname{dist}(y, y') < R_n^\alpha$, or else

(ii) $\operatorname{dist}(y, y') \gg R_n^\gamma$.

Typically the choice must be that $0 < \alpha \ll \gamma < 1$. In words, the requirement is that within the nth annulus $B_{n+1} \backslash B_n$, the singular sites can be divided into clusters, with slow growth of the cardinality of each cluster, namely $R_n^{\alpha m}$, and with large distances R_n^γ between clusters.

It is principally in this second property that the specific details of the particular partial differential equation and its nonlinearity and dispersion relation come into the analysis in an important way. In particular, the Schrödinger equation in one space dimension has singular sites which are automatically well separated, due to the nondegeneracy of the dispersion relation. The nonlinear wave equation in one dimension achieves the required degree of separation throug a diophantine condition on the frequency Ω^0. Problems in higher space dimensions address the two conditions with increasingly sophisticated techniques, which can be seen in more detail in [B95, B96]. It is however beyond the scope of the present article to pursue this in any further detail.

References

[BB97] T.B. Benjamin and T.J. Bridges. Reappraisal of the Kelvin-Helmholtz problem. I. Hamiltonian structure. *J. Fluid Mech.* **333** pp. 301–325, (1997).

[B95] J. Bourgain. Construction of periodic solutions of nonlinear wave equations in higher dimension. *Geom. Funct. Anal.* **5** (1995), no. 4, pp. 629–639.

[B96] J. Bourgain. Construction of approximative and almost periodic solutions of perturbed linear Schrödinger and wave equations. *Geom. Funct. Anal.* **6** (1996), no. 2, pp. 201–230.

[B98] J. Bourgain. Quasi-periodic solutions of Hamiltonian perturbations of 2D linear Schrödinger equations. *Ann. of Math. (2)* **148** (1998), no. 2, pp. 363–439.

[B00] J. Bourgain. Problems in Hamiltonian PDE's. GAFA 2000 (Tel Aviv, 1999). *Geom. Funct. Anal.* (2000), Special Volume, Part I, pp. 32–56.

[C00] W. Craig. Problèmes de petits diviseurs dans les équations aux dérivées partielles. (French) [Small divisor problems in partial differential equations] *Panoramas et Synthèses* **9**. Société Mathématique de France, Paris, 2000. viii+120 pp. ISBN: 2-85629-095-7

[CGK03] W. Craig, P. Guyenne & H. Kalisch. A new model for large amplitude long internal waves. *C. R. Acad. Sci. Paris* **332** (2004), pp. 525 - 530.

[CGK04] W. Craig, P. Guyenne & H. Kalisch. Hamiltonian long wave expansions for free surfaces and interfaces. *Commun. Pure Applied Math.* **LLVIII** (2005).

[CN00] W. Craig & D. Nicholls. Traveling two and three dimensional capillary gravity waves. *SIAM: Math. Analysis* **32** (2000), pp. 323–359.

[CS93] W. Craig and C. Sulem. Numerical simulation of gravity waves. *J. Comp. Phys.***108**, pp. 73–83, (1993).

[CW93] W. Craig and C.E. Wayne. Newton's method and periodic solutions of nonlinear wave equations. *Commun. Pure Applied Math.* **XLVI** pp. 1409-1501, (1993).

[CW94] W. Craig and C.E. Wayne. Periodic solutions of nonlinear Schrdinger equations and the Nash-Moser method. *Hamiltonian mechanics* (Toruń, 1993), pp. 103–122, NATO Adv. Sci. Inst. Ser. B Phys., 331, Plenum, New York, (1994).

[FS83] J. Fröhlich and T. Spencer. Absence of diffusion in the Anderson tight binding model for large disorder or low energy. *Comm. Math. Phys.* **88** (1983), no. 2, pp. 151–184.

[H10] J. Hadamard. *Leçons sur le calcul des variations.* Hermann, Paris, 1910; paragraph §249.

[H16] J. Hadamard. Sur les ondes liquides. *Rend. Acad. Lincei* **5**, (1916) no. 25, pp. 716–719.

[K87] S. B. Kuksin. Hamiltonian perturbations of infinite-dimensional linear systems with imaginary spectrum. (Russian) *Funktsional. Anal. i Prilozhen.* **21** (1987), no. 3, pp. 22–37, 95.

[K00] S. B. Kuksin. Analysis of Hamiltonian PDEs. *Oxford Lecture Series in Mathematics and its Applications,* **19.** Oxford University Press, Oxford, 2000. xii+212 pp. ISBN: 0-19-850395-4

[KP96] S. Kuksin and J. Pöschel. Invariant Cantor manifolds of quasi-periodic oscillations for a nonlinear Schrödinger equation. *Ann. of Math. (2)* **143** (1996), no. 1, pp. 149–179.

[M76] J. Moser. Periodic orbits near an equilibrium and a theorem by Alan Weinstein. *Comm. Pure Appl. Math.* **29** (1976), no. 6, pp. 724–747.

[P74] I. C. Percival. *Variational principles for the invariant tori of classical dynamics* J. Phys. A **7** (1974), pp. 794–802.

[P96] J. Pöschel. A KAM-theorem for some nonlinear partial differential equations. *Ann. Scuola Norm. Sup. Pisa Cl. Sci. (4)* **23** (1996), no. 1, pp. 119–148.

[W90] E. Wayne. Periodic and quasi-periodic solutions of nonlinear wave equations via KAM theory. *Comm. Math. Phys.* **127** (1990), no. 3, pp. 479–528.

[W73] A. Weinstein. Normal modes for nonlinear Hamiltonian systems. *Invent. Math.* **20** (1973), pp. 47–57.

[Z68] V.E. Zakharov. Stability of periodic waves of finite amplitude on the surface of a deep fluid. *J. Appl. Mech. Tech. Phys.* **9** (1968), pp. 190–194.

Fields Institute Communications
Volume **48**, 2006

Stable and Not Too Unstable Solutions on R^n for Small Diffusion

Norman Dancer
School of Mathematics and Statistics
The University of Sydney
NSW 2006, Australia
normd@maths.usyd.edu.au

The purpose of this survey is to survey and discuss some open questions for problems with small diffusion. More precisely our main interest is in studying problems.

$$-\epsilon^2 \Delta u = f(u) \text{ in } \Omega \tag{1}$$

where Ω is a smooth bounded domain in $\mathbb{R}^n$, and we have an appropriate homogeneous boundary condition on $\partial\Omega$ (usually Dirichlet or Neumann) and ϵ is small. Here $f : \mathbb{R} \longrightarrow \mathbb{R}$ is a given C^1 function. We are interested in how many solutions our problem has and their asymptotic shape. At the same time, we want results that apply to a wide variety of f's. Usually, we restrict to positive solutions though not always. Our aim is to obtain results which do not depend on domain shape. Our results are only for $n = 2$ (and sometimes $n = 3$) so that many interesting questions remain open.

The question is almost certainly impossible for all solutions so that our aim is to study solutions which are stable (in the natural sense and defined precisely in the text) or "not too unstable". By the latter, we mean solutions whose Morse index stays bounded as ϵ tends to zero.

The basis of the method are blow up arguments and a careful study of our problem when $\Omega = \mathbb{R}^n$ or a half space. Here we make use of important results of Berestycki, Caffarelli and Nirenberg [5] and ideas of Gui-Ghoussoub [40] and Alberti, Ambrosio and Cabre [1]. Here there are also a number of important open problems. We improve a number of the original results.

We also discuss results for generalizations of (1), for some systems and for some other geometries. We ignore the case where $n = 1$ where results are much easier to obtain. We also improve many of the original results. It should be noted that many important equations in catalysis, combustion and population models are included in (1). Thus results for (1) are of considerable importance in applications. Note also that in these applications $n = 2$ or $n = 3$.

We will not give complete proofs of results but aim to give the key ideas of the proofs of the main results.

2000 *Mathematics Subject Classification.* Primary 35J65; Secondary 35K55, 35B25.

Part 1 Stable and finite Morse index solutions on unbounded domains

§1. Linearized stable solutions on R^n

In this section, we study the problem

$$-\Delta u = f(u) \tag{2}$$

on $\mathbb{R}^n$. Note that the ϵ is irrelevant here because we can rescale so that $\epsilon = 1$ (if $\epsilon \neq 0$). A solution u of (2) is said to be linearized stable if

$$J(\phi) = \int (\nabla\phi)^2 - f'(u)\phi^2 \geq 0 \tag{3}$$

for all $\phi \in C_0^\infty(\mathbb{R}^n)$. Note that this makes sense even if u is unbounded. If $f'(u)$ is bounded, (3) continues to hold for $u \in W^{1,2}(\mathbb{R}^n)$. It is not difficult to show that in this case (3) is equivalent to assuming the spectrum of $-\Delta - f'(u)I$ on $L^2(\mathbb{R}^n)$ is contained in the right half plane.

Our main result is the following. It is the key to most of our later results.

Theorem 1. *Assume that (i) u is bounded and linearized stable and either $n = 3$ and $f \geq 0$ or $n = 2$ or (ii) u is linearized stable and $\int_{B_R} |\nabla u|^2 \leq CR^2$ for large R. Then after a rotation of axes, $u = u(x_1)$ and u is monotone in x_1.*

Remarks.

(1) If u is bounded, then such bounded ordinary differential equation solutions which are not constant are rare. For example, they can only occur if there exist $a, b \in \mathbb{R}$ such that $f(a) = f(b) = 0, F(a) = F(b)$ (where $F' = f$) and $F(x) < F(a)$ on (a, b). This follows from the first integral of the ordinary differential equation. The converse also holds if the zeros of f are isolated.

(2) In (ii), it suffices to assume the inequality holds for a sequence R_i tending to infinity. This is sometimes useful. Moreover if $n = 3$, it suffices to assume f has fixed sign on the range of u.

Sketch of proof. It suffices to prove (ii). Indeed if u is bounded then $f(u)$ is bounded and we can apply standard local estimates [41] for elliptic equation on balls of radius 1 to deduce that u is bounded in C^1. Thus if $n = 2$ the bound in (ii) automatically holds. If $n = 3$ the divergence theorem implies that

$$\int_{B_R} f(u) = -\int_{B_R} \Delta u = -\int_{\partial B_R} \frac{\partial u}{\partial R} \leq KR^2$$

since u is bounded in C^1 and ∂B_R has surface area of order R^2. Since f has fixed sign and u is bounded, it follows that

$$\int_{B_R} uf(u) \leq K_1 R^2 . \tag{4}$$

By using the divergence theorem again, we see that

$$\int_{B_R} uf(u) = -\int_{B_R} u\Delta u = \int_{B_R} |\nabla u|^2 - \int_{\partial B_R} u\frac{\partial u}{\partial R} .$$

Since u is bounded and ∇u is bounded, the second term in the right hand is of order R^2. Hence (4) implies the required estimate for $\int_{B_R} |\nabla u|^2$.

Thus it suffices to prove (ii). We first use the Harnack inequality to prove that there is a positive function ψ and $\overline{\lambda} \geq 0$ such that

$$-\Delta\psi = f'(u)\psi + \overline{\lambda}\psi \ \text{ on } \ \mathbb{R}^n \,.$$

To see this, we note that by the variational characterization of eigenvalues, our assumption ensures that the principal eigenvalue λ_R of

$$\begin{aligned} -\Delta\widetilde{\psi} &= f'(u)\widetilde{\psi} + \lambda\widetilde{\psi} \ \text{ in } \ B_R \\ \psi &= 0 \quad \text{on } \partial B_R \end{aligned}$$

satisfies $\lambda_R \geq 0$. Let ψ_R be the corresponding positive eigenfunction normalized so that $\psi_R(0) = 1$. Moreover, by the variational characterization of eigenvalues, λ_R decreases in R. Hence $\overline{\lambda} = \lim_{R \longrightarrow \infty} \lambda_R$ exists and $\overline{\lambda} \geq 0$. If $k > 0$ and $R \geq k+1$, the Harnack inequality [41] ensures a bound in $L^\infty(B_k)$ for ψ_R independent of R. (This uses that $\psi_R > 0$ and $\psi_R(0) = 1$). Thus, as before, we have a bound for ψ_R in C^1 on $B_{\frac{1}{2}k}$ where the bound is independent of R. Hence we can use a diagonalisation argument to find a subsequence of ψ_{R_i} (where $R_i \longrightarrow \infty$ as $i \longrightarrow \infty$) converging uniformly on compact sets to $\overline{\psi}$ where $\overline{\psi} \geq 0, \overline{\psi}(0) = 1$, $-\Delta\overline{\psi} = f'(u)\overline{\psi} + \overline{\lambda\psi}$ on $\mathbb{R}^n$. This is the required ψ. The difficulty is that we know little about the behaviour of ψ at infinity.

Now note that $\frac{\partial u}{\partial x_i}$ satisfies the same equation as ψ except that $\overline{\lambda} = 0$. Let $\sigma_i = \frac{\partial u}{\partial x_i}/\psi$. Then, by an elementary computation,

$$\text{div } (\psi^2 \nabla\sigma_i) = \overline{\lambda}\psi^2\sigma_i \,.$$

We will now use a clever test function argument of Ambrosio and Cabre [2] to prove that $\nabla\sigma_i = 0$. This implies that $\sigma_i =$ constant and hence $\frac{\partial u}{\partial x_i} = C_i\psi$ where C_i is a constant. By a rotation of axes, we can assume $C_i = 0$ for $i \geq 2$. Thus $\frac{\partial u}{\partial x_i} = 0$ for $i \geq 2$ and hence $u = u(x_1)$. If $C_1 \neq 0$, since $\psi > 0$, $\frac{\partial u}{\partial x_1}$ has fixed sign and u is monotone, as required. (If $C_1 = 0, u$ is constant).

Thus the proof reduces to using the test function argument to show that σ_i is constant. Choose ℓ smooth of compact support in B_1 such that $\ell \geq 0$ and $\ell = 1$ on $B_{\frac{1}{2}R}$ and let $\ell_R(x) = \ell(\frac{x}{R})$.

We multiply the equation for σ_i by $\ell_R^2\sigma_i$, integrate by parts and use that $\overline{\lambda} \geq 0$. We find after some easy calculations that

$$\begin{aligned} \int_{B_R} \psi^2\ell_R^2(\nabla\sigma_i)^2 &\leq -\int_{B_R} \ell_R\nabla\ell_R \cdot \nabla\sigma_i\psi^2\sigma_i \\ &\leq (\int_{B_R} (\psi\ell_R\nabla\sigma_i)^2)^{\frac{1}{2}} (\int_{B_R} (\psi\sigma_i\nabla\ell_R)^2)^{\frac{1}{2}} \end{aligned} \tag{5}$$

by the Cauchy-Schwartz inequality.

Since $\nabla\ell$ is bounded on B_1, $|\nabla\ell_R| \leq C/R$ on $\mathbb{R}^n$ were C is independent of R. Hence we find that

$$\int_{B_R} (\psi\ell_R\nabla\sigma_i)^2 \leq C(\int_{B_R} (\psi\ell_R\nabla\sigma_i)^2)^{\frac{1}{2}} (R^{-2}\int_{B_R} (\sigma_i\psi)^2)^{\frac{1}{2}} \,. \tag{6}$$

Now $\sigma_i\psi = \frac{\partial u}{\partial x_i}$ so that our assumptions on $\frac{\partial u}{\partial x_i}$ imply that $R^{-2}\int_{B_R}(\sigma_i\psi)^2 \leq K$ for all large R. Hence (6) implies that for large R

$$(\int_{B_R} (\psi\ell_R\nabla\sigma_i)^2)^{\frac{1}{2}} \leq C_1 \,.$$

Hence by taking the limit as R tends to infinity

$$\int_{R^n} (\psi \nabla \sigma_i)^2 \leq C_1^2 \, .$$

(Remember that ℓ_R is uniformly bounded and equals 1 on $B_{\frac{1}{2}R}$.) Thus $\psi \nabla \sigma_i \in L^2$. Hence $\lim_{R \longrightarrow \infty} \int_{B_R \backslash B_{\frac{1}{2}R}} (\psi \nabla \sigma_i)^2 = 0$. We now note that $\nabla \ell_R$ is only non-zero if $\frac{1}{2}R \leq \|x\| \leq R$. Thus we can refine (5) by replacing both integrals on the right hand side by integrals over $B_R \backslash B_{\frac{1}{2}R}$ and repeating the argument above we now find

$$\left(\int_{B_R} (\psi \ell_R \nabla \sigma_i)^2 \right)^{\frac{1}{2}} \leq C_1 \left(\int_{B_R - B_{\frac{1}{2}R}} (\psi \nabla \sigma_i)^2 \right)^{\frac{1}{2}} \longrightarrow 0$$

as $R \longrightarrow \infty$, Hence in the limit $\int_{R^n} (\psi \nabla \sigma_i)^2 = 0$ which implies that $\nabla \sigma_i = 0$ and hence σ_i is constant, as required.

Remarks.

(1) The most important open question is whether stable bounded solutions are always ordinary differential equation solutions if $n = 3$. (Note that the cases $n = 2$ or $n = 3$ are the ones of most physical interest for these problems). We discuss some partial results on this problem below.

(2) This question arose from work on the De Giorgi conjecture [40], [1],[2], [56]. Here, they replace linearized stability by the stronger condition that u is monotone in one direction. In this case much better results are known. In particular, monotone bounded solutions are ordinary differential equation solutions if $n \leq 7$ [56] (but for a restricted class of nonlinearities). For general nonlinearities it is proved in [1] for $n = 3$. Unfortunately, it is very unclear how to apply the techniques in [56] to our case. The techniques we use are based on earlier work on the De Giorgi conjecture in lower dimensions in [1], [2], [40]. It seems to us that in applications it is much more natural to assume stability rather than monotonicity. The monotone case is intuitively the most difficult case. However, it does not seem easy to prove this mathematically. We will sketch below an example in $n \geq 11$ of a linearized stable solution which is not a function of a single x_i. No such examples are known in the De Giorgi case. Note that there are examples of solutions of (1) with $n = 3$ where our gradient estimate fails. (These solutions are periodic in all variables).

(3) Our methods can be modified [25] to prove that there are no linearized stable negative solutions of $-\Delta u = e^u$ on $\mathbb{R}^3$. (Note that such solutions could not be bounded.) Remark 2 after the statement of Theorem 1 is used here.

We now discuss partial results for the case $n = 3$. Our conclusion is true if there is an i such that for some component T_i of

$$\{x \in \mathbb{R}^3, \frac{\partial u}{\partial x_i} \neq 0\}, \int_{T_i \cap B_R} \left| \frac{\partial u}{\partial x_i} \right|^2 \leq CR^2 \tag{7}$$

for large R. Note that this is a much weaker assumption. To prove this claim, we follow part of the proof of Theorem 1 to show that $\sigma_i = \frac{\partial u}{\partial x_i} / \psi$ is constant on T_i. (We restrict and only consider σ_i on T_i and note that $\sigma_i = 0$ on ∂T_i so that we do not introduce boundary terms in the test function argument). If $T_i \neq \mathbb{R}^3$, by considering the points on ∂T_i, we deduce that $\frac{\partial u}{\partial x_i} \equiv 0$ on T_i and hence on $\mathbb{R}^n$ by using that $\frac{\partial u}{\partial x_i}$ solves a homogeneous elliptic linear equation and then use the vanishing theorem [9]. Thus we reduce to the case $n = 2$. Note that the case $T_i = \mathbb{R}^n$ reduces to the De Giorgi case. This condition (7) is much weaker than the assumption in Theorem 1. For example, it always holds if u is bounded and if $m(T_i \cap B_R) \leq CR^2$ for large R. Note that (7) holds if u is bounded and $\int_{T_i \cap B_R} \left|\frac{\partial u}{\partial x_i}\right| \leq CR^2$ for large R (since ∇u is bounded). This holds if each line parallel to e_i meets T_i in at most k components where k is finite and independent of the line. In particular this holds if $\frac{\partial u}{\partial x_i}$ changes sign at most k times on any line in $\mathbb{R}^3$ parallel to e_i. These remarks suggest that any counter examples will have to be rather complicated (if they exist). Moreover, with care one can prove some such results if we only require the conditions for x_i large.

Lastly for this section, we sketch a counterexample to Theorem 1 in dimensions $n \geq 11$. By Joseph and Lungren [48], the positive solutions of

$$\begin{aligned} -\Delta u &= \lambda(1+u)^p \text{ in } B_1 \\ u &= 0 \qquad \text{on } \partial B_1 \end{aligned}$$

form a smooth curve $(u(\lambda), \lambda) : 0 < \lambda < a\}$ if p is large enough and $\|u(\lambda)\|_\infty \longrightarrow \infty$ as $\lambda \longrightarrow a^-$. ($a$ depends on p). Moreover, by [48] or by general theory, u is stable in the sense that the principal eigenvalue of the eigenvalue problem (for μ)

$$\begin{aligned} -\Delta h - \lambda p(1+u_\lambda)^{p-1} h &= \mu h \text{ in } B_1 \\ h &= 0 \text{ on } \partial B_1 \end{aligned}$$

is positive. Let $v_\lambda = (\|u_\lambda\|_\infty)^{-1} u_\lambda$. Then after a suitable rescaling of x, u_λ (rescaled) is a stable solution of

$$\begin{aligned} -\Delta v_\lambda &= (v_\lambda + o(1))^p \text{ on } B_{\tau(\lambda)} \\ v_\lambda &= 0 \text{ on } \partial B_{\tau(\lambda)} \end{aligned}$$

where $\tau(\lambda) \longrightarrow \infty$ as $\lambda \longrightarrow a$. Since v_λ is bounded in C^1 uniformly in λ (by using [41] again), it is not difficult to prove that a subsequence of u_λ converges uniformly on compact sets to a linearized stable positive radial solution of $-\Delta v = v^p$ on $\mathbb{R}^n$ which is our required example. It is close to examples in [42] and [8]. Note that our example is still in a sense one-dimensional but not in our earlier sense.

Lastly for this section, we state a very useful variant of Theorem 1 which is particularly useful in dimension 3.

Theorem 2. *Assume that u is a bounded positive linearized stable solution of (1) on $\mathbb{R}^3$ such that $u \longrightarrow 0$ as $x_1 \longrightarrow -\infty$ uniformly in x_2, x_3. Assume either that there is a $\delta > 0$ such that $f'(y) \leq 0$ on $(0, \delta)$ or $f'(0) > 0$ or f is convex on $[0, \delta_1]$ for some $\delta > 0$ and $f'(y) \sim ay^\tau$ as $y \longrightarrow 0^+$ where $a > 0$ and $0 < \tau < 4$. Then $u = u(x_1)$ and u is increasing in x_1.*

Remarks. It is easy to see that our assumptions imply that $f(0) = 0$ and that no such positive solution could exist if $f'(0) > 0$ (by the method of sub and supersolutions). It would be very interesting to remove the condition that $\tau < 4$. (If $\tau = 4$, it can be shown that there is no solution such that $u \longrightarrow 0$ as $\|x\| \longrightarrow \infty$). Note also that Theorem 2 is only of interest if f changes sign on the range of u.

This theorem is proved by combining moving plane ideas with the ideas in the proof of Theorem 1 (cp [22]). We actually obtain a slightly better result but it is convenient to defer this till the next section.

Note that we can easily weaken the decay condition at $-\infty$ to $u(x) \leq m$ if $x_1 \leq -a$ if we know that there is no positive solution of (1) with $0 \leq u \leq m$ on $\mathbb{R}^3$ (by a simple limiting argument).

§2. Finite Morse index solutions on R^n

In this section, we study a slightly larger class of solutions, the solutions of finite Morse index. We make use of the theory in §1.

A solution u of (1) is said to have Morse index at most k if whenever W is a finite dimensional subspace of $C_o^\infty(\mathbb{R}^n)$ and $J(w) < 0$ on $W \backslash \{0\}$ then dim $W \leq k$. It has finite Morse index if it has Morse index at most k for some k and the Morse index is the sup of such k. There are many other ways to express this definition. In particular, if $f'(u)$ is bounded on $\mathbb{R}^n$, it is not difficult to show that this condition is equivalent to assuming that the natural self-adjoint operator $-\Delta - f'(u)I$ on $L^2(R^n)$ has a finite dimensional spectral projection corresponding to the interval $(-\infty, 0)$. Note that we need to be a little careful in the formulation because the spectrum of $-\Delta - f'(u)I$ need not consist of eigenvalues. If u is a solution of Morse index k on $\mathbb{R}^n$, then there is an $R > 0$ such that u is linearized stable on $\mathbb{R}^n \backslash B_R$ (in the sense that $J(\phi) \geq 0$ for $\phi \in C_0^\infty(\mathbb{R}^n \backslash B_R)$ (where this denotes the smooth functions of compact support on $\mathbb{R}^n \backslash B_R$). To see this, we choose $\{\phi_i\}_{i=1}^k$ in $C_o^\infty(\mathbb{R}^n)$ such that J is negative definite on span $\{\phi_i\}$. Choose $\widetilde{R}$ so that supp $\phi_i \subseteq B_{\widetilde{R}}$ for each i. Then $J(\phi) \geq 0$ if $\phi \in C_0^\infty(\mathbb{R}^n \backslash B_{\widetilde{R}})$. To see this, note that if $J(\phi) < 0$ for some such ϕ, then since ϕ and $\sum \alpha_i \phi_i$ have disjoint support

$$\begin{aligned} J(\alpha_o \phi + \sum \alpha_i \phi_i) &= J(\alpha_o \phi) + J(\sum \alpha_i \phi_i) \\ &= \alpha_o^2 J(\phi) + J(\sum \alpha_i \phi_i) \\ &< 0 \end{aligned}$$

when at least one $\alpha_i, i = 0, \cdots, k$ is non-zero. This contradicts that k is the Morse index which proves our claim. We suspect that there is a converse to this result though it does not seem easy to prove. It is true if $-\Delta - f'(u)I$ is Freedholm on $L^2(\mathbb{R}^n)$.

Our main result in this case is the following simple theorem. We say that condition nw holds if there does not exist any non constant bounded monotone solutions of $-u'' = f(u)$ on $\mathbb{R}$. By our earlier remarks, this certainly holds if whenever $a \neq b$ and $f(a) = f(b) = 0$ then $F(a) \neq F(b)$. The importance of this assumption is that it ensures that the linearized stable solutions are constant (under the assumptions of Theorem 1).

Theorem 3. *Assume that condition nw holds, that $n = 2$ or $n = 3$, that f has fixed sign, that the zeros of f do not contain a non-trivial interval and that u is a bounded solution of (1) of finite Morse index. Then there is a $C \in R$ such that $u(x) \longrightarrow C$ as $\|x\| \longrightarrow \infty$. Moreover $f(C) = 0$ and $f'(C) \leq 0$.*

Sketch of proof. If $\|x_i\| \longrightarrow \infty$ as $i \longrightarrow \infty$, then $\{u(x + x_i)\}$ are a sequence of solutions of (1) bounded in C^1 and thus a subsequence converges uniformly on compact sets to a bounded solution v of (1). Moreover, v is stable. To see this suppose $R > 0$ and $\phi \in C_0^\infty(B_R)$. By our remarks before the statement of the theorem, $\int_{\mathbb{R}^n} |\nabla \phi_i|^2 - f'(u(x))\phi_i^2 \geq 0$ where $\phi_i(x) = \phi(x - x_i)$. (This follows since ϕ_i has support outside $B_{\widetilde{R}}$ for large i). Hence $\int_{B_R} (\nabla \phi)^2 - f'(u(x + x_i))\phi^2 \geq 0$ and thus in the limit $\int_{B_R} (\nabla\phi)^2 - f'(v)\phi^2 \geq 0$. Thus v is stable and hence, by Theorem 1, v is constant.

Hence we see the $u(x+x_i) \longrightarrow C$ uniformly on compact sets where C is constant and is a stable solution of (1). Thus $f(C) = 0$ and $f'(C) \leq 0$. Note that if $B > 0, -\Delta - BI$ is not positive. Hence by a simple compactness argument, if R is large, $\{u(x) : \|x\| \geq R\}$ is uniformly close to $\{c : f(c) = 0, f'(c) \leq 0\}$. However $\{x : \|x\| \geq R\}$ is connected and so $\{u(x) : \|x\| \geq R$ is connected $\}$ and hence $\{u(x) : \|x\| \geq R\}$ must be close to a connected subset of $\{c : f(c) = 0, f'(c) \leq 0\}$ and the result follows.

Remark.

(1) If condition *nw* fails the finite Morse index solutions may be more complicated. See the example in [58] which is obtained by reflection from a quarter space solution in $\mathbb{R}^2$. It would be interesting to understand more generally which of these solutions have finite Morse index (for a more general nonlinearity than in [58]).

(2) In Theorem 3, we frequently also know that $u \geq C$ on $\mathbb{R}^n$ (or $u \leq C$ on $\mathbb{R}^n$). In this case, one can frequently use moving plane arguments to prove that u is radial. Thus in many cases the finite Morse index solutions are quite simple (solutions of an ordinary differential equation).

(3) If u is a solution of (1) such that $u \longrightarrow C$ as $\|x\| \longrightarrow \infty$ and $f'(C) < 0$, it is easy to show that u has finite Morse index (since $-\Delta - f'(C)I$ is invertible and has spectrum in $(0, \infty)$). In general, the problem seems more complicated if $f'(C) = 0$. (It depends on the decay of $f'(u(x))$. However, if u is radial and $u - C$ has fixed sign, then it can be shown that the Morse index is the same as in the space of radial functions and as in [39] this can be determined by oscillation theory. (We use here that $u'(r) \neq 0$ for $r \neq 0$). If $u(r) - C$ changes sign, it can be shown that the Morse index on $\mathbb{R}^n$ is strictly greater than the Morse index in the radial space though the difference is finite if $u(r) - C$ has finitely many zeros. It is also possible to give a formula for the difference in terms of oscillation theory.

(4) It is unclear how frequently there are non-radial changing sign solutions with $u \longrightarrow C$ as $\|x\| \longrightarrow \infty$ and in particular ones with finite Morse index. This seems an interesting question. The changing sign solutions (found in [38]) of $-\Delta u = |u|^{p-1}u$ on $\mathbb{R}^n$ where $n \geq 3$ and $p = (n+2)/(n-2)$ can be shown to be non-radial and of finite Morse index. (They can not be radial by Pokojaev's identity and they are of finite Morse index because they have fast decay at infinity since they are obtained from smooth solutions on a sphere). The only other examples known to the author are the ones for $n \geq 6$ or $n = 4$ in Bartsch and Willem [1]. It seems very likely that some others can be constructed by perturbations from Ding's examples.

(5) There is a variant of Theorem 2 where all the conditions there hold except that u has finite Morse index rather than stable and u is assumed not to be stable and we assume condition *nw* holds. Then it can be proved that $u \longrightarrow 0$ as $\|x\| \longrightarrow \infty$ and that u is radial (up to translation). The first part is proved by sliding plane arguments in [22] while the second then follows by combining this result with results in Li and Ni [51] and Zou [65].

§3. The half space case

We consider the half space T which we take to be $\{x \in \mathbb{R}^n : x_1 \geq 0\}$ and we consider

$$-\Delta u = f(u) \text{ on } T \tag{8}$$

plus a boundary condition $u = 0$ on ∂T or $\frac{\partial u}{\partial x_1} = 0$ on ∂T. In the first case we denote it by (8_D) while in the second case we denote it by (8_N). We first consider the Neumann case. We say that u is linearized stable if $J(\phi) \geq 0$ for all $\phi \in C_0^\infty(\overline{T})$, that is the C^∞ functions of compact support in $\overline{T}$.

Theorem 4. *Assume that in u is a bounded linearized stable solution of (8_N) and that $n = 2$ or $\int_{B_R \cap T} |\nabla u|^2 \leq CR^2$ for large R. Then after a rotation of axes in $(x_2, \cdots, x_n)$, $u = u(x_2)$ and u is monotone.*

Sketch of Proof. We can extend u to be a solution on $\mathbb{R}^n$ by extending it evenly across $x_1 = 0$ by reflection. (This uses the Neumann boundary condition). If we can prove that this solution is stable on $\mathbb{R}^n$, the result will follow easily from Theorem 1 and the boundary condition. By the monotonicity of the principal eigenvalue, it suffices to consider a ball B_R centre zero. Since $f'(u)$ is even in x_1, the simplicity and positivity of the principal eigenfunction w of

$$\begin{aligned} -\Delta v - f'(u)v &= \lambda v \text{ in } B_R \\ v &= 0 \text{ on } \partial B_R \end{aligned}$$

ensures that w is even in x_1. Thus $\inf\{J(v) : v \in C_o^\infty(B_R), \|v\|_2 = 1\} = 2J(w)$ (for w suitably normalized). Thus J is linearized stable on B_R and our claim follows.

Remark. Note that positivity is not very helpful for Neumann boundary conditions because it can always be obtained by translating u without affecting the boundary condition .

Analogously to earlier we can define a solution u of (8_N) to have finite Morse index if there is an integer k such that whenever W is subspace of $C_0^\infty(\overline{T})$ such that $J(\phi) < 0$ on $W \backslash \{0\}$, then $\dim W \leq k$.

Theorem 5. *If u is a bounded solution of (8_N) of finite Morse index, if $n = 2$, if condition nw holds and if the zeros of f are isolated, then there is a $C \in R$ such that $f(C) = 0, f'(C) \leq 0$ and $u(x) \longrightarrow C$ as $\|x\| \longrightarrow \infty$.*

Proof. Much as in the proof of Theorem 3, we find that there is an $\widetilde{R} > 0$ such that $J(\phi) \geq 0$ if $\phi \in C_0^\infty(T)$ and if $\phi(x) = 0$ for $\|x\| \leq \widetilde{R}$, and then prove that if $|x_i| \longrightarrow \infty, x_i \in T, u(x + x_i) \longrightarrow C$ as $i \longrightarrow \infty$ uniformly on compact sets where $f(C) = 0$ and $f'(C) \leq 0$ (where C may depend on $\{x_i\}$). Since $\{x \in T : \|x\| \geq \widetilde{R}\}$ is connected, we can argue as there deduce that C is independent of $\{x_i\}$.

Note that this gets us back very much to the situation of Theorem 3.

We now consider the Dirichlet case. This is somewhat different. We have two main theorems.

Theorem 6. (Berestycki, Caffarelli and Nirenberg [5]). *Assume that u is a non-negative bounded solution of (8_D) where $f(0) \geq 0$ and $n \leq 3$. Then $u = u(x_1)$.*

Remark. Note that we do not assume that u is linearized stable though this is a consequence of the theorem. They prove a slightly stronger theorem. Note also that the proof of Theorem 6 in [5] introduced a number of the ideas we have used above. In fact, the ideas here could be used to obtain a slightly different proof of Theorem 6. We sketch this very briefly. One uses moving plane ideas as in [16] or [5] to prove that $\frac{\partial u}{\partial x_1} > 0$ if $x_1 \in T$ (if u is non-trivial) and hence $lim_{x_1 \longrightarrow \infty} u = \widetilde{u}$ exists. One then proves that $\widetilde{u}$ is a linearized stable solution of $-\Delta' v = f(v)$ on $\mathbb{R}^{n-1}$. Thus, by Theorem 1, $v = v(x_2)$ (after a rotation in $x_2, \cdots, x_n$) and v is monotone in x_2. One then uses sub and supersolutions to prove that $\sup_{x_i, i\geq 2} u(x_1, \cdots, x_n)$ is a subsolution of $-y'' = f(y)$ and hence there is a positive solution of $-y'' = f(y), y(0) = 0, y(\infty) = \sup v$. We then use this to construct a subsolution of compact support and hence prove that v is constant (by sweeping families of subsolutions). Finally, if v is constant and if we construct a subsolution $\widetilde{u}$ as above, then $\sup_{x_i, i\geq 2} \widetilde{u}(x_1, \cdots, x_n)$ is a subsolution below our solution u (by sweeping families) and is a function of one variable. Hence by iteration there is a one dimensional solution below u with limit v. However, the first integral ensures the ordinary differential equation solution is unique and our claim follows. (Note that the maximal solution is also an ordinary differential equation solution).

Theorem 7. *Assume that u is a bounded linearized stable sign changing solution of (5_D) and $n = 2$ or $\int_{B_R \cap T} |\nabla u|^2 \leq CR^2$ for large R and f has isolated zeros. Then $u = u(x_1, x_2)$ after a rotation of coordinates in $x_2, \cdots, x_n$, u is monotone in $x_2, \lim_{x_2 \longrightarrow \pm\infty} u(x_1, x_2) = u\pm$ are monotone solutions of $-u'' = f(u)$ with opposite signs such that $u_+'(0) = -u_-'(0) > 0$ and there exists a monotone solution of $-u'' = f(u)$ joining $u_-(\infty)$ and $u_+(\infty)$ and in particular $F(u_-(\infty)) = F(u_+(\infty))$.*

Sketch of Proof. Much as before the stability of u and the Harnack inequality (or rather the version up to the boundary in [6]) guarantees the existence of $\overline{\lambda} \geq 0$ and a positive ψ such that $-\Delta\psi = f'(u)\psi + \overline{\lambda}\psi$ on $T, \psi = 0$ on ∂T. One uses the maximum principle to prove that $\frac{\partial u}{\partial x_i} \ / \ \psi$ is locally bounded near ∂T for $i \geq 2$. (Remember that $\frac{\partial u}{\partial x_i} = 0$ on ∂T if $i \geq 2$). We consider $\sigma_i = \frac{\partial u}{\partial x_i}/\psi$ and note as before that div $(\psi^2 \nabla \sigma_i) = \overline{\lambda}\psi^2\sigma_i$. We then repeat part of the proof of Theorem 1 to prove that σ_i is constant. (To justify the integration by parts we first integrate over $\{x \in B_R : x_1 \geq \epsilon\}$ and then let ϵ tend to zero. That $\psi \longrightarrow 0$ as $x_1 \longrightarrow 0$ ensures the boundary terms vanish in the limit and hence the proof is valid). Hence we find that $\frac{\partial u}{\partial x_i}/\psi = C_i$ for $i \geq 2$. By a rotation in $x_2, \cdots, x_n$, we can ensure that $C_i = 0$ for $i \geq 3$. Thus $\frac{\partial u}{\partial x_i} = 0$ for $i \geq 3$ and hence $u = u(x_1, x_2)$. Moreover $\frac{\partial u}{\partial x_2} = C_2\psi$ and u is monotone in x_2. (That $C_2 = 0$ is impossible since then $u = u(x_1)$.) It is easy to prove that any linearized stable ordinary differential equation solution on $[0, \infty)$, is monotone and thus has fixed sign). Hence u must depend on x_2. Hence u_+ and u_- exist and are distinct linearized stable bounded solutions of $-y'' = f(y), y(0) = 0$. One must be positive and one must be negative because u changes sign. Suppose $u_-(\infty) < \ell < u_+(\infty)$ such that no other zeros of f lies in $[\ell, u_+(\infty))$. For each x_1 large enough there exists $\widetilde{x}_2$ such that $u(x_1, \widetilde{x}_2) = \ell$. Then, if we shift the origin to (x_1, x_2) and then let x_1 tend to infinity, a subsequence of these solutions will converge uniformly on compact sets to a linearized stable solution $\widetilde{u}$ on $\mathbb{R}^2$ of $-\Delta u = f(u), u(0,0) = \ell, u \leq u_+(\infty)$. By Theorem 1, u must be a monotone ordinary differential equation solution joining two zeros s, t of $\widetilde{u}$ with $s < \ell < t \leq u_+(\infty)$. ($\widetilde{u}$ cannot be constant by the choice of ℓ). By the choice of ℓ, $t = u_+(\infty)$ and $F(y) < F(u_+(\infty))$ on $(s, u_+(\infty))$ by the first integral. Set $a_2 = s$. We now repeat the construction with ℓ chosen less than s such that $[\ell, s)$ contains no zeros of f. We obtain a monotone solution joining s_1, s_2 where $s_1 < \ell < s_2, s \leq s_2 \leq u_+(\infty)$ and $F(y) < F(s_2)$ on (s_1, s_2) by the first integral. We prove that $s_2 = s$. If not, $s_2 > s$ and then $F(s_2) < F(u_+(\infty))$ (since $F(y) < F(u_+(\infty))$ on $(s, u_+(\infty))$ when $s \in (s_1, s_2)$) and thus $F(s) < F(s_2) < F(u_+(\infty))$ which gives a contradiction. Thus $s = s_2$ and we can continue the argument and eventually reach $u_-(\infty)$. We obtain that $u_+'(0) = -u_-'(0)$ from the various first integrals. Since $u_+ \neq u_-, u'_+(0) \neq u'(0)$ by uniqueness and hence $u'_+(0) > 0$. Thus by the first integral for $u_+, F(y) < F(u_+(\infty)$ on $[0, u_+(\infty))$. Similarly, $F(y) < F(u_-(\infty)) = F(u_+(\infty))$ on $(u_-(\infty), 0]$. Thus $F(y) < F(u_+(\infty))$ on $(u_-(\infty), u_+(\infty))$ and hence by the construction of $s, s = u_-(\infty)$. We have completed the proof.

Thus we see that if $n = 2$, it is rather difficult to have bounded changing sign linearized stable solutions of (8_D). Indeed there are none unless condition nw fails. The simplest way to construct such solutions is to assume f is odd and $f(a) = 0$ where $a > 0$ and $F(a) > F(0)$. One uses sub and supersolutions to construct a positive solution on a quarter space for Dirichlet boundary conditions and reflects it as an odd function to obtain a solution on a half space. Indeed sub and supersolutions can be used to construct these solutions always under the assumptions of Theorem 7. We suspect they have good uniqueness properties. One proves existence by first using a variant of results on infinite strips in [23] and then passing to the limit.

There is one special case that is worthy of mention. Suppose that $n = 3, M > 0, f(M) = 0, f'(M) < 0$. Then there can be no linearized stable solution of (8_D) such that $u \longrightarrow M$ uniformly as $x_1 \longrightarrow \infty$ except if $u = u(x_1)$. To see this, one proves that $u \longrightarrow M$ exponentially and hence $\frac{\partial u}{\partial x_i} \longrightarrow 0$ exponentially as $x_1 \longrightarrow \infty$. Thus our condition on the gradient in Theorem 7 holds and the result follows from Theorems 6 and 7.

Lastly note that the estimate on ∇u can sometimes be obtained from local energy estimates, monotonicity and energy bounds.

Finally we mentioned one result on finite Morse index solutions of Dirichlet problems on half spaces. Note that by the result of Berestycki, Caffarelli and Nirenberg, the only interesting case (for bounded solutions, $f(0) \geq 0$ and $n \leq 3$) is where u changes sign. (Because a non-trivial non-negative solution satisfies $\frac{\partial u}{\partial x_1} > 0$ and hence is linearized stable).

Theorem 8. *Assume that $n = 2, u$ is a bounded finite Morse index solution of $(8_D), u$ changes sign, condition nw holds and the zeros of f are discrete. Then there is a non-negative (or non-positive) bounded linearized stable solution $\widetilde{u} = \widetilde{u}(x_1)$ of (8_D) such that $u(x) - \widetilde{u}(x) \longrightarrow 0$ as $\|x\| \longrightarrow \infty, x \in \overline{T}$. Here we also allow $\widetilde{u} \equiv 0$.*

Sketch of Proof. This is very similar to the proof of Theorem 5. If $|x^i| \longrightarrow \infty$ as $i \longrightarrow \infty, x_i \in \overline{T}, u(x + x_i)$ must converge through a subsequence uniformly on compact sets to a linearized stable solution of (8) on $\overline{T}$ or on $\mathbb{R}^2$ which must be a constant or a solution $\widetilde{u}(x_1)$ on $\overline{T}$. (The limit might depend on the sequence x^i). Suppose $x^i, \widetilde{y}^i \in \overline{T}, d(x^i, \partial T) \longrightarrow \infty$ as $i \longrightarrow \infty, d(y^i, \partial T) \longrightarrow \infty$ as $i \longrightarrow \infty$ and $u(x^i) \longrightarrow c_1$ as $i \longrightarrow \infty$ while $u(y^i) \longrightarrow c_2$ as $i \longrightarrow \infty$ where $c_1 \neq c_2$. Choose $\tau \in (c_1, c_2)$ with $\tau \neq 0$ and $f(\tau) \neq 0$. By considering the line segment joining x^i and y^i, we find z^i such that $d(z^i, \partial T) \longrightarrow \infty$ as $i \longrightarrow \infty$ and $u(z_i) = \tau$. Since τ is not a solution, this contradicts what we have already proved. Thus $u(x^i) \longrightarrow C$ as $i \longrightarrow \infty$ if $d(x^i, \partial T) \longrightarrow \infty$ as $i \longrightarrow \infty$ where $f(C) = 0, f'(C) \leq 0$ and C is independent of the choice of x^i. It then follows that each such one-dimensional solution $\widetilde{u}$ on T must also have the property that $\widetilde{u} \longrightarrow C$ as $x_1 \longrightarrow \infty$ and this uniquely determines $\widetilde{u}$ (by the first integral).

Remark. Note that $\widetilde{u} \equiv 0$ may be possible if $f(0) = 0$ and $f'(0) \leq 0$ though we do not have an example. Such a solution would have to change sign. One would expect them to be rare. Note that such a solution u which is not stable and $\widetilde{u} \neq 0$ must have a peak down on a positive solution or vice versa (by Theorem 6). An example appears in [31].

§4. The case of strips

We consider very briefly analogous results in $\Omega \times \mathbb{R}^k$ where Ω is a smooth bounded domain in $\mathbb{R}^n$. For simplicity, we only consider Dirichlet boundary conditions. We give one example why we might be interested in this problem in §8 though there are number of others. We consider

$$\begin{aligned} -\Delta u &= f(u) \text{ on } \Omega \times \mathbb{R}^k \\ u &= 0 \text{ on } \partial\Omega \times \mathbb{R}^k . \end{aligned} \tag{9}$$

Theorem 9. *Assume that u is a bounded linearized stable solution of (6) and $k \leq 2$. We write $X = (x, y)$ with $x \in \Omega, y \in R^k$. Then, after a rotation of the y coordinates, $u = u(x, y_1)$ and u is monotone in y_1.*

Sketch of Proof. As before, we use the stability to find ψ positive and $\overline{\lambda} \geq 0$ such that $-\Delta\psi = f'(u)\psi + \overline{\lambda}\psi$ on $\Omega \times \mathbb{R}^k, \psi = 0$ on $\partial\Omega \times \mathbb{R}^k$.

Then, much as in the proof of Theorem 7 we find that $\frac{\partial u}{\partial y_i}/\psi$ satisfies a nice equation and by a similar test function argument to that in the proof of Theorem 7 $\frac{\partial u}{\partial y_i}/\psi =$ constant. We can then argue much as before. The condition on k is used to guarantee that the volume of the ball of radius R intersected with $\Omega \times \mathbb{R}^k$ grows no faster than R^2.

Remarks.

(1) We obtain more information if u depends on y_1. By the results in [23], $v_\pm = \lim_{y_1 \longrightarrow \pm\infty} u(x, y_1)$ must be stable solutions of the problem on Ω and have the same energy. Thus these linearized stable solutions depending on y_1 are rather rare.

(2) The analogous results hold for Neumann boundary conditions (where we can allow $k = 3$ if $f \geq 0$).

(3) If the problem on Ω does not have two linearized stable solutions with the same energy and if solutions of $-\Delta u = f(u)$ in Ω with the appropriate boundary condition are isolated, we can argue much as before to prove that if u is a bounded finite Morse index solution on $\Omega \times \mathbb{R}^k$ and $k \leq 2$, then $u(x, y) \longrightarrow w(x)$ as $|y| \longrightarrow \infty$ where w is a linearized stable solution of $-\Delta w = f(w)$ on Ω with the appropriate boundary condition. (If $k = 1$, the limits as $y \longrightarrow \pm\infty$ could possibly be different but then the integral in [23] shows they must have the same energy and hence must be the same by our assumptions). Note that if f is real analytic on $\mathbb{R}$ it can be shown that positive linearized stable solutions are isolated if $f(0) \geq 0$ when f is not linear (cp [18]) or if f is sublinear or superlinear with subcritical power growth (or possibly supercritical if $\dim \Omega = 3$) at ∞ (or $-\infty$) if we do not assume positivity. Note that, if $u - w$ has fixed sign, one can frequently use moving plane techniques to obtain further properties of u.

Part 2 Problems on bounded domains with small diffusion

§5. Stable solutions of Neumann problems

In this section, we discuss stable solutions of the problem

$$\begin{aligned} -\epsilon^2 \Delta u &= f(u) \text{ in } \Omega \\ \frac{\partial u}{\partial \nu} &= 0 \text{ on } \partial\Omega\,. \end{aligned} \tag{10}$$

Here Ω is a smooth bounded domain in $\mathbb{R}^n$ and ν is the outward normal to $\partial\Omega$.

We show under weak assumptions that such a solution for small ϵ is constant if $n = 2$. We can actually handle more general problems than the above including non self-adjoint problems but we leave this to the remarks. Our strategy will be to use Theorem 1 and blow ups to show that u is nearly constant on Ω and then use classical techniques. Note that if we could prove that Theorem 1 holds for $n = 3$, we could also prove the theorem below for $n = 3$.

Firstly we give a formal definition that a solution u of (10) is stable. Here for simplicity we assume that $f'(y)$ has polynomial growth at infinity. (In higher dimensions, we would have to assume a suitable subcritical growth that is $f'(y)$ grows at most like $|y|^{p-1}$ where $p < (n+2)/(n-2)$.) We can then define stability by requiring that u is a local minimum of $E_\epsilon(v) = \int_\Omega \frac{1}{2}\epsilon^2|\nabla v|^2 - F(v)$ on $W^{1,2}(\Omega)$. Here I should clarify this definition because several are used in the literature. We mean that there is a neighbourhood W_ϵ of u in $W^{1,2}(\Omega)$ show that $E_\epsilon(v) \geq E_\epsilon(u)$ for v in W_ϵ. Note that W_ϵ is allowed to depend on ϵ. If we wish to avoid the growth restriction we only require this for v in $W_\epsilon \cap L^\infty(\Omega)$ (assuming u_ϵ is bounded). By using perturbations $v = u + \delta w$ where $w \in C_0^\infty(\Omega)$, we see that in either case $J_\epsilon(\phi) \geq 0$ for $\phi \in C_0^\infty(\Omega)$ where $J_\epsilon(\phi) = \int_\Omega \epsilon^2(\nabla\phi)^2 - f'(u)\phi^2$. This is actually what we will use. In fact our definition is equivalent to required that u is a stable solution of the parabolic flow of

$$\begin{aligned} u_t &= \epsilon^2\Delta u + f(u) \text{ in } \Omega \\ \frac{\partial u}{\partial \nu} &= 0 \text{ on } \partial\Omega \end{aligned}$$

in $L^s(\Omega)$ for large s. See [27].

If we do not assume the growth condition on f', this is still true but we need to consider the flow on $L^\infty(\Omega)$ where we consider solutions which are continuous in t for $t > 0$ but only weak $*$ continuous at zero.

Theorem 10. *Assume that $\epsilon_i \longrightarrow 0$ as $i \longrightarrow \infty, u_{\epsilon_i}$ are stable solutions of (10) for $\epsilon = \epsilon_i$ such that $\|u_{\epsilon_i}\|_\infty \leq k$ for all i. In addition, assume that $n = 2$, that condition nw holds, that the zeros of f are isolated and that, if α is a zero of f where f changes from negative to positive, then $f' \geq 0$ in a neighbourhood of α. Then u_{ϵ_i} is a constant c for large i where $f(c) = 0, f'(c) \leq 0$.*

Sketch of Proof. This proof as do many others below depend on using Theorem 1 to obtain the asymptotics of u. In this case, we first show that there is a zero c of f such that $\|u_{\epsilon_i} - c\|_\infty \longrightarrow 0$ as $i \longrightarrow \infty$ (after choosing subsequences). The result then follows from more classical techniques.

We sketch the proof of our claim above. Since the zeros of f are isolated, it suffices to prove that if $\epsilon_i \longrightarrow 0$ as $i \longrightarrow \infty$ and $x_i \in \Omega$ then $d(u_{\epsilon_i}(x_i), f^{-1}(0)) \longrightarrow 0$ as $i \longrightarrow \infty$. It is now convenient to use the rescaled equation

$$-\Delta u = f(u) \text{ on } \Omega_{\epsilon_i} = \epsilon_i^{-1}\Omega \tag{11}$$

with the Neumann boundary condition. Note that we are assuming $0 \in \Omega$ (which we can do without loss of generality). Suppose by way of contradiction that $\widetilde{x}_i \in \Omega_{\epsilon_i}$ such that $d(\widetilde{u}_i(\widetilde{x}_i), f^{-1}(0)) \geq \alpha > 0$ for all large i. We also assume that $d(\widetilde{x}_i, \partial\Omega_{\epsilon_i}) \longrightarrow \infty$ as $i \longrightarrow \infty$. We return to the other case later. Here $\widetilde{u}_i$ is u_i rescaled (to satisfy (11)). Now $\widetilde{u}_i(x + \widetilde{x}_i)$ is bounded in C^1 on compact sets in $\mathbb{R}^n$ (by local estimates as earlier) and hence we can choose a subsequence which converges uniformly on compact sets to a bounded solution v of $-\Delta u = f(u)$ on $\mathbb{R}^n$. (We use that $d(\widetilde{x}_i, \partial\Omega_{\epsilon_i}) \longrightarrow \infty$ as $i \longrightarrow \infty$ to ensure that, given $R > 0, \widetilde{u}_i(x + \widetilde{x}_i)$ is defined on B_R for large i). Hence, since $d(\widetilde{u}_i(\widetilde{x}_i), f^{-1}(0)) \geq \alpha, d(v(0), f^{-1}(0)) \geq \alpha$. Moreover, since $\widetilde{u}_i$ are linearized stable, v is linearized stable (cp part of the proof of Theorem 3). Thus, by Theorem 1, v is constant. (We use condition nw here). Hence, since v is a solution of $-\Delta v = f(v), f(v) = 0$ which contradicts that $d(\widetilde{u}_i(\widetilde{x}_i), f^{-1}(0)) \geq \alpha$. In the case when $d(\widetilde{u}_i(\widetilde{x}_i), f^{-1}(0)) \geq \alpha$ but $d(\widetilde{x}_i, \partial\Omega_{\epsilon_i}) \leq K_1$ for all i, we use similar blow up argument except that in the limit we obtain a solution of $-\Delta v = f(v)$ on a half space T and satisfying the Neumann boundary condition on ∂T. (A detailed proof of this appears in [21]. Note that locally $\partial\Omega_{\epsilon_i}$ is rather flat so that in the limit we obtain a half space). Otherwise the argument is the same except we use Theorem 4 rather than Theorem 1.

We now complete the proof of Step 1. $\{u_{\epsilon_i}(x) : x \in \overline{\Omega}\}$ is a connected set which is uniformly close to the discrete set $f^{-1}(0)$ for large i and hence must be close to a single member of $f^{-1}(0)$ by connectedness.

The rest of the proof is rather classical. We consider linearized stable solutions u uniformly close to a constant c where $f(c) = 0$. If the maximum of u occurs at $x_{\max} \in \Omega, \Delta u(x_{\max}) \geq 0$ and hence $f(u(x_{\max})) \geq 0$. If the maximum occurs on $\partial\Omega$ (and then $x_{\max} \in \partial\Omega$), it is not difficult to use the Neumann boundary condition to check that the same result holds. (We use that $\nabla u(x_{\max}) = 0$ and that $\frac{\partial^2 u}{\partial t^2}(x_{\max}) = 0$ if t is tangent to $\partial\Omega$). If $f(u)(x_{\max}) = 0$, one can apply the strong maximum principle and the Neumann boundary conditions to $u(x_{\max}) - u$ to deduce a contradiction unless u is constant. Hence, unless u is constant, $f(u(x_{\max})) > 0$. Similarly $f(u(x_{\min})) < 0$ if u attains its minimum on $\overline{\Omega}$ at $x_{\min}$. Hence c must be a nodal zero of f where f changes from negative to positive. Hence, by our assumptions, $f'(u(x)) \geq 0$ on Ω (since $u(x)$ is uniformly close to c). Now $J_\epsilon(1) = -\int_R f'(u(x))$ and hence, if u is stable, $\int_\Omega f'(u(x)) = 0$ and hence $f'(u(x)) = 0$ on Ω (since $f' \geq 0$ near c). Hence $f'(y) = 0$ on $[u(x_{\min}), u(x_{\max})]$ which is an interval with c in its interior (since f has different signs at $x_{\min}$ and $x_{\max}$). This contradicts that c is an isolated zero of f.

This completes the sketch of the proof.

Remarks.

1. This result admits many variants. We always assume $n = 2$. Note that the main part of the proof (the first part) shows that there can be no solutions of $-\epsilon^2\Delta u = f(x, u)$ in Ω with sharp transitions if for each $x_0 \in \overline{\Omega}$ the map $y \longrightarrow f(x_0, y)$ satisfies condition nw. In fact $\{(x, u(x)) : x \in \overline{\Omega}\}$ must be uniformly close to $\{(x, y) : x \in \overline{\Omega}, f(x, y) = 0\}$. In particular, if $A \subseteq \overline{\Omega}$ is closed and connected, if for each $x_0 \in A$ the map $x \longrightarrow f(x_0, y)$ satisfies condition nw and if $\{y \in [-K, K] : f(x, y) = 0\}$ is $\{y = \ell_i(x) :$ where the l_i are continuous and never intersect for $x \in A\}$ then any stable solution (1) with $\|u\| \leq K$ is uniformly close to some $l_i(x)$ on A for some i (independent of x) and $f_2'(x, l_i(x)) \leq 0$ on A. (One could also formulate a theorem for the case where $l_i(x)$ intersect on some surfaces).

We can then use classical techniques to look for such solutions and frequently prove local uniqueness. As a very particular example, the conclusion of Theorem 10 continues to hold for $\epsilon^2\Delta u = g(x)f(u)$ if f satisfies the assumptions of Theorem 10 and g is continuous and strictly positive on $\overline{\Omega}$. Another example is that we could prove the uniqueness of the positive stable solution of $-\epsilon^2\Delta u = u^2(1-\ell(x)u)$ if ℓ is continuous and positive on $\overline{\Omega}$ and ϵ is small.

2. Our theory continues to hold if $-\Delta$ is replaced by $-\Delta - b\cdot\nabla$ where $b : \overline{\Omega} \longrightarrow \mathbb{R}^2$ is continuous. This situation frequently arises when the ϵ^2 occurs as a large parameter multiplying the nonlinear term $f(u)$. We explain this briefly. Firstly, when one rescales as earlier, the drift term (that is, the ∇u term) becomes $\epsilon b(\epsilon^{-1}x)\cdot\nabla u$ and this will go to zero in the limit. Secondly, our problem is non-self adjoint on Ω. We say that a solution u is linearized stable if the principal eigenvalue of

$$-\epsilon^2(\Delta h + b(x)\cdot\nabla h) - f'(u)h = \lambda h \text{ in } \Omega$$
$$\frac{\partial h}{\partial\nu} = 0 \text{ on } \partial\Omega$$

is non-negative. It is well known (cp [43]) that the stability of the corresponding parabolic flow implies this. Note that we need to use this as the definition of stability in the non self-adjoint case. Note that this assumption implies that the principal eigenvalue for Dirichlet boundary conditions is also non-negative (cp Hess [44], Proposition 17.7) and hence the principal eigenvalue on Ω_1 (where Ω_1 is an open subset of Ω) for Dirichlet boundary conditions on $\partial\Omega_1$ is also non-negative by monotonicity. The monotonicity is a simple application of subsolution results. We use this result on balls $x_0 + \epsilon B_R(0)$ and use that on compact domains the principal eigenvalue depends continuously on the coefficients (as is easily proved) to prove that the blow up limit of linearized stable solutions on Ω is linearized stable on $\mathbb{R}^2$ and hence we can proceed as before. Note that our definition of linearized stability on $\mathbb{R}^n$ is equivalent to requiring the principal eigenvalue on each B_R for Dirichlet boundary conditions is non-negative (by our comments in §1.) In fact these arguments can still be used for a drift term $g(x,u)\cdot\nabla u$ is g is continuous on $\overline{\Omega}\times\mathbb{R}$ provided f has simple zeros. (This extra condition is only used in the last step of the proof).

3. Note that the proof of Theorem 10 does not use the full strength of the linearized stability. If f has simple zeros, we could consider solutions u_ϵ such that the principal eigenvalue λ_ϵ of

$$\begin{aligned} -\epsilon^2\Delta h - f'(u_\epsilon)h &= \lambda h \text{ in } \Omega \\ \frac{\partial h}{\partial\nu} &= 0 \text{ on } \partial\Omega \end{aligned}$$

satisfies $\liminf_{\epsilon\longrightarrow 0}\lambda_\epsilon = 0$. This is a meta stability condition on u. Under the same condition on f, we could also consider nonlinearities $f(y)+\epsilon g(y,\epsilon)$ with similar results.

4. If $f(c) = 0$ and $f'(c) = 0$, the stability of the constant solution c of (10) reduced to that of c as a solution of $y' = f(y)$ by the centre manifold theorem (cp Henry [43, Theorem 6.4.1]) and this is easily determined.

5. The first part of the proof could also be applied for a Robin boundary condition $\frac{\partial u}{\partial\nu} + \widetilde{c}u = 0$ to deduce that a linearized stable solution is uniformly close to a zero of f. The point is that when we blow up the boundary condition becomes $\frac{\partial u}{\partial\nu} + \epsilon cu = 0$ which becomes a Neumann boundary condition in the limit. We will return to this problem in the next section. In particular, if $\widetilde{c}$ is non-negative, we can use the methods in the proof of Theorem 11 in the next section to obtain an exact count of the number of stable positive solutions with $\|u\|_\infty \leq K$. If $f \geq 0$ and $\widetilde{c}$ is non-negative, we can also obtain results for $n = 3$.

6. We discuss the boundedness assumption on the stable solutions. In many cases, this holds. For example, if $yf(y) \leq 0$ for $|y|$ large, this follows easily from the maximum principle. If we can bound solutions above or below for example by the maximum principle or if we are only interested in positive solutions, we can frequently use blow up arguments to bound stable solutions. For example if $f'(y) \sim ay^{p-1}$ as $y \longrightarrow \infty$ where $a, p > 0$, then the bound follows if we know that the only bounded linearized stable positive solution of $-\Delta u = u^p$ on $\mathbb{R}^2$ is zero. This follows easily from Theorem 1. By [25], we could also handle the case where $e^{-y}f'(y) \longrightarrow a > 0$ as $y \longrightarrow \infty$. If $f'(y) \sim |y|^{p-2}$ as $|y| \longrightarrow \infty$ where $1 < p < \infty$, we could apply the ideas in [3]. (Here we do not first need to establish a one sided bound.)

7. If Ω is convex, Theorem 10 is true in all dimensions without the assumption that ϵ is small cp [10] and [52]. A result of Kohn and Sternberg [50] shows that Theorem 10 is false for some domains if condition nw fails. It remains unclear for exactly which domains it fails. Note that by results of Matano and Mimura and the author [20] Theorem 10 may fail for star shaped domains if ϵ is not small even if f satisfies condition nw.

8. Note that our assumption of f near its zeros always holds if f is real analytic and not identically zero or if f is C^∞ and $f^k(\alpha) \neq 0$ for some k at each zero α of f (where k may depend upon α).

§6. Stable solutions of the Dirichlet problem

We discuss analogues of the results of §5 for the Dirichlet problem

$$\begin{aligned} -\epsilon^2 \Delta u &= f(u) \text{ in } \Omega \\ u &= 0 \text{ on } \Omega\,. \end{aligned} \tag{12}$$

We mainly consider positive solutions where we obtain good results for $n = 2, 3$ but, if $n = 2$, we also consider changing sign solutions. In a sense, our results are as close as they can be to the results of the previous section in that the stable solutions are almost constant away from $\partial\Omega$.

We can define stable and linearized stable solutions in the same way as the previous section (except that our space is $\dot{W}^{1,2}(\Omega)$ not $W^{1,2}(\Omega)$).

We state the simplest version of the main theorem and leave improvements to the remarks.

Theorem 11. *Assume that $n \leq 3$, f has simple zeros, condition nw holds, $f(0) \geq 0$ and $K > 0$ such that $f(K) \neq 0$. Let*

$$\widetilde{Z} = \{C \in (0, K) : f(C) = 0, f \text{ is negative on } (C - \delta, C)$$

$$\int_0^t f < \int_0^C f \text{ if } 0 \leq t \leq C\,.$$

Then the number of non-trivial non-negative stable solutions u of (12) with $\|u\|_\infty \leq K$ for small positive ϵ is the number of elements of $\widetilde{Z}$. Moreover, if $C \in \widetilde{Z}$, define $\phi_\epsilon^C(x) = C$ if $d(x, \partial\Omega) \geq \delta$ (where δ is small and fixed) and $\phi_\epsilon^C(x) = u_0(\epsilon^{-1}, t)$ if $d(x, \partial\Omega) \leq \delta$. Here u_0 is the unique solution of $-u''(t) =$

$f(u(t)), u(0) = 0, u(\infty) = C$. (Note that each point x near $\partial\Omega$ can be uniquely written in the form $x = x_0 - t\nu(x_0)$ where $x_0 \in \partial\Omega$ and $\nu(x_0)$ is the outward normal to $\partial\Omega$ at x_0.) Then any stable solution for small ϵ is uniformly close to a ϕ_ϵ^C on Ω and there is a unique positive solution for each such C and this solution is non degenerate.

Sketch of Proof. First assume that $n = 2$. We can use Theorem 1 and a blow up argument rather similar to the first part of the proof of Theorem 10 to show that a stable positive solution u is uniformly close to a zero C of f with $f'(C) \leq 0$ except when $\epsilon^{-1}d(x, \partial\Omega)$ is bounded. (We use that $\{x \in \Omega : d(x, \partial\Omega) \geq \mu\}$ is connected if μ is small). On the other hand, if we blow up a solution of our problem at a point within order ϵ of the boundary, we easily see that we obtain a non-negative bounded solution of

$$\begin{aligned} -\Delta u &= f(u) \text{ on } T \\ u &= 0 \text{ on } \partial T \end{aligned}$$

where T is the half space $\{x \in R^2 : x_1 \geq 0\}$. Since $f(0) \geq 0$, we can use the maximum principle to deduce that either $u \equiv 0$ on $u > 0$ in T. By a result of Berestycki, Caffarelli and Nirenberg [5], $u = u(x_1)$ and u is monotone in x_1. (For future reference, note that this part of the argument is also valid if $n = 3$). Let $C = \lim_{x_1 \longrightarrow \infty} u(x_1)$. At this stage, C may depend on the choice of boundary point. Since $\frac{1}{2}(u'(x_1))^2 + F(x_1) = \text{constant} = d = F(u(\infty))$, we see that this solution is uniquely determined by $u(\infty)$ (for $u \geq 0$) and $F(y) < F(u(\infty))$ for $y \in (0, u(\infty))$. Note that this implies that $f(y) < 0$ for y slightly less than $u(\infty)$ if $u(\infty) > 0$. Since the boundary blow up must match with the interior blow up, $u(\infty) = C$ and thus, by our comments above, this implies that the boundary blow up is the same at each point of $\partial\Omega$. This gives the asymptotics in the theorem. Note that $C = 0$ is impossible for stable positive solutions. (Such positive solutions would be uniformly small. This is obviously impossible if $f'(0) < 0$ since $f(u_{\max}) \geq 0$. If $f'(0) > 0$, the linearization is $-\epsilon^2\Delta - g(x)I$ on Ω with Dirichlet boundary conditions on Ω where $g(x)$ is uniformly close to $f'(0)$ and this clearly has negative eigenvalues. Hence small positive solution are not stable. This shows that any positive stable solution has the required asymptotics if $n = 2$ (or if $n = 3$ and $f \geq 0$ or more generally $f \geq 0$ on $(0, a), f \leq 0$ on (a, ∞)). If $n = 3$, we have to argue somewhat differently. We make more use of the boundary. We sketch this very briefly. We can argue as above to obtain the asymptotic behaviour near the boundary except we also use sweeping family of subsolutions as in Sweers [57] to show that the boundary blow up solution is independent of the point $x_0 \in \partial\Omega$. (Here we use the half space solution is uniquely determined by $u(\infty) = C_1$. We construct the subsolutions from solutions on large balls.) This argument also shows that $u \geq C_1 - \delta$ on Ω except within order ϵ of the boundary. If $\mu > 0$ is small, we choose the point $x_\epsilon \in \{x \in \Omega : u(x) \geq C_1 + \mu\}$ closest to $\partial\Omega$. By blowing up we then obtain a bounded linearized stable solution $\widetilde{u}$ of $-\Delta u = f(u)$ on $\mathbb{R}^3$ such that $\widetilde{u} \geq C_1$ on $\mathbb{R}^3$ and $\widetilde{u} \leq C_1 + \mu$ if $x_1 \leq a, \widetilde{u}(0) = C_1 + \mu$. (This requires a little care). By considering subsequences of $\widetilde{u}(x - \alpha_i e_1)$ where $\alpha_i \longrightarrow \infty$ as $i \longrightarrow \infty$, we see that either $\widetilde{u} \longrightarrow C_1$ as $x_1 \longrightarrow -\infty$ uniformly in (x_2, x_3) or there exists a solution $\hat{u}$ of $-\Delta\hat{u} = f(\hat{u})$ such that $C_1 < \hat{u} \leq C_1 + \mu$ on $\mathbb{R}^3$. Since $f(y) < 0$ for $y > C_1$ but close to C_1 (if $C_1 > 0$) and since $f(\hat{u}_{\max}) \geq 0$ (even if u does not achieve its maximum), we obtain a contradiction if μ is small. (If $C_1 = 0$, and $f'(0) > 0$, we have to argue differently. We use that there is no bounded positive solution of $-\Delta u = f(u)$ if

$f(y) \geq \alpha y$ for $y > 0$, where $\alpha > 0$ (cp Toland [59],). Thus $\hat{u} \longrightarrow 0$ as $x_1 \longrightarrow -\infty$ uniformly in (x_2, x_3) and then Theorem 2 (applied to $\widetilde{u} - C_1$) implies that $\widetilde{u}(x_1)$ is monotone in x_1 and, since condition nw holds, $\widetilde{u}$ is constant $\equiv C_1$. Hence we have a contradiction unless $u \leq C_1 + \mu$ on Ω. Since μ was arbitrary, this gives the required asymptotics.

The rest of the argument is rather classical. We use that $f(u_{\max}) \geq 0$ and is not zero (as earlier) to deduce that $\|u\|_\infty < C$ (but close to C). To construct such a solution, we use that C is a supersolution. It then suffices to show that there is a subsolution ϕ of our problem with $\|\phi\|_\infty < C$ but close to C and larger than C_1 which is the largest zero of f less than C. Set $C_1 = 0$ if $f(y) > 0$ on $[0, C)$. Then for each ϵ small, there is a stable solution between ϕ and C (cp Dancer and Hess [28]) and by the first part of the theorem, this has sup close to C. Moreover, the theory in Sweers [57] ensures the uniqueness of this solution. Thus the proof reduces to the construction of ϕ. By a result of Hess [45], there is a positive solution $\hat{u}$ of $-\Delta u = f(u)$ in B_{R_0} with $u = 0$ on ∂B_{R_0} and $C_1 < \|\hat{u}\|_\infty < C$ if R_0 is large. We fix R_0. Then, for ϵ small, $B_{R_0} \subseteq \epsilon^{-1}\Omega$ and we obtain a subsolution of $-\Delta u = f(u)$ on $\epsilon^{-1}\Omega$ by defining ϕ to be $\hat{u}$ on B_{R_0} and zero elsewhere. Here we are using that $f(0) \geq 0$ and Proposition 1 in Berestycki and Lions [7] to ensure it is a subsolution on $\epsilon^{-1}\Omega$. We obtain a subsolution of (12) on Ω by simply rescaling ϕ. This completes the sketch of the proof.

Remarks.

1. The most important remark is that if $n = 2$, we can delete the assumptions that condition nw holds and that $f(0) \geq 0$. To see this, first note that, if $0 < a < b$, if a and b are zeros of f and if there exists a monotone heteroclinic orbit joining a and b, then any solution of (12) with $\|u\|_\infty < b$ satisfies $\|u\|_\infty \leq a$. This follows from the proof of Theorem 1 in Clement and Sweers [11] (by sweeping families of supersolutions). It follows from this and our interior blow up argument in the proof of Theorem 11 that u is always uniformly close to a zero $\hat{c}$ of f in the interior of Ω. (Note that our blow up argument implies that in the interior of Ω, u can never take a value close to $\widetilde{c}$ where $f(\widetilde{c}) \neq 0$ and $\widetilde{c}$ does not lie on a non-negative monotone heteroclinic solution.) On the other hand the boundary blow up argument and Theorem 7 implies that near the boundary u rescaled must be uniformly close to the solution of $-u'' = f(u(t)), u(0) = 0, u(\infty) = \widetilde{c}$ which is positive. (Note that, if $f(0) < 0, \widetilde{c} = 0$ is impossible). This proves the asymptotics of the stable solutions. We can then complete the proof much as before except that as in Clement and Sweers [11], we must be a little more careful in the construction of the subsolution. We do not know if these results are true for $n = 3$ when f changes sign on $[0, \infty)$.

2. Once again, we can weaken greatly the simplicity of the zeros in Theorem 11. We firstly assume that the non-negative zeros of f are isolated, that whenever f is positive just to the left of a positive zero $\bar{c}$ of f, then $f'(x) \leq 0$ on $(\bar{c} - \delta, \bar{c})$. Moreover, if $f'(\bar{c}) = 0$ and $f > 0$ just to the right of such a zero, we assume that f is convex on $[\bar{c}, \bar{c} + \delta]$ and $f'(x + \bar{c}) = ax^{r-1} + o(x^{r-1})$ on $[0, \delta]$ where $a > 0$ and $1 \leq r < (n+2)/(n-2)$. Finally if $f(0) = f'(0) = 0$ we need to assume that either $f'(x) \leq 0$ on $(0, \delta)$ or f is convex on $[0, \delta]$ and $f'(x) = ax^{r-1} + o(x^{r-1})$ on $(0, \delta)$ with a and r as above. Note that these assumptions are weak assumptions near the zeros and allow non-nodal zeros. We could actually do slightly better if $n = 2$. We could also allow r to be critical or supercritical if there is a $\tau > \bar{c}$ such that $f(x) > 0$ on $(\bar{c}, \tau)$ and $f(x) < 0$ if $x > \tau$. These are proved by combining the ideas in Remark 1 with the ideas in the proofs of Theorem 1 in the cases $n = 2$ and $n = 3$ and noting there they can be no small positive stable solutions since $yf'(y) > f(y)$ for small positive if $f(0) = f'(0) = 0$ and $f > 0$ on $(0, \delta)$ (under our assumptions above).

3. Once again our ideas could also be used if f dependent on x. The blow up arguments can be used to obtain good asymptotics for stable positive solutions if the map $y \longrightarrow f(x_0, y)$ satisfies condition nw (at least for $y \geq 0$) for each $x_0 \in \overline{\Omega}$. In particular, there are no sharp transitions in this case. We can then use similar techniques to the theorem to obtain existence and uniqueness, especially if in addition $f_y(x_0, y) \neq 0$ whenever $f(x_0, y) = 0$ and $x_0 \in \overline{\Omega}$. As to the simplest example, we obtain good results for $g(x)f(y)$ where f satisfies the assumptions of Theorem 11 and g is continuous and positive on $\overline{\Omega}$.

4. With care, our results in this section still hold if f is locally Lifschitz, f is C^1 except at isolated points and f has continuous right and left derivatives at x_0 if $f(x_0) = 0$. This generalization is frequently useful for problem in mathematical biology. The key point is that if $z \in \mathbb{R}\backslash\{0\}, \{x \in \Omega : u(x) = z\}$ has measure zero and hence the linearization makes sense. There is a similar result for the Neumann problem though one has difficulties if $f(x_0) = 0$ and f is not differentiable at x_0.

5. Most of the other remarks at the end of §5 hold here (in particular, the counterexamples with ϵ not small). Note that in the non-self adjoint case, we must use some of the ideas near the end of [17] (with some minor corrections) to construct subsolutions.

Lastly for this section, we consider changing sign stable solutions for $n = 2$. It is proved by combining the ideas in the proof of Theorem 11 with some of the ideas in Remark 1. (In particular we use Theorem 7 and Sweers subsolutions). Note that it is easy to check that there is no small stable changing sign solutions.

Theorem 12. *Assume that $n = 2$, condition nw holds, f has simple zeros and $K > 0$. There are no stable changing sign solutions u if (12) with $\|u\|_\infty \leq K$ and ϵ small.*

Remarks.

1. By examining the proof here and our remarks above we see that, if $n = 2$ and f has simple zeros, we have a rather complete understanding of the uniformly bounded stable solutions for small ϵ unless there exist zeros a, b of f with $a < 0 < b$ and a monotone heteroclinic solution of $-u''(t) = f(u(t))$ joining a and b. Conversely a result of Kohn and Sternberg [50] shows that if this condition occurs, then for certain domains there always exist stable solutions with sharp interior transition layers for all small ϵ. However, it is unclear for which domains this can occur (even for strongly convex domains). It can occur for strongly convex domains if ϵ is not small (cp Sweers [57]).

2. Once again, we could weaken the condition that the zeros of f are simple.

We can do better for global maxima when they exist if $n = 3$. The difference occurs here because it is easy to see that when we blow up at an interior point, we obtain a solution u on $\mathbb{R}^3$ which is a local minimum in the much stronger sense of Alberti, Ambriosio, Cabré [1] that $E_R(v) \geq E_R(\overline{u})$ for every smooth function v such that $v - \overline{u}$ has support in B_R where $E_R(s) = \int_{B_R} \frac{1}{2}|\nabla s|^2 - F(s)$.

The results in [1] then ensure that $\int_{B_R} |\nabla \overline{u}|^2 \leq CR^2$ for large R and hence Theorem 1 applies. In particular if condition nw holds, we deduce that $\overline{u}$ is constant and much as before, we can deduce that a global minimum is nearly constant in the interior of Ω. In in addition the zeros of f are simple, we can use a half space result mentioned after the proof of Theorem 7 to deduce that, in the Dirichlet case that either u is trivial or has fixed sign and hence we are back to the case in Theorem 11. (In the Neumann case we deduce u is constant.)

§7. Finite Morse index solutions

We want to indicate rather more briefly how we can obtain information on the finite Morse index solutions for small ϵ. The interest here is that we can frequently prove that the finite Morse index solutions are exactly the solutions with finitely many sharp peaks. These have been very extensively studied in recent years.

Recall that the Morse index of a solution u of (12) is the number of negative eigenvalues counting multiplicity of

$$-\epsilon^2 \Delta h - f'(u)h = \lambda h \text{ in } \Omega \tag{13}$$

with the appropriate boundary condition. (For a non self-adjoint problem, we would define it to be the number of eigenvalues with negative real part counting multiplicity).

Our main result is the following. This is for positive solutions of the Dirichlet problem. We will briefly discuss analogues for other cases when $n = 2$ in the remarks afterwards.

Theorem 13. *Assume that $n = 2$ or $3, f(0) \geq 0$, f has simple zeros, condition nw holds and $K > 0$. Suppose that u_i are positive solutions of (12) with $\|u_i\|_\infty \leq K$ and Morse index at least 1 and at most k for all i where $\epsilon_i \longrightarrow 0$ as $i \longrightarrow \infty$. Then after taking subsequences, if necessary, there exist $C_1 \geq 0$ and a finite number of distinct points $\{x_i^j\}_{j=1}^m$ in Ω where $1 \leq m \leq k$ such that $f(C_1) = 0, f'(C_1) \leq 0$ and u_i is uniformly close to $\phi_{\epsilon_i}^{C_1}(x)$ on Ω except within order ϵ_i of $x_i^j, \epsilon_i^{-1} d(x_i^j, \partial\Omega) \longrightarrow \infty$ as $i \longrightarrow \infty$, $\epsilon_i^{-1} d(x_i^j, x_i^r) \longrightarrow \infty$ as $i \longrightarrow \infty$ if $r \neq j$ and $u_i(\epsilon_i^{-1}(x - x_i^j))$ converges uniformly to a solution $\widetilde{u}_j$ of finite Morse index of $-\Delta u = f(u)$ on $\mathbb{R}^n$ such that $\widetilde{u}_j \longrightarrow C_1$ uniformly as $\|x\| \longrightarrow \infty$ and $\widetilde{u}_j > C_1$ on $\mathbb{R}^n$. Moreover $\widetilde{u}_j$ is not linearized stable.*

Remark. I do not exclude that x_i^j are close to each other or to the boundary, though this can often be excluded by peak solution ideas. There appears to be no simple relationship between the Morse index and the number of peaks. We could weaken the assumption to u_i has finite meta Morse index, that is, there exist μ_i with $\mu_i \longrightarrow 0$ as $i \longrightarrow \infty$ such that the sum of the multiplications of the eigenvalues of the linearization in $(-\infty, \mu_i)$ is at most k for all i.

Sketch of Proof. We only sketch the proof for $n = 2$. The generalization to $n = 3$ uses similar ideas to the proof of Theorem 11 for $n = 3$. (In particular, it makes more uses of the half space results and subsolutions).

We define $\widetilde{Z}$ as in the proof of Theorem 11. Suppose that $u_i(x_i)$ is bounded away from the finite set $\widetilde{Z}$ and $\epsilon_i^{-1} d(x_i, \partial\Omega) \longrightarrow \infty$. Then, much as earlier, $u_i(\epsilon_i^{-1}(x - x_i))$ converges uniformly on compact sets to a solution u_1 of $-\Delta u = f(u)$ on $\mathbb{R}^n$ of finite Morse index. Moreover $u_1(0)$ is not close to $\widetilde{Z}$. Thus, by Theorem 1 and condition nw, u_1 is not linearly stable and hence has Morse index at least 1. In particular, there exists $\phi \in C_0^\infty(\mathbb{R}^n)$ such that $\int_{R^n} |\nabla\phi|^2 - f'(u_1)\phi^2 < 0$. It follows easily that, for large i, $\widetilde{\phi}_i(x) = \phi(\epsilon_i^{-1}(x - x_i))$ has the property that $T_i(\widetilde{\phi}) = \int_\Omega \epsilon_i^2 (\nabla\widetilde{\phi}_i)^2 - f'(u_i)(\widetilde{\phi}_i)^2 < 0$ and $\widetilde{\phi}_i$ has support within order ϵ_i of x_i. If $y_i \in \Omega$

such that $\epsilon_i^{-1}d(y_i, \partial\Omega) \longrightarrow \infty, \epsilon_i^{-1}d(x_i, y_i) \longrightarrow \infty$ and $u_i(y_i)$ is not close to $\widetilde{Z}$, we see that we have two functions of disjoint support for which $T_i(\phi) < 0$ and hence a two dimensional subspace on which $T_i(\phi) < 0$ (except for zero). Continuing this process, we see that we will have a contradiction unless there exist $m_i \leq k$ points $\{x_i^j\}_{j=1}^{m_i}$ in Ω such that $\epsilon_i^{-1}d(x_i^j, \partial\Omega) \longrightarrow \infty$ as $i \longrightarrow \infty, \epsilon_i^{-1}d(x_i^{j_i}, x_i^{k_i}) \longrightarrow \infty$ as $i \longrightarrow \infty$ if $j_i \neq k_i$ and $u_{\epsilon_i}(x)$ is uniformly close to $\widetilde{Z}$ if $x \in \Omega$, and if $\epsilon_i^{-1}d(x, \partial\Omega)$ is large unless $\epsilon_i^{-1}d(x, x_i^j)$ is bounded for some j with $1 \leq j \leq m_i$. Since Ω minus a finite number of points is connected, we see that, for fixed large $i, u_{\epsilon_i}(x)$ must be uniformly close to a single element C_1 of $\widetilde{Z}$ except within order ϵ_i of the boundary or within order ϵ_i of some x_i^j. On the other hand, if we blow up within order ϵ_i of the boundary, we obtain a bounded non-negative solution of $-\Delta u = f(u)$ on a half space. By Theorem 6, this must be a function of x_1 only and monotone in x_1. Since by using the first integral (and by our assumptions), these solutions are isolated and uniquely determined by their value at infinity, we see the blow-up half space solution must be the same on all of $\partial\Omega$ and the convergence is uniform near the boundary. Suppose we blow up at a point x_i^j for fixed j and obtain a solution $\hat{u}$. By what we have proved, $\hat{u} \longrightarrow C_1$ uniformly as $|x| \longrightarrow \infty$. This proves our claim except to show that $\hat{u} \geq C_1$. To prove this, we use sweeping families of subsolutions much as in the proof of Theorem 11 for $n = 3$. Hence $\hat{u} \geq C_1$ as required. This completes the proof.

Remarks.

1. The importance of this result is that it shows that, under suitable conditions, that the solutions of finite Morse index are exactly the solutions with sharp peaks. These have been very extensively studied by many authors. See [32], [33], [42], [63] where many other references can be found. In particular, by combining with results of Wei [60], [62], [63] on solutions with 1 sharp peak, we can largely classify the uniformly bounded positive solutions of Morse index 1 if condition *nw* holds and a non-degeneracy condition is satisfied (for Dirichlet boundary conditions if $n = 2, 3$ and for Neuman boundary conditions if $n = 2$). However our information on peak solutions is far from complete. Our result is often very useful for showing that a solution obtained by variational methods (mountain pass, saddle point etc.) is a peak solution for small ϵ. See [26], for example.

2. If $n = 2$, there are analogous theorems for changing sign solutions or for solutions under Neumann boundary conditions. However, there are some differences. The peaks can now be up or down on the stable solutions and indeed the peaks could be solutions which oscillate about the limit at infinity. In addition, the peaks can be within order ϵ of the boundary (as in [31]) and so appear in the boundary blow up solutions (as in Theorems 5 and 8). Moreover, in the Neumann case, the peaks could be on the boundary.

3. As before, our methods could be used to study the case where f depends on x, mainly when the map $y \longrightarrow f(x_0, y)$ satisfies condition *nw* for each $x_0 \in \overline{\Omega}$ and $n = 2$. As a simple example of this, our techniques imply that every positive solution of bounded Morse index of

$$\begin{aligned} -\epsilon^2 \Delta u &= b(x)u^p - u \text{ on } \Omega \\ u &= 0 \text{ on } \partial\Omega \end{aligned}$$

for small ϵ, consists of finitely many sharp peaks on the zero solutions if $n \leq 3, 1 < p < \frac{n+2}{n-2}$ and b is continuous and positive is $\overline{\Omega}$. This case is a little easier because one can show that the boundary blow up solution is always trivial. (In fact, if b is C^1, one can then use the theory of peak solutions to show that the peaks must occur near critical points of b on Ω or of $b|_{\partial\Omega}$ on $\partial\Omega$.

4. As in the remarks after Theorem 12, one can greatly weaken the assumption that the zeros of f are simple. However, there is one change. We must change the statement if we have a positive zero α of f such that $f > 0$ in some deleted neighbourhood of α. If $f(y) \geq c|y-\alpha|^q$ in this neighbourhood where $c > 0$ and $1 < q \leq n/(n-2)$ and if f has a zero α_1 larger than α for which the boundary equation also has a solution, there will be a solution of Morse index 1 uniformly close to the stable solution ϕ_ϵ^α (and thus uniformly close to α in the interior of Ω). To see this, note that as in [29] there is an unstable mountain pass solution $\widetilde{u}_\epsilon$ between $\phi_\epsilon^\alpha(x)$ and $\phi_\epsilon^{\hat{\alpha}}(x)$. We can repeat the argument in the proof of Theorem 13 to show that either $\widetilde{u}_\epsilon$ is uniformly close to $\phi_\epsilon^\alpha(x)$ or is a one peak positive solution on top of $\phi_\epsilon^\alpha(x)$. We show the second case is impossible, which will prove our claim. If the second case occurred, the proof of Theorem 13 implies that there is a solution of $-\Delta u = f(u), u \geq \alpha, \alpha < \|u\|_\infty < \hat{\alpha}$ and $u \longrightarrow \alpha$ as $\|x\| \longrightarrow \infty$. However a theorem of Toland [59] and our assumption on q implies that no such solution exists. It is unclear what happens if $n = 3$ and $3 < q < 5$ except the behaviour is similar when $f(y)/(y-\alpha)^5$ is decreasing on $(\alpha, \hat{\alpha})$. It would be interesting to understand better these unstable solutions uniformly close to ϕ_ϵ^α (both their existence and multiplicity). This and the discussion in the following remark seem to suggest that the finite Morse index solution results are less robust than the stable solution results.

5. Analogous results hold if we add a term $\epsilon^2 b(x)\cdot\nabla u$ to equation (12) provided we assume that there is a C^1 function ϕ on $\overline{\Omega}$ such that $b = \nabla\phi$. The reason we need this assumption is to control the spectrum of the linearized equation. (Our assumption ensures that the spectral problem for the linearized problem on $\overline{\Omega}$ can be written as a self-adjoint problem with a weight and hence all its eigenvalues are real with good monotonicity properties). We can then proceed much as before. In the general case, the difficulties seem to centre on understanding the behaviour of the eigenvalues (which may not be real).

§8. Some simple applications

In this section, we want to consider very briefly a few simple applications of the results in §6-7 other then the obvious ones. We also raise a number of open problems.

Firstly assume that $a \leq 0$, f is C^1, $f(0) = 0$, $f'(0) < 0$, f is negative on $(0, a)$, f is positive on (a, b), $f(b) = 0$ $f'(x) \leq 0$ on $(b-\delta, b)$ and $\int_0^b f > 0$. Then, by a result of Sweers [57] for small $\epsilon > 0$, there is a positive solution u_ϵ of

$$\begin{aligned} -\epsilon^2\Delta u &= f(u) \text{ in } \Omega \\ u &= 0 \text{ on } \partial\Omega \end{aligned} \tag{14}$$

with $\|u_\epsilon\| < b$ but close to b. This is locally unique. This solution is non-degenerate and stable (by [57] and [13]) and, by Theorem 11, u_ϵ is the only stable positive solution with $\|u\|_\infty < b$. As in [29], it is then easy to construct for small positive ϵ a positive mountain pass solution v_ϵ with $v_\epsilon(x) < u_\epsilon(x)$ in Ω. By Theorem 13, $v_\epsilon(x)$ must be a one positive peak (on the zero solution). By using some ideas of Del Pino and Felmer [37], it is possible to prove that v_ϵ has its peak close to a point of Ω of maximal distance from the boundary. This is proved in [26]. It was proved earlier for a restricted class of domains and under a non-degeneracy hypothesis in [30] (for all n). One would expect a similar result in the case that there exist a, b, c such that $0 < a < b < c$, $f > 0$ on $(0, a) \cup (b, c)$, $f < 0$ on (a, b), $f'(a) < 0$, $f'(x) \leq 0$ on $(c - \delta, c)$ and $\int_a^c f > 0$. More precisely, for ϵ small, there are stable positive solutions $u_{\epsilon,1}$ and $u_{\epsilon,2}$ with $u_{\epsilon,1} < a$ on Ω, $\|u_{\epsilon,1}\|_\infty$ close to a, $u_{\epsilon,2} < c$ on Ω and $\|u_{\epsilon,2}\|_\infty$ is close to c. One can construct a mountain peak v_ϵ between them (in the natural order). Once again, by Theorem 13, v_ϵ must be a positive peak on top of $u_{\epsilon,1}$ and one would expect the location of the peak to be the same as before. This is proved under additional hypotheses in [26], [29], and [30] but the general case remains open.

Secondly, we consider (14) for Dirichlet boundary conditions where $n \leq 3$, $f(y) = y^p - y$ and $1 < p < (n+2)/(n-2)$. If Ω is strictly convex it is proved in [34] as a consequence of Theorem 13 and the theory of peak solutions (especially Theorem 1.2 in [32]) that for small positive ϵ, the only positive solution with bounded Morse index is the positive one peak solution (which can be obtained as a mountain pass). The open question is whether there are any other positive solutions. When Ω is also rather symmetric, it is proved in [15] that there are no other positive solutions for small positive ϵ (and all n). Note that some conditions on the geometry of Ω are needed for these results to be true.

Thirdly, assume that $n \leq 3$, f is C^1, $f > 0$ on $[0, \infty)$, and $y^{1-p} f'(y) \longrightarrow C > 0$ as $y \longrightarrow \infty$ where $p > 1$. Then it is proved in [25] that the branch of positive solutions for $\lambda > 0$ emanating from $(0, 0)$ must have a bifurcation point (possibly a turning point). A key role in this is played by the result that

$$-\Delta u = u^p \tag{15}$$

on $\mathbb{R}^n$ has no positive bounded linearized stable solution (which follows from Theorem 1). The main result in [25] can also be generalized to some functions which grow exponentially. It would be interesting to prove that for $n = 3$, p supercritical and Ω convex the branch has infinitely many bifurcations. Partial results appear in [18]. This is related to studying finite Morse index solutions of (15).

Fourthly if, $n = 2$, $yf(y)$ is negative for large y, condition nw holds, $f(0) = 0$, and $f'(0) > 0$, then our results imply that any unbounded branch of C non-trivial solutions of $-\Delta u = \lambda f(u)$ in $\Omega, u = 0$ on $\partial\Omega$ which has only finitely many bifurcation points on it must approach a peak solution as $\lambda \longrightarrow \infty$. (The problem here is to prove that the branches are unbounded).

As a simple example of the application of or results on infinite strips, we consider the expanding domains D_R in [34] for $n \leq 3$. (The simplest example is the domains $\{x \in \mathbb{R}^n : R \leq \|x\| \leq R + k\}$ for large R where k is positive and fixed). Then we can combine the remarks at the end of §4 with the proof of Theorem 13 to prove that if $1 < p < \frac{n+2}{n-2}$ and $n \leq 3$ a positive solution of $-\Delta u = u^p - u$ on D_R, with Dirichlet boundary conditions and bounded Morse index is asymptotically for large R on a finite sum of humps where the humps are asymptotically a positive decaying solution of $-\Delta u = u^p - u$ on $\mathbb{R}^{n-1} \times [0, k]$ with Dirichlet boundary conditions. We call these humps rather than peaks because they have a peak but it is not a sharp peak.

§9. Some simple systems

It would be very interesting to generalize our results to various classes of systems. We sketch two very simple results very briefly here, though it is clear other results can be obtained.

Firstly, we consider the Fithugh-Nagumo type system

$$\begin{aligned}-\epsilon^2\Delta u &= f(u) - v\\ -\Delta v + \lambda v &= \delta u\end{aligned}$$

on Ω with Neumann boundary conditions (though we could also study Dirichlet boundary conditions.) We assume $\gamma, \delta > 0$ and $n = 2$. It is easy to show that solutions are critical points of the functional $H_\epsilon(u) = \int_\Omega \frac{1}{2}\epsilon^2|\nabla u|^2 + F(u) - \frac{1}{2}\delta \int_\Omega uG_\gamma u$ on $W^{1,2}(\Omega)$) where $G_\gamma(u) = (-\Delta + \gamma I)^{-1}u$ and $v = G_\gamma(u)$ (for Neumann boundary conditions) provided f grows slowly enough at infinity. We look for solutions u for which $H_\epsilon^{''}(u)$ is positive semi-definite. It can be easily shown solutions which are either local minima of H_ϵ or which are linearized stable (in the second case, for the natural parabolic system) satisfy this (though the converse is false in the second case).

Now suppose u_{ϵ_i} is a sequence of solutions of the above type such that $\epsilon_i \longrightarrow 0$ as $i \longrightarrow \infty$ and $\|u_{\epsilon_i}\|_\infty \leq M$ for all i. By the second equation $\|v_{\epsilon_i}\|_\infty$ is also uniformly bounded. In rescaled variables,

$$\begin{aligned}-\Delta\widetilde{u}_i &= f(\widetilde{u}_i) - \widetilde{v}_i\\ -\Delta\widetilde{v}_i + \epsilon_i^2\gamma\widetilde{v}_i &= \epsilon_i^2\delta\widetilde{u}_i\end{aligned}$$

with Neumann boundary condition on $\partial\Omega_i$ where $\Omega_i = \epsilon_i^{-1}\Omega$ and $\widetilde{u}_i, \widetilde{v}_i$ are u_i, v_i rescaled. If $x_0 \in \Omega_i$ such that $d(x_0, \partial\Omega_i) \longrightarrow \infty$ as $i \longrightarrow \infty$, we can then argue much as before to prove that, after taking subsequences, $\widetilde{u}_i(x_0 + x) \longrightarrow \overline{u}$ and $\widetilde{v}_i(x_0 + x) \longrightarrow \overline{v}$ uniformly on compact sets where $\overline{u}$ and $\overline{v}$ are uniformly bounded and

$$\begin{aligned}-\Delta\overline{u} &= f(\overline{u}) - \overline{v}\\ -\Delta\overline{v} &= 0\end{aligned}$$

on $\mathbb{R}^n$. Hence $\overline{v}$ is a constant $\overline{c}$. Moreover, if $f(\overline{u}) - \overline{c}$ satisfies condition nw and if we prove that $\overline{u}$ is a linearly stable solution of $-\Delta u = f(u) - \overline{c}$, Theorem 1 implies that $\overline{u}$ is constant and $f(\overline{u}) = \overline{c} = \overline{v}$. Hence provided that v stays away from values $\widetilde{v}$ such that $f(u) - \widetilde{v}$ fails condition nw, then $f(u_{\epsilon_i}) - v_{\epsilon_i} \longrightarrow 0$ uniformly on compact subsets of Ω. In fact, this is easily generalized to uniform convergence up to the boundary, while if we use Dirichlet boundary conditions we need to assume $f(0) = 0$. In particular, if f is invertible (whence $f(y) - c$ satisfies condition nw for every c), we find that the limiting equation for v is

$$-\Delta v + \gamma v = \delta f^{-1}(v) \text{ on } \Omega$$

with Neumann boundary condition.

It remains to be proved that $\overline{u}$ is linearized stable. If $h \in C_0^\infty(\mathbb{R}^2)$, we know that

$$\int_{R^2} |\nabla h|^2 - f'(\widetilde{u}_{\epsilon_i})h^2 - \delta h\widetilde{G}_{\gamma_i}(h) \geq 0 \tag{16}$$

by our assumptions, where $\widetilde{G}_i(h) = (-\epsilon_i^{-2}\Delta + \gamma I)^{-1}h$ on Ω_i with the Neumann boundary conditions. Thus we can easily pass to the limit and deduce the linearized stability of $\overline{u}$ if we prove that $\widetilde{G}_i(h) \rightharpoonup 0$ in $L^2(\mathbb{R}^2)$ weakly because the last term in (16) will then tend to zero as $i \longrightarrow \infty$. To prove that $w_i = \widetilde{G}_i(h)$ converges weakly to zero, we first multiply the equation for w_i by w_i to deduce that $\|w_i\|_2 \leq \delta\|h\|_2$ and hence a subsequence of $\{w_i\}$ converges weakly to $\overline{w}$ in $L^2(\mathbb{R}^2)$. On the other hand we can easily pass to the (weak) limit in the w_i equation and deduce that

$\overline{w}$ is harmonic. Since $\overline{w} \in L^2(\mathbb{R}^2), \overline{w} = 0$ and our claimed linearized stability of $\overline{u}$ follows.

Finally, note that for this example, it can be shown [35] that for some f's even the global minimum of H (when it exists) may have very complicated behaviour in regions where v is close to s if $f(y) - s$ fails condition nw.

As a second example, we briefly consider the Geirer-Meinhardt system on $\Omega \subseteq \mathbb{R}^2$. More precisely, we consider the shadow system (which corresponds to one diffusion going to infinity). Our system is

$$\begin{aligned} \dot{u} &= \epsilon^2 \Delta u - u^p h^{-q} \text{ in } \Omega \\ \tau \dot{h} &= -h + |\Omega|^{-1} \int_\Omega a^r h^{-s} \end{aligned}$$

where $u = u(x,t), h = h(t), \epsilon, \tau, q, r > 0, s \geq 0, \gamma_0 = \frac{qr}{(p-1)(s+1)} > 1, \Omega$ is a smooth bounded domain in $\mathbb{R}^2$ and u satisfies a Neumann boundary condition on $\partial\Omega$. It arises in the theory of shells in biology (cp [63]) except that this is a limit equation (the shadow system) as a diffusion coefficient tends to infinity. Wei [63] shows that stationary solutions where u has one sharp peak can sometimes be meta stable for the shadow system for small positive ϵ. We prove conversely that the only stable stationary positive solutions are ones where u has at most one sharp peak, at least in the special case where $r = p + 1$ or $r = 2$ (which were the main cases studied by Wei and which include the original application). In fact, with care, one can prove the one peak solutions are the only possible meta stable positive solutions. (One uses the idea of finite meta Morse index solutions). Thus Wei's solution is the only possible positive stable solution. Firstly, note that as in [63], one easily sees that the positive stationary solutions are given by $u = tw$ where t is constant and positive and

$$-\epsilon^2 \Delta w = -w + w^p \text{ in } \Omega \tag{17}$$

where there are simple formulae for t and h in terms of integrals of w^m. Now the stability of this solution is determined by the eigenvalues of

$$-\epsilon^2 \Delta z + z - p w^{p-1} z + \frac{qr}{s + 1 - \tau\lambda} \frac{\int_\Omega w^{r-1} z}{\int_\Omega w^r} w^p = \lambda z \tag{18}$$

on Ω with Neumann boundary conditions. (This requires some tedious but easy calculations.) Here we have solved the second equation of the linearized eigenvalue problem to reduce to a scalar equation. We prove that, if (18) has no negative real eigenvalues (in particular if our solution is stable), then $Z_\epsilon = -\epsilon^2 \Delta z + z - p w^{p-1} z$ on Ω with Neumann boundary conditions has at most 1 negative real eigenvalue counting multiplicity. If we prove this, w is a solution of (17) of Morse index at most 1 since the eigenvalues of Z_ϵ are real). Hence we can apply Theorem 13 to show that, for small ϵ, u is a one peak solution of the type in Wei. (It is easy to use the convexity of $y^p - y$ for $y \geq 0$ to show that (17) has no stable positive solution).

Thus it remains to prove that if (18) has no negative real eigenvalue then w has Morse index at most 1. We sketch this in the much easier case where $r = p + 1$. In this case, it is easy to the left hand side of (18) is self adjoint for λ real. Let $W_\lambda = \widetilde{W}_\lambda - \frac{1}{2}\lambda \|x\|_2^2$ where $\widetilde{W}_\lambda$ is the quadratic form corresponding to the left hand side of (18). It is easy to see that W_λ is positive definite if λ is large negative and hence by continuity and our assumption that (18) has no negative real eigenvalues, $W_0 = \widetilde{W}_0$ is positive semidefinite. Similar arguments appear in [24]. Note that we need to use that the operators corresponding to W_λ is Fredholm for $\lambda \leq 0$ to justify this. On the other hand, if the operator H defined by $Hz = -\Delta z + z - pw^{p-1}z$ had 2 negative eigenvalues counting multiplicity, there would be a two-dimensional subspace T of $\mathcal{D}(H)$ such that $(Hz, z) < 0$ on $T \backslash \{0\}$. Since T is two dimensional, there exists $z_1 \in T \backslash \{0\}$ such that $(w^{r-1}, z_1) = 0$. Hence $(W_0 z_1, z_1) < 0$ and we have a contradiction as required. In the case where $r = 2$, we need to use ideas in Kato [49, Chapter 4.6] and Ni, Takagi and Yana[54]. Note that the reason we do not use the blow up limit as in [63] or [24] is that it is unclear if one of the terms in the limit has a reasonably large domain of definition and that our methods could also be used for the case where $\Omega \subseteq \mathbb{R}^3$.

Lastly, for this section, we can frequently, obtain information for systems by blow-up techniques in cases where the blow up systems are diagonal.

Added in Proof: The author has now solved affirmatively the question after eqn. (15).

References

[1] G. Alberti, L. Ambriosio and X. Cabré, "On a long standing conjecture of E. de Giorgi : symmetry in 3d for general nonlinearities and a local minimum property", Acta Applicandae Math. 65 (2001), 9-33.

[2] L. Ambrosio and X. Cabré, "Entire solutions of semilinear elliptic problems in R^3 and a conjecture of de Giorgi", J. Amer. Math Soc 13 (2000), 725-739.

[3] A. Bahri and P. Lions, "Solutions of superlinear elliptic problems and their indices", Comm Pure Appl Math 45 (1992), 1205-1215.

[4] T. Bartsch and M. Willem, "Infinitely many non-radial solutions of an Euclidean scalar field equation", J. Funct Anal 117 (1993), 447-460.

[5] H. Berestycki, L. Caffarelli and L. Nirenberg, "Further qualitative properties for elliptic equations in unbounded domains", Ann Scuola Norm Sup Pisa 25 (1997), 69-94.

[6] H. Berestycki, L. Caffarelli and L. Nirenberg, "Inequalities for second order elliptic equations with applications to unbounded domains", Duke Math J. 81 (1966), 467-494.

[7] H. Berestycki and P. Lions, "Some applications of the methods of sub and supersolutions" pp 16-41 in Bifurcation and Nonlinear Eigenvalue Problems, Lecture Notes in Mathematics vol 782, Springer, Berlin, 1980.

[8] X. Cabré and A. Capella, "On the stability of radial solutions of semilinear elliptic equation in all of R^n", C.R. Acad. Sci Paris, 338 (2004), 769-774.

[9] L. Caffarelli and A. Friedman, "Partial regularity of the zero set of solutions of linear and superlinear elliptic equations", J. Diff Eqns 60 (1985), 420-433.

[10] R. Casten and C. Holland, "Instability results for reaction diffusion equations with Neumann boundary conditions", J. Diff Eqns 27 (1978), 266-273.

[11] P. Clement and G. Sweers, "Existence and multiplicity results for a semilinear elliptic eigenvalue problem", Ann Scuola Norm Sup Pisa 14 (1987), 97-121.

[12] E.N. Dancer, "New solutions of equations on R^n", Ann Scuola Norm Sup Pisa 30 (2001), 535-563.

[13] E.N. Dancer, "On the number of positive solutions of weakly nonlinear elliptic equations when a parameter is large", Proc London Soc 53 (1986), 429-452.

[14] E.N. Dancer, "The effect of domain shape on the number of positive solutions of certain nonlinear equations", J. Diff Eqns 74 (1988), 120-156.

[15] E.N. Dancer, "On positive solutions of some singularly perturbed problems where the nonlinearity changes sign", Top Methods Nonlinear Anal 5 (1995), 141-175.

[16] E.N. Dancer, "A note on the method of moving planes", Bull Austral Math Soc 46 (1992), 425-434.

[17] E.N. Dancer, "Multiple fixed points of positive maps", J. Reine Ang Math 371 (1986), 46-66.

[18] E.N. Dancer, "Infinitely many turning points for some supercritical problems", Annali di Matematica 98 (2000), 225-233.

[19] E.N. Dancer, "Superlinear problems on domains with holes of asymptotic shape and exterior problems", Math Zeit 229 (1988), 475-491.

[20] E.N. Dancer, "The effect of domain shape on the number of positive solutions of certain nonlinear equations II", J. Diff Eqns 87 (1990), 316-339.
[21] E.N. Dancer, "Stable and finite Morse index solutions on R^n or on bounded domains with small diffusion", Trans Amer Math Soc 357 (2005), 1225-1243.
[22] E.N. Dancer, "Stable and finite Morse index solutions on R^n on bounded domains with small diffusion II", Indiana U. Math J. 53 (2004), 97-108.
[23] E.N. Dancer, Weakly nonlinear Dirichlet problems on long or thin domains, Memoirs Amer Math Soc 501, (1993).
[24] E.N. Dancer, "On stability and Hopf bifurcations for chemotaxis systems", Methods and Applications of Analysis 8 (2001), 245-256.
[25] E.N. Dancer, "Stable solutions on R^n and the primary branch of some non-self-adjoint convex problems", Diff Int Eqns 17 (2004), 961-970.
[26] E.N. Dancer, "Some mountain pass theorems for small diffusion", Diff Int Eqns 16 (2003), 1013-1024.
[27] E.N. Dancer and Z.M. Guo, "Some remarks on the stability of sign changing solutions", Tohoku Math J. 47 (1995), 199-225.
[28] E.N. Dancer and P. Hess, "Stability of fixed points for order preserving discrete time dynamical systems", J. Reine Ang Math 419 (1991), 125-139.
[29] E.N. Dancer and J. Wei, "On the profile of solutions with two sharp layers to a singularly perturbed semilinear Dirichlet problem", Proc Royal Soc Edinburgh A 127 (1997), 691-701.
[30] E.N. Dancer and J. Wei, "On the location of spikes of solutions with two sharp layers for a similarly perturbed semilinear Dirichlet problem", J. Diff Eqns 157 (1999), 82-101.
[31] E.N. Dancer and S. Yan, "On the profile of the sign changing mountain pass solutions for an elliptic problem", Trans Amer Math Soc. 354 (2002), 3573-3600.
[32] E.N. Dancer and S. Yan, "Effect of the domain geometry on the existence of multipeak solutions for an elliptic problem", Top. Methods in Nonlinear Analysis 14 (1999), 1-38.
[33] E.N. Dancer and S. Yan, "Interior and boundary peak solutions for a mixed boundary value problem", Indiana Math J. 48 (1999) 1177-1212.
[34] E.N. Dancer and S. Yan, "Multibump solutions for an elliptic problem in expanding domains", Comm Partial Diff Eqns 27 (2002), 23-53.
[35] E.N. Dancer and S. Yan, "A minimization problem associated with elliptic systems of Fitzhugh-Nagumo type", Analyse Nonlinéaire 21 (2004), 237-253.
[36] E.N. Dancer and K. Schmitt, "On positive solutions of semilinear elliptic problems", Proc Amer Math Soc 101 (1987), 445-452.
[37] M. Del Pino and P. Felmer, "Spike layered solutions of singularly perturbed elliptic problems", Indiana Math J. 48 (1999), 883-898.
[38] W.Y. Ding, "On a conformally invariant elliptic equation on R^n", Comm Math Phys 107 (1986), 331-335.
[39] N. Dunford and J. Schwartz, Linear operators Vol II, Interscience, New York, 1963.
[40] N. Ghoussoub and C. Gui, "On a conjecture of de Giorgi and some related problems", Math Ann 311 (1998), 481-491.
[41] D. Gilbarg and N. Trudinger, Elliptic partial differential equations of second order, Springer Verlag, Berlin, 1977.
[42] C. Gui and J. Wei, "Multiple interior peaks for some singularly perturbed Neumann problems", J. Diff Eqns 158 (1999), 1-27.
[43] D. Henry, Geometric theory of semilinear parabolic equations, Lecture Notes in Mathematics 840, Springer, Berlin, 1981.
[44] P. Hess, Periodic - Parabolic boundary value problems and positivity, Pitman, Harlow, 1991.
[45] P. Hess, "On multiple solutions of nonlinear elliptic eigenvalue problems", Comm Partial Diff Eqns 6 (1981), 951-961.
[46] H. Hofer, "A note on the topological degree at a critical point of mountain pass type", Proc Amer Math Soc 90 (1984), 309-315.
[47] J. Jang, "On spike solutions of singularly perturbed semilinear Dirichlet problems", J. Diff Eqns 114 (1994), 370-395.
[48] D. Joseph and T. Lungren, "Quasilinear Dirichlet problems driven by positive sources", Arch Rat Mech Anal 49 (1973), 241-269.
[49] T. Kato, Perturbation theory for linear operators, Springer-Verlag, Berlin, 1966.

[50] R. Kohn and P. Sternberg, "Local minimizers and singular perturbations", Proc Royal Soc Edinburgh A 111 (1989),69-84.

[51] Y. Li and W. Ni, "Radical symmetry of positive solutions of nonlinear elliptic equations on R^n", Comm Partial Diff Eqns 18 (1993), 1043-1054.

[52] H. Matano, "Asymptotic behaviour and stability of solutions of semilinear diffusion equations", Publ Res Int Math Sci 15 (1977), 401-454.

[53] W.M. Ni, I. Takagi and J. Wei, "On the location and profile of spike layer solutions to a singularly perturbed semilinear Dirichlet problem : intermediate solutions", Duke Math J. 94 (1998), 597-618.

[54] W.M. Ni, I. Takagi and E. Yanagida, "Stability of least energy patterns of the shadow system for an activator-inhibitor model", Japanese J. Industrial Appl Math 18 (2001), 259-272.

[55] W.M. Ni and J. Wei, "On the location and profile of spike layer solutions to singularly perturbed semilinear Dirichlet problems", Comm Pure App Math 48 (1995), 731-768.

[56] O. Savin, "Phase transitions, regularity of flat level sets", preprint.

[57] G. Sweers, "On the maximum of solutions for a semilinear elliptic problem", Proc Royal Soc Edinburgh A 108 (1988), 357-370.

[58] M. Schatzman, "On the stability of the saddle solution of Allen Cahn's equation", Proc Royal Soc Edinburgh 125A (1995), 1241-1275.

[59] J. Toland, "On positive solutions of $-\Delta u = F(x, u)$", Math Zeit 182 (1983), 351-357.

[60] J. Wei, "On the construction of single peaked solutions to a singularly perturbed semilinear Dirichlet problem", J. Diff Eqns 129 (1996), 315-333.

[61] J. Wei, "On the interior spike solutions for some singular perturbation problems", Proc Royal Soc Edinburgh A 128 (1998), 849-874.

[62] J. Wei, "On the effect of domain geometry in singular perturbation problems", Diff Integral Eqns 13 (2000), 15-45.

[63] J. Wei, "On single interior spike solutions of Gierer-Meinhardt system : Uniqueness and spectrum estimates", Europ J. App Math 10 (1998), 353-378.

[64] S. Yan, "On the number of interior multipeak solutions for singularly perturbed Neumann problems", Top Methods Nonlinear Anal 12 (1999), 61-78.

[65] H. Zou, "Symmetry of ground states of semilinear elliptic equations with mixed Sobolev growth", Indiana U Math J. 45 (1996), 221-240.

Fields Institute Communications
Volume **48**, 2006

Some Recent Results on Diffusive Predator-prey Models in Spatially Heterogeneous Environment

Yihong Du
School of Mathematics
Statistics and Computer Sciences
University of New England
Armidale, NSW2351, Australia
ydu@turing.une.edu.au

Junping Shi
Department of Mathematics
College of William and Mary
Williamsburg, VA 23187-8795, USA
and
School of Mathematics
Harbin Normal University
Harbin, Heilongjiang, 150080, P.R.China
shij@math.wm.edu

Abstract. We present several recent results obtained in our attempts to understand the influence of spatial heterogeneity in the predator-prey models. Two different approaches are taken. The first approach is based on the observation that the behavior of many diffusive population models is very sensitive to certain coefficient functions becoming small in part of the underlying spatial region. We apply this observation to three predator-prey models to reveal fundamental differences from the classical homogeneous case in each model, and demonstrate the essential differences of these models from each other. In the second approach, we examine the influence of a protection zone in a Holling type II diffusive predator-prey model, which introduces different mathematical problems from those in the first approach, and reveals important impacts of the protection zone.

2000 *Mathematics Subject Classification.* Primary 35J55, 92D40; Secondary 35J60, 35K50, 35K57, 92D25.

The first author is partially supported by the Australia Research Council.

The second author is partially supported by United States NSF grants DMS-0314736 and EF-0436318, and Heilongjiang Province oversea scholar grant, China.

1 Introduction

The natural environments for most biological species are inhomogeneous in space. Therefore it is reasonable to expect the dynamics of their populations to be influenced by the heterogeneity of the environments, apart from the interactions between the species. Many scientific experiments support this view; for example, biological experiments of Huffaker [Hu] found that a predator-prey system consisting of two species of mites could collapse to extinction quickly in small homogeneous environments, but would persist longer in suitable heterogeneous environments. However, it is generally not easy to capture the influence of heterogeneous environment mathematically. Traditionally the main focus in the study of population models was on the effects of interactions between the species; so even if diffusive population models are used, the coefficients appearing in the models are chosen to be positive constants, representing a homogeneous environment. One natural way to include spatial variations of the environment in these models is to replace the constant coefficients by positive functions of the space variable x. But the mathematical techniques developed to study these models are typically either not sensitive to this change, in which case the effects of heterogeneous spatial environment are difficult to observe in the mathematical analysis, or the techniques are too sensitive to this change and become unapplicable when the constant coefficients are replaced by functions. (We will be more specific below.)

In this paper, we will discuss several results obtained in our recent attempts to better understand the influence of spatial heterogeneity in predator-prey models, where new ideas and techniques are used to overcome the aforementioned difficulties. Two different approaches are taken. One approach is based on the observation that the behavior of many diffusive population models is very sensitive to certain coefficient functions becoming small in part of the underlying spatial region. With the hope of providing some insights in improving existing biological environments or designing new ones, we will show how this observation can be applied to several predator-prey models to reveal essential differences from the classical homogeneous case for each model, and demonstrate interesting differences between the models. Our second approach examines the influence of a protection zone in a diffusive predator-prey model, which introduces different mathematical problems from those in the first approach, and reveals important impacts of the protection zone; we hope our results here may shed some light on the creation and design of nature reserves.

To be more realistic, the coefficients in the diffusive population models should also be dependent on time t, but we will not consider this case here. A systematic discussion of time-periodic diffusive population models can be found in Hess [He]. It would be interesting to see how our approaches discussed here can be extended to the time-periodic case.

Much research in the directions discussed in this paper is still on-going. Moreover, what we consider here is only a small proportion of a vast area of study in population models. To put this paper into perspective, we now briefly recall some history and related research.

The first population models are ordinary differential equations. It has long been accepted that the interaction between a pair of predator and prey influences the population growth of both species. The first evidence of the interaction was provided by the population data of Canadian lynx and snowshoe hare. Hudson Bay company kept careful records of all furs from the early 1800s into the 1900s. It is often assumed that the number of furs are representative of the populations in the wild, so these data can be used to represent the relative populations of the lynx and hares in the wild. The records for the furs collected by the Hudson Bay company showed distinct oscillations (approximately 9 year periods), suggesting that these species caused almost periodic fluctuations of each other's populations. Many other ecological examples can be found in, for example, [MBN].

To describe the interactions between the species, the first ordinary differential equations (ODEs) of predator-prey type were formulated by American chemist and biologist Alfred James Lotka in 1920 [Lo1, Lo2, Lo3], and Italian mathematician Vito Volterra in 1926 [V1, V2]. The simplest form of the equation is

$$\frac{du}{dt} = \lambda u - buv, \quad \frac{dv}{dt} = -\mu v + cuv, \tag{1.1}$$

where $\lambda, \mu, b, c > 0$. Volterra's model of biological interaction is motivated by the statistical studies of the populations of various species of fish in the Adriatic Sea conducted by biologist U. d'Ancona (see [V1, V2]). In [V2], he proposed a more general model:

$$\frac{du}{dt} = \lambda u - au^2 - buv, \quad \frac{dv}{dt} = \mu v - dv^2 + cuv, \tag{1.2}$$

where $\lambda, a, b, c, d \geq 0$, and $\mu \in \mathbf{R}$.

The Lotka-Volterra predator-prey model (1.2) possesses a unique coexistence steady state solution (u_*, v_*) when $d \geq 0$, and (u_*, v_*) is globally asymptotically stable when $d > 0$ (a complete phase plane analysis of (1.2) can be found in, for example, [BC] or [HS]). However biologists have observed that while some predator-prey interactions lead to a system with a stable equilibrium, some others do not (such as the data from the above mentioned Hudson Bay company show), and there exists ecological process which destabilizes the coexistence equilibrium. Various mechanisms are introduced by ecologists for destabilization of the equilibrium [MBN]. In consideration of the limited ability of a predator to consume its prey, a general functional response of the predator $\phi(u)$ was introduced by Solomon [So] and Holling [H1, H2], and the classical Lotka-Volterra model is modified to

$$\frac{du}{dt} = \lambda u - au^2 - b\phi(u)v, \quad \frac{dv}{dt} = \mu v - dv^2 + c\phi(u)v. \tag{1.3}$$

Here $\phi(u)$ is a positive and nondecreasing function of u (prey density). Among many possible choices of $\phi(u)$, the Holling type II functional response is most commonly used in the ecological literature, which is defined by

$$\phi(u) = \frac{u}{1+mu}, \tag{1.4}$$

where m is a positive constant: $m = bT_h$ and T_h is the handling time for a generic predator to kill and consume a generic prey ([H1, H2]). Hence a more realistic Holling type II predator-prey model takes the form

$$\frac{du}{dt} = \lambda u - au^2 - \frac{buv}{1+mu}, \quad \frac{dv}{dt} = \mu v - dv^2 + \frac{cuv}{1+mu}. \tag{1.5}$$

When $\mu < 0$ and $d = 0$, (1.5) becomes the Rosenzweig-MacArthur model widely used in real-life ecological applications [RM]; when $\mu < 0$ and $d > 0$, the model was used by Bazykin [Ba, Tur].

Generally speaking, the dynamical behavior of the predator-prey models is very sensitive to the changes of the reaction functions. To accommodate the various different situations in ecological applications, many other functional responses and more general interactions have been introduced, and excellent surveys can be found in recent monographs [MBN, Tur]. We list below some more examples without quoting the original references (which can be retrieved from [MBN, Tur]):

$$\text{(Yodzis)} \quad \begin{cases} \dfrac{du}{dt} = \lambda u - au^2 - \dfrac{bu^2 v}{1+mu^2}, \\ \dfrac{dv}{dt} = \mu v - dv^2 + \dfrac{cu^2 v}{1+mu^2}; \end{cases} \tag{1.6}$$

$$\text{(Beddington-DeAngelis)} \quad \begin{cases} \dfrac{du}{dt} = \lambda u - au^2 - \dfrac{buv}{1+mu+nv}, \\ \dfrac{dv}{dt} = \mu v - dv^2 + \dfrac{cuv}{1+mu+nv}; \end{cases} \tag{1.7}$$

$$\text{(Variable Territory)} \quad \begin{cases} \dfrac{du}{dt} = \lambda u - au^2 - \dfrac{buv}{1+mu}, \\ \dfrac{dv}{dt} = \mu v - \dfrac{dv^2}{u} + \dfrac{cuv}{1+mu}; \end{cases} \tag{1.8}$$

$$\text{(Leslie)} \quad \begin{cases} \dfrac{du}{dt} = \lambda u - au^2 - buv, \\ \dfrac{dv}{dt} = \mu v - \dfrac{dv^2}{u}; \end{cases} \tag{1.9}$$

$$\text{(Leslie-May-Tanner)} \quad \begin{cases} \dfrac{du}{dt} = \lambda u - au^2 - \dfrac{buv}{1+mu}, \\ \dfrac{dv}{dt} = \mu v - \dfrac{dv^2}{u}; \end{cases} \tag{1.10}$$

$$\text{(Ratio-dependent)} \quad \begin{cases} \dfrac{du}{dt} = \lambda u - au^2 - \dfrac{buv}{u+nv}, \\ \dfrac{dv}{dt} = \mu v + \dfrac{cuv}{u+nv}. \end{cases} \tag{1.11}$$

These ODE models are usually less difficult to understand than their reaction-diffusion counterparts to be discussed below, but there is considerable biological evidence that space can affect the dynamics of populations and the structure of communities, and such effects are not captured in this kind of ODE models. In 1937, Fisher [Fis] and Kolmogoroff, Petrovsky, and Piscounoff [KPP] used a reaction-diffusion equation to study the spread of an advantageous form (allele) of a single gene in a population of diploid individuals. In the early 1950s' Skellam [Sk], Kierstead and Slobodkin [KS] studied the application of reaction-diffusion models in population persistence when spatial dispersal is involved, and Turing [Tu] used reaction-diffusion equations to study the formation of patterns for his model of morphogenesis. These pioneering works marked the beginning of reaction-diffusion models for biological problems. Since the 1970's, the corresponding reaction-diffusion equations of an ODE population model of the form

$$\frac{du}{dt} = f(u,v), \ \frac{dv}{dt} = g(u,v),$$

namely,

$$\frac{\partial u}{\partial t} - d_1 \Delta u = f(u,v), \ \frac{\partial v}{\partial t} - d_2 \Delta u = g(u,v),$$

together with suitable boundary conditions on the boundary of the underlying spatial domain, have been widely used to study the dynamical behavior of the same species over the given spatial domain ([St1, St2, L, Mu2, OL, CC2].) These reaction-diffusion systems are usually called diffusive population models. Such models have been successfully used to study the waves of invasion and persistence of species in mathematical ecology, but for the purpose of this paper, we only describe below their use in understanding the long-time dynamical behavior on a bounded spatial domain Ω, especially the set of steady-state solutions and their spatial patterns.

Much progress on the study of reaction-diffusion systems had been made in the 1970's and 1980's. For example, "invariant rectangle" techniques ([CCS, W]) and Lyapunov function techniques ([A, dMR]) were successfully used to understand the long-time dynamical behavior for such systems, especially for the case of Neumann boundary conditions. Remarkably, if the diffusion rates d_1 and d_2 are both very large, it was shown that the only steady-state solutions of a general reaction-diffusion system with Neumann boundary conditions are constant ones, and the attractor for the reaction-diffusion system is the same as for the corresponding ordinary differential equations ([CHS]). If the underlying domain of a single reaction-diffusion equation with Neumann boundary conditions is convex, then any stable steady-state solution must be a constant ([CH, Ma]), but stable non-constant solutions can exist if Ω is not convex ([MaM]). Similar results were also proved for competition reaction-diffusion systems with Neumann boundary conditions ([KW]), but it is unclear whether a homogeneous diffusive predator-prey system with Neumann boundary conditions can possess stable non-constant steady states since the predator-prey system does not generate a monotone dynamical system. Considerable progress has also been made on the pattern formation and bifurcation of diffusive predator-prey systems (see, *e.g.*, [Mi, MiM, MNY, Mu1, SJ, SL]). See [Co1, Co2] for surveys of some of the earlier works on diffusive predator-prey systems.

In the 1980's various analytical and topological techniques emerged for the investigation of the set of coexistence steady-states of diffusive population models, such as local bifurcation theory (see [CGS]), monotone iteration schemes (see [Pao, KL]), global bifurcation theory [BB1, BB2], and fixed point index theory [Da1, Da2] (see also [Li1, Li2]). These methods were further refined and used from the 1990's to understand the question of uniqueness, multiplicity and stability of the steady-state solutions (see, e.g., [Da3, Da4], [LP], [DD1], [LN], [DL1, DL2, DL3], [PaW, PeW], [KY]).

The techniques described in the last paragraph can be largely extended to cover the case that the constant coefficients are replaced by positive functions of the space variable x. Therefore, it seems difficult to capture the effects of heterogeneous environment on diffusive population models by exploring these techniques alone. It is also worthwhile to point out that, on the other hand, some powerful techniques involving the use of a Lyapunov functional (as used in [dMR]), and the "contracting rectangle" method (as used in [Br]), may collapse when the reaction terms are space-dependent, as they rely heavily on the existence of constant solutions of the system.

In [D1, D2], based on earlier work on the single logistic equation (see, *e.g.*, [Ou, dP, FKLM, DHu]), a degenerate competition model was examined, revealing fundamental changes of dynamical behavior when the crowding effect coefficient function for one of the competing species vanishes on part of the underlying domain. This observation was further developed in [D3] to study the influence of heterogeneity on the formation of sharp spatial patterns and the stability of the sharp patterned steady-state solutions in the classical (but near degenerate) competition model.

In this paper, we will examine, along a similar line of thinking, the effects of spatial heterogeneity on several diffusive predator-prey models. We will present some recent results in [DD2, DHs, DW, DS1]. Since the predator-prey models do not have the monotonicity property enjoyed by the competition model, very different techniques from those used in [D1, D2, D3] are needed, and due to the fundamental difference between the predator-prey and competition models, very different phenomena will be revealed. We will also describe our recent results in [DS2] on the effects of a protection zone in a diffusive predator-prey model. Much work is still on-going in this direction.

The rest of this paper is organized as follows. In Section 2, we consider the Lotka-Volterra predator-prey model, where after a brief discussion of the homogeneous case, we present some resent results of Dancer and Du ([DD2, DD3]) on the inhomogeneous case. In Section 3, we consider the predator-prey model with a Holling type II response function, and present some recent results of Du and Shi ([DS1, DS2]). Section 4 is concerned with a diffusive Leslie model and some recent results of Du, Hsu and Wang ([DHs, DW]) are discussed there.

Before ending this section, we would like to mention some related research on the effect of heterogeneous environment in diffusive population models. In a series of recent papers, Hutson, Lou, Mischaikow and Polacik studied various perturbations of the special competition model

$$\begin{cases} u_t - \mu\Delta u = \alpha(x)u - u^2 - uv, & x \in \Omega,\ t > 0, \\ v_t - \mu\Delta v = \alpha(x)v - v^2 - uv, & x \in \Omega,\ t > 0, \\ \partial_\nu u = \partial_\nu v = 0, & x \in \partial\Omega,\ t > 0, \end{cases}$$

and obtained interesting results revealing some fundamental effects of heterogeneous environment on the competition model. We refer to [HMP, HLM1, HLM2, HLMP] for details. Other related results can be found in [CC2] and [Lo, LS]. In recent years, much in-depth research on diffusive predator-prey systems has been carried out by numerical simulations; a review in that direction can be found in [MPT].

2 The Lotka-Volterra model

2.1 The homogeneous case. After some suitable rescalings, the classical homogeneous diffusive Lotka-Volterra predator-prey model over a bounded smooth domain $\Omega \subset \mathbf{R}^n$ may be expressed in the following form:

$$\begin{cases} u_t - d_1\Delta u = \lambda u - u^2 - buv, & x \in \Omega,\ t > 0, \\ v_t - d_2\Delta v = \mu v - v^2 + cuv, & x \in \Omega,\ t > 0, \\ B_1 u = B_2 v = 0, & x \in \partial\Omega,\ t > 0, \\ u(x,0) = u_0(x) \geq 0,\ v(x,0) = v_0(x) \geq 0, & x \in \Omega. \end{cases} \tag{2.1}$$

where $d_1, d_2, \lambda, \mu, b, c$ are constants, and d_1, d_2, b, c are always assumed to be positive; B_1 and B_2 are boundary operators of the form

$$B_1 u(x,t) = \alpha_1 \partial_\nu u(x,t) + \beta_1 u(x,t),\ B_2 v(x,t) = \alpha_2 \partial_\nu v(x,t) + \beta_2 v(x,t),$$

where ν denotes the unit outward normal of $\partial\Omega$, and α_i, β_i are nonnegative constants with $\alpha_i^2 + \beta_i^2 > 0$, $i = 1, 2$. Therefore B_1 and B_2 are the Neumann boundary operators if $(\alpha_i, \beta_i) = (1, 0)$, and they are Dirichlet if $(\alpha_i, \beta_i) = (0, 1)$.

When the Neumann boundary conditions are used, the understanding of (2.1) is rather complete. It is proved by Leung [Le] that all positive solutions of (2.1), regardless of the initial data, converge to a constant steady-state solution as time goes to infinity. This implies that (2.1) behaves similarly to the corresponding ODE model. In sharp contrast, much less is known when Dirichlet boundary conditions are used. The techniques in [Le] rely heavily on the fact that (2.1) has constant steady-states under Neumann boundary conditions. It has been conjectured that the dynamics of (2.1) under Dirichlet boundary conditions behaves similarly to the Neumann case, *i.e.*, all the positive solutions converge to the steady-state solutions as time goes to infinity, but the conjecture remains open so far.

From now on in this section, we will only consider the Dirichlet case, and will focus on the steady-state solutions. The techniques below carry over easily to the case of general boundary conditions. For convenience of comparison to the inhomogeneous case to be discussed later, we first briefly recall some well-known results on the nonnegative steady-state solutions of (2.1), namely, nonnegative solutions of the system

$$\begin{cases} -\Delta u = \lambda u - u^2 - buv, & x \in \Omega, \\ -\Delta v = \mu v - v^2 + cuv, & x \in \Omega, \\ u = v = 0, & x \in \partial\Omega. \end{cases} \tag{2.2}$$

where for convenience of notation, we have assumed that $d_1 = d_2 = 1$. (One could also rescale u, v and the coefficients to reduce the original elliptic system to the form (2.2).)

In the following sections, linear eigenvalue problems will play important roles in our results. We define $\lambda_1^D(\phi, O)$ and $\lambda_1^N(\phi, O)$ to be the first eigenvalues of $-\Delta + \phi$ over the region O, with Dirichlet or Neumann boundary conditions respectively. If the region O is omitted in the notation, then we understand that $O = \Omega$. If the potential function ϕ is omitted, then we understand that $\phi = 0$. We recall some well-known properties of $\lambda_1^D(\phi, O)$ and $\lambda_1^N(\phi, O)$:

1. $\lambda_1^D(\phi, O) > \lambda_1^N(\phi, O)$;
2. $\lambda_1^B(\phi_1, O) > \lambda_1^B(\phi_2, O)$ if $\phi_1 \geq \phi_2$ and $\phi_1 \not\equiv \phi_2$, for $B = D, N$;
3. $\lambda_1^D(\phi, O_1) \geq \lambda_1^D(\phi, O_2)$ if $O_1 \subset O_2$.

The understanding of (2.2) relies on the diffusive logistic equation

$$-\Delta w = \lambda w - w^2, \; x \in \Omega, \quad w = 0, \; x \in \partial\Omega. \tag{2.3}$$

It is well-known that (2.3) has no positive solution when $\lambda \leq \lambda_1^D \equiv \lambda_1^D(\Omega)$, and it has a unique positive solution θ_λ when $\lambda > \lambda_1^D$. Therefore, if $\lambda > \lambda_1^D$, then (2.2) has a unique semitrivial solution of the form $(u, 0)$, namely $(\theta_\lambda, 0)$. Similarly, (2.2) has a unique semitrivial solution of the form $(0, v)$ when $\mu > \lambda_1^D$, namely, $(0, \theta_\mu)$. Clearly $(0, 0)$ is always a solution to (2.2), and we call it the trivial solution. By the strong maximum principle, any other nonnegative solution (u, v) of (2.2) must be a positive solution, *i.e.*, $u > 0, v > 0$ in Ω. It is proved in [Da2] that (2.2) has a positive solution if and only if

$$\lambda > \lambda_1^D(b\theta_\mu) \text{ and } \mu > \lambda_1^D(-c\theta_\lambda), \tag{2.4}$$

where we understand that $\theta_\eta = 0$ if $\eta \leq \lambda_1^D$. Clearly (2.4) implies $\lambda > \lambda_1^D$. If we fix $b, c > 0$ and $\lambda > \lambda_1^D$ and use μ as a parameter, then from the monotonicity of $\lambda_1^D(b\theta_\mu)$ in μ it is easily seen that there exists a unique $\mu^0 > \lambda_1^D$ such that

$$\lambda = \lambda_1^D(b\theta_{\mu^0}).$$

Denote $\mu_0 = \lambda_1^D(-c\theta_\lambda)$. We find that (2.4) is equivalent to

$$\mu_0 < \mu < \mu^0. \tag{2.5}$$

It is possible to use a bifurcation argument to show that a branch of positive solutions $\Gamma = \{(\mu, u, v)\}$ bifurcates from the semitrivial solution branch $\Gamma_u := \{(\mu, \theta_\lambda, 0) : -\infty < \mu < \infty\}$ at $\mu = \mu_0$, and Γ then joins the other semitrivial solution branch $\Gamma_v := \{(\mu, 0, \theta_\mu) : \mu > \lambda_1^D\}$ at $\mu = \mu^0$, see [BB2]. These results are illustrated by Figure 1 (left). It has been conjectured that (2.2) has at most one positive solution, but this is only proved for the case Ω is an interval, *i.e.*, when the space dimension is 1 (see [LP], see also [DLO] for some partial uniqueness results for the radial case in high dimensions). Even in that case, it is not known whether the unique positive solution is asymptotically stable as a steady-state of the corresponding parabolic system, though it is expected that the unique positive steady-state is globally attractive, as in the case of Neumann boundary conditions.

2.2 The heterogeneous case. When the spatial environment is inhomogeneous, (2.1) with Dirichlet boundary conditions may be modified to the following form

$$\begin{cases} u_t - \operatorname{div}\big(d_1(x)\nabla u\big) = \lambda a_1(x)u - a(x)u^2 - b(x)uv, & x \in \Omega,\ t > 0, \\ v_t - \operatorname{div}\big(d_2(x)\nabla v\big) = \mu a_2(x)v - d(x)v^2 + c(x)uv, & x \in \Omega,\ t > 0, \\ u = v = 0, & x \in \partial\Omega,\ t > 0, \\ u(x,0) = u_0(x) \geq 0,\ v(x,0) = v_0(x) \geq 0, & x \in \Omega, \end{cases} \tag{2.6}$$

where λ, μ are constants and $d_1, d_2, a_1, a_2, a, b, c, d$ are nonnegative continuous functions on $\overline{\Omega}$. If all the coefficient functions are strictly positive over Ω, it is easy to see that (2.6) behaves similarly to the homogeneous case (2.1). Thus, we may call (2.6) a *classical* predator-prey model when all the coefficient functions are strictly positive over Ω. In order to capture the influence of the heterogeneity of environment on (2.6), we consider a limiting case, where some of the coefficient functions in (2.6) vanish partially over Ω, which we call a degeneracy henceforth. We will show that the dynamical behavior of (2.6) with certain degeneracies may change drastically from the classical model.

Note that though a degenerate model as described above is not very realistic as a population model, it is a natural limiting problem for the classical model when some of its coefficients are very small on part of the underlying domain. Hence a better understanding of the degenerate cases provides important insight to the classical model with variable coefficients.

As a first step in this direction, we examine the effects of heterogeneous spatial environment on the set of steady-state solutions of (2.6). To be more specific, we consider such effects caused by the partial vanishing of $a(x)$. Let us recall that $a(x)$ describes the intro-specific pressure of u. The partial vanishing of $a(x)$ implies that, in the absence of v, the growth of u is governed by a degenerate logistic law, or more precisely, a mixture of the logistic and Malthusian laws over Ω.

To avoid unnecessary complications and to simplify our notations, we assume that all the other coefficient functions are positive constants. Through some simple rescalings of u, v and the coefficients, one sees that for the steady-state solutions, we need only consider the following system,

$$\begin{cases} -\Delta u = \lambda u - a(x)u^2 - buv, & x \in \Omega, \\ -\Delta v = \mu v - v^2 + cuv, & x \in \Omega, \\ u = v = 0, & x \in \partial\Omega, \end{cases} \tag{2.7}$$

where b, c are positive constants. We assume that $a(x) \equiv 0$ on the closure of some smooth domain Ω_0 satisfying $\overline{\Omega}_0 \subset \Omega$ and $a(x) > 0$ over $\overline{\Omega} \setminus \overline{\Omega}_0$. We will show that there exists a critical value $\lambda^* > 0$ such that (2.7) behaves largely as the homogeneous case $a(x) \equiv 1$ when $\lambda < \lambda^*$, but fundamental changes occur once $\lambda \geq \lambda^*$.

From now on, we fix b, c and regard λ and μ as varying parameters. Clearly $v \equiv 0$ satisfies the second equation in (2.7). In this case u satisfies the so called degenerate logistic equation

$$-\Delta u = \lambda u - a(x)u^2, \ x \in \Omega, \ \ u = 0, \ x \in \partial\Omega. \tag{2.8}$$

It is well known ([Ou, dP, FKLM]) that (2.8) has only the trivial nonnegative solution $u \equiv 0$ when $\lambda \notin (\lambda_1^D(\Omega), \lambda_1^D(\Omega_0))$, and it has a unique positive solution u_λ when λ belongs to this open interval. It is easily seen that $u_\lambda \to 0$ in $L^\infty(\Omega)$ when $\lambda \to \left(\lambda_1^D(\Omega)\right)^+$. Moreover, by [DHu], as $\lambda \to \left(\lambda_1^D(\Omega_0)\right)^-$,

$$u_\lambda \to \infty \ \text{ uniformly on } \overline{\Omega}_0,$$

$$u_\lambda \to U_{\lambda_1^D(\Omega_0)} \text{ locally uniformly on } \overline{\Omega} \setminus \overline{\Omega}_0,$$

where U_λ denotes the minimal positive solution of the following boundary blow-up problem

$$-\Delta U = \lambda U - a(x)U^2, \ x \in \Omega \setminus \overline{\Omega}_0; \ U|_{\partial\Omega} = 0, \ U|_{\partial\Omega_0} = \infty. \tag{2.9}$$

Here $U|_{\partial\Omega_0} = \infty$ means $\lim_{d(x,\partial\Omega_0)\to 0} U(x) = \infty$. By [DHu], (2.9) has a minimal and maximal positive solution for each $\lambda \in (-\infty, \infty)$.

To summarize, for each $\lambda \in (\lambda_1^D(\Omega), \lambda_1^D(\Omega_0))$, (2.7) has a unique semitrivial solution of the form $(u, 0)$ with $u > 0$, namely, $(u_\lambda, 0)$; there is no such semitrivial solution for other λ values. Similar to (2.2), (2.7) has a unique semitrivial solution $(0, \theta_\mu)$ of the form $(0, v)$ if $\mu > \lambda_1^D(\Omega)$ and there is no such semitrivial solution for other μ values. As before, the obvious solution $(u, v) = (0, 0)$ of (2.7) is called the trivial solution.

To analyze the set of positive solutions for (2.7) we will need the following *a priori* estimates. (Similar result also holds for Neumann boundary value problems.)

Theorem 2.1 *Given an arbitrary positive constant M we can find another positive constant C, depending only on M and $a(x), b, c, \Omega$ in* (2.7), *such that if (u, v) is a positive solution of* (2.7) *with $|\lambda| + |\mu| \leq M$, then*

$$\|u\|_\infty + \|v\|_\infty \leq C.$$

Here $\|\cdot\|_\infty = \|\cdot\|_{L^\infty(\Omega)}$.

We refer to [DD2] for the proof of Theorem 2.1. An important step in this proof is the following useful result.

Lemma 2.2 *Suppose $\{u_n\} \subset C^2(\overline{\Omega})$ satisfies*

$$-\Delta u_n \leq \lambda u_n, \ u_n|_{\partial\Omega} = 0, \ u_n \geq 0, \ \|u_n\|_\infty = 1,$$

where λ is a positive constant. Then, there exists $u_\infty \in L^\infty(\Omega) \cap H_0^1(\Omega)$ such that, subject to a subsequence, $u_n \to u_\infty$ weakly in $H_0^1(\Omega)$, strongly in $L^p(\Omega)$, $\forall p \geq 1$, and $\|u_\infty\|_\infty = 1$.

To study the positive solution set of (2.7), we will make use of the comparison principles and adapt the bifurcation approach used by Blat and Brown in [BB2] by fixing λ and using μ as the main bifurcation parameter. It turns out that our results will be fundamentally different in the following two cases:

$$\text{(i) } \lambda_1^D(\Omega) < \lambda < \lambda_1^D(\Omega_0) \quad \text{and} \quad \text{(ii) } \lambda \geq \lambda_1^D(\Omega_0).$$

In the first case, (2.7) can be analyzed as in the classical case (2.2). It has a unique semitrivial solution of the form $(u, 0)$, namely, $(u_\lambda, 0)$. If (u, v) is a positive solution to (2.7), then u satisfies

$$-\Delta u \leq \lambda u - a(x)u^2, \; u|_{\partial\Omega} = 0.$$

A simple comparison argument shows

$$0 < u \leq u_\lambda, \; \forall x \in \Omega. \tag{2.10}$$

From the equation for v we find

$$-\Delta v > \mu v - v^2, \; v|_{\partial\Omega} = 0,$$

which implies that

$$v \geq \theta_\mu, \; \forall x \in \Omega, \tag{2.11}$$

where we have used the convention that $\theta_\mu \equiv 0$ whenever $\mu \leq \lambda_1^D(\Omega)$.

From the equation for v we also obtain

$$\mu = \lambda_1^D(v - cu) \equiv \lambda_1^D(v - cu, \Omega).$$

Therefore, by (2.10) and the well-known monotonicity property of $\lambda_1^D(\phi)$, we easily deduce

$$\mu > \lambda_1^D(-cu_\lambda). \tag{2.12}$$

By the equation for u and (2.11), we deduce

$$\lambda = \lambda_1^D(au + bv) > \lambda_1^D(bv) \geq \lambda_1^D(b\theta_\mu),$$

that is,

$$\lambda > \lambda_1^D(b\theta_\mu). \tag{2.13}$$

Summarizing, we have the following result.

Theorem 2.3 *In the case that $\lambda_1^D(\Omega) < \lambda < \lambda_1^D(\Omega_0)$, a necessary condition for* (2.7) *to possess a positive solution is that both* (2.12) *and* (2.13) *hold.*

We will see in the following that (2.12) and (2.13) are also sufficient conditions for the existence of positive solutions.

In the (μ, u, v)-space $X := \mathbf{R} \times C^1(\overline{\Omega}) \times C^1(\overline{\Omega})$, we have two semitrivial solution curves

$$\Gamma_u := \{(\mu, u_\lambda, 0) : \mu \in (-\infty, \infty)\} \text{ and } \Gamma_v := \{(\mu, 0, \theta_\mu) : \lambda_1^D < \mu < \infty\}.$$

A local bifurcation analysis along Γ_u shows that from $(\lambda_1^D(-cu_\lambda), u_\lambda, 0) \in \Gamma_u$ bifurcates a smooth curve of positive solutions $\Gamma' = \{(\mu, u, v)\}$. A global bifurcation consideration, together with an application of the maximum principle, shows that Γ' is contained in a global branch (*i.e.*, connected set) of positive solutions

$\Gamma = \{(\mu, u, v)\}$ which is either unbounded or joins the semitrivial curve Γ_v at exactly $(\mu_0, 0, \theta_{\mu_0}) \in \Gamma_v$, where $\mu_0 > \lambda_1^D(\Omega)$ is determined uniquely by

$$\lambda = \lambda_1^D(b\theta_{\mu_0}). \tag{2.14}$$

It follows from (2.13) that $\mu < \mu_0$ whenever $(\mu, u, v) \in \Gamma$. Therefore, we find that $(\mu, u, v) \in \Gamma$ implies

$$\lambda_1^D(-cu_\lambda) < \mu < \mu_0. \tag{2.15}$$

From this, and applying Theorem 2.1, we conclude that Γ is bounded in the space $\mathbf{R} \times L^\infty(\Omega) \times L^\infty(\Omega)$. By standard L^p theory for elliptic operators, we conclude that Γ is also bounded in X. Hence Γ must join Γ_v. A local bifurcation analysis near $(\mu_0, 0, \theta_{\mu_0})$ shows that near this point, Γ consists of a smooth curve.

To summarize, we have proved the following result.

Theorem 2.4 *When $\lambda_1^D(\Omega) < \lambda < \lambda_1^D(\Omega_0)$, there is a bounded connected set of positive solutions $\Gamma = \{(\mu, u, v)\}$ in the space X which joins the semitrivial solutions branches Γ_u and Γ_v at $(\lambda_1^D(-cu_\lambda), u_\lambda, 0)$ and $(\mu_0, 0, \theta_{\mu_0})$, respectively; moreover, near these two points, Γ consists of smooth curves. (See Figure 1 (left).)*

Clearly, (2.15) is equivalent to (2.12) and (2.13) combined. It now follows from Theorems 2.3 and 2.4 the following result.

Corollary 2.5 *When $\lambda_1^D(\Omega) < \lambda < \lambda_1^D(\Omega_0)$,* (2.7) *has a positive solution if and only if* (2.15) *holds.*

The above discussion shows that our results for case (i) are similar to that of the classical case $b(x) \equiv 1$.

Let us now consider the second case where $\lambda \geq \lambda_1^D(\Omega_0)$. A fundamental difference to the first case now is that we no longer have a semitrivial solution of the form $(u, 0)$. However, the semitrivial solution curve Γ_v is unchanged, and the bifurcation analysis of [BB2] along Γ_v can still be adapted. Again, a local bifurcation analysis shows that a smooth curve of positive solutions $\Gamma' = \{(\mu, u, v)\}$ bifurcates from $(\mu_0, 0, \theta_{\mu_0}) \in \Gamma_v$ where μ_0 is determined by (2.14). As before, a global bifurcation analysis, together with an application of the maximum principle, shows that Γ' is contained in a global branch of positive solutions Γ which is either unbounded in X or joins a semitrivial solution of the form $(u, 0)$. But we already know that there is no semitrivial solution of the form $(u, 0)$. Therefore, Γ must be unbounded.

One easily sees that the arguments leading to (2.13) still work for our present situation. Hence $\mu < \mu_0$ whenever (2.7) has a positive solution. We now apply Theorem 2.1 and conclude that

$$proj_\mu \Gamma = (-\infty, \mu_0). \tag{2.16}$$

Summarizing the above discussion, we obtain the following result.

Theorem 2.6 *When $\lambda \geq \lambda_1^D(\Omega_0)$,* (2.7) *has a positive solution if and only if $\mu < \mu_0$. Moreover, there is an unbounded connected set of positive solutions $\Gamma = \{(\mu, u, v)\}$ in X which joins the semitrivial solution branch Γ_v at $(\mu_0, 0, \theta_{\mu_0})$ and satisfies* (2.16). *(See Figure 1 (right).)*

The fact that (2.7) has a positive solution for arbitrarily large negative μ is strikingly different from the classical case. Biologically, this implies that the prey species can support a predator species of arbitrarily negative growth rate. This is due to the fact that the population of the prey would blow up in the region Ω_0 in

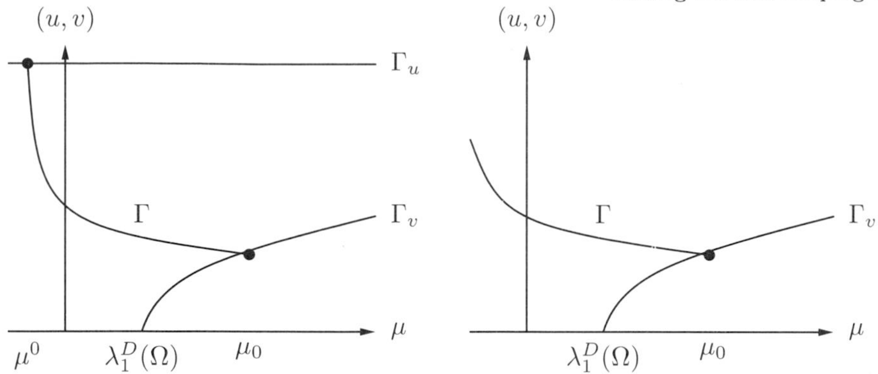

Figure 1 Bifurcation diagrams of positive steady state solutions of (2.7). $\Gamma_u = \{(\theta_\lambda, 0) : \mu \in \mathbf{R}\}$; $\Gamma_v = \{(0, \theta_\mu) : \mu \geq 0\}$. The branch Γ connecting Γ_u and Γ_v consists of coexistence states. (left): $\lambda_1^D(\Omega) < \lambda < \lambda_1^D(\Omega_0)$; (right): $\lambda \geq \lambda_1^D(\Omega_0)$.

the absence of the predator and hence one might think Ω_0 as a region where food is abundant for the predator. On the other hand, our above result indicates that the blow-up of the prey population can be avoided by introducing a predator with rather arbitrary growth rate.

We now consider the asymptotic behavior of the positive solutions of (2.7) as $\mu \to -\infty$. For this purpose, we consider a decreasing sequence of negative numbers μ_n which converges to $-\infty$, and let (u_n, v_n) be an arbitrary positive solution of (2.7) with $\mu = \mu_n$. We have the following result.

Theorem 2.7 *Let (μ_n, u_n, v_n) be as above. Then the following conclusions hold:*

(i) $\lim_{n\to\infty} \|u_n\|_\infty / |\mu_n| = 1/d$, $\lim_{n\to\infty} \|v_n\|_\infty / |\mu_n| = 0$;
(ii) $u_n/|\mu_n| \to 0$ *and* $v_n \to 0$ *uniformly on any compact subset of* $\overline{\Omega} \setminus \overline{\Omega}_0$*;*
(iii) $\underline{\lim}_{n\to\infty} \|u_n\|_{L^1(\Omega)}/|\mu_n| > 0$, $\overline{\lim}_{n\to\infty} \|v_n\|_{L^1(\Omega)} < \infty$, *and when* $\lambda > \lambda_1^D(\Omega_0)$, $\underline{\lim}_{n\to\infty} \|v_n\|_{L^1(\Omega)} > 0$*;*
(iv) $u_n/\|u_n\|_\infty \to \hat{u}$ *weakly in* $H_0^1(\Omega)$, v_n *weak* converges in* $C(\Omega)^*$ *to* $(\lambda/c)\chi_{\{\hat{u}=1\}}$, *where* $\hat{u} = 0$ *a.e. in* $\Omega \setminus \Omega_0$ *and* $\hat{u}|_{\Omega_0}$ *is the unique positive solution of*

$$-\Delta u = \lambda \chi_{\{u<1\}} u, \; u|_{\partial\Omega_0} = 0, \; \|u\|_\infty = 1. \tag{2.17}$$

We refer to [DD2, DD3] for the proof of this theorem.

Remark 2.8 (i) As $\int_\Omega u dx$ and $\int_\Omega v dx$ represent the total population of u and v, respectively, the conclusions in part (i) and part (iii) of Theorem 2.7 imply that, as $\mu \to -\infty$, the total population of u blows up at the rate of $|\mu|$, while the total population of v stays bounded. Moreover, when $\lambda > \lambda_1^D(\Omega_0)$, the total population of v is bounded from below by a positive constant independent of μ.

(ii) Note that (2.17) is a kind of free boundary problem. It was proved in [DD3] that this problem has no positive solution when $\lambda < \lambda_1^D(\Omega_0)$, and it has a unique positive solution when $\lambda \geq \lambda_1^D(\Omega_0)$. Moreover, if $\lambda > \lambda_1^D(\Omega_0)$, it is easily seen from equation (2.17) that $|\{\hat{u} = 1\}| > 0$ in part (iv) of Theorem

2.7; hence $\hat{u}$ has a "flat core". If $\lambda = \lambda_1^D(\Omega_0)$, then the flat core has measure zero and $\hat{u}$ is the first normalized eigenfunction on Ω_0.

(iii) If $\{\|v_n\|_{L^q}\}$ is bounded for some $q > 1$, then it is easy to show that $v_n \to 0$ uniformly on any compact subset of the set $\{\hat{u} < 1\}$. However, we were unable to determine whether $\{\|v_n\|_{L^q}\}$ is always bounded for some $q > 1$.

3 The Holling type II model

As mentioned in Section 1, it is generally believed that a Holling type II functional response is more reasonable than the unbounded functional response used in the Lotka-Volterra model. In this section we will examine a Holling type II diffusive predator-prey model with Neumann boundary conditions. We will see that in both the homogeneous case and the heterogeneous case, the Holling type II model exhibits much richer dynamics than the Lotka-Volterra model. In particular, we will reveal some fundamental differences caused by the spatial heterogeneity of the environment. Moreover, we will see that a similar degeneracy to that for the Lotka-Volterra model will result in very different effects for the Holling type II model. Furthermore, we will investigate the impact of a protection zone for the prey species on this model, and reveal some interesting results and problems.

3.1 The homogeneous case. In a homogeneous environment with no-flux boundary conditions, a Holling type II diffusive model has the form

$$\begin{cases} u_t - d_1\Delta u = \lambda u - au^2 - \dfrac{buv}{1+mu}, & x \in \Omega,\ t > 0, \\ v_t - d_2\Delta v = \mu v - dv^2 + \dfrac{cuv}{1+mu}, & x \in \Omega,\ t > 0, \\ \partial_\nu u = \partial_\nu v = 0, & x \in \partial\Omega,\ t > 0, \\ u(x,0) = u_0(x) \geq 0,\ v(x,0) = v_0(x) \geq 0, & x \in \Omega. \end{cases} \tag{3.1}$$

When $m = 0$, (3.1) reduces to (2.1). In fact, it follows from the Lyanunov functional analysis in [dMR] that the dynamics of (3.1) is essentially the same as that of (2.1) (with Neumann boundary conditions) when $0 < m < a/\lambda$; namely, $(\lambda/a, 0)$ is globally attractive when $\mu \leq -c\lambda/(a + m\lambda)$, the unique constant positive steady-state is globally attractive when $-c\lambda/(a+m\lambda) < \mu < \lambda d/b$, and $(0, \mu/d)$ is globally attractive when $\mu \geq \lambda d/b$. Therefore, in such a case, the diffusive model has the same dynamical behavior as the ODE model. When m is increased, multiple constant positive steady-states can arise and periodic solutions may occur through Hopf bifurcation. We refer to [Hz] for an analysis of Hopf bifurcation for the ODE model. The constant steady-states are described in the following results (which is essentially results in [DL3] Section 2):

Theorem 3.1 *Suppose that $d_1, d_2, a, b, c, d, \lambda, m > 0$ are fixed and $\mu \in \mathbf{R}$. Then the following hold:*

1. *The constant steady state solutions of* (3.1) *consist of* $(0,0)$, $(\lambda/a, 0)$, $(0, \mu/d)$ *and a branch of coexistence states $(\mu(u), u, v(u))$ given by*

$$\mu(u) = \frac{d}{b}(\lambda - au)(1 + mu) - \frac{cu}{1+mu}, \quad v(u) = \frac{(\lambda - au)(1+mu)}{b}, \tag{3.2}$$

where $0 < u < \lambda/a$ (see Figure 2); the bifurcation point along $\Gamma_u = \{(\lambda/a, 0) : \mu \in \mathbf{R}\}$ is $\mu^0 = -c\lambda/(1 + m\lambda)$, and the bifurcation point along $\Gamma_v = \{(0, \mu/d) : \mu \geq 0\}$ is $\mu_0 = \lambda d/b$.

2. *Define*

$$q = \frac{bc}{ad}, \quad m_2 = \frac{a}{\lambda}(q+1), \quad m_1 = \begin{cases} m_2 & \text{if } q \le 1; \\ \frac{a}{\lambda}(3q^{1/3}-1) & \text{if } q > 1. \end{cases} \tag{3.3}$$

Let

$$\overline{\mu}_* = \min_{u \in [0,\lambda/a]} \mu(u), \quad \overline{\mu}^* = \max_{u \in [0,\lambda/a]} \mu(u). \tag{3.4}$$

Then for all $m \ge 0$, $\overline{\mu}_* = \mu^0$; *when* $0 \le m \le m_2$, $\overline{\mu}^* = \mu_0$, *and when* $m > m_2$, $\overline{\mu}^* > \mu_0$. *Moreover when* $0 \le m \le m_1$, $\mu(u)$ *is strictly decreasing in* $(0, \lambda/a)$; *when* $m_1 < m < m_2$, $\mu(u)$ *has exactly one local minimum and one local maximum point in* $(0, \lambda/a)$; *and when* $m \ge m_2$, $\mu(u)$ *has exactly one local maximum point in* $(0, \lambda/a)$ *(see Figure 2 left.)*

3. *For all* $m \ge 0$, *when* $\mu \le \overline{\mu}_*$, *there is no any other non-negative steady state solutions, and* $(\lambda/a, 0)$ *is globally asymptotically stable. If in addition,* $0 \le m \le a/\lambda$, *then* $(0, \mu/d)$ *is globally asymptotically stable when* $\mu > \overline{\mu}^*$, *and the unique positive steady state is globally asymptotically stable when* $\overline{\mu}_* < \mu < \overline{\mu}^*$.

4. $\overline{\mu}^* = \dfrac{\lambda^2 dm}{4a^2 b} + o(m)$ *as* $m \to \infty$.

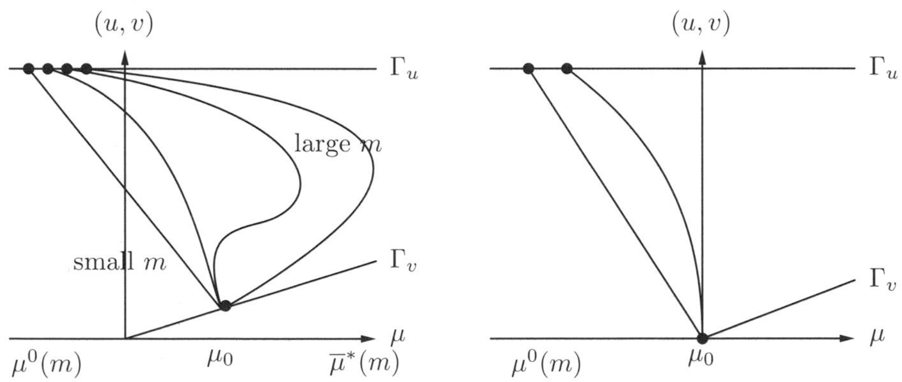

Figure 2 Bifurcation diagrams of constant non-negative steady state solutions of (3.1). $\Gamma_u = \{(\lambda/a, 0) : \mu \in \mathbf{R}\}$; $\Gamma_v = \{(0, \mu/d) : \mu \ge 0\}$. The branch Γ connecting Γ_u and Γ_v consists of coexistence states. (left): $d > 0$, with four different values of m (0, small positive, intermediate, large), the corresponding coexistence curves move from left to right when these values are taken by m successively; (right): $d = 0$, with $m = 0$ and $m > 0$ respectively.

The existence of space dependent positive steady-state solutions was considered by Du and Lou [DL3] when m is sufficiently large, and they proved the following results (see Theorems 1.1, 1.2 and Corollary 4.6):

Theorem 3.2 *Suppose that* $d_1, d_2, a, b, c, d, \lambda > 0$ *are fixed,* $\mu \in \mathbf{R}$. *Then there exists a large* $M > 0$ *such that for* $m \ge M$, *the following hold:*

1. *If* $\mu < \mu_0 \equiv \lambda d/b$ *or* $\mu > \overline{\mu}^*$, *then* (3.1) *has no non-constant steady state solutions.*

2. *If* $\mu \in (\mu_0 + \varepsilon_0, \overline{\mu}^* - \varepsilon_0)$, *and the equation*

$$-\Delta w + w\left(\lambda - \frac{b}{1/\mu + w}\right) = 0, \quad x \in \Omega, \quad \partial_\nu w = 0, \quad x \in \partial\Omega, \tag{3.5}$$

has a non-degenerate solution w, *then* (3.1) *has a non-constant positive solution* (u, v).

Except for some special cases, the stability of the constant and non-constant positive steady state solutions of (3.1) are not known. The secondary bifurcations (including Turing and Hopf type bifurcations) from the branch of constant steady state solutions seem to be complicated, and further investigation is needed.

In Theorems 3.1 and 3.2, we have assumed that $d > 0$. An ecologically important case is $d = 0$ and $\mu < 0$ (so the predator cannot survive without the prey u):

$$\begin{cases} u_t - d_1\Delta u = \lambda u - au^2 - \dfrac{buv}{1+mu}, & x \in \Omega,\ t > 0, \\ v_t - d_2\Delta v = \mu v + \dfrac{cuv}{1+mu}, & x \in \Omega,\ t > 0, \\ \partial_\nu u = \partial_\nu v = 0, & x \in \partial\Omega,\ t > 0, \\ u(x,0) = u_0(x) \geq 0,\ v(x,0) = v_0(x) \geq 0, & x \in \Omega. \end{cases} \tag{3.6}$$

In (3.6), if $\mu > 0$, then it is clear that the predator will have an exponential growth and wipe out the prey population, thus the only interesting case is when $\mu < 0$. When $\mu \leq \mu^0 \equiv -c\lambda/(a + m\lambda)$, the steady state $(\lambda/a, 0)$ is globally asymptotically stable. When $\mu \in (\mu^0, 0)$, (3.6) has a unique positive constant steady state solution for any positive $m > 0$ (see Figure 2 right). Thus bistability for the ODE system never happens. However it is known that for a certain range of m and μ, the unique positive constant steady state is not stable in the ODE model, and there exists a stable limit cycle. Numerical experiments on the diffusive model (3.6) have shown that various complicated spatiotemporal patterns can be generated in some parameter ranges (see [MPT] for a survey in that direction), but no rigorous treatments are available yet.

Finally we mention that the Dirichlet counterpart of (3.1) has been studied in [BB2, CEL, DL1, DL2]. Similar bifurcation diagrams for small m were obtained in [BB2, CEL], and the multiplicity and exact multiplicity of the steady state solutions when m is large were studied in [DL1, DL2]. In particular an exact S-shaped bifurcation diagram with large m (and under some conditions on the other parameters) was shown in [DL2].

3.2 The heterogeneous case. In heterogeneous environment, the diffusive predator-prey system becomes:

$$\begin{cases} u_t - \operatorname{div}(d_1(x)\nabla u) = \lambda(x)u - a(x)u^2 - \dfrac{b(x)uv}{1+m(x)u}, & x \in \Omega,\ t > 0, \\ v_t - \operatorname{div}(d_2(x)\nabla v) = \mu(x)v - d(x)v^2 + \dfrac{c(x)uv}{1+m(x)u}, & x \in \Omega,\ t > 0, \\ \partial_\nu u = \partial_\nu v = 0, & x \in \partial\Omega,\ t > 0, \\ u(x,0) = u_0(x) \geq 0,\ v(x,0) = v_0(x) \geq 0, & x \in \Omega. \end{cases} \tag{3.7}$$

To focus our interest, again we only consider a more special model with all parameter functions in (3.7) being constant except $a(x)$. Moreover, for the convenience of notations, we assume that the diffusion constants $d_1 = d_2 = 1$ and

$d \equiv 1$:

$$\begin{cases} u_t - \Delta u = \lambda u - a(x)u^2 - \dfrac{buv}{1+mu}, & x \in \Omega,\ t > 0, \\ v_t - \Delta v = \mu v - v^2 + \dfrac{cuv}{1+mu}, & x \in \Omega,\ t > 0, \\ \partial_\nu u = \partial_\nu v = 0, & x \in \partial\Omega,\ t > 0, \\ u(x,0) = u_0(x) \geq 0,\ v(x,0) = v_0(x) \geq 0, & x \in \Omega, \end{cases} \tag{3.8}$$

and for the equation of steady state solutions, now we have

$$\begin{cases} -\Delta u = \lambda u - a(x)u^2 - \dfrac{buv}{1+mu}, & x \in \Omega, \\ -\Delta v = \mu v - v^2 + \dfrac{cuv}{1+mu}, & x \in \Omega, \\ \partial_\nu u = \partial_\nu v = 0, & x \in \partial\Omega. \end{cases} \tag{3.9}$$

As in subsection 2.1, we assume that the function $a(x)$ is a nonnegative continuous function on $\overline{\Omega}$, and there exists a subdomain Ω_0 such that $\overline{\Omega_0} \subset \Omega$, and

$$a(x) \equiv 0 \ \ \forall x \in \overline{\Omega_0}, \text{ and } a(x) > 0 \ \ \forall x \in \overline{\Omega} \backslash \overline{\Omega_0}. \tag{3.10}$$

Similar to the classical Lotka-Volterra case considered in subsection 2.1, the eigenvalue $\lambda_1^D(\Omega_0)$ plays an important role in our analysis, and our results will be fundamentally different in the following two cases:

$$\text{(a) weak prey growth: } 0 < \lambda < \lambda_1^D(\Omega_0), \ \ \text{(b) strong prey growth: } \lambda > \lambda_1^D(\Omega_0). \tag{3.11}$$

As in the Dirichlet case, it is well-known (see [Ou]) that the boundary value problem

$$-\Delta u = \lambda u - a(x)u^2,\ x \in \Omega,\ \ \partial_\nu u = 0,\ x \in \partial\Omega, \tag{3.12}$$

has a unique positive solution $u_\lambda(x)$ when $0 < \lambda < \lambda_1^D(\Omega_0)$, and it has no positive solution when $\lambda \geq \lambda_1^D(\Omega_0)$. Moreover, $u_\lambda \to 0$ uniformly in Ω as $\lambda \to 0^+$, and $u_\lambda \to \infty$ uniformly on $\overline{\Omega_0}$ as $\lambda \to [\lambda_1^D(\Omega_0)]^-$, and $u_\lambda \to \underline{U}^\lambda$ uniformly on any compact subset of $\overline{\Omega}\backslash\overline{\Omega_0}$ as $\lambda \to [\lambda_1^D(\Omega_0)]^-$, where $\underline{U}^\lambda$ is the minimal positive solution of the boundary blow-up problem (see [DHu]):

$$-\Delta u = \lambda u - a(x)u^2,\ x \in \Omega\backslash\overline{\Omega_0},\ \ \partial_\nu u = 0,\ x \in \partial\Omega,\ \ u = \infty,\ x \in \partial\Omega_0. \tag{3.13}$$

It is shown in [DHu] that for any $\lambda \in \mathbf{R}$, (3.13) has a minimal positive solution $\underline{U}^\lambda$ and a maximal positive solution $\overline{U}^\lambda$. Problems (3.12) and (3.13) will play essential roles in our analysis below.

3.2.1 *Weak prey growth rate.* We now consider the case that $0 < \lambda < \lambda_1^D(\Omega_0)$. In contrast to the homogeneous case, the set of steady-state solutions of (3.9) are now much more difficult to analyze, and the dynamical behavior of (3.8) is almost out of reach. We will use a bifurcation approach to study the steady-state solutions, and more importantly, we will make use of an auxiliary single equation (see (3.17) below) to obtain some in-depth results for the global bifurcation branch of the positive solutions of (3.9), which will enable us to obtain some partial results on the dynamical behavior of (3.8). Since a single reaction-diffusion equation enjoys the monotonicity property, we are able to obtain rather detailed understanding of our auxiliary equation, which in turn helps us to understand our diffusive predator-prey model. So we are indirectly using the monotonicity property to study a system which lacks such a property.

Under our assumptions for λ, for any $\mu > 0$, (3.9) has two semi-trivial solutions: $(u_\lambda, 0)$ and $(0, \mu)$. They form two smooth curves in the (μ, u, v)-space:

$$\Gamma_u = \{(\mu, u_\lambda, 0) : -\infty < \mu < \infty\}, \quad \Gamma_v = \{(\mu, 0, \mu) : 0 < \mu < \infty\}. \tag{3.14}$$

Despite the different boundary conditions, the bifurcation analysis for the Dirichlet boundary value problem can be carried over to (3.9). More precisely, along Γ_u, there is a bifurcation point $\mu^0 = \lambda_1^N(-cp(u_\lambda))$ such that a smooth curve Γ_1' of positive solutions to (3.9) bifurcates from Γ_u at $(\mu, u, v) = (\mu^0, u_\lambda, 0)$. Similarly, there is another bifurcation point $\mu_0 = \lambda/b$ such that a smooth curve of positive solutions Γ_2' to (3.9) bifurcates from Γ_v at $(\mu, u, v) = (\mu_0, 0, \mu_0)$. Moreover one can show that (3.9) has only the trivial or semi-trivial solutions when $|\mu|$ is sufficiently large. Hence Γ_1' and Γ_2' are connected to each other, and we denote by Γ the maximal connected component of the set of positive steady state solutions which contains Γ_1' and Γ_2'. The direction of bifurcation of Γ at $\mu = \mu_0$ can be determined by m. In summary we have the following rough global bifurcation picture (see subsection 3.1 of [DS1] for details):

Theorem 3.3 *When $0 < \lambda < \lambda_1^D(\Omega_0)$, there exists a continuum Γ of positive solutions of (3.9) satisfying*

$$proj_\mu \Gamma = (\overline{\mu}_*, \overline{\mu}^*] \text{ or } (\overline{\mu}_*, \overline{\mu}^*), \tag{3.15}$$

where

$$\overline{\mu}_* = \lambda_1^N(-cp(u_\lambda)), \quad \lambda/b \leq \overline{\mu}^* \leq \lambda b^{-1}(1 + m||u_\lambda||_\infty).$$

Moreover, Γ connects the branches Γ_u and Γ_v at $(\mu, u, v) = (\mu^0, u_\lambda, 0)$ and $(\mu, u, v) = (\mu_0, 0, \mu_0)$ respectively. Furthermore, the bifurcation at $\mu = \mu^0$ is always supercritical, the bifurcation at μ_0 is supercritical if $m > m_0$ and it is subcritical if $0 \leq m < m_0$, where m_0 is given by

$$m_0 = \lambda^{-1}[\overline{a} + bc], \quad \text{where } \overline{a} = \frac{1}{|\Omega|}\int_\Omega a(x)dx. \tag{3.16}$$

While Theorem 3.3 provides useful information on the set of positive steady states, a more detailed description can be obtained by making use of the following auxiliary equation:

$$-\Delta u = \lambda u - a(x)u^2 - b\mu\frac{u}{1+mu}, \; x \in \Omega, \; \partial_\nu u = 0, \; x \in \partial\Omega. \tag{3.17}$$

To see the relevance of (3.17) to (3.9), let us observe that if (u, v) is a positive solution of (3.9), then a simple comparison argument shows $\mu < v < \mu + c/m$. Therefore u is a sub-solution to (3.17), and it is a super solution to (3.17) with μ replaced by $\mu + c/m$.

An in-depth study of the solution set of (3.17) will enable us to partially overcome the lack of comparison principle for the full system (3.8). By making extensive use of the monotonicity property of (3.17) and a global bifurcation argument, we have the following result:

Proposition 3.4 *Suppose that $0 < \lambda < \lambda_1^D(\Omega_0)$, and $b, m > 0$ are fixed. Then $\mu = \mu_0 \equiv \lambda/b$ is a bifurcation point for (3.17) such that a global unbounded continuum Σ of positive solutions of (3.17) emanates from $(\mu, u) = (\mu_0, 0)$, and*

$$proj_\mu \Sigma = (-\infty, \hat{\mu}^*] \text{ or } (-\infty, \hat{\mu}^*), \tag{3.18}$$

where $\hat{\mu}^* = \sup\{\mu > 0 :$ (3.17) has a positive solution$\} \geq \mu_0$. Moreover Σ satisfies the following:

1. *Near* $(\mu, u) = (\mu_0, 0)$, Σ *is a curve.*
2. *When* $\mu \le 0$, (3.17) *has a unique positive solution* $\overline{U}_\mu(x)$, *and* $\{(\mu, \overline{U}_\mu) : \mu \le 0\}$ *is a smooth curve.*
3. *For* $\mu \in (-\infty, \hat{\mu}^*)$, (3.17) *has a maximal positive solution* $\overline{U}_\mu(x)$, $\overline{U}_\mu$ *is strictly decreasing with respect to* μ.
4. *For* $\mu \in (-\infty, \lambda/b)$, (3.17) *has a minimal positive solution* $\underline{U}_\mu(x)$, $\underline{U}_\mu(x) \equiv \overline{U}_\mu(x)$ *when* $\mu \le 0$, $\underline{U}_\mu$ *is strictly decreasing with respect to* μ.
5. *If* $\hat{\mu}^* > \mu_0$, *then* (3.17) *has a maximal positive solution for* $\mu = \hat{\mu}^*$, *and has at least two positive solutions for* $\mu \in (\mu_0, \hat{\mu}^*)$.
6. *If* $\hat{\mu}^* > \mu_0$ *and* $0 < m < m_0$, *then there exists* $\hat{\mu}_* \in (0, \mu_0)$ *such that* (3.17) *has at least three positive solutions for* $\mu \in (\hat{\mu}_*, \mu_0)$ *and* $\underline{U}_\mu(x) < \overline{U}_\mu(x)$ *for* $\mu \in [\hat{\mu}_*, \mu_0)$. *Moreover,* $\lim_{\mu \to (\mu_0)^-} \underline{U}_\mu = 0$ *uniformly for* $x \in \overline{\Omega}$.

All these solutions mentioned above can be chosen from the unbounded continuum Σ.

The proof of Proposition 3.4 can be found in [DS1]. Estimates of $\hat{\mu}^*$ in terms of the parameter m can also be deduced. Indeed, let $M_a = \max_{x \in \overline{\Omega}} a(x)$ and $\overline{a} = |\Omega|^{-1} \int_\Omega a(x) dx$. Then

1. $\hat{\mu}^*(m) \ge \dfrac{\lambda M_a + m\lambda^2}{4bM_a}$, and in particular, $\lim\limits_{m \to \infty} \hat{\mu}^*(m) = \infty$.
2. $\hat{\mu}^*(m) > \lambda/b$ if $m > 3M_a/\lambda$; $\hat{\mu}_*(m) < \lambda/b$ if $m < \overline{a}/\lambda$.

We remark that $\hat{\mu}^*(m) > \lambda/b > \hat{\mu}_*(m)$ is only possible for certain patterned $a(x)$ (see [DS1] for a concrete example). With that possibility, Proposition 3.4 suggests three possible minimal bifurcation diagrams for the set of positive steady states of (3.17) as in Figure 3.

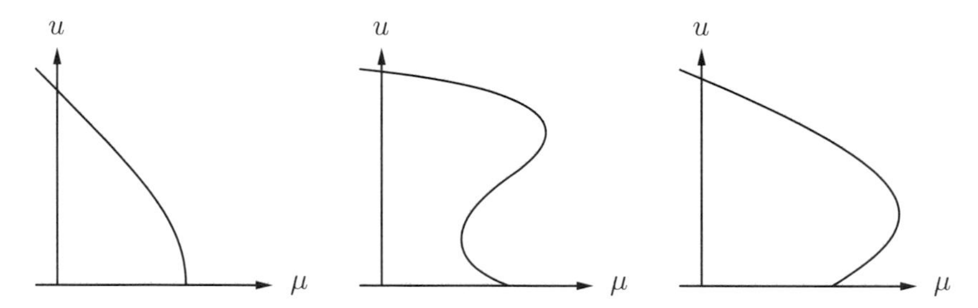

Figure 3 Possible bifurcation diagrams for (3.17)

The global bifurcation branch Σ of (3.17) somehow provides an estimate for the global bifurcation branch Γ of (3.9). A loose relationship between Σ and Γ can be rigorously established through some topological arguments, and we have the following results (see [DS1] for the proof):

Theorem 3.5 *Suppose that* $0 < \lambda < \lambda_1^D(\Omega_0)$, *and* $b, c, m > 0$ *are fixed. Let* $\hat{\mu}^*$ *and* $\hat{\mu}_*$ *be defined as in Proposition 3.4. Then the following hold:*

1. *Define*

$$\mu^* = \sup\{\mu > 0 : \text{ (3.9) } \textit{has a positive solution}\}, \tag{3.19}$$

and

$$\mu_* = \inf\{\mu > 0 : \text{(3.9) } \textit{has a positive solution } (u, v), \textit{ and } u \not\geq \overline{U}_{\hat{\mu}^*}\}. \tag{3.20}$$

Then $\hat{\mu}^* - c/m \leq \mu^* \leq \hat{\mu}^*$, *and* $\hat{\mu}_* \leq \mu_* \leq \hat{\mu}_* + c/m$.

2. *If* $\lambda_1^N(-cp(u_\lambda)) < \hat{\mu}^* - c/m$, *then for* $\mu \in (\lambda_1^N(-cp(u_\lambda)), \hat{\mu}^* - c/m]$, (3.9) *has a positive solution* (u_μ^1, v_μ^1) *satisfying*

$$\begin{aligned} &\min\{\overline{U}_\mu(x), \overline{U}_0(x)\} > u_\mu^1(x) > \overline{U}_{\mu+c/m}(x),\ x \in \Omega, \\ &\mu + \frac{c}{m} > v_\mu^1(x) > \max\{\mu, 0\},\ x \in \Omega. \end{aligned} \tag{3.21}$$

3. *If* $\hat{\mu}_* + c/m < \mu_0$, *then for* $\mu \in [\hat{\mu}_* + c/m, \mu_0)$, (3.9) *has a positive solution* (u_μ^2, v_μ^2) *satisfying*

$$\begin{aligned} &\underline{U}_\mu(x) > u_\mu^2(x) > \max\{\underline{U}_{\mu+c/m}(x), 0\},\ x \in \Omega, \\ &\mu + \frac{c}{m} > v_\mu^2(x) > \mu,\ x \in \Omega. \end{aligned} \tag{3.22}$$

4. *If* $\hat{\mu}^* > \mu_0 + c/m$, *then* (3.9) *has at least two positive solutions for* $\mu_0 < \mu \leq \hat{\mu}^* - c/m$.
5. *If* $\hat{\mu}^* > \mu_0 + c/m$ *and* $\hat{\mu}_* < \mu_0 - c/m$, *then* (3.9) *has at least three positive solutions for* $\hat{\mu}_* + c/m < \mu < \mu_0$.

All these solutions above can be chosen from the continuum Γ.

(Notice that $\underline{U}_{\mu+c/m}$ is not always defined in Part 3. In that case we assume $\underline{U}_{\mu+c/m} = 0$. Similarly, if $\overline{U}_{\hat{\mu}^*}$ is not defined, we understand that it equals 0.)

In Figure 4, three possible minimal bifurcation diagrams for (3.9) are shown, and the solid curves represent the bifurcation branches in the case $0 < \lambda < \lambda_1^D(\Omega_0)$ considered in Theorem 3.5. The diagrams in (a) and (c) are similar to those of homogeneous cases considered in subsection 3.1. The diagram in (b) shows a reversed S-shaped branch, which can happen for all small m, all $b, c > 0$ and some particularly designed $a(x)$. In the homogeneous case, such reversed S-shaped curve happens when m takes intermediate values.

With the above results for the steady state solutions, and by making use of the corresponding parabolic problem of (3.17), we can prove the following partial classification of the dynamics of (3.8) (see [DS1]):

Theorem 3.6 *Suppose that* $0 < \lambda < \lambda_1^D(\Omega_0)$, *and* $b, c, m > 0$ *are fixed. Then all solutions* $(u(x,t), v(x,t))$ *of* (3.8) *are globally bounded, and* $v(x,t)$ *satisfies*

$$\max\{\mu, 0\} \leq \underline{\lim}_{t\to\infty} v(x,t) \leq \overline{\lim}_{t\to\infty} v(x,t) \leq \max\{\mu + \frac{c}{m}, 0\}. \tag{3.23}$$

Moreover, the following hold:

1. *If* $\mu < \mu^0 \equiv \lambda_1^N(-cp(u_\lambda))$, *then* $(u_\lambda, 0)$ *is globally asymptotically stable.*
2. *If* $\mu > \hat{\mu}^*$, *then* $(0, \mu)$ *is globally asymptotically stable.*
3. *If* $\lambda_1^N(-cp(u_\lambda)) < \mu < \hat{\mu}^* - c/m$, *and* $u_0(x) \geq \overline{U}_{\hat{\mu}^*}(x)$, *then*

$$V_{\mu+c/m,1}(x) \leq \underline{\lim}_{t\to\infty} u(x,t) \leq \overline{\lim}_{t\to\infty} u(x,t) \leq \min\{\overline{U}_\mu(x), \overline{U}_0(x)\}, \tag{3.24}$$

where $V_{\mu,1}$ *is unique positive solution of*

$$-\Delta u = \left(\lambda - \frac{b\mu}{1+mU}\right) u - a(x)u^2,\ x \in \Omega,\ \ \partial_\nu u = 0,\ x \in \partial\Omega, \tag{3.25}$$

with $U = \overline{U}_{\hat{\mu}^*}$.

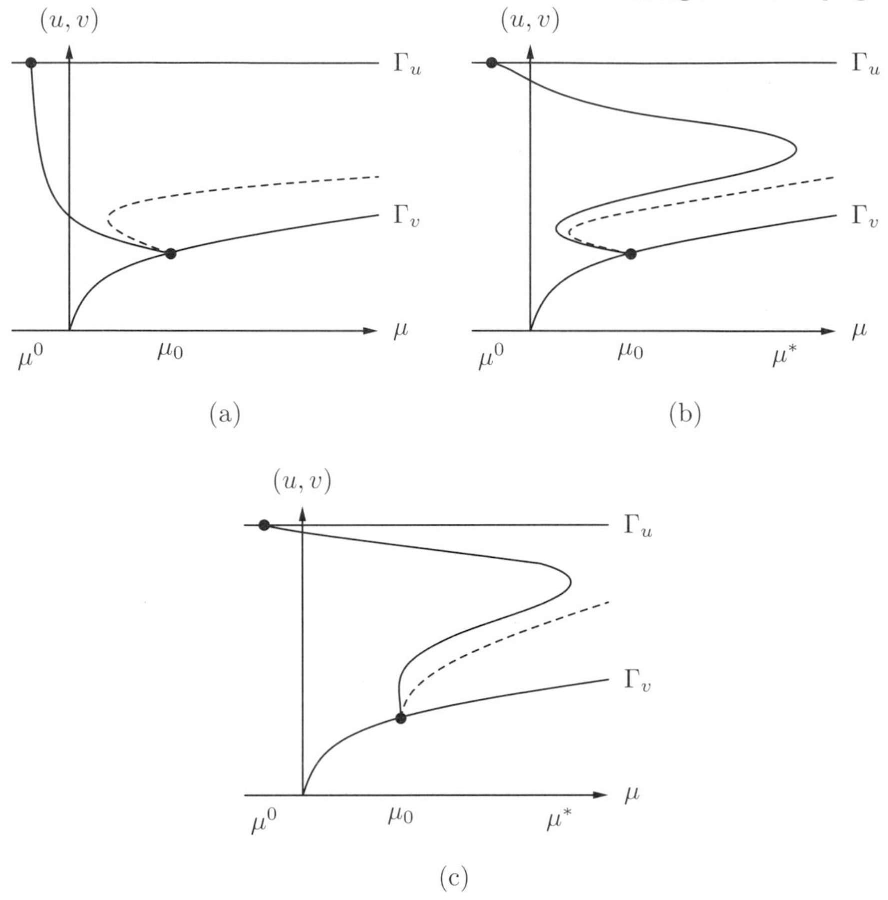

Figure 4 Bifurcation diagram of positive steady state solutions of (3.9); solid line: $\lambda < \lambda_1^D(\Omega_0)$, dashed line: $\lambda > \lambda_1^D(\Omega_0)$. (a): m is small; (b): m small but special $a(x)$; (c): m is large.

4. *Suppose that* $\hat{\mu}_* < \mu_0$, *then there exists* $\hat{\mu}^\sharp \in (\lambda/b, \hat{\mu}^*)$ *such that* $\lambda = \lambda_1^N(b\hat{\mu}^\sharp/(1+m\underline{U}_{\hat{\mu}_*}))$, *and for* $\hat{\mu}_* + c/m \le \mu < \hat{\mu}^\sharp$, *if* $u_0(x) \le \underline{U}_{\hat{\mu}_*}(x)$, *then*

$$\max\{\underline{U}_{\mu+c/m}(x), 0\} \le \underline{\lim}_{t\to\infty} u(x,t) \le \overline{\lim}_{t\to\infty} u(x,t) \le V_{\mu,2}(x), \tag{3.26}$$

where $V_{\mu,2}$ *is the unique positive solution of* (3.25) *with* $U = \underline{U}_{\hat{\mu}_*}$, *and for* $\mu \ge \hat{\mu}^\sharp$, *if* $u_0(x) \le \underline{U}_{\hat{\mu}_*}(x)$, *then* $\lim_{t\to\infty} u(x,t) = 0$ *and* $\lim_{t\to\infty} v(x,t) = \mu$ *uniformly for* $x \in \overline{\Omega}$.

The above results reveal that when the prey has a weak growth rate $\lambda < \lambda_1^D(\Omega_0)$, both populations stay bounded, and the bifurcation diagram and the dynamics are similar to the homogenous case. In biological terms, the above mathematical results imply that in the weak prey growth rate case, the dynamics of (3.8) has three possibilities according to the predator growth rate μ:

1. (Weak predator growth rate) When $\mu < \lambda_1^N(-cp(u_\lambda)) < 0$, the predator will become extinct, while the prey will reach its carrying capacity. Thus

we have unconditional persistence for the prey and unconditional extinction for the predator.

2. (Strong predator growth rate) This is when $\mu > \hat{\mu}^*$. Opposite to the previous case, the prey will become extinct while the predator will persist unconditionally for any initial population distribution.
3. (Intermediate predator growth rate) If $\hat{\mu}_* + c/m < \mu < \hat{\mu}^* - c/m$ with all these constants well defined and well ordered, then a bistable phenomenon appears, where two attracting regions exist, one is defined in (3.24) for u and (3.23) for v, and the other is defined in (3.26) for u and (3.23) for v. The former one contains a coexistence steady state, but it is unclear whether that coexistence steady state is the global attractor; the latter attracting region contains a coexistence steady state with a smaller u component when $\hat{\mu}_* + c/m < \mu < \lambda/b$, but that coexistence steady state becomes $(0, \mu)$ (prey extinction and predator persistence) when $\mu > \hat{\mu}^\sharp$. Though the predator population will settle in between μ and $\mu + c/m$, but the fate of the prey will depend on its initial population. We notice that it is possible that $\hat{\mu}_* = \hat{\mu}^* = \lambda/b$, then a coexistence steady state exists for $\lambda_1^N(-cp(u_\lambda)) < \mu < \lambda/b$, and this is a case that the system exhibits permanence. However, bistability can happen for certain ranges of the parameters if $a(x)$ is chosen properly.

Unfortunately, a full description of the dynamics of (3.8) is still a formidable job.

3.2.2 *Strong prey growth rate.* Now we turn to the case of $\lambda > \lambda_1^D(\Omega_0)$. We will only describe the results; their proofs can be found in [DS1]. Compared to the weak prey growth case, there is only one semi-trivial branch Γ_v, and there is still a bifurcation point $\mu_0 = \lambda/b$ such that a smooth curve of positive solutions Γ_2' (contained in a global branch Γ) to (3.9) bifurcates from Γ_v at $(\mu, u, v) = (\mu_0, 0, \mu_0)$. Since this is the only possible bifurcation point for positive solutions, Γ is unbounded in the space of (μ, u, v). Similar to the situation considered in subsection 2.2, we can prove the following *a priori* bound for the steady states:

Proposition 3.7 *Let $\lambda > \lambda_1^D(\Omega_0)$ be fixed and $\mu_n \le M$. Then there exists a positive constant C independent of n such that any positive solution (u_n, v_n) of* (3.9) *(with $\mu = \mu_n$) satisfies*

$$||u_n||_\infty + ||v_n||_\infty \le C. \tag{3.27}$$

Combining the bifurcation analysis and the above *a priori* estimate, we have

Theorem 3.8 *Suppose that $\lambda > \lambda_1^D(\Omega_0)$ is fixed. Then there exists a continuum Γ of positive solutions of* (3.9) *such that*

$$proj_\mu \Gamma = [\mu_*, \infty) \text{ or } (\mu_*, \infty), \quad -\frac{c}{m} < \mu_* \le \frac{\lambda}{b} \tag{3.28}$$

and satisfies the following:

1. *Γ bifurcates from $(\mu, u, v) = (\lambda/b, 0, \lambda/b)$, and the bifurcation there is supercritical if $m > m_0$ and it is subcritical if $0 \le m < m_0$, where m_0 is defined by* (3.16).
2. *For any $\mu > \mu_*$,* (3.9) *has at least one positive solution.*
3. *If $\mu_* < \lambda/b$, then for $\mu = \mu_*$,* (3.9) *has a positive solution, and thus $proj_\mu \Gamma = [\mu_*, \infty)$.*

4. *All these solutions mentioned above can be chosen from the unbounded continuum* Γ.

An illustration of the bifurcation diagrams can be seen in Figure 4 (dotted lines).

Remark 3.9 (i) From Theorem 3.8, one sees that a drastic change occurs when the prey growth rate λ increases across the threshold value $\lambda_1^D(\Omega_0)$. In the strong prey growth case, the system has a positive steady-state for all $\mu > \mu_0$; in the weak prey growth case, the corresponding range of μ is a bounded interval.

(ii) This is also in sharp contrast to the $m = 0$ case (the Lotka-Volterra model) described in Theorem 2.6, where positive steady-states exists if and only if μ belongs to an interval of the form $(-\infty, \tilde{\mu}_0)$.

(iii) Note that in the homogeneous case, we have seen that the dynamical behavior of (3.1) is similar for all $m \in [0, a/\lambda]$, but in the heterogeneous case, when $\lambda > \lambda_1^D(\Omega_0)$, a fundamental difference exists between the cases $m = 0$ (Theorem 2.6) and $m > 0$ (Theorem 3.8).

Next, we make use of the scalar equation (3.17) to obtain a more detailed description of the solutions of (3.8). It is clear that $\mu = \lambda/b$ is a bifurcation point for (3.17) such that a global continuum Σ_0 of positive solutions of (3.17) emanates from $(\mu, u) = (\lambda/b, 0)$, and

$$proj_\mu \Sigma_0 = [\mu_*^0, \infty) \text{ or } (\mu_*^0, \infty), \quad -\frac{c}{m} < \mu_*^0 \leq \frac{\lambda}{b}. \tag{3.29}$$

By making extensive use of the monotonicity property, one can show that $\mu_*^0 = \inf\{\mu : (3.17) \text{ has a positive solution}\}$. Moreover if $\mu_*^0 < \lambda/b$, then for $\mu \in [\mu_*^0, \lambda/b)$, (3.17) has a minimal positive solution $\underline{U}_\mu$, which is strictly decreasing with respect to μ, and for $\mu \in (\mu_*^0, \lambda/b)$, (3.17) has at least two positive solutions.

By using μ_*^0 and $\underline{U}_\mu$, we are able to partially identify the following bistable dynamics:

Theorem 3.10 *Suppose that* $\lambda > \lambda_1^D(\Omega_0)$, *and* $b, c, m > 0$ *are fixed. Let* $(u(x,t), v(x,t))$ *be the solution of* (3.8). *Then for any non-negative* (u_0, v_0) *with* $u_0, v_0 \not\equiv 0$ *and any* $\mu \in \mathbf{R}$, $v(x,t)$ *satisfies*

$$\max\{\mu, 0\} \leq \underline{\lim}_{t\to\infty} v(x,t) \leq \overline{\lim}_{t\to\infty} v(x,t) \leq \max\{\mu + \frac{c}{m}, 0\}, \tag{3.30}$$

and the asymptotic behavior of $u(x,t)$ *is as follows:*

1. (Blow-up) *If* $\mu < \mu_*^0 - c/m$, *then* $u(x,t)$ *satisfies*

$$\begin{aligned} &\lim_{t\to\infty} u(x,t) = \infty \text{ uniformly on } \overline{\Omega_0}, \\ &\overline{U}^\lambda(x) \geq \overline{\lim}_{t\to\infty} u(x,t) \geq \underline{\lim}_{t\to\infty} u(x,t) \geq 0, \quad x \in \overline{\Omega}\backslash\overline{\Omega_0}. \end{aligned} \tag{3.31}$$

If in addition, $\mu < -c/m$, *then* $\lim_{t\to\infty} v(x,t) = 0$ *uniformly on* $\overline{\Omega}$.

2. (Extinction) *If* $\mu_*^0 + c/m < \lambda/b$, *and* $u_0(x) \leq \underline{U}_{\mu_*^0}(x)$, *then for* $\mu \in [\mu_*^0 + c/m, \lambda/b]$,

$$\underline{U}_{\mu+c/m}(x) \leq \underline{\lim}_{t\to\infty} u(x,t) \leq \overline{\lim}_{t\to\infty} u(x,t) \leq V_{\mu,5}(x), \tag{3.32}$$

where $V_{\mu,5}$ *is the solution of of* (3.25) *with* $U = \underline{U}_{\mu_*^0}$, *and for* $\mu > \lambda/b$,

$$0 \leq \underline{\lim}_{t\to\infty} u(x,t) \leq \overline{\lim}_{t\to\infty} u(x,t) \leq \max\{V_{\mu,5}(x), 0\}. \tag{3.33}$$

Note that Part 2 of the above theorem demonstrates a bistable behavior for the dynamics of (3.8), since in that parameter range, the semitrivial steady-state $(u_\lambda, 0)$ is asymptotically stable. It is then important to further identify the basins of attraction of the persistence and extinction behaviors. To this end, we describe in the following two criteria for the persistence (and blow-up) or extinction of u. The strategy is to first find suitable sub(sup)-solutions of (3.17), then construct suitable sub(sup)-solutions for the full system.

Suppose that U_{μ_a} is a positive solution of (3.17) for $\mu = \mu_a$; then one can show that there exists $\mu_b < \mu_a$ such that U_{μ_a} is a sub-solution of (3.17) for all $\mu < \mu_b$; on the other hand, there exists $\mu_c > \mu_a$ such that U_{μ_a} is a super-solution of (3.17) for all $\mu > \mu_c$. To be more precise, we have the following results regarding the dynamics of the scalar equation

$$\begin{cases} w_t - \Delta w = \lambda w - a(x)w^2 - b\mu \dfrac{w}{1+mw}, & x \in \Omega,\ t > 0, \\ \partial_\nu w = 0, & x \in \partial\Omega,\ t > 0, \\ w(x,0) = w_0(x) \geq 0, & x \in \Omega, \end{cases} \tag{3.34}$$

and the full system (3.8):

Theorem 3.11 *Suppose that $\lambda > \lambda_1^D(\Omega_0)$, and $b, m > 0$ are fixed. Let $w(x,t)$ be a solution of (3.34), and let $(u(x,t), v(x,t))$ be a solution of (3.9). Then the following hold:*

1. (Blow-up) *For any given $\widehat{\mu} > 0$ we define*

$$C_{\widehat{\mu}} = \sup\{||u||_\infty : \ u \text{ is a positive solution of } (3.17) \text{ with } \mu \leq \widehat{\mu}\} \tag{3.35}$$

(the supremum exists from Proposition 3.7). Let $\mu_a > \widehat{\mu}$ be such that

$$C_{\widehat{\mu}} < (b\mu_a - \lambda)/(m\lambda)$$

and let $U_{\mu_a}(x)$ be a positive solution of (3.17) with $\mu = \mu_a$. Then for $\mu \leq \widehat{\mu}$ and $w_0(x) \geq U_{\mu_a}(x)$, the solution $w(x,t) \to \infty$ uniformly for $x \in \overline{\Omega_0}$, and $\overline{\lim}_{t\to\infty} w(x,t) \leq \overline{U}^\lambda(x)$ for $x \in \overline{\Omega}\backslash\overline{\Omega_0}$; for $\mu < \widehat{\mu} - c/m$ and $u_0(x) \geq U_{\mu_a}(x)$, (3.31) holds.

2. (Extinction) *Let $U_{\mu_a}(x)$ be a positive solution of (3.17) with $\mu = \mu_a > 0$. Define $\widetilde{\mu}^*$ to be the unique positive number such that $\lambda = \lambda_1^N(b\widetilde{\mu}^*/(1 + mU_{\mu_a}))$. Then for $\mu \geq \widetilde{\mu}^*$ and $w_0(x) \leq U_{\mu_a}(x)$, $\lim_{t\to\infty} w(x,t) = 0$ uniformly for $x \in \Omega$; and for $\mu > \widetilde{\mu}^*$, if $u_0(x) \leq U_{\mu_a}(x)$, $\lim_{t\to\infty} u(x,t) = 0$ and $\lim_{t\to\infty} v(x,t) = \mu$ uniformly for $x \in \overline{\Omega}$.*

Theorem 3.11 provides a criterion in terms of a generic solution U_{μ_a} of (3.17) for the persistence or extinction of u, but the profile of the steady state U_μ is not known. For large μ, we could determine the asymptotic behavior of U_μ, and the following result was proved in [DS1].

Proposition 3.12 *Suppose that $\lambda > \lambda_1^D(\Omega_0)$, and $b, m > 0$ are fixed. Let (μ_n, u_n) be a sequence of positive solutions of (3.17) and $\mu_n \to \infty$ as $n \to \infty$. Then subject to a subsequence,*

$$\lim_{n\to\infty} \mu_n ||u_n||_\infty^{-1} = \sigma \in \left(0, \frac{m\lambda}{b}\right), \tag{3.36}$$

and $\sigma u_n/\mu_n \to \widehat{u}$ *weakly in* $H^1(\Omega)$ *and strongly in* $L^p(\Omega_0)$ *for any* $p > 1$, *where* $\widehat{u}$ *is a nonnegative function satisfying* $\widehat{u}(x) = 0$ *for* $x \in \Omega\backslash\overline{\Omega_0}$, *and* $\widehat{u}|_{\Omega_0} \in H_0^1(\Omega_0)$ *is a weak solution of the (free boundary) problem*

$$-\Delta u = \left(\lambda u - \frac{b\sigma}{m}\right)\chi_{\{u>0\}}, \;\; x \in \Omega_0, \;\; u(x) = 0, \;\; x \in \partial\Omega_0, \;\; ||u||_\infty = 1. \qquad (3.37)$$

We can now use the solution of the free boundary problem as a candidate for the sub-solution which induces blow-up:

Proposition 3.13 *Suppose that* $\lambda > \lambda_1^D(\Omega_0)$, *and* $b, m > 0$ *are fixed. Suppose that* $U_0(x)$ *is a nontrivial solution of*

$$-\Delta u = \left(\lambda_a u - \frac{b}{m}\right)\chi_{\{u>0\}}, \;\; x \in \Omega_0, \;\; u(x) = 0, \;\; x \in \partial\Omega_0, \qquad (3.38)$$

where $\lambda_1^D(\Omega_0) < \lambda_a < \lambda$. *We extend* U_0 *by* $U_0(x) \equiv 0$ *for* $x \in \overline{\Omega}\backslash\overline{\Omega_0}$. *Then there exists* $\sigma > 0$ *such that for* $w_0(x) \geq \sigma U_0(x)$, *the solution of* (3.34) *satisfies* $w(x,t) \to \infty$ *uniformly for* $x \in \overline{\Omega_0}$, *and* $\overline{\lim}_{t\to\infty} w(x,t) \leq \overline{U}^\lambda(x)$ *for* $x \in \overline{\Omega}\backslash\overline{\Omega_0}$; *and for* $u_0(x) \geq \sigma' U_0(x)$, *where* $\sigma' = \max\{\mu + c/m, C_{\mu+c/m}/\|U_0\|_\infty\}$, (3.31) *holds.*

For a general domain Ω_0, the solution of the free boundary problem is not completely understood. However for a one dimensional domain, a complete solution to the free boundary problem (3.38) can be given. Indeed for $\Omega_0 = (-\pi, \pi)$, it can be shown (see [DS1]) that if $1/4 < \lambda \leq 1$, (3.38) has a unique solution

$$u(x) = \frac{b\sigma}{\lambda m}\big(1 - \frac{\cos(\sqrt{\lambda}\,x)}{\cos(\sqrt{\lambda}\,\pi)}\big),$$

where

$$\sigma = \frac{\lambda m}{b}\big(1 - \frac{1}{\cos(\sqrt{\lambda}\pi)}\big)^{-1};$$

if $\lambda > 1$, then the solutions are given by

$$u(x) = \sum_{i=1}^{k} \frac{\lambda m}{2b}\phi_\lambda(x_i + x), \; 1 \leq k \leq [\sqrt{\lambda}\,],$$

where

$$\phi_\lambda(x) = \frac{b}{\lambda m}[\cos(\sqrt{\lambda}x) + 1], \; x \in (-\pi/\sqrt{\lambda}, \pi/\sqrt{\lambda}).$$

and x_i satisfying

$$2|x_i - \pi|, \, 2|x_i + \pi|, \, |x_i - x_j| \geq \frac{2\pi}{\sqrt{\lambda}}, \qquad (3.39)$$

where $[\sqrt{\lambda}\,]$ denotes the largest positive integer no bigger than $\sqrt{\lambda}$.

Let us now explain the biological implications of our mathematical results. In the strong prey growth case ($\lambda > \lambda_1^D(\Omega_0)$), no matter how large the predator growth rate is, the prey population can blow up for certain initial population distributions. Moreover from our discussions in Theorem 3.11 and Proposition 3.13, even if the prey is initially restricted to a very narrow region in the habitat (but perhaps with a high density), an unbounded growth of this prey population is still possible if its growth rate λ is high enough. This is of particular interest for understanding the biological invasion of the prey species, since it shows that the presence of a very strong predator cannot stop the invasion if there exists an ideal environment Ω_0 which sufficiently nutrients the growth of the prey population.

Theorem 3.11 has another interesting biological explanation. These results show that for the initial population distribution $u_0 = U_{\mu_0}$, the prey population will become extinct when the predator is strong ($\mu > \widetilde{\mu}^*$), but for the same initial distribution, the prey population can also blow up if the predator is weak ($\mu < \widehat{\mu} - c/m$). Hence no any initial distribution is guaranteed to predict extinction or blow-up of the prey population. We remark that this idea can also be applied to the weak prey growth case considered in subsection 3.1, to show that the same initial distribution can lead to either extinction or persistence depending on the predator growth rate.

Finally we notice that in the weak prey growth case, the possibility of Allee effect (that is, bistability) depends on the value of m: when m is large, two attracting regions exist for the system, but when m is small, a unique attracting region will absorb all initial states. In sharp contrast, in the strong prey growth case, bistability is always possible no matter how small m is. Hence in the latter case the Allee effect is mainly caused by the degeneracy of $a(x)$, not the saturation (*i.e.* $m > 0$) of the predation rate.

3.3 Impact of protect zones. In this subsection, we use a rather different approach to understand the effects of inhomogeneous environment on the Holling type II diffusive predator-prey model. We will examine a situation that the heterogeneity of the environment is mainly caused by the creation of a protection zone for the prey species. We refer to [DS2] for the proofs of the results to be presented below.

From our analysis in subsection 3.1, we know that in a homogeneous environment, the prey population would extinguish if the growth rate of the predator is too large, or the predation rate is too high. It is not difficult to see that this remains the case if the coefficients in the model are replaced by positive functions of x. To save an otherwise endangered prey species, a natural idea is to set up one or several protection zones for the prey, where the prey species can enter and leave freely but the predator is kept out. This create a spatial environment felt very differently by the prey and predator species. We may ask several related biological questions: Are such protection zones effective to save an endangered prey population? What are the effects of such protection zones on the predator species? Could such protection zones induce unexpected new dynamics for the species involved?

We now attempt to address these questions by examining the Holling type II diffusive predator-prey model with a single protection zone. Our model can be described by the following system of equations:

$$\begin{cases} u_t - \operatorname{div}(d_1(x)\nabla u) = \lambda(x)u - a(x)u^2 - \dfrac{b(x)uv}{1+m(x)u}, & x \in \Omega,\ t > 0, \\ v_t - \operatorname{div}(d_2(x)\nabla v) = \mu(x)v - d(x)v^2 + \dfrac{c(x)uv}{1+m(x)u}, & x \in \Omega\backslash\overline{\Omega}_0,\ t > 0, \\ \partial_\nu u = 0, \ x \in \partial\Omega, \ t > 0, \quad \partial_\nu v = 0, \ x \in \partial\Omega \cup \partial\Omega_0, \ t > 0, \\ u(x,0) = u_0(x) \geq 0, \ x \in \Omega, \quad v(x,0) = v_0(x) \geq 0, \ x \in \Omega_0. \end{cases} \tag{3.40}$$

Here the protection zone is represented by Ω_0, a subdomain of Ω with smooth boundary $\partial\Omega_0$. The larger region Ω is the habitat of the prey, but the predator species can only exist in $\Omega\backslash\overline{\Omega}_0$. All the coefficient functions are nonnegative in $\overline{\Omega}$. The function $b(x)$ is zero when $x \in \Omega_0$, representing the assumption that the prey population enjoys predation-free growth in Ω_0; this also makes the interaction term in the equation for u well-defined over Ω. The boundary of the protection zone does not affect the dispersal of prey, but it works as a barrier to block the predator from entering Ω_0; thus a no-flux boundary condition should be imposed for the predator on $\partial\Omega_0$. For technical reasons, we assume further that $\overline{\Omega}_0 \subset \Omega$. Therefore, we may call Ω_0 an interior protection zone.

To focus on the impact of the protection zone on the dynamics, we will assume that all the parameter functions in (3.40) are constant except $b(x)$, and $a = d = d_1 = d_2 = 1$. Thus we have the following system:

$$\begin{cases} u_t - \Delta u = \lambda u - u^2 - \dfrac{b(x)uv}{1+mu}, & x \in \Omega,\ t > 0, \\ v_t - \Delta v = \mu v - v^2 + \dfrac{cuv}{1+mu} & x \in \Omega\backslash\overline{\Omega}_0,\ t > 0, \\ \partial_\nu u = 0,\ \ x \in \partial\Omega,\ t > 0, & \partial_\nu v = 0,\ \ x \in \partial\Omega \cup \partial\Omega_0,\ t > 0, \\ u(x,0) = u_0(x) \geq 0,\ \ x \in \Omega, & v(x,0) = v_0(x) \geq 0,\ \ x \in \Omega_0. \end{cases} \tag{3.41}$$

The steady state solutions satisfy

$$\begin{cases} -\Delta u = \lambda u - u^2 - \dfrac{b(x)uv}{1+mu}, & x \in \Omega, \\ -\Delta v = \mu v - v^2 + \dfrac{cuv}{1+mu}, & x \in \Omega\backslash\overline{\Omega}_0, \\ \partial_\nu u = 0,\ \ x \in \partial\Omega, & \partial_\nu v = 0,\ \ x \in \partial\Omega \cup \partial\Omega_0. \end{cases} \tag{3.42}$$

Here λ, μ, c, m are positive constants, $b(x) \in L^\infty(\Omega)$, $b(x) \geq 0$ in $\overline{\Omega}$, $b(x) \equiv 0$ on $\overline{\Omega}_0$ and for any compact subset A of $\overline{\Omega} \setminus \overline{\Omega}_0$, there exists $\delta_A > 0$ such that

$$b(x) \geq \delta_A \ \ \forall x \in A. \tag{3.43}$$

For simplicity, we may think of $b(x)$ as given by $b(x) = 0$ on $\overline{\Omega}_0$ and $b(x) = 1$ otherwise.

We will consider λ and μ as varying parameters while fixing the other parameters. It turns out that the ranges of λ where the dynamics will be drastically different are as in subsection 3.2:

$$\text{(a) } 0 < \lambda < \lambda_1^D(\Omega_0), \ \ \text{(b) } \lambda > \lambda_1^D(\Omega_0). \tag{3.44}$$

Let us now make an interesting comparison between (3.8) and (3.41): In (3.8), Ω_0 is the region where the prey could have an unbounded growth as the crowding effect is zero, but the predator can be present in Ω_0; in (3.41), the prey has the same or similar living condition in Ω_0 as in other parts of the habitat Ω, but the predator is blocked out. Hence the models (3.8) and (3.41) can be viewed as two different ways to create some ideal subregions for the prey. Indeed, our mathematical results will show that under a weak prey growth rate λ (which means $0 < \lambda < \lambda_1^D(\Omega_0)$), the two systems (3.8) and (3.41) exhibit very similar dynamical behavior. However, these systems behave very differently when the prey growth rate is strong ($\lambda > \lambda_1^D(\Omega_0)$), implying that the different strategies of designing favorable regions for the prey could lead to very different consequences.

In the following, it is more convenient for us to regard λ as fixed while interpreting the cases $0 < \lambda < \lambda_1^D(\Omega_0)$ and $\lambda > \lambda_1^D(\Omega_0)$ as caused by the change of sizes of the protect zone Ω_0. Such a view may be best explained by the well-known diffusive logistic equation with hostile boundary condition:

$$\begin{cases} u_t - \Delta u = \lambda u - u^2, & x \in \Omega_0,\ t > 0, \\ u(x,t) = 0, & x \in \partial\Omega_0,\ t > 0, \\ u(x,0) = u_0(x) \geq 0, & x \in \Omega_0. \end{cases} \tag{3.45}$$

It is known that there exists a unique $\lambda_* = \lambda_1^D(\Omega_0) > 0$ such that when $\lambda \leq \lambda_*$, the population u would eventually extinguish, and when $\lambda > \lambda_*$, the population u will persist and settle at a unique positive steady state. Thus λ_* is the threshold growth rate for persistence/extinction. If one regards the growth rate λ as fixed, then there is a minimal patch size S, such that when the domain is smaller than S, the population will become extinct, but otherwise the population will persist. The minimal patch size is determined by the principal eigenvalue of the associated linear operator, and it depends on the geometry of the habitat (see [CC2]). (The concept of minimal patch size first appeared in the pioneering work [Sk] and [KS] as mentioned in Section 1.) Therefore, we may call $0 < \lambda < \lambda_1^D(\Omega_0)$ the small protect zone case, and $\lambda > \lambda_1^D(\Omega_0)$ the large protect zone case.

For the small protect zone case, the dynamical behavior of (3.41) is essentially the same as that of (3.8), hence we will not go into details for this case. In the remaining part of this subsection, we shall concentrate on the large protect zone case: $\lambda > \lambda_1^D(\Omega_0)$.

We start our analysis by a standard local bifurcation argument. We fix $\lambda, c, m > 0$, and take μ as the bifurcation parameter. For any $\mu > 0$, (3.42) has two semi-trivial solutions: $(\lambda, 0)$ and $(0, \mu)$. They form two smooth curves in the (μ, u, v)-space:

$$\Gamma_u = \{(\mu, \lambda, 0) : -\infty < \mu < \infty\}, \quad \Gamma_v = \{(\mu, 0, \mu) : 0 < \mu < \infty\}. \tag{3.46}$$

Similar to before, a supercritical bifurcation occurs along the semi-trivial branch Γ_u. But the bifurcation along Γ_v now depends on the equation

$$\lambda = \lambda_1^N(b(x)\mu, \Omega). \tag{3.47}$$

It turns out that (2.1) can be satisfied by some $\mu > 0$ only if $0 < \lambda < \lambda_1^D(\Omega_0)$; when $\lambda \geq \lambda_1^D(\Omega_0)$, $\lambda > \lambda_1^N(b(x)\mu, \Omega)$ holds for all $\mu > 0$. Thus when $\lambda > \lambda_1^D(\Omega_0)$, the semi-trivial state $(0, \mu)$ is unstable for every $\mu > 0$, and there is no bifurcation of positive solutions occurring along Γ_v. Hence we have the following global bifurcation picture (see Figure 5):

Theorem 3.14 *If $\lambda \geq \lambda_1^D(\Omega_0)$, then*

1. $\mu^0 = -c\lambda/(1 + m\lambda)$ *is a bifurcation point where an unbounded continuum* Γ *of positive solutions to* (3.42) *bifurcates from* Γ_u *at* $(\mu, u, v) = (\mu^0, \lambda, 0)$.
2. *Near* $(\mu^0, \lambda, 0)$, Γ *is a smooth curve* $(\mu(s), u(s), v(s))$ *with* $s \in (0, \delta)$, *such that* $(\mu(0), u(0), v(0)) = (\mu^0, \lambda, 0)$ *and* $\mu'(0) > 0$.
3. $\mathrm{proj}_\mu \Gamma = (\mu^0, \infty)$, *and so* (3.42) *has at least one positive solution for any* $\mu > \mu^0$, *but* (3.42) *has no positive solution for* $\mu \leq \mu^0$.

The instability of the semi-trivial state $(0, \mu)$ implies that the system is permanent for all $\mu > 0$ (see [CCH]), but a better description of the dynamics can be obtained, especially when $\mu > 0$ is large. For (3.41), the population of the predator $v(x,t)$ still asymptotically lies between the constants μ and $\mu + c/m$, *i.e.* $v(x,t)$ satisfies (3.23). Hence similar to the analysis in subsection 3.2, more information

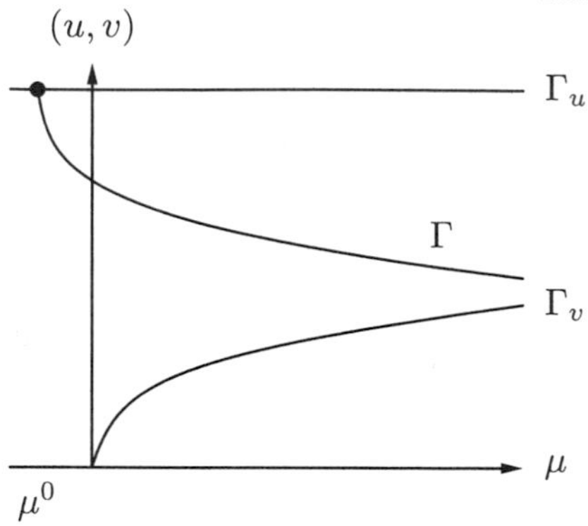

Figure 5 Bifurcation diagram of positive steady state solutions of (3.42) when $\lambda > \lambda_1^D(\Omega_0)$; the diagrams when $0 < \lambda < \lambda_1^D(\Omega_0)$ are similar to the ones with solid lines in Fig. 4.

on the set of coexistence states of (3.41) can be obtained if we can extract more information from the the scalar equation:

$$-\Delta u = \lambda u - u^2 - b(x)\mu \frac{u}{1+mu}, \quad x \in \Omega, \quad \partial_\nu u = 0, \quad x \in \partial\Omega. \tag{3.48}$$

The analysis of (3.48) relies on a well-known comparison lemma (see [SY] Lemma 2.3, and a more general version can be found in [DM] Lemma 2.1):

Lemma 3.15 *Suppose that* $f : \Omega \times \mathbf{R}^+ \to \mathbf{R}$ *is a continuous function such that* $f(x, s)$ *is decreasing for* $s > 0$ *at almost all* $x \in \overline{\Omega}$. *Let* $w, v \in C(\overline{\Omega}) \cap C^2(\Omega)$ *satisfy*

1. $\Delta w + wf(x, w) \leq 0 \leq \Delta v + vf(x, v)$ *in* Ω,
2. $w, v > 0$ *in* Ω *and* $w \geq v$ *on* $\partial\Omega$,
3. $\Delta v \in L^1(\Omega)$.

Then $w \geq v$ *in* $\overline{\Omega}$.

By using Lemma 3.15, one can construct a positive solution of (3.48) by the sub- and super-solution method: $u = \lambda$ is a super-solution, and W_λ is a sub-solution, where W_λ is defined by $W_\lambda(x) = 0$ in $\Omega \backslash \overline{\Omega}_0$, $W_\lambda = w_\lambda$ in Ω_0, where w_λ denotes the unique positive solution of

$$-\Delta w = \lambda w - w^2, \quad x \in \Omega_0, \qquad w = 0, \quad x \in \partial\Omega_0. \tag{3.49}$$

When $\mu \to \infty$, we could prove that each solution U of (3.48) uniformly converges to W_λ. Therefore for large μ, $U \approx w_\lambda$ for $x \in \Omega_0$ and $U \approx 0$ for $x \notin \Omega_0$. With this explicit spatial profile of the solution, we are able to conclude that any positive solution of (3.48) is linearly stable, and we can then apply a standard fixed point index argument to show that (3.48) has a unique positive solution when μ is sufficiently large. More precisely, we have

Proposition 3.16 *Suppose that* $\lambda > \lambda_1^D(\Omega_0)$. *Then*

1. *For each* $\mu \leq 0$, (3.48) *has a unique positive solution* U_μ, *which is strictly decreasing in* μ, *and* $\{(\mu, U_\mu) : \mu \leq 0\}$ *is a smooth curve.*
2. *For each* $\mu > 0$, (3.48) *has a minimal positive solution* $\underline{U}_\mu$ *and a maximal positive solution* $\overline{U}_\mu$, *and they satisfy*

$$W_\lambda(x) < \underline{U}_\mu(x) \leq \overline{U}_\mu(x) < \lambda, \quad \forall x \in \Omega. \tag{3.50}$$

3. *There exists* $\overline{\mu}^* > 0$ *such that for* $\mu > \overline{\mu}^*$, $\overline{U}_\mu = \underline{U}_\mu$, *and* (3.48) *has a unique positive solution, which we denote by* U_μ, *and* $\{(\mu, U_\mu) : \mu > \overline{\mu}^*\}$ *is a smooth curve. Moreover, as* $\mu \to \infty$, $U_\mu \to W_\lambda$ *in* $C(\overline{\Omega})$.

Regarding the last convergence result in Proposition 3.16, we remark that it is not difficult to show that $U_\mu \to W_\lambda$ in $L^p(\Omega)$ for any $p > 1$ as $\mu \to \infty$ from energy estimates, but the convergence in $C(\overline{\Omega})$ requires more delicate interior and boundary estimates, see [DS2] Proposition 3.3. The uniform convergence is crucial for the later results.

The uniqueness result in Proposition 3.16 enables us to prove the uniqueness and global asymptotical stability of the positive steady state for the full system (3.41):

Theorem 3.17 *Suppose that* $\lambda > \lambda_1^D(\Omega_0)$. *Then*

1. *There exists* $\delta > 0$ *such that* (3.42) *has a unique positive solution when* $\mu \in (\mu^0, \mu^0 + \delta)$.
2. *For any* $\mu > \mu^0$, *if* (u, v) *is a positive solution of* (3.42), *then*

$$\underline{U}_{\mu+c/m}(x) \le u(x) \le \overline{U}_\mu(x), \quad \max\{\mu, 0\} \le v(x) \le \mu + c/m, \tag{3.51}$$

where $\underline{U}_\mu$ *and* $\overline{U}_\mu$ *are the minimal and maximal solutions of* (3.48), *respectively.*

3. *There exists* $\mu^* > 0$ *such that* (3.42) *has a unique positive solution* (u_μ, v_μ) *when* $\mu \ge \mu^*$, *and* (u_μ, v_μ) *is linearly stable in the sense that* $Re(\eta) > 0$ *if* η *is an eigenvalue of the linearized eigenvalue problem at* (u_μ, v_μ). *Moreover, when* $\mu \to \infty$, $u_\mu \to W_\lambda$ *uniformly in* $\overline{\Omega}$, *and* $v_\mu - \mu \to 0$ *uniformly in* $\overline{\Omega \setminus \Omega_0}$.
4. *There exists* $\mu_* > \mu^*$ *such that if* $\mu \ge \mu_*$, *and if* $(u(x,t), v(x,t))$ *is a solution of* (3.41), *then* $\lim_{t\to\infty} u(x,t) = u_\mu(x)$ *and* $\lim_{t\to\infty} v(x,t) = v_\mu(x)$ *uniformly for* $x \in \overline{\Omega}$ *and* $x \in \overline{\Omega \setminus \Omega_0}$, *respectively.*

The proof of the uniqueness of steady state is based on the linear stability analysis. For the linearized eigenvalue system, we use Kato's inequality ($-\Delta|\phi| \le -Re\left(\overline{\phi}|\phi|^{-1}\Delta\phi\right)$) to establish second order differential inequalities satisfied by any eigenfunction (ϕ, ψ) with large μ (thanks to the explicit asymptotic profile of the steady state), then we use energy estimates to conclude that $Re(\eta) > 0$ for any eigenvalue η, which implies the linear stability of the steady state. Again the uniqueness conclusion follows from the linear stability and a standard fixed point index argument. The proof of the global asymptotic stability follows the same line, but the estimates are more involved: We first obtain careful estimates of the L^2 norms of the solution $(u(x,t), v(x,t))$ for large time t, then we apply Gronwall type inequalities to show the global asymptotic stability.

Mathematically, it is usually difficult to prove not only the uniqueness but also the global asymptotical stability of the coexistence steady state for a predator-prey model. For reaction-diffusion systems without a proper order structure, rigorous proof of global asymptotical stability has been rarely achieved. For diffusive predator-prey models, we recall that global asymptotical stability of the constant coexistence state was obtained in [dMR, Le] for the classical Lotka-Volterra system with Neumann boundary conditions. For even just slightly more complicated systems, the uniqueness is only known when the spatial dimension is 1 ([LP], see also [DLO] for partial uniqueness results for the radial case in high dimensions). Even in this special case, the local stability is still unknown (Hopf bifurcation has not been ruled out).

In biological terms, the most significant feature of our study in this subsection is the existence of a critical patch size described by the principal eigenvalue $\lambda_1^D(\Omega_0)$ for the protection zone Ω_0. If the protection zone if above that size (*i.e.*, if $\lambda_1^D(\Omega_0)$ is less than λ, the prey growth rate), then the dynamics of the model is fundamentally changed from the usual predator-prey dynamics; in such a case, the prey population can survive regardless of the level of predation, and if the predator is strong, then the two populations stabilize at a unique coexistence state. If the protection zone is below the critical patch size, then the dynamics of the model is qualitatively similar to the usual case without protection zone, but the chances of survival of the prey species increase with the size of the protection zone, as generally expected.

The value of $\lambda_1^D(\Omega_0)$ depends on the size as well as the shape of Ω_0. The smaller the value of $\lambda_1^D(\Omega_0)$, the more protection Ω_0 provides to the prey species. For the interior protection zone case discussed here, if the volume of the zone is fixed, then it is well-known that a spherical protection zone has the smallest eigenvalue $\lambda_1^D(\Omega_0)$ from the classical Rayleigh-Faber-Krahn inequality (see [P]). Thus if a ball of the given size can be inscribed into Ω, then the optimal interior protection zone should be a ball. If fencing is needed to create the protection zone (for example, nets with suitable mesh sizes could be used if the prey has considerably smaller body size than its predator), then a ball shaped protection zone also uses the least fencing material as a ball has the least surface area among all regions of the same volume. This also suggests that having one big protection zone is usually better than having several protection zones that add to the same size (but this is not always so as the shape of the protection zones matters).

It is natural to have a boundary protection zone which is built along part or all the boundary of Ω. If Ω_0 is a ring shaped domain which has $\partial\Omega$ as its outer boundary, that is, $\partial\Omega \subset \partial\Omega_0$ and $\Gamma := \partial\Omega_0 \setminus \partial\Omega$ is nonempty and is contained in Ω, then the techniques and results here carry over easily. The critical patch size for this case is determined by $\lambda_1^M(\Omega_0) = \lambda$, where $\lambda_1^M(\Omega_0)$ denotes the principal eigenvalue of the problem

$$-\Delta\phi = \lambda\phi,\ x \in \Omega_0, \quad \partial_\nu\phi = 0,\ x \in \partial\Omega, \quad \phi = 0,\ x \in \Gamma.$$

From the variational characterization of eigenvalues, we have

$$\lambda_1^D(\Omega_0) > \lambda_1^M(\Omega_0) > \lambda_1^N(\Omega_0),$$

for any given region Ω_0. Therefore, a boundary protection zone is better than a same shaped interior protection zone. We should note, however, that there are more choices for the shape of interior zones than boundary zones.

A more natural boundary protection zone is one where $\partial\Omega_0$ splits into two parts Γ_1 and Γ_2, with $\Gamma_1 \subset \partial\Omega$, $int(\Gamma_2) \subset \Omega$, and $\Gamma_2 \cap \partial\Omega$ is an $(N-2)$-dimensional manifold. This is a mathematically challenging case, and great technical difficulties will be involved to extend our results here to this case. But we believe that similar results hold, and the critical patch size is determined by $\lambda_1^{M'}(\Omega_0) = \lambda$, where $\lambda_1^{M'}(\Omega_0)$ denotes the principal eigenvalue of the problem

$$-\Delta\phi = \lambda\phi,\ x \in \Omega_0, \quad \partial_\nu\phi = 0,\ x \in \Gamma_1, \quad \phi = 0,\ x \in \Gamma_2.$$

Again we have

$$\lambda_1^D(\Omega_0) > \lambda_1^{M'}(\Omega_0) > \lambda_1^N(\Omega_0).$$

If the boundary has a flat part, then a half ball H along the flat part of the boundary has the same principal eigenvalue $\lambda_1^{M'}(H)$ as that of a whole ball of the same radius in the interior. We conjecture that if the area is fixed, then the optimal protection zone is always achieved by a boundary one. This is apparently true for some special domains. A related optimization problem is considered in [KuS].

4 The Leslie model

In this section, we closely examine the diffusive Leslie model. The corresponding ODE model is given by (1.9), where all the coefficients are positive constants. It is known that (1.9) has simple dynamics: the unique positive equilibrium (u^*, v^*) attracts all the positive solutions as $t \to \infty$; in contrast, the slightly different Leslie-May-Tanner model (1.10) may have stable limiting cycles. See [Hz, HH1, HH2] for details.

We will consider the diffusive Leslie model with Neumann boundary conditions. Therefore in the case of homogeneous environment, the ODE solutions are also solutions to the diffusive model. We will employ a Lyapunov function technique to show that the homogeneous diffusive model has similar dynamics as the ODE model under some restriction of the parameters. We suspect that the restriction on the parameters is not necessary but are not able to remove it. In sharp contrast, we will show that when the environment is heterogeneous, the dynamics can be drastically changed.

4.1 The homogeneous case. If we replace u by u/d, a by αd and b by β, then the corresponding diffusive model of (1.9) with Neumann boundary conditions can be written in the following simpler form

$$\begin{cases} u_t - d_1\Delta u = u(\lambda - \alpha u - \beta v), & x \in \Omega, t > 0, \\ v_t - d_2\Delta v = \mu v(1 - \dfrac{v}{u}), & x \in \Omega, t > 0, \\ \partial_\nu u = \partial_\nu v = 0, & x \in \partial\Omega, t > 0, \end{cases} \tag{4.1}$$

where d_1, d_2, λ, μ, α, β are positive constants. Clearly,

$$(u^*, v^*) = (\frac{\lambda}{\alpha + \beta}, \frac{\lambda}{\alpha + \beta})$$

is the only constant positive equilibrium of (4.1).

Let $(u(x,t), v(x,t))$ be a positive solution of (4.1). A simple comparison argument yields $0 < u(x,t) < U(x,t)$ for all $t > 0$ and $x \in \Omega$, where U is the unique solution of

$$U_t - d_1\Delta U = \lambda U - \alpha U^2 \text{ in } \Omega \times (0,\infty),\ \partial U_\nu|_{\partial\Omega\times(0,\infty)} = 0,\ U(x,0) = u(x,0).$$

It is well known that $U(x,t) \to \lambda/\alpha$ as $t \to \infty$ uniformly in x. From these facts, it follows by standard comparison arguments that $u(x,t)$ and $v(x,t)$ exist and remain positive for all $t > 0$, and

$$\overline{\lim}_{t\to\infty} u(x,t) \le \lambda/\alpha,\ \overline{\lim}_{t\to\infty} v(x,t) \le \lambda/\alpha.$$

Adapting the Lyapunov function in [HH1], we define

$$V(u,v) = \int \frac{u - u^*}{u^2} du + c \int \frac{v - v^*}{v} dv,$$

$$W(t) = \int_\Omega V(u(x,t), v(x,t)) dx,$$

where $c > 0$ is a constant to be determined later, and $(u(x,t), v(x,t))$ is an arbitrary positive solution of (4.1).

Denote

$$f(u,v) = u(\lambda - \alpha u - \beta v), \; g(u,v) = \mu v(1 - v/u).$$

We have

$$\begin{aligned} & V_u(u,v)f(u,v) + V_v(u,v)g(u,v) \\ = \; & \frac{u-u^*}{u}(\lambda - \alpha u - \beta v) + c\mu(v - v^*)(1 - v/u) \\ = \; & \frac{u-u^*}{u}(\alpha u^* + \beta v^* - \alpha u - \beta v) + c\mu(v - v^*)\frac{u - u^* + v^* - v}{u} \\ = \; & -\alpha\frac{(u-u^*)^2}{u} + (c\mu - \beta)\frac{(u-u^*)(v-v^*)}{u} - c\mu\frac{(v-v^*)^2}{u}. \end{aligned}$$

We now choose $c = \beta/\mu$ and obtain

$$V_u(u,v)f(u,v) + V_v(u,v)g(u,v) = -\alpha\frac{(u-u^*)^2}{u} - \beta\frac{(v-v^*)^2}{u}.$$

It follows that

$$\begin{aligned} & W'(t) \\ = \; & \int_\Omega \big(V_u(u(x,t),v(x,t))u_t + V_v(u(x,t),v(x,t))v_t\big)dx \\ = \; & \int_\Omega \big(\frac{u-u^*}{u^2}d_1\Delta u + c\frac{v-v^*}{v}d_2\Delta v\big)dx - \int_\Omega \big(\alpha\frac{(u-u^*)^2}{u} + \beta\frac{(v-v^*)^2}{u}\big)dx \\ = \; & -\int_\Omega \big(d_1\frac{2u^*-u}{u^3}|\nabla u|^2 + d_2\frac{v^*}{v^2}|\nabla v|^2 + \alpha\frac{(u-u^*)^2}{u} + \beta\frac{(v-v^*)^2}{u}\big)dx. \end{aligned}$$

Suppose that $\alpha > \beta$. Then $2u^* = 2\lambda/(\alpha+\beta) > \lambda/\alpha$. Since $u(x,t) < U(x,t)$ and $U(x,t) \to \lambda/\alpha$ as $t \to \infty$, we can find $T > 0$ large such that $U(x,t) < 2\lambda/(\alpha+\beta)$ for all $t \geq T$ and all $x \in \Omega$. Therefore, $w'(t) \leq 0$ for all $t > T$ and the equality holds only if $(u,v) \equiv (u^*, v^*)$. Together with some standard arguments based on the boundedness of (u,v) and parabolic regularity, this proves the following result.

Proposition 4.1 *When $\alpha > \beta$, (u^*, v^*) attracts every positive solution of* (4.1).

The restriction $\alpha > \beta$ can be relaxed by replacing $V(u,v)$ by the following Lyapunov function

$$V^*(u,v) = \int \frac{u^2 - (u^*)^2}{u^2}du + c\int \frac{v-v^*}{v}dv,$$

with a suitable choice of $c > 0$, and the following result is proved in [DHs]:

Theorem 4.2 *There exists a constant $s_0 \in (1/5, 1/4)$ such that if $\alpha/\beta > s_0$, then (u^*, v^*) attracts every positive solution of* (4.1).

Remark 4.3 From the proof in [DHs], s_0 is the unique positive zero of the polynomial $h(s) = 32s^3 + 16s^2 - s - 1$. We conjecture that the conclusion of Theorem 4.2 is valid for all positive constants α and β. Like in [dMR], the above Lyapunov function technique collapses when the diffusive system fails to have a positive constant steady-state.

4.2 The heterogeneous case. As before, to simplify our presentation, we only consider the case that α in (4.1) is replaced by a positive continuous function $\alpha(x)$, while the other coefficients remain positive constants. However, with the loss of the Lyapunov function technique, we can say much less than in the homogeneous case about the global dynamical behavior, except that all the positive solutions remain bounded for all $t > 0$, which follows from a simple comparison argument. We will instead concentrate on the understanding of the positive steady-state solutions, namely positive solutions of the elliptic system

$$\begin{cases} -d_1\Delta u = \lambda u - \alpha(x)u^2 - \beta uv, & x \in \Omega, \\ -d_2\Delta v = \mu v(1 - \dfrac{v}{u}), & x \in \Omega, \\ \partial_\nu u = \partial_\nu v = 0, & x \in \partial\Omega. \end{cases} \tag{4.2}$$

By some simple change of scales, (4.2) can be reduced to the following simpler form:

$$\begin{cases} -\Delta u = \lambda u - \alpha(x)u^2 - \beta uv, & x \in \Omega, \\ -\Delta v = \mu v(1 - \dfrac{v}{u}), & x \in \Omega, \\ \partial_\nu u = \partial_\nu v = 0, & x \in \partial\Omega. \end{cases} \tag{4.3}$$

Here Ω is a bounded smooth domain in $\mathbf{R}^n$, λ, μ, β are positive constants, and $\alpha(x)$ is a continuous positive function over $\overline{\Omega}$.

Making use of a standard continuation and topological degree argument, it is shown in [DHs] that (4.3) always has a positive solution.

Theorem 4.4 *Suppose that λ, μ, β are positive constants, and $\alpha(x)$ is a continuous positive function over $\overline{\Omega}$. Then* (4.3) *has at least one positive solution.*

In the one spatial dimension case, that is, if Ω is a bounded interval in $\mathbf{R}^1$, then the techniques in [LP] can be easily adapted to show that any possible positive solution of (4.3) is non-degenerate, namely, the linearized problem of (4.3) at any positive solution (u, v),

$$\begin{cases} -\Delta\phi = \lambda\phi - 2\alpha(x)u\phi - \beta v\phi - \beta u\psi, & x \in \Omega, \\ -\Delta\psi = \mu(\psi - \dfrac{1}{u}\phi + \dfrac{v}{u^2}\phi), & x \in \Omega, \\ \partial_\nu\phi = \partial_\nu\psi = 0, & x \in \partial\Omega, \end{cases} \tag{4.4}$$

has only the zero solution $(\phi, \psi) = (0, 0)$. It follows from this fact and a topological degree argument that (4.3) has a unique positive solution in this one dimension case. We suspect that this is true in high dimension and the unique positive solution is globally attractive to all the positive solutions of the corresponding parabolic system, but cannot find a proof for this. In fact, even in the one dimension case where the uniqueness is known, we are not able to show that the unique positive solution of (4.3) is globally attractive for the corresponding parabolic system.

4.2.1 *The degenerate case.* In order to capture the influence of the heterogeneity of the spatial environment on the behavior of (4.3), we consider in this subsection a degenerate case, where $\alpha(x)$ is allowed to vanish on part of Ω. For simplicity, we suppose that $\alpha(x) = 0$ on $\overline{\Omega}_0 \subset \Omega$, where Ω_0 is a connected set with smooth boundary, and $\alpha(x) > 0$ in the rest of Ω. We will show a crucial difference in the behavior of (4.3) between the case $\lambda < \lambda_1^D(\Omega_0)$ and the case $\lambda > \lambda_1^D(\Omega_0)$. Here, as before, $\lambda_1^D(\Omega_0)$ denotes the principal eigenvalue of

$$-\Delta\phi = \lambda\phi,\ x \in \Omega_0, \quad \phi(x) = 0,\ x \in \partial\Omega_0.$$

Let us first observe that if $\lambda \in (0, \lambda_1^D(\Omega_0))$, then by the main result in [Ou] (see also [FKLM]), the problem

$$-\Delta u = \lambda u - \alpha(x)u^2,\ x \in \Omega, \quad \partial_\nu u = 0,\ x \in \partial\Omega,$$

still has a unique positive solution u_λ^*. Using this fact, one easily sees that the proof of Theorem 4.4 carries over to the present degenerate case. Therefore we have the following result.

Theorem 4.5 *Let $\alpha(x)$ be a continuous function as described above, $\lambda \in (0, \lambda_1^D(\Omega_0))$, $\mu > 0$ and $\beta > 0$. Then (4.3) has at least one positive solution.*

In sharp contrast, we will show in the following that if $\lambda > \lambda_1^D(\Omega_0)$, then (4.3) will no longer have a positive solution for certain $\mu > 0$ and $\beta > 0$.

Let us fix $\mu \in (0, \lambda_1^D(\Omega_0))$ and suppose $\lambda > \lambda_1^D(\Omega_0)$. By Lemma 2.6 in [DL], the boundary blow-up problem

$$-\Delta u = \lambda u - \alpha(x)u^2,\ x \in \Omega\backslash\overline{\Omega_0}, \quad \partial_\nu u = 0,\ x \in \partial\Omega, \quad u = \infty,\ x \in \partial\Omega_0, \tag{4.5}$$

has a minimal positive solution U_λ. Applying Lemma 2.3 in [DL], we find that if (u, v) is a positive solution of (4.3), then

$$u(x) \le U_\lambda(x), \quad x \in \Omega \setminus \overline{\Omega}_0.$$

Define

$$\alpha_\lambda(x) = \begin{cases} 0, & x \in \overline{\Omega}_0, \\ 1/U_\lambda(x), & x \in \Omega \setminus \overline{\Omega}_0. \end{cases}$$

Clearly α_λ is continuous on $\overline{\Omega}$ and $\alpha_\lambda > 0$ on $\Omega \setminus \overline{\Omega}_0$.

By our choice of μ and the main result of [Ou], the problem

$$-\Delta V = \mu V(1 - \alpha_\lambda(x)V),\ x \in \Omega, \quad \partial_\nu V = 0,\ x \in \partial\Omega, \tag{4.6}$$

has a unique positive solution V_λ.

We want to show that $v \le V_\lambda$ if (u, v) is a positive solution of (4.3). Indeed, we already know that $u(x) \le U_\lambda(x)$ on $\Omega \setminus \overline{\Omega}_0$. Hence

$$1/u(x) \ge \alpha_\lambda(x), \quad x \in \Omega.$$

It follows that

$$-\Delta v = \mu v(1 - v/u) \le \mu v(1 - \alpha_\lambda(x)v), \quad x \in \Omega.$$

Thus, v is a lower solution of (4.6). It is easily checked that for any constant $M > 1$, MV_λ is an upper solution of (4.6), and $MV_\lambda > v$ if M is large enough. Therefore, $v \le V_\lambda \le MV_\lambda$ in Ω.

We recall that we use $\lambda_1^N(\phi, \omega)$ to denote the first eigenvalue of the operator $-\Delta + \phi$ over ω under Neumann boundary conditions, and $\lambda_1^D(\phi, \omega)$ to denote the first eigenvalue under Dirichlet boundary conditions.

If (u, v) is a positive solution of (4.3), then from the equation for u we obtain

$$\lambda = \lambda_1^N(\alpha u + \beta v, \Omega) < \lambda_1^D(\alpha u + \beta v, \Omega) < \lambda_1^D(\alpha u + \beta v, \Omega_0) = \lambda_1^D(\beta v, \Omega_0).$$

Since $v \le V_\lambda$, we obtain

$$\lambda < \lambda_1^D(\beta V_\lambda, \Omega_0). \tag{4.7}$$

By the properties of the principle eigenvalue, we see that $f(\beta) \equiv \lambda_1^D(\beta V_\lambda, \Omega_0)$ is a continuous, strictly increasing function of β, and $f(0) = \lambda_1^D(\Omega_0)$, $f(\infty) = \infty$.

Since $\lambda > \lambda_1^D(\Omega_0)$, we can find a unique $\beta_0 = \beta_0(\lambda) > 0$ such that $f(\beta_0) = \lambda$. Therefore, we have

$$\lambda = \lambda_1^D(\beta_0 V_\lambda, \Omega_0), \quad \text{and } \lambda \geq \lambda_1^D(\beta V_\lambda, \Omega_0),\ \forall \beta \leq \beta_0. \tag{4.8}$$

Comparing (4.7) with (4.8), we immediately obtain the following result.

Theorem 4.6 *Suppose $\mu \in (0, \lambda_1^D(\Omega_0))$ and $\lambda \in (\lambda_1^D(\Omega_0), \infty)$. Let β_0 be as in* (4.8). *Then* (4.3) *has no positive solution if $0 < \beta \leq \beta_0$.*

Our next result shows that for the case $\lambda > \lambda_1^D(\Omega_0)$, (4.3) can still have a positive solution for every $\beta > 0$ if μ is large enough; precisely, if $\mu > \max\{\lambda_1^D(\Omega_0), \lambda\}$. Thus, existence of a positive solution is regained when μ becomes large.

Theorem 4.7 *Suppose that $\mu > \lambda_1^D(\Omega_0)$, then* (4.3) *has a positive solution for every $\lambda \in (0, \mu)$ and $\beta > 0$.*

We refer to [DHs] for the proof of Theorem 4.7, which uses a degree argument, and a key step is to obtain a priori bounds for all possible solutions of (4.3).

Remark 4.8 The existence problem of (4.3) in the degenerate case is not completely understood. For example, we do not know whether (4.3) has a positive solution for every $\beta > 0$ when $\lambda \geq \mu > \lambda_1^D(\Omega_0)$.

4.2.2 *Perturbation and sharp spatial patterns.* Suppose that $\alpha(x)$ is as in the last subsection, that is, it is continuous in $\overline{\Omega}$, vanishes on $\overline{\Omega}_0 \subset \Omega$, and is positive elsewhere. We now fix $\mu \in (0, \lambda_1^D(\Omega_0))$, $\lambda > \lambda_1^D(\Omega_0)$ and $\beta \in (0, \beta_0)$, where β_0 is determined by (4.8). For $\epsilon > 0$ we consider the following perturbation of (4.3):

$$\begin{cases} -\Delta u = \lambda u - [\alpha(x) + \epsilon]u^2 - \beta uv, & x \in \Omega, \\ -\Delta v = \mu v(1 - \dfrac{v}{u}), & x \in \Omega, \\ \partial_\nu u = \partial_\nu v = 0, & x \in \partial\Omega. \end{cases} \tag{4.9}$$

By Theorem 4.4, (4.9) always has a positive solution. Denote by (u_ϵ, v_ϵ) an arbitrary positive solution of (4.9). Since (4.9) does not have a positive solution when $\epsilon = 0$ (Theorem 4.6), it is interesting to see how (u_ϵ, v_ϵ) evolves as ϵ shrinks to 0. We will show that as $\epsilon \to 0$, (u_ϵ, v_ϵ) exhibits a sharp spatial pattern determined by the inhomogeneity of $\alpha(x)$. To this end, let $\{\epsilon_n\}$ be an arbitrary sequence of positive numbers decreasing to 0 as $n \to \infty$, and denote $(u_n, v_n) = (u_{\epsilon_n}, v_{\epsilon_n})$.

We have the following result.

Theorem 4.9 *$\{(u_n, v_n)\}$ has a subsequence, still denoted by (u_n, v_n), such that*

(i) *$u_n \to \infty$ uniformly on $\overline{\Omega}_0$,*
(ii) *$u_n \to \tilde{u}$ in $C^1(\overline{\omega})$ for any subdomain ω satisfying $\overline{\omega} \subset \overline{\Omega} \setminus \overline{\Omega}_0$,*
(iii) *$v_n \to \tilde{v}$ in $C^1(\overline{\Omega})$,*

where $(\tilde{u}, \tilde{v})$ is a positive solution to the problem

$$\begin{cases} -\Delta \tilde{u} = \lambda \tilde{u} - \alpha(x)\tilde{u}^2 - \beta \tilde{u}\tilde{v}, \quad x \in \Omega \setminus \overline{\Omega}_0, \\ -\Delta \tilde{v} = \mu \tilde{v}(1 - \dfrac{\tilde{v}}{\tilde{u}}), \quad x \in \Omega, \\ \tilde{u} = \infty,\ x \in \partial\Omega_0, \quad \partial_\nu \tilde{u} = 0,\ x \in \partial\Omega, \quad \partial_\nu \tilde{v} = 0,\ x \in \partial\Omega, \end{cases} \tag{4.10}$$

where we use the convention that $1/\tilde{u} = 0$ in $\overline{\Omega}_0$.

From Theorem 4.9 we see that for large n, u_n exhibits a clear spatial pattern: its value is big over Ω_0, but stays bounded away from $\overline{\Omega}_0$. The spatial pattern of u_n can be better observed through a rescaling.

Theorem 4.10 *Suppose that (u_n, v_n) converges to $(\tilde{u}, \tilde{v})$ as in Theorem 4.9. Then $\epsilon_n u_n \to w$ in $C(\overline{\Omega})$, where $w = 0$ on $\overline{\Omega} \setminus \Omega_0$, and on Ω_0, w is the unique positive solution of*

$$-\Delta w = \lambda w - w^2 - \beta \tilde{v} w, \; x \in \Omega_0, \quad w = 0, \; x \in \partial\Omega_0.$$

We refer to [DHs] for the proof of Theorems 4.9 and 4.10, where a more general situation was considered.

Remark 4.11 In the statement of Theorem 3.16 in [DHs], equation (3.32) should be deleted. In fact, (3.32) is never satisfied. Since $\lambda > \lambda_1^{D_j}(\beta V_\lambda)$, we always have $\lambda > \lambda_1^{D_j}(\beta \tilde{v})$. Theorem 3.16 remains valid when (3.32) is deleted.

Remark 4.12 (i) If $\lambda \in (0, \lambda_1^D(\Omega_0))$, then it is easy to show that, for small $\epsilon > 0$, any positive solution (u_ϵ, v_ϵ) of (4.9) is close to a positive solution (u_0, v_0) of (4.3). Therefore, (u_ϵ, v_ϵ) does not develop a sharp spatial pattern as $\epsilon \to 0$.

(ii) Likewise, if we perturb (2.7) by replacing $b(x)$ with $b(x)+\epsilon$, then the positive solution (u_ϵ, v_ϵ) of the perturbed equation converges to a positive solution of the original unperturbed problem, so no sharp spatial patterns for the solutions of (2.7) can be observed through this perturbation.

(iii) The above point, when compared with Theorems 4.9 and 4.10, reveals a significant difference between the diffusive Lotka-Volterra predator-prey model and the diffusive Leslie model. A key observation is that when a degeneracy occurs, the range of parameters where a positive solution exists for (2.7) is **enlarged** from the non-degenerate case, while such a range for (4.3) is **reduced** when degeneracy occurs. It is interesting to note that this difference between (2.7) and (4.3) can only be observed in the heterogeneous case.

By Theorem 4.7, (4.3) has a positive solution for every $\beta > 0$ if $\mu > \lambda \geq \lambda_1^D(\Omega_0)$. In [DW], to better understand the effect of the degeneracy, the asymptotic behavior of the positive solutions of (4.3) was examined in each of the following cases:
(i) $\beta \to 0$, (ii) $\beta \to \infty$, (iii) $\mu \to \infty$,
and comparison was made with the perturbed system (4.9). We refer to [DW] for the proofs of the following results.

Theorem 4.13 (Limiting behavior as $\beta \to 0^+$) *Suppose that $\mu > \lambda \geq \lambda_1^D(\Omega_0)$ and $\beta > 0$. Let (u_β, v_β) be a positive solution of (4.3). Then we have the following conclusions:*

(a) $\lim_{\beta\to 0^+}(u_\beta(x), v_\beta(x)) = (\infty, \infty)$ *uniformly on $\bar{\Omega}_0$;*

(b) *Along any sequence of β decreasing to 0, there is a subsequence $\{\beta_n\}$ such that,*

$$\lim_{n\to\infty} (u_{\beta_n}, v_{\beta_n}) = (u, v) \textit{ uniformly on any compact subset of } \bar{\Omega} \setminus \bar{\Omega}_0,$$

where (u, v) is a positive solution of the following system

$$\begin{cases} -\Delta u = \lambda u - a(x)u^2, & x \in \Omega \setminus \bar{\Omega}_0, \\ -\Delta v = \mu v\left(1 - \dfrac{v}{u}\right), & x \in \Omega \setminus \bar{\Omega}_0, \\ \partial_\nu u = \partial_\nu v = 0, \quad x \in \partial\Omega, & u = v = \infty, \quad x \in \partial\Omega_0. \end{cases} \tag{4.11}$$

(c) *If there exists $\xi > 0$ such that $a(x)d(x, \Omega_0)^{-\xi}$ is bounded for all $x \in \Omega \setminus \overline{\Omega}_0$ close to $\partial\Omega_0$, then (4.11) has a unique positive solution (u, v) and the convergence in part* (b) *holds for $\beta \to 0^+$.*

(d) *If (u_β, v_β) is a positive solution of (4.9), then* $\lim_{\beta\to 0^+}(u_\beta(x), v_\beta(x)) = (U_\epsilon, V_\epsilon)$ *uniformly over $\overline{\Omega}$, where U_ϵ and V_ϵ are the unique positive solutions to*

$$-\Delta U = \lambda U - [a(x) + \epsilon]U^2, \quad \partial_\nu U = 0, \ x \in \partial\Omega,$$

and

$$-\Delta V = \mu V\left(1 - \frac{V}{U_\epsilon}\right), \quad \partial_\nu V = 0, \ x \in \partial\Omega$$

respectively.

Theorem 4.14 (Limiting behavior as $\beta \to +\infty$) *Suppose that $\mu > \lambda \geq \lambda_1^D(\Omega_0)$ and $\beta > 0$. Let (u_β, v_β) be a positive solution of* (4.3). *Then* $\lim_{\beta\to\infty}(u_\beta, v_\beta) = (0, 0)$ *uniformly on $\bar{\Omega}$. The same holds for positive solutions of (4.9).*

Theorem 4.15 (Limiting behavior as $\mu \to +\infty$) *Suppose that $\mu > \lambda \geq \lambda_1^D(\Omega_0)$ and $\beta > 0$. Let (u_μ, v_μ) be a positive solution of* (4.3). *Then $u_\mu \to w$ in $C^1(\bar{\Omega})$ and $v_\mu \to w$ uniformly on any compact subset of Ω, where w is the unique positive solution of*

$$-\Delta w = \lambda w - [a(x) + \beta]w^2, \ x \in \Omega, \quad \partial_\nu w = 0, \ x \in \partial\Omega. \tag{4.12}$$

A similar conclusion holds for the positive solutions of (4.9) except that the limiting function is the unique positive solution of

$$-\Delta w = \lambda w - [a(x) + \epsilon + \beta]w^2, \ x \in \Omega, \quad \partial_\nu w = 0, \ x \in \partial\Omega. \tag{4.13}$$

Clearly Theorem 4.13 reveals a striking different between (4.3) and (4.9): The positive solutions of (4.3) exhibit a sharp spatial pattern for small β while those of (4.9) do not. On the other hand, the cases considered in Theorems 4.14 and 4.15 do not seem to reveal any significant difference between (4.3) and (4.9).

Remark 4.16 If $0 < \lambda < \lambda_1^D(\Omega_0)$, then it follows from Theorem 4.5 that (4.3) has a positive solution for every $\mu > 0$ and $\beta > 0$. Using the techniques in [DW], it can be shown that the corresponding limiting behaviors of (4.3) and (4.9) are the same in each of the three cases: $\beta \to 0^+$, $\beta \to \infty$, and $\mu \to \infty$.

References

[A] Alikakos, Nicholas D., A Liapunov functional for a class of reaction-diffusion systems. Modeling and differential equations in biology, pp. 153–170, *Lecture Notes in Pure and Appl. Math.*, **58**, Dekker, New York, 1980.

[Ba] Bazykin, A.D., Sistema Volterra i uravnenie Mihaelisa-Menton. *Voprosy matematicheskoy genetiki*, Nauka, Novosibirsk, Russia, 103–143, 1974.

[BL] Berestycki, H.; Lions, P.-L., Some applications of the method of super and subsolutions. *Lecture Notes in Math.*, **782**, Springer, Berlin, 1980, pp. 16–41.

[BB1] Blat, J.; Brown, K.J. Bifurcation of steady-state solutions in predator-prey and competition systems. *Proc. Roy. Soc. Edinburgh Sect. A*, **97** (1984), 21–34.

[BB2] Blat, J.; Brown, K.J., Global bifurcation of positive solutions in some systems of elliptic equations. *SIAM J. Math. Anal.*, **17** (1986), 1339–1353.

[BC] Brauer, Fred; Castillo-Chávez, Carlos, Mathematical models in population biology and epidemiology. *Texts in Applied Mathematics*, **40**. Springer-Verlag, New York, 2001.

[Br] Brown, Peter N. Decay to uniform states in ecological interactions. *SIAM J. Appl. Math.* **38** (1980), no. 1, 22–37.

[CC1] Cantrell, Robert Stephen; Cosner, Chris, Diffusive logistic equations with indefinite weights: population models in disrupted environments. *Proc. Roy. Soc. Edinburgh Sect. A* **112** (1989), no. 3-4, 293–318.

[CC2] Cantrell, Robert Stephen; Cosner, Chris, Spatial ecology via reaction-diffusion equation. *Wiley series in mathematical and computational biology*, John Wiley & Sons Ltd, 2003.

[CCH] Cantrell, Robert Stephen; Cosner, Chris; Hutson, Vivian, Permanence in ecological systems with spatial heterogeneity. *Proc. Roy. Soc. Edinburgh Sect. A* **123** (1993), no. 3, 533–559.

[CEL] Casal, A.; Eilbeck, J. C.; López-Gómez, J. Existence and uniqueness of coexistence states for a predator-prey model with diffusion. *Differential Integral Equations* **7** (1994), no. 2, 411–439.

[CH] Casten, Richard G.; Holland, Charles J., Instability results for reaction diffusion equations with Neumann boundary conditions. *J. Differential Equations* **27** (1978), no. 2, 266–273.

[CCS] Chueh, K. N.; Conley, C. C.; Smoller, J. A. Positively invariant regions for systems of nonlinear diffusion equations. *Indiana Univ. Math. J.* **26** (1977), no. 2, 373–392.

[Co1] Conway, E. D., Diffusion and predator-prey interaction: Steady states with flux at the boundaries. *Contemporary Mathematics*, **17** (1983), 217–234.

[Co2] Conway, E. D., Diffusion and predator-prey interaction: pattern in closed systems. Partial differential equations and dynamical systems, 85–133, *Res. Notes in Math.*, 101, Pitman, Boston, Mass.-London, 1984.

[CGS] Conway, E.; Gardner, R.; Smoller, J., Stability and bifurcation of steady-state solutions for predator-prey equations. *Adv. in Appl. Math.* **3** (1982), no. 3, 288–334.

[CHS] Conway, E.; Hoff, D.; Smoller, J., Large time behavior of solutions of systems of nonlinear reaction-diffusion equations. *SIAM J. Appl. Math.* **35** (1978), no. 1, 1–16.

[Da1] Dancer, E. N., On positive solutions of some pairs of differential equations. *Trans. Amer. Math. Soc.* **284** (1984), no. 2, 729–743.

[Da2] Dancer, E.N., On positive solutions of some pairs of differential equations, II. *J. Diff. Eqns.* **60** (1985), 236-258.

[Da3] Dancer, E. N., On the existence and uniqueness of positive solutions for competing species models with diffusion. *Trans. Amer. Math. Soc.***326** (1991), no. 2, 829–859.

[Da4] Dancer, E. N. On uniqueness and stability for solutions of singularly perturbed predator-prey type equations with diffusion. *J. Differential Equations* **102** (1993), no. 1, 1–32.

[DD1] Dancer, E. N.; Du, Yihong, Competing species equations with diffusion, large interactions, and jumping nonlinearities. *J. Differential Equations* **114** (1994), 434-475.

[DD2] Dancer, E. N.; Du, Yihong, Effects of certain degeneracies in the predator-prey model. *SIAM J. Math. Anal.* **34** (2002), no. 2, 292–314.

[DD3] Dancer, E.N.; Du, Yihong, On a free boundary problem arising from population biology, *Indiana Univ. Math. J.* **52** (2003), 51-68.

[DLO] Dancer, E. N.; López-Gómez, J.; Ortega, R., On the spectrum of some linear noncooperative elliptic systems with radial symmetry. *Differential Integral Equations* **8** (1995), no. 3, 515–523.

[dP] del Pino, M.A., Positive solutions of a semilinear elliptic equation on a compact manifold, *Nonl. Anal. TMA.* **22**(1994), 1423-1430.

[dMR] deMotoni, P.; Rothe, F., Convergence to homogeneous equilibrium state for generalized Volterra-Lotka systems. *SIAM J. Appl. Math.* **37** (1979), 648-663.

[D1] Du, Yihong, Effects of a degeneracy in the competition model. I. Classical and generalized steady-state solutions. *J. Differential Equations* **181** (2002), no. 1, 92–132.

[D2] Du, Yihong, Effects of a degeneracy in the competition model. II. Perturbation and dynamical behaviour. *J. Differential Equations* **181** (2002), no. 1, 133–164.

[D3] Du, Yihong, Realization of prescribed patterns in the competition model. *J. Differential Equations* **193** (2003), no. 1, 147–179.

[D4] Du, Yihong, Spatial patterns for population models in a heterogeneous environment. *Taiwanese J. Math.* **8** (2004), no. 2, 155–182.

[D5] Du, Yihong, Bifurcation and related topics in elliptic problems. Handbook of Partial Diff. Eqns., Vol. 2, Edited by M. Chipot and P. Quittner, Elsevier, 2005, 127–209.

[D6] Du, Yihong, Asymptotic behavior and uniqueness results for boundary blow-up solutions. *Diff. Integral Eqns.*, **17**(2004), 819-834.

[DG] Du, Yihong; Guo, Zongming, The degenerate logistic model and a singularly mixed boundary blow-up problem. *Discrete and Continuous Dynamical Systems* **14** (2006), no. 1, 1-29.

[DHs] Du, Yihong; Hsu, Sze-Bi, A diffusive predator-prey model in heterogeneous environment. *J. Differential Equations* **203** (2004), no. 2, 331–364.

[DHu] Du, Yihong; Huang, Qingguang, Blow-up solutions for a class of semilinear elliptic and parabolic equations. *SIAM J. Math. Anal.* **31** (1999), no. 1, 1–18.

[DL] Du, Yihong; Li, Shujie, Positive solutions with prescribed patterns in some simple semilinear equations, *Diff. Integral Eqns.***15** (2002), 805–822.

[DL1] Du, Yihong; Lou, Yuan, Some uniqueness and exact multiplicity results for a predator-prey model. *Trans. Amer. Math. Soc.* **349** (1997), no. 6, 2443–2475.

[DL2] Du, Yihong; Lou, Yuan, S-shaped global bifurcation curve and Hopf bifurcation of positive solutions to a predator-prey model. *J. Differential Equations* **144** (1998), no. 2, 390–440.

[DL3] Du, Yihong; Lou, Yuan, Qualitative behaviour of positive solutions of a predator-prey model: effects of saturation. *Proc. Roy. Soc. Edinburgh Sect. A* **131** (2001), no. 2, 321–349.

[DM] Du, Yihong; Ma, Li, Logistic type equations on $\mathbf{R}^N$ by a squeezing method involving boundary blow-up solutions. *J. London Math. Soc.* **64** (2001), no. 1, 107–124.

[DS1] Du, Yihong; Shi, Junping, Allee effect and bistability in a spatially heterogeneous predator-prey model. Submitted, (2005).

[DS2] Du, Yihong; Shi, Junping, A diffusive predator-prey model with a protect zone. Submitted, (2005).

[DW] Du, Yihong; Wang, Mingxin, Asymptotic behavior of positive steady-states to a predator-prey model, *Proc. Roy. Soc. Edinburgh Sect. A*, to appear.

[Fis] Fisher, R.A., The wave of advance of advantageous genes. *Ann. Eugenics*, **7**, (1937), 353–369.

[FKLM] Fraile, J.M.; Koch Medina, P.; López-Gómez, J.; Merino, S., Elliptic eigenvalue problems and unbounded continua of positive solutions of a semilinear elliptic problem. *J. Diff. Eqns.* **127** (1996), 295-319.

[Hz] Hainzl, J., Multiparameter bifurcation of a predator-prey system, *SIAM J. Math. Anal.***23**(1992), 150-180.

[He] Hess, Peter, Periodic -Parabolic Boundary Value Problems and Positivity, Longman Scientific and Technical, Pitman *Research Notes in mathemaics Series* **247**, Harlow, Essex, 1991.

[HS] Hofbauer, Josef; Sigmund, Karl, Evolutionary games and population dynamics. Cambridge University Press, Cambridge, 1998.

[H1] Holling, C. S., Some characteristics of simple types of predation and parasitism. *Canadian Entomologist* **91** (1959), 385–398.

[H2] Holling, C. S., The functional response of predators to prey density and its role in mimicry and population regulation. *Mem. Entom. Soc. Can.* **45** (1965), 1–60.

[HH1] Hsu S.B.; Hwang,T.W., Global stability for a class of predator-prey systems, *SIAM J. Appl. Math.* **55** (1995), 763–783.

[HH2] Hsu S.B.; Hwang,T.W., Uniqueness of limit cycles for a predator-prey system of Holling and Leslie type, *Canad. Appl. Math. Quart.* **6** (1998), no. 2, 91–117.

[Hu] Huffaker, C.B., Exerimental studies on predation: Despersion factors and predator-prey oscilasions, *Hilgardia* **27** (1958), 343–383.

[HLM1] Hutson, V; Lou, Y.; Mischaikow, K, Spatial heterogeneity of resources versus Lotka-Volterra dynamics, *J. Diff. Eqns.* **185** (2002), 97-136.

[HLM2] Hutson, V.; Lou, Y.; Mischaikow, K., Convergence in competition models with small diffusion coefficients, *J. Differential Equations* **211** (2005), no. 1, 135–161.

[HLMP] Hutson, V; Lou, Y; Mischaikow, K; Polacik, P, Competing species near a degenerate limit, *SIAM J. Math. Anal.* **35** (2003), 453-491.

[HMP] Hutson, V.; Mischaikow, K.; Polacik, P., The evolution of dispersal rates in a heterogeneous time-periodic environment. *J. Math. Biol.* **43** (2001), 501-533.

[KS] Kierstead, H. ; Slobodkin, L.B., The size of water masses containing plankton bloom. *J. Mar. Res.* **12**, (1953), 141–147.

[KW] Kishimoto, Kazuo; Weinberger, Hans F. The spatial homogeneity of stable equilibria of some reaction-diffusion systems on convex domains. *J. Differential Equations* **58** (1985), no. 1, 15–21.

[KPP] Kolmogoroff, A., Petrovsky, I, Piscounoff, N, Study of the diffusion equation with growth of the quantity of matter and its application to a biological problem. (French) *Moscow Univ. Bull. Math.* **1**, (1937), 1–25.

[KL] Koman, P; Leung, A.W., A general monotone scheme for elliptic systems with applications to ecological models. *Proc. Roy. Soc. Edin. A* **102** (1986), 315-325.

[KuS] Kurata, Kazuhiro; Shi, Junping, Optimal spatial harvesting strategy and symmetry-breaking. Submitted, (2005).

[KY] Kuto, Kousuke; Yamada, Yoshio, Multiple coexistence states for a prey-predator system with cross-diffusion. *J. Differential Equations* **197** (2004), no. 2, 315–348.

[LM] Lazer, A. C.; McKenna, P. J. On steady state solutions of a system of reaction-diffusion equations from biology. *Nonlinear Anal.* **6** (1982), no. 6, 523–530.

[Le] Leung, Anthony, Limiting behaviour for a prey-predator model with diffusion and crowding effects. *J. Math. Biol.* **6** (1978), no. 1, 87–93.

[L] Levin, Simon A., Population models and community strcture in heterogeneous environments. Studies in mathematical biology. Part II. Populations and communities. pp439–476. Edited by Simon A. Levin. *MAA Studies in Mathematics*, **16**. Mathematical Association of America, Washington, D.C., 1978.

[Li1] Li, Lige, Coexistence theorems of steady states for predator-prey interacting systems. *Trans. Amer. Math. Soc.* **305** (1988), no. 1, 143–166.

[Li2] Li, Lige, On the uniqueness and ordering of steady states of predator-prey systems. *Proc. Roy. Soc. Edinburgh Sect. A* **110** (1988), no. 3-4, 295–303.

[Lo] López-Gómez, J., On the structure of the permanence region for competing species models with general diffusivities and transport effects. *Discrete and Continuous Dynamical Systems* **2** (1996), 525–542.

[LP] López-Gómez, J.; Pardo, R.M., Invertibility of linear noncooperative elliptic systems, *Nonlinear Anal.* **31** (1998), 687–699.

[LS] López-Gómez, J.; and Sabina de Lis, J., Coexistence states and global attractivity for some convective diffusive competing species models, *Trans. Amer. Math. Soc.* **347**(1995), 3797-3833.

[Lo1] Lotka, A. J., Analytical note on certain rhythmic relations in organic systems. *Proc. Natl. Acad. Sci.* **6**, (1920), 410–415.

[Lo2] Lotka, A. J., Updated oscillations derived from the law of mass action. *J. Am. Chem. Soc.* **42**, (1920), 1595–1599.

[Lo3] Lotka, A. J., Elements of Physical Biology. Baltimore: Williams & Wilkins Co., 1925.

[LN] Lou, Yuan; Ni, Wei-Ming, Diffusion vs cross-diffusion: an elliptic approach. *J. Differential Equations* **154** (1999), no. 1, 157–190.

[Ma] Matano, Hiroshi, Asymptotic behavior and stability of solutions of semilinear diffusion equations. *Publ. Res. Inst. Math. Sci.* **15** (1979), no. 2, 401–454.

[MaM] Matano, H.; Mimura, M., Pattern formation in competition-diffusion systems in nonconvex domains, *Publ. Res. Inst. Math. Sci.* **19** (1983), 1049-1079.

[MPT] Medvinsky, Alexander B.; Petrovskii, Sergei V.; Tikhonova, Irene A.; Malchow, Horst; Li, Bai-Lian, Spatiotemporal complexity of plankton and fish dynamics. *SIAM Rev.* **44** (2002), no. 3, 311–370.

[Mi] Mimura, Masayasu Asymptotic behaviors of a parabolic system related to a planktonic prey and predator model. *SIAM J. Appl. Math.* **37** (1979), no. 3, 499–512.

[MiM] Mimura, M.; Murray, J. D., On a diffusive prey-predator model which exhibits patchiness. *J. Theoret. Biol.* **75** (1978), no. 3, 249–262.

[MNY] Mimura, Masayasu; Nishiura, Yasumasa; Yamaguti, Masaya, Some diffusive prey and predator systems and their bifurcation problems. Bifurcation theory and applications in scientific disciplines (Papers, Conf., New York, 1977), pp. 490–510, *Ann. New York Acad. Sci.*, **316**, New York Acad. Sci., New York, 1979.

[MBN] Murdoch, William W.; Briggs, Cheryl J.; Nisbert, Roger, M., Consumer-resource dynamics. *Monographs in population biology*, **36**, Edited by Simon A. Levin and Henry S. Horn, Princeton University Press, 2003.

[Mu1] Murray, J. D. Non-existence of wave solutions for the class of reaction-diffusion equations given by the Volterra interacting-population equations with diffusion. *J. Theoret. Biol.* **52** (1975), no. 2, 459–469.

[Mu2] Murray, J. D. Mathematical biology. Third edition. I. An introduction. *Interdisciplinary Applied Mathematics*, **17**. Springer-Verlag, New York, 2002; II. Spatial models and biomedical applications. *Interdisciplinary Applied Mathematics*, **18**. Springer-Verlag, New York, 2003.

[OL] Okubo, Akira; Levin, Simon, Diffusion and ecological problems: modern perspectives. Second edition. *Interdisciplinary Applied Mathematics*, **14**. Springer-Verlag, New York, 2001.

[Ou] Ouyang, Tiancheng, On the positive solutions of semilinear equations $\Delta u + \lambda u - hu^p = 0$ on the compact manifolds. *Trans. Amer. Math. Soc.* **331** (1992), no. 2, 503–527.

[PaW] Pang, P.Y.H.; Wang, Mingxin, Qualitative analysis of a ratio-dependent predator–prey system with diffusion. *Proc. Roy. Soc. Edinburgh Sect. A* **133** (2003), 919–942.

[Pao] Pao, C.V., On nonlinear reaction-diffusion systems, *J. Math. Anal. Appl.* **87** (1982), 165-198.

[P] Payne, L. E., Isoperimetric inequalities and their applications. *SIAM Rev.* **9** (1967) 453–488.

[PeW] Peng, Rui; Wang, Mingxin, Positive steady states of the Holling-Tanner prey-predator model with diffusion. *Proc. Roy. Soc. Edinburgh Sect. A* **135** (2005), no. 1, 149–164.

[RM] Rosenzweig, M.L.; MacArthur, R., Graphical representation and stability conditions of predator-prey interactions. *Amer. Natur.* **97** (1963), 209-223.

[SJ] Segel, L.A.; Jackson, J.L., Dissipative structure: An explaination and an ecological example. *J. Theor. Biol.*, **37**, (1975), 545–559.

[SL] Segel, Lee A.; Levin, Simon A. Application of nonlinear stability theory to the study of the effects of diffusion on predator-prey interactions. *AIP Conf. Proc.*, **27**, Amer. Inst. Phys., New York, (1976), 123–156.

[SS] Shi, Junping; Shivaji, Ratnasingham, Persistence in reaction diffusion models with weak Allee effect. Submitted, (2005).

[SY] Shi, Junping; Yao, Miaoxin, On a singular nonlinear semilinear elliptic problem. *Proc. Royal Soc. Edin. A*, **128**, (1998), no. 6, 1389–1401.

[Sk] Skellam, J. G., Random dispersal in theoritical populations. *Biometrika* **38**, (1951), 196–218.

[So] Solomon, M. E., The natural control of animal populations. *Journal of Animal Ecology* **18**, (1949), 1-35.

[St1] Steele, J. H., Spatial heterogeneity and population stability. *Nature*, **83**, (1974), 248.

[St2] Steele, J. H., Stability of plankton ecosystems. Ecological Stability, M.B. Usher and M.H. Williamson, eds., Chapman and Hall, London, 1974, 179–191.

[Tu] Turing, A., The chemical basis of morphogenesis. *Phil. Trans. Royal Soc. Lond* **B237**, (1952), 37–72.

[Tur] Turchin, Peter, Complex Population Dynamics: A Theoretical/Empirical Synthesis. *Monographs in Population Biology* **35**, Edited by Simon A. Levin and Henry S. Horn, Princeton University Press, 2003.

[V1] Volterra, V., Variazioni e fluttuazioni del numero d'individui in specie animali conviventi. *Mem. R. Accad. Naz. dei Lincei. Ser. VI*, **2**, (1926), 31–113. (Variations and fluctuations of the number of individuals in animal species living together. Translation in: R.N. Chapman: Animal Ecology. New York: McGraw Hill, 1931, 409–448.)

[V2] Volterra, V. Fluctuations in the abundance of a species considered mathematically. *Nature*, **118**, (1926), 558–560.

[W] Weinberger, Hans F. Invariant sets for weakly coupled parabolic and elliptic systems. *Rend. Mat. (6)* **8** (1975), 295–310.

Fields Institute Communications
Volume **48**, 2006

Delayed Non-local Diffusive Systems in Biological Invasion and Disease Spread

S. A. Gourley
Department of Mathematics and Statistics
University of Surrey, Guildford
Surrey GU2 7XH, England
S.Gourley@surrey.ac.uk

J. Wu
Laboratory for Industrial and Applied Mathematics
Department of Mathematics and Statistics
York University
Toronto, Ontario, Canada, M3J 1P3
wujh@mathstat.yorku.ca

Abstract. We survey some recent progress towards modeling and analysis of long-term behaviors of biological and epidemiological systems for which individuals move randomly and the feedback nonlinearity involves time lags. Simultaneous consideration of both spatial diffusion and time delay leads to a new class of infinite dimensional dynamical system: delayed reaction diffusion equations with global interaction, where the delayed non-local nonlinearity is either explicitly or implicitly given by the population density at a previous time and in all possible spatial locations. We show how the interweaving diffusion and delay results in some interesting and challenging mathematical problems such as the possibility of periodic oscillations and traveling wave solutions from a spatially homogeneous equilibrium to a periodic solution even in the mono-stable case. We illustrate that this seemingly complicated formulation of the biological reality common to many systems may facilitate and simplify the development of a qualitative theory, such as the exponential ordering and order-preserving property in the case where the nonlinearity is not monotone. We also suggest some open problems and issues for future research.

2000 *Mathematics Subject Classification.* Primary 35R10; Secondary 34K05, 34K60, 35K55, 35L60, 35M10, 92D25, 92D30.

Research was partially supported by Natural Sciences and Engineering Research Council of Canada, by Canada Research Chairs Program, and by Network of Centers of Excellence Program: Mathematics for Information Technology and Complex Systems.

1 Introduction

There has been a long history in modeling either time delay or spatial diffusion in population biology, spatial ecology and disease spread. The 1970s saw the first efforts to model time delay and spatial diffusion together and there has been some intensive study of the corresponding mathematical formulation in the format of partial functional differential equations. See [91] and the references therein.

Although incorporating both time delay and spatial diffusion into a single deterministic model represents a very natural attempt to bring the model closer to the biological reality, only limited progress has been made so far since, as we will explain, describing the implications of individual movement on the formulation of the time delay term seems to be a highly non-trivial task and yields some mathematically challenging non-local delayed problems. The earliest studies of nonlocal delay differential equations in mathematical biology seem to date back about a quarter of a century, although it is only in much more recent years that there has been a real explosion of interest. Investigators who published papers on such problems in the early 1980s include (alphabetically) M. Pozio, R. Redlinger and Y. Yamada. Many of these earlier papers were very insightful and provided important mathematical tools which are in regular use today and from which other investigators have gained inspiration.

There are at least two complementary approaches developed to derive the time delay term. Both approaches lead to non-local diffusive systems (reaction diffusion or lattice equations) with delay. One approach is proposed in Britton [8] where the delay term involves a spatio-temporal weight, derived using a probabilistic argument, to account for the drift of individuals to their present positions from all possible positions at previous times. Another approach was developed by Smith and Thieme in [73] and is based on structured population models coupled with the technique of integration along characteristics. This approach is particularly useful when the individuals have two clearly distinguishable biological age classes (mature and immature), and the resulting model for the matured population becomes a system of ordinary delay differential equations or a lattice delay differential system when the spatial domain consists of discrete patches (see also [80]), or a reaction-diffusion equation with the nonlinearity being delayed and non-local when the spatial domain is continuous (see [81]). When spatial movement involves a time lag, we will also obtain a system of neutral functional differential equations [88] or a hyperbolic parabolic non-local functional differential equation [65, 59, 56].

In [25] the authors provide a survey of results on the biological modeling, dynamics analysis and numerical simulation of nonlocal spatial effects induced by time delays in reaction diffusion models for a single species population confined to either a finite or an infinite domain. There has been rapid development over the last few years since the publication of this survey article in both Russian and English, and it seems thus necessary to have a report to update the progress and to address some unresolved problems. It is hoped that this article will stimulate further interest in the subject that may advance the theory of infinite dimensional dynamical systems and find more applications to issues important for the study of biological invasion and disease spread.

In this survey, we start with the derivation of a reaction-diffusion equation with non-local delayed nonlinearity using structured population models and Fourier transformations and show how the model reduces to the local delayed reaction diffusion model when the immature becomes immobile. We then discuss variations of this model, which include systems of ordinary delay differential equations and lattice delay differential equations with global interaction when space is discrete, and systems of neutral functional differential equations and hyperbolic parabolic equations with non-local delayed reaction when time lags for spatial movement are considered. More complicated models are described in the later sections, along with the results for the asymptotic behaviors of the solutions to these model equations.

We must emphasize that there have been intensive research activities for non-local partial differential equations arising from a number of applications including phase transitions and population biology that do not involve time lags. These activities should definitely be included in a comprehensive survey on diffusive systems with non-local interactions. We will however, in the current survey, focus on the non-locality arising exclusively due to the interaction of time lag of feedback and the spatial movement of biological species.

2 A prototype delayed non-local reaction diffusion equation

We start with the work of [81] to illustrate the approach of Smith and Thieme [73] based on structured population models and their reduction, and we shall also discuss some extensions in a number of directions.

2.1 Single species population in 1-D unbounded domain. Let $u(t,a,x)$ denote the density of the population of the species under consideration at time $t \geq 0$, age $a \geq 0$ and location $x \in \mathbb{R}$. It is natural to assume

$$|u(t,a,\pm\infty)| < \infty \quad \text{for} \quad t \geq 0, \quad a \geq 0.$$

A standard argument on population with age structure and diffusion [53] gives

$$\frac{\partial u}{\partial t} + \frac{\partial u}{\partial a} = D(a)\frac{\partial^2 u}{\partial x^2} - d(a)u, \tag{2.1}$$

where $D(a)$ and $d(a)$ are the age-dependent diffusion rate and death rate respectively. If $\tau \geq 0$ is the maturation time for the species (assumed, at this moment, to be a constant), then the total matured population at time t and location x is given by

$$w(t,x) = \int_\tau^\infty u(t,a,x)\,da,$$

and satisfies, under the biologically realistic assumption $u(t,\infty,x) = 0$, that

$$\begin{aligned}\frac{\partial w}{\partial t} &= \int_\tau^\infty \frac{\partial}{\partial t}u(t,a,x)\,da = \int_\tau^\infty \left[-\frac{\partial u}{\partial a} + D(a)\frac{\partial^2 u}{\partial x^2} - d(a)u\right] da \\ &= u(t,\tau,x) + \int_\tau^\infty \left[D(a)\frac{\partial^2 u}{\partial x^2} - d(a)u\right] da.\end{aligned} \tag{2.2}$$

To proceed further, we assume that only the mature can reproduce and the reproduction rate depends on the matured population only, i.e.

$$u(t,0,x) = b(w(t,x)), \tag{2.3}$$

where $b(\cdot)$ is the birth function. We also assume the diffusion and death rates for the mature population are age independent, that is, $D(a) = D_m$ and $d(a) = d_m$ for

$a \in [\tau, \infty)$, where D_m and d_m are constants. Then from equation (2.2) it follows that

$$\frac{\partial w}{\partial t} = u(t, \tau, x) + D_m \frac{\partial^2 w}{\partial x^2} - d_m w. \tag{2.4}$$

To obtain an equation for w, we need to evaluate $u(t, \tau, x)$: the maturation rate at time t and spatial location x. For this purpose, we fix $s \geq 0$ and let $V^s(t, x) = u(t, t-s, x)$ for $s \leq t \leq s + \tau$. Then

$$\begin{aligned} \frac{\partial}{\partial t} V^s(t, x) &= \frac{\partial u}{\partial t}(t, a, x)\bigg|_{a=t-s} + \frac{\partial u}{\partial a}(t, a, x)\bigg|_{a=t-s} \\ &= D(t-s)\frac{\partial^2}{\partial x^2} V^s(t, x) - d(t-s) V^s(t, x) \end{aligned} \tag{2.5}$$

with

$$|V^s(t, \pm\infty)| < \infty. \tag{2.6}$$

Applying the method of separation of variables to (2.5)-(2.6) and using Fourier transformation, we obtain

$$V^s(t, x) = \int_{-\infty}^{\infty} k(s, \omega) e^{-\int_0^{t-s} (\omega^2 D(a) + d(a))\, da} e^{-i\omega x}\, d\omega,$$

with

$$k(s, \omega) = \frac{1}{2\pi} \int_{-\infty}^{\infty} b(w(s, y)) e^{i\omega y}\, dy.$$

From this it follows that

$$u(t, \tau, x) = V^{t-\tau}(t, x) = \frac{e^{-\int_0^\tau d_I(a)\, da}}{\sqrt{4\pi\alpha}} \int_{-\infty}^{\infty} b(w(t-\tau, y)) e^{\frac{-(x-y)^2}{4\alpha}}\, dy,$$

and hence $w(t, x)$ satisfies

$$\frac{\partial w}{\partial t} = D_m \frac{\partial^2 w}{\partial x^2} - d_m w + \varepsilon \int_{-\infty}^{\infty} b(w(t-\tau, y)) f_\alpha(x-y)\, dy, \tag{2.7}$$

where $\alpha := \int_0^\tau D(a)\, da > 0$, $\varepsilon = e^{-\int_0^\tau d(a)\, da} \in [0, 1]$ and $f_\alpha(x) = \frac{1}{\sqrt{4\pi\alpha}} e^{\frac{-x^2}{4\alpha}}$.

Each term in (2.7) can be explained naturally. The growth rate of the mature population at time t and spatial location $x \in \mathbb{R}$ is the balance of the spatial diffusion, the death rate and the maturation rate. To calculate the latter we integrate over all possible birth locations y. The birth rate at time $t - \tau$ at spatial location y is $b(w(t-\tau, y))$; this has to be multiplied by the probability that an individual born at y moves to location x to mature, and further multiplied by the probability of actually surviving (the quantity ε) up to the age of maturation.

Note that (2.7) becomes a reaction diffusion equation with time delays and nonlocal effects. It is interesting to remark that the survival rate ε and the net diffusion rate α for the immature enter the model in a completely different fashion than the death rate and diffusion rate of the matured. How this impacts the global dynamics remains to be an interesting question.

This observation is also important if we use the same idea to model disease spread when the age involved means disease age, i.e. the time since the individual became exposed to the disease. In this case, the "matured" population is the compartment of infectious individuals and the "immature" population is the compartment of the latent individuals. The spatial movement patterns of the asymptomatic and symptomatic individuals will certainly be different due to different control and treatment measures; how this difference of individual movement patterns affects the spatio-temporal patterns of disease spread is a critical issue for public health.

It is easy to see that when $\alpha \to 0$, that is as the immature become immobile, equation (2.7) reduces to the local problem

$$\frac{\partial w}{\partial t} = D_m \frac{\partial^2 w}{\partial x^2} - d_m w + \varepsilon b(w(t-\tau, x)),$$

so much of the earlier work, presented in [91], on modeling time delay and spatial diffusion using reaction diffusion equations with local effects does apply to the case where the mobility of the immature can be ignored: much of the effort on partial functional differential equations over the last 30 years is applicable to this special case!

2.2 Patch models and delayed lattice equations. The approach of Smith and Thieme [73] is utilized in [80] to formulate the model of a single species living in a two identical patches with two age classes. The model for the density of the matured population $(w_1(t), w_2(t))$ in two patches is shown to take the following form

$$\begin{cases} \dfrac{dw_1(t)}{dt} = -d_{1,m}w_1(t) + D_{2,m}w_2(t) - D_{1,m}w_1(t) \\ \qquad +e^* \left[1 - \displaystyle\int_0^\tau e^{-\int_\theta^\tau \hat{D}(a)\,da} D_1(\theta)\,d\theta\right] b_1(w_1(t-\tau)) \\ \qquad +e^* \left[\displaystyle\int_0^\tau e^{-\int_\theta^\tau \hat{D}(a)\,da} D_2(\theta)\,d\theta\right] b_2(w_2(t-\tau)), \\ \dfrac{dw_2(t)}{dt} = -d_{2,m}w_2(t) + D_{1,m}w_1(t) - D_{2,m}w_2(t) \\ \qquad +e^* \left[1 - \displaystyle\int_0^\tau e^{-\int_\theta^\tau \hat{D}(a)\,da} D_2(\theta)\,d\theta\right] b_2(w_2(t-\tau)) \\ \qquad +e^* \left[\displaystyle\int_0^\tau e^{-\int_\theta^\tau \hat{D}(a)\,da} D_1(\theta)\,d\theta\right] b_1(w_1(t-\tau)). \end{cases} \tag{2.8}$$

In the above system, $e^* = \exp(-\int_0^\tau d_I(a)\,da)$ is the surviving probability, and $d_{i,m}$, b_i and $D_{i,m}$ are the death rate, birth rate and dispersal rate (from the ith patch to another). The constant term $e^* \int_0^\tau e^{-\int_\theta^\tau \hat{D}(a)\,da} D_j(\theta)\,d\theta$ denotes the fraction of the mature population which was born at time $t-\tau$ in the j–th patch and is in the i–th patch at the current time t. This factor was usually ignored in the literature where a delay differential equation is used to model the dynamics of interacting species in patchy environment with time delay. It was observed in [80] that this non-local delayed coupling can give rise to some transient spatially heterogeneous periodic solutions: the so-called phase-locked oscillations.

In the reaction diffusion model (2.7) and the two patch model (2.8), the dynamics during the maturation period is quite simple: it is governed by a system involving linear dispersal/diffusion and linear death rates, and hence the maturation rate is a linear combination of the birth rates in all other possible spatial locations at an earlier time: the nonlinearity comes from the birth function. This simple dynamics makes it possible to formulate the maturation rate explicitly in terms of the density of the matured population at the earlier time. Otherwise, we will have a coupled system of reaction-diffusion equations involving maturation rates that are implicitly determined by a hyperbolic equation, as shall be shown in later sections on epidemiological models.

We should mention that the general multiple patchy case is considered in [73] where the dynamics of the population during the maturation period is incorporated. Also the situation where there is some spread of the maturation period around some mean value was considered in [36] by replacing the constant delay τ with a distributed term.

Finally, the case of infinitely many discrete patches is considered in [90] where the authors obtain a lattice delay differential equation using discrete Fourier transformations. More precisely, they consider a single species population living in infinitely many patches in a one dimensional space connecting locally for diffusion, and with age structure and a fixed maturation period. For the matured population density in the ith patch, they obtain

$$\frac{dw_j(t)}{dt} = u_j(t,\tau) + D_m[w_{j+1}(t) + w_{j-1}(t) - 2w_j(t)] - d_m w_j(t) \quad \text{for } t > 0. \quad (2.9)$$

Using discrete Fourier transformations and the aforementioned technique of integration along characteristics, they derive

$$u_j(t,\tau) = \frac{\mu}{2\pi} \sum_{k=-\infty}^{\infty} \beta(j,k) b(w_k(t-\tau)),$$

where

$$\beta(j,k) = \int_{-\pi}^{\pi} \exp\{i[(j-k)\omega] - 4\alpha \sin^2(\frac{\omega}{2})\}\, d\omega.$$

Note that, although the patches are connected only locally through nearest neighborhood dispersal, the resulting equation for the matured population involves a global interaction term. Note also that the discrete Laplace operator $D_m[w_{j+1}(t) + w_{j-1}(t) - 2w_j(t)]$ can be replaced by any algebraic sum of the dispersal to and from the i-th patch, so that equation (2.9) does not necessarily come from the spatial discretization of a non-local reaction diffusion equation with delay.

2.3 Non-local delayed hyperbolic-parabolic equations. In [65], a deterministic model is formulated as an approximation for a single species population distributed in an unbounded one dimensional domain where individuals move around following the classical Fick's diffusion law with a time lag, and where the population has two age classes: immature and mature, with a constant maturation time. The resulting model is a scalar hyperbolic-parabolic differential equation with a non-local time-delayed nonlinearity for the matured population density.

As before, let $u(t,a,x)$ denote the population density at time t, age $a \geq 0$ and spatial location $x \in \mathbb{R}$. Let $\tau > 0$ be the maturation time and let

$$w(t,x) = \int_{\tau}^{\infty} u(t,a,x)\, da$$

denote the adult population at time t and spatial location x.

Assuming the spatial movement of the mature individual follows the Fick's diffusion law, but with a time delay $r > 0$, and using the standard age-structured population model, we have

$$\Big(\frac{\partial}{\partial t} + \frac{\partial}{\partial a}\Big) u(t+r, a+r, x) = D(a) \frac{\partial^2}{\partial x^2} u(t,a,x) - d(a+r) u(t+r, a+r, x), \quad (2.10)$$

where $D(a)$ and $d(a)$ are the diffusion and death rates.

We now assume the delay r for the spatial diffusion is sufficiently small, so that we can use the following approximation to the above equation:

$$\begin{aligned}&\Big(\frac{\partial}{\partial t}+\frac{\partial}{\partial a}\Big)u(t,a,x)+r\Big(\frac{\partial}{\partial t}+\frac{\partial}{\partial a}\Big)^2u(t,a,x)\\&=D(a)\frac{\partial^2}{\partial x^2}u(t,a,x)-[d(a)+rd'(a)]\Big[u(t,a,x)+r\Big(\frac{\partial}{\partial t}+\frac{\partial}{\partial a}\Big)u(t,a,x)\Big].\end{aligned}$$

Therefore,

$$\begin{aligned}&\{1+r[d(a)+d'(a)r]\}\Big(\frac{\partial}{\partial t}+\frac{\partial}{\partial a}\Big)u(t,a,x)+r\Big(\frac{\partial}{\partial t}+\frac{\partial}{\partial a}\Big)^2u(t,a,x)\\&=D(a)\frac{\partial^2}{\partial x^2}u(t,a,x)-[d(a)+rd'(a)]u(t,a,x).\end{aligned}\tag{2.11}$$

In what follows, we assume the death and diffusion rates of the mature population are independent of the age, and let $d_m=d(a), D_m=D(a)$ for all $a\geq\tau$. Then for $a\geq\tau$, we have

$$\begin{aligned}&[1+rd_m]\Big(\frac{\partial}{\partial t}+\frac{\partial}{\partial a}\Big)u(t,a,x)+r\Big(\frac{\partial}{\partial t}+\frac{\partial}{\partial a}\Big)^2u(t,a,x)\\&=D_m\frac{\partial^2}{\partial x^2}u(t,a,x)-d_mu(t,a,x),\end{aligned}$$

and hence we have, noting that $u(t,\infty,x)=0$, the following:

$$\begin{aligned}\frac{\partial}{\partial t}w(t,x)&=\int_\tau^\infty\frac{\partial}{\partial t}u(t,a,x)\,da\\&=\int_\tau^\infty\Big[-\frac{r}{1+rd_m}\Big(\frac{\partial}{\partial t}+\frac{\partial}{\partial a}\Big)^2u(t,a,x)\\&\quad+\frac{D_m}{1+rd_m}\frac{\partial^2}{\partial x^2}u(t,a,x)-\frac{d}{1+rd}u(t,a,x)-\frac{\partial}{\partial a}u(t,a,x)\Big]da\\&=-\frac{r}{1+rd_m}\{\frac{\partial^2}{\partial t^2}w(t,x)-\Big(2\frac{\partial}{\partial t}+\frac{\partial}{\partial a}\Big)u(t,\tau,x)\}\\&\quad+\frac{D_m}{1+rd_m}\frac{\partial^2}{\partial x^2}w(t,x)-\frac{d}{1+rd_m}w(t,x)+u(t,\tau,x).\end{aligned}$$

So, we obtain

$$\begin{aligned}&\frac{r}{1+rd_m}\frac{\partial^2}{\partial t^2}w(t,x)+\frac{\partial}{\partial t}w(t,x)\\&=\quad\frac{D_m}{1+rd}\frac{\partial^2}{\partial x^2}w(t,x)-\frac{d_m}{1+rd_m}w(t,x)+u(t,\tau,x)\\&\quad+\frac{r}{1+rd_m}\Big(2\frac{\partial}{\partial t}+\frac{\partial}{\partial a}\Big)u(t,\tau,x).\end{aligned}\tag{2.12}$$

It remains to derive $u(t,\tau,x)$, the maturation rate at time t and spatial location x, and this should be determined from the consideration of the immature population. In particular, in the case where the diffusion of the immature population is rapid, we can use the formula obtained before:

$$u(t,\tau,x)=\varepsilon\int_{-\infty}^\infty b(w(t-\tau,y))f_\alpha(x-y)\,dy.$$

Therefore, the model finally becomes

$$
\begin{aligned}
&(1+rd_m)\frac{\partial}{\partial t}w(t,x)+r\frac{\partial^2}{\partial t^2}w(t,x)\\
&=D_m\frac{\partial^2}{\partial x^2}w(t,x)-dw(t,x)+\epsilon(1+rd)\int_{-\infty}^{\infty}f_\alpha(x-y)b(w(t-\tau,y))\,dy\\
&\quad+r\frac{\partial}{\partial\theta}\big[\epsilon\int_{-\infty}^{\infty}f(x-y)b(w(\theta,y))\,dy\big]\big|_{\theta=t-\tau}.
\end{aligned}
\tag{2.13}
$$

We should note that if the immature population also moves with delay, we will obtain a linear hyperbolic-parabolic partial differential equation that determines $u(t,\tau,x)$, and as such we are led to a hyperbolic-parabolic partial differential equation with non-local delayed nonlinear term appearing in the highest (2nd) temporal derivative for $w(t,x)$. Much remains to be done for this type of model equation. Furthermore, in the case where the environment for the species is discrete (patchy environment), then we would obtain a system of neutral functional differential equations, as shown in the recent work [88].

We should mention that delayed diffusion (with instantaneous reaction) has been studied intensively using hyperbolic equations, see the excellent survey [17] for more details. The type of equations considered there seems to provide analytic estimation of the velocity for the spread of agriculture that is in good agreement with archaeological evidence leading to the conclusion that European farming originated in the Near East, from where it spread across Europe with a certain speed. It would be interesting to see how time lags in both diffusion and reaction can be incorporated into a more comprehensive model.

2.4 Nonlocal equations as limits of singularly perturbed systems. Nonlocal equations do not have to arise as a consequence of having a time delay. They may arise as a formal simplification of a singularly perturbed local system. Raquepas and Dockery [64] studied the following nonlocal problem on the interval $x\in(0,1)$:

$$
u_t=d^2u_{xx}+f(u)+\gamma\int_0^1 g(x,s)u(s,x)\,ds \tag{2.14}
$$

on homogeneous Neumann boundary conditions. This equation can be considered as a model of the dynamics of the singularly perturbed activator-inhibitor system in the limit $\epsilon\to 0$:

$$
\begin{aligned}
u_t&=d^2u_{xx}+f(u)-w,\\
\epsilon w_t&=Dw_{xx}+\gamma u-w,
\end{aligned}
\tag{2.15}
$$

subject to the same boundary conditions. Indeed, if one sets $\epsilon=0$ in (2.15) and then solves the second equation to obtain w in terms of u using the Green's function

$$
g(x,s)=\begin{cases}
-\dfrac{\sqrt{D}\cosh\left(\frac{x}{\sqrt{D}}\right)\cosh\left(\frac{s-1}{\sqrt{D}}\right)}{\sinh\left(\frac{1}{\sqrt{D}}\right)}, & 0\le x<s\le 1,\\[2ex]
-\dfrac{\sqrt{D}\cosh\left(\frac{x-1}{\sqrt{D}}\right)\cosh\left(\frac{s}{\sqrt{D}}\right)}{\sinh\left(\frac{1}{\sqrt{D}}\right)}, & 0\le s<x\le 1,
\end{cases}
$$

then one obtains (2.14). Raquepas and Dockery [64] conducted an extensive bifurcation study of (2.14), showing for example that stable heterogeneous equilibria having any number of zeros on the spatial domain can be obtained by choosing the parameters appropriately.

Gourley [22] similarly derived a nonlocal scalar equation as a limit of a singularly perturbed local system. His analysis was for the unbounded domain $x \in (-\infty, \infty)$, but can easily be modified for a finite domain by using a Greens function as just described. He studied the nonlocal Fisher equation

$$\frac{\partial u}{\partial t} = \frac{\partial^2 u}{\partial x^2} + u(x,t)\left(1 - \int_{-\infty}^{\infty} g(x-y)u(t,y)\,dy\right) \tag{2.16}$$

for $x \in (-\infty, \infty)$, $t > 0$, with kernel g as general as possible. If viewed as a model of resource-consumer dynamics, in a situation in which the individuals are competing for a resource which can redistribute itself, then some simple modelling to be described below leads us to the view that the choice

$$g(x) = \frac{1}{2}\lambda e^{-\lambda|x|}, \qquad \lambda > 0 \tag{2.17}$$

is especially important ecologically. Furter and Grinfeld [18] considered a similar model on a finite domain. Gourley [22] derived equation (2.16) with (2.17) as follows. Begin with an equation of the form

$$\frac{\partial u}{\partial t} = \frac{\partial^2 u}{\partial x^2} + u(t,x)\,r(t,x) \tag{2.18}$$

where $r(x,t)$ measures a local resource level ($r < 0$ means that the population is declining locally and $r > 0$ means it is growing locally).

We now need an equation for r, which we take to be

$$\epsilon\frac{\partial r}{\partial t} = K - r - u + \alpha\frac{\partial^2 r}{\partial x^2}. \tag{2.19}$$

This assumes the resource diffuses, is consumed by u and is replaced at a rate proportional to its distance from its carrying capacity K. Assuming fast time dynamics ($\epsilon \to 0$), (2.19) becomes

$$\alpha\frac{\partial^2 r}{\partial x^2} - r = u - K, \qquad x \in (-\infty, \infty)$$

for which a bounded solution is sought. Solving this equation gives

$$r(t,x) = K - \int_{-\infty}^{\infty} g(x-y)u(t,y)\,dy$$

with

$$g(x) = \frac{1}{2\sqrt{\alpha}}e^{-|x|/\sqrt{\alpha}}$$

which is (2.17) with $\lambda = 1/\sqrt{\alpha}$. Insertion of this solution into (2.18) then yields the form of equation (2.16) with (2.17). Thus, in the application to resource-consumer dynamics, $1/\lambda$ is proportional to the square root of the diffusivity of the resource.

An equation similar to (2.16) was studied by Billingham [6] who considered in detail the situation when the equation is very strongly nonlocal (i.e. the parameter λ in (2.17) is small).

3 The reaction-advection case

For the case of an infinite one-dimensional domain, this section will consider what changes result when a species is subject to advection as well as Fickian diffusion. Let $w(t,x)$ be the density of mature adults, i.e. those of age at least τ, given by

$$w(t,x) = \int_\tau^\infty u(t,a,x)\,da.$$

This time, assume that u satisfies

$$\frac{\partial u}{\partial t} + \frac{\partial u}{\partial a} = U_i \frac{\partial u}{\partial x} + d_i \frac{\partial^2 u}{\partial x^2} - \gamma u, \qquad 0 < a < \tau \tag{3.1}$$

so that the juveniles are subject to advection from right to left with speed $U_i > 0$. Suppose for simplicity that the mature population $w(t,x)$ has a (possibly different) advection speed U_m and satisfies

$$\frac{\partial w}{\partial t} = U_m \frac{\partial w}{\partial x} + d_m \frac{\partial^2 w}{\partial x^2} + u(t,\tau,x) - d(w), \qquad x \in (-\infty,\infty), \tag{3.2}$$

and suppose that

$$\begin{aligned} u(t,0,x) &= b(w(t,x)), \qquad t>0, \\ u(0,a,x) &= u_0(a,x), \qquad a > 0. \end{aligned} \tag{3.3}$$

We will adopt any reasonable set of assumptions concerning the birth and death functions $b(\cdot)$ and $d(\cdot)$, which obviously will include the requirement that $b(0) = 0$ and $d(0) = 0$. The first condition in (3.3) is the birth law and we have assumed that the birth rate is a function of the total number of adults, while $u_0(a,x)$ is the prescribed initial age distribution at each location x. Again it is necessary to solve (3.1) subject to (3.3) to calculate $u(t,\tau,x)$. The formula for $u(t,\tau,x)$ depends on whether $t > \tau$ or $t < \tau$. For $t > \tau$ it can be shown, similarly to calculations presented earlier in this paper, that

$$u(t,\tau,x) = e^{-\gamma\tau} \int_{-\infty}^{\infty} \frac{1}{\sqrt{4\pi d_i \tau}} e^{\frac{-(U_i\tau + x - y)^2}{4 d_i \tau}} b(w(t-\tau,y))\,dy. \tag{3.4}$$

See also [37]. For $t < \tau$ the expression for $u(t,\tau,x)$ is different and its derivation will be discussed, partly to show the reader clearly why there are two different expressions for the cases $t > \tau$ and $t < \tau$. Set

$$v_\xi(t,x) = u(t,t+\xi,x).$$

Then

$$\frac{\partial v_\xi}{\partial t} = U_i \frac{\partial v_\xi}{\partial x} + d_i \frac{\partial^2 v_\xi}{\partial x^2} - \gamma v_\xi. \tag{3.5}$$

Applying the Fourier transform

$$\hat{v}_\xi(t,s) = \int_{-\infty}^{\infty} v_\xi(t,x) e^{-isx}\,dx$$

to (3.5) gives

$$\frac{d\hat{v}_\xi}{dt} = (U_i is - d_i s^2 - \gamma)\hat{v}_\xi$$

so that

$$\hat{v}_\xi(t,s) = \hat{v}_\xi(0,s) e^{(U_i is - d_i s^2 - \gamma)t}.$$

Also

$$\hat{v}_\xi(0,s) = \widehat{u(0,\xi,x)} = \widehat{u_0(\xi,x)}.$$

Thus

$$\begin{aligned}\hat{v}_\xi(t,s) &= \widehat{u_0(\xi,x)}e^{(U_i is-d_i s^2-\gamma)t}\\ &= \widehat{u_0(\xi,x)}\mathcal{F}\left\{\frac{1}{\sqrt{4\pi d_i t}}e^{-\gamma t}e^{-(x+U_i t)^2/4d_i t};\ t\to s\right\}\end{aligned}$$

and so

$$u(t,t+\xi,x)=\int_{-\infty}^{\infty}u_0(\xi,y)\frac{e^{-\gamma t}}{\sqrt{4\pi d_i t}}e^{-(x-y+U_i t)^2/4d_i t}\,dy.$$

Setting $\xi=\tau-t$ gives (provided $t<\tau$)

$$u(t,\tau,x)=\int_{-\infty}^{\infty}u_0(\tau-t,y)\frac{e^{-\gamma t}}{\sqrt{4\pi d_i t}}e^{-(x-y+U_i t)^2/4d_i t}\,dy. \tag{3.6}$$

In the above calculation, ξ always has to be positive because the initial data $u_0(a,x)$ is prescribed only for $a>0$. This is why the above calculation is only for $t<\tau$.

We may now state that the initial value problem consisting of (3.1), (3.3) and (3.2) leads to the following *reduced problem*. For $0<t<\tau$,

$$\begin{aligned}\frac{\partial w}{\partial t} &= U_m\frac{\partial w}{\partial x}+d_m\frac{\partial^2 w}{\partial x^2}-d(w)\\ &\quad+\int_{-\infty}^{\infty}u_0(\tau-t,y)\frac{e^{-\gamma t}}{\sqrt{4\pi d_i t}}e^{-(x-y+U_i t)^2/4d_i t}\,dy, \qquad x\in\mathbb{R},\end{aligned} \tag{3.7}$$

and, for $t>\tau$,

$$\begin{aligned}\frac{\partial w}{\partial t} &= U_m\frac{\partial w}{\partial x}+d_m\frac{\partial^2 w}{\partial x^2}-d(w)\\ &\quad+e^{-\gamma\tau}\int_{-\infty}^{\infty}\frac{1}{\sqrt{4\pi d_i\tau}}e^{-(x-y+U_i r)^2/4d_i r}b(w(t-\tau,y))\,dy, \qquad x\in\mathbb{R}.\end{aligned} \tag{3.8}$$

The reduced problem, consisting of (3.7) and (3.8) above, would be solved subject to the initial data

$$w(0,x)=\int_{\tau}^{\infty}u_0(a,x)\,dx \tag{3.9}$$

with $u_0(a,x)$ from (3.3).

Expression (3.6) has a clear ecological interpretation. It represents the rate of adult recruitment at time t, position x, for times $t<\tau$. It should be contrasted with the corresponding expression (3.4) for times $t>\tau$. The latter states that adult recruitment for $t>\tau$ is essentially the birth rate at time $t-\tau$, corrected to allow for juvenile mortality (hence the factor $e^{-\gamma\tau}$). The convolution integral accounts for drift (due here to both advection and diffusion) from all possible birth locations at time $t-\tau$ to position x at time t. However, if we are solving the initial value problem consisting of (3.1), (3.3) and (3.2), then (3.4) no longer makes sense for $t<\tau$ because it refers to the adult population at a negative time. For times $t<\tau$ the correct expression for adult recruitment is (3.6), which counts the number of juveniles that were present at time $t=0$, when these individuals would have been of age $\tau-t$, hence the presence of the quantity $u_0(\tau-t,y)$. Juvenile mortality is now represented by $e^{-\gamma t}$, and all possible locations of these age $\tau-t$ juveniles at time $t=0$ are taken care of through the convolution integral.

As the above analysis has shown, the reason why the cases $t>\tau$ and $t<\tau$ have to be dealt with separately is to do with the fact that the partial differential equation (3.1) is to be solved for $a>0$, $t>0$. The characteristics of the hyperbolic partial differential equation (3.1) are, of course, the straight lines $t = a + \text{const}$. We solve for $a > 0$, $t > 0$ and so one of these characteristics passes through the origin. Characteristics belonging to the region $a > t$ will intersect the a-axis and therefore when the partial differential equation is integrated along these characteristics it is necessary to use the condition $u(0,a,x) = u_0(a,x)$. On the other hand, characteristics belonging to the region $t > a$ intersect only the t-axis and so integration along these characteristics involves using $u(t,0,x) = b(w(t,x))$. Thus there are two expressions for $u(t,\tau,x)$ according as to whether $t > \tau$ or $t < \tau$. In the literature it seems to be common for authors to derive the model for $w(t,x)$ only for the case when $t > \tau$, but then to assume that the resulting reduced model is, in fact, valid for all times $t > 0$ subject to an arbitrary prescribed initial function specifying the initial data for $t \in [-\tau, 0]$. This is probably a reasonable assumption in general but it does raise delicate issues regarding initial data and positivity which have been treated in detail in Bocharov and Hadeler [7] for problems without diffusion. If, instead of solving (3.7), (3.8) and (3.9) above, we were to regard equation (3.8) as valid for all $t > 0$, and then proceed to solve this equation subject to prescribed non-negative initial data on $[-\tau, 0]$, we would find that only certain initial data for this latter problem is related to the original problem which has its initial data (see (3.3)) prescribed only at $t = 0$. Also, while positive solutions of the original problem lead to positive solutions of the modified problem just described, the cone of positive solutions of the latter problem is larger in general, since it would be studied for arbitrary non-negative initial data prescribed on $[-\tau, 0]$, and not just for the subset of these initial data that are properly related to the original problem.

Notwithstanding the above remarks, our comments shall henceforth be confined to the study of the delay equation that applies beyond the initial transient period, i.e. equation (3.8). If we let $d_i \to 0$ then it can be shown that (3.8) becomes, in the limit,

$$\frac{\partial w}{\partial t} = U_m \frac{\partial w}{\partial x} + d_m \frac{\partial^2 w}{\partial x^2} - d(w) + e^{-\gamma\tau} b(w(t-\tau, x+U_i\tau)). \tag{3.10}$$

Equation (3.10) describes the evolution of the adult population $w(t,x)$ in the case when the immatures do not diffuse but are still subject to advection. The last term represents adult recruitment and involves the birth rate $b(w(t-r, x+U_i\tau))$ at time $t-\tau$ and position $x + U_i\tau$. Since the advection of juveniles is from right to left with speed U_i, the location $x + U_i r$ is of course the exact birth location of every individual that becomes mature at position x at time t.

When the immature of a species are subject to advection but not diffusion we know, as explained above, the exact birth location of every individual that becomes mature at position x at time t. This is in contrast with the situation for immatures that diffuse according to Fickian diffusion. Given only the location of an individual when it reaches maturity, its exact birth location cannot be known exactly although the probability of birth at a particular location can be specified. Adult recruitment terms in such situations therefore involve an integral over all possible birth locations.

It may, of course, be the case that immature individuals are subject to both advection and diffusion. Adult recruitment is then still given by an integral, as in (3.8), but here we are making the simplifying assumption that the advection velocity is constant. This might not always be realistic. The study of problems with both advection and diffusion is very important in hydrology, physiology and chemical engineering and the theory of Taylor diffusion is often discussed. See, for example, Young and Jones [100] or the original article by Taylor [82]. In these disciplines the scenario might well be one of diffusion of a passive tracer in a situation where the advection velocity is not constant, for example where it is given by Poiseuille flow in a pipe. The problem is then one of solving the advection diffusion equation

$$c_t + uc_x = D\Delta c$$

with c being the concentration of the tracer and $u(y,z) = 2(U/a^2)(a^2 - y^2 - z^2)$ being the axial velocity along the pipe. Here, a is the radius of the pipe and U is the mean of the advection speed over the cross section, while the x-axis is that of the pipe. The velocity is zero where the fluid touches the pipe wall (the no-slip boundary condition of viscous flow). Taylor [82] showed that, at large times, the sectionally averaged concentration $C(x,t)$ satisfies a simpler advection-diffusion equation, namely

$$C_t + UC_x = D_{\text{eff}} C_{xx} \tag{3.11}$$

where

$$C(x,t) = \frac{1}{A}\int c(x,y,z,t)\,dA, \qquad U = \frac{1}{A}\int u(y,z)\,dA$$

and D_{eff}, which is called the "effective diffusivity", is given by

$$D_{\text{eff}} = D + \frac{a^2U^2}{48D}.$$

At the heart of the derivation of (3.11) by Taylor is the assumption that, at large times, molecular diffusivity would mix the tracer in the transverse direction while advection would stretch it out in the axial direction. Since Taylor's work there have been further developments, for example, improvements to the approximation by adding additional correction terms to (3.11) (see Young and Jones [100]). It might be interesting to use these ideas in the study of populations that both diffuse and advect, but with a non uniform advection speed.

The situation when the species is close to extinction and the immatures are subject to advection but are not diffusing can be studied by replacing (3.10) by its linearized approximation for small w:

$$\frac{\partial w}{\partial t} = U_m \frac{\partial w}{\partial x} + d_m \frac{\partial^2 w}{\partial x^2} - d'(0)w + b'(0)e^{-\gamma\tau} w(t-\tau, x + U_i\tau). \tag{3.12}$$

If the immatures are neither advecting nor diffusing (though the adults may still advect and/or diffuse) then (3.12) further simplifies to

$$\frac{\partial w}{\partial t} = U_m \frac{\partial w}{\partial x} + d_m \frac{\partial^2 w}{\partial x^2} - d'(0)w + b'(0)e^{-\gamma\tau} w(t-\tau, x) \tag{3.13}$$

and, if one poses this equation on the finite one-dimensional interval $x \in (0, L)$ with homogeneous Dirichlet boundary conditions, a condition can be found on the domain size L to avoid extinction. We shall assume here that $b'(0) > 0$ and that $d'(0) \geq 0$. Note that (3.12) does not make sense on the finite domain $(0, L)$, as it comes from a model the derivation of which assumed the domain to be the whole real line. Non-trivial solutions of the form $w = e^{\sigma t}\psi(x)$ exist to equation (3.13) under homogeneous Dirichlet boundary conditions whenever σ satisfies

$$b'(0)e^{-\gamma\tau}e^{-\sigma\tau} = \sigma + d'(0) + \frac{d_m n^2\pi^2}{L^2} + \frac{U_m^2}{4d_m} \tag{3.14}$$

with n a positive integer playing the role of a wavenumber. Therefore, a sufficient condition for extinction of the species is that all roots σ of (3.14) satisfy $\mathrm{Re}\,\sigma < 0$ for every $n = 1, 2, 3, \ldots$. Suppose, for a contradiction, that there exists an n and a σ with $\mathrm{Re}\,\sigma \geq 0$. Then, taking the modulus of (3.14), we find that

$$\begin{aligned}\left|\sigma + d'(0) + \frac{d_m n^2 \pi^2}{L^2} + \frac{U_m^2}{4d_m}\right| &= b'(0)e^{-\gamma r}|e^{-\sigma r}| \\ &= b'(0)e^{-\gamma\tau}e^{-(\mathrm{Re}\,\sigma)\tau} \\ &\leq b'(0)e^{-\gamma\tau}\end{aligned}$$

so that σ is in a certain disk in the complex plane centered at a point (depending on n) on the negative real axis. This entire disk is in the left half plane (giving a contradiction) if

$$b'(0)e^{-\gamma\tau} < d'(0) + \frac{d_m n^2 \pi^2}{L^2} + \frac{U_m^2}{4d_m}. \tag{3.15}$$

We would like this contradiction for every $n = 1, 2, 3, \ldots$, therefore by imposing the condition

$$b'(0)e^{-\gamma\tau} < d'(0) + \frac{d_m \pi^2}{L^2} + \frac{U_m^2}{4d_m} \tag{3.16}$$

we are assured that $w(t, x) \to 0$ as $t \to \infty$ in the linearised approximation. It can be shown that if the strict inequality in (3.16) is reversed then an exponentially growing mode can be found, and so survival of the species is predicted in this case. Note, however, that (3.16) is only a condition for extinction if the immatures neither advect nor diffuse.

Condition (3.16) makes sense in every obvious respect; there is perhaps one point that needs a little explaining in ecological terms, and that is the fact that, for any given adult advection speed $U_m > 0$ (no matter how small), the condition is satisfied for sufficiently small d_m, resulting in extinction. The explanation for this is that, if adult diffusion is sufficiently small, then there is no adequate mechanism for sufficient movement of individual adults to the right, to compensate for the continual leftward movement by advection. As a consequence, an uninhabited gap develops at the right of the domain, and this gap continually widens until the whole domain is uninhabited, resulting in extinction. Recall that we are assuming at the moment that immatures do not move at all, otherwise they might in principle be able to recolonize the gap if they had sufficient diffusivity and sufficiently low advection.

4 Non-local delayed equations in bounded domains

In this section we concentrate on delayed non-local reaction diffusion equations in bounded domains. As previously mentioned there were some earlier efforts on this aspect and these have had significant impact on the development of the field.

4.1 Background. From modeling point of view, the case of a bounded domain is often viewed as more difficult than the case of a domain with no boundaries. The reason why the delay term is usually nonlocal is because individuals have moved during the period of the delay. For example, if the delay is a maturation time and if the immatures can move, then an individual will be born at one location and (if it survives the juvenile phase) become a mature adult at another. If the domain is finite then during the maturation phase the individual might reach one of the domain's boundaries during the period of the delay.

On a finite domain it is usually still possible to solve the von-Foerster equations and hence to derive the time delay term rigorously. The difference is that we will no longer be working with the fundamental solution of the heat equation on the whole real line as in the calculations presented thus far, but instead with the Green's function of a certain differential operator subject to the boundary conditions. In some cases it may be possible to find this Green's function explicitly. For example, in Gourley & So [24] the Green's function is just a Fourier series like (4.4) below. In a square or rectangular domain it will be a double Fourier series. In a circular domain it will be expressible in terms of Bessel functions. However, an explicit expression for the Green's function is usually unnecessary as all that is usually needed for the analysis is to know the structure of the nonlocal term and the general properties of the Green's function. Thus, it seems that it should usually be possible to carry out theoretical analysis on general finite domains (see, for example, Thieme and Zhao [84]).

In the rigorous study of nonlocal delay equations on finite domains an important early study was that of Yamada [98]. In fact, Yamada worked on a fairly general class of equations written in the abstract form

$$\frac{du(t)}{dt} = -A_p u(t) + Bu(t) + \int_{-\infty}^{t} C(t-s)u(s)\,ds + f(u_t), \qquad t>0$$

with initial data specified for $-\infty < t \leq 0$. Here, $-A_p$ is a closed linear operator generating an analytic semigroup $\{\exp(-tA_p)\}$ on $L^p(\Omega;\mathbb{R}^N)$, $B=(B_{ij})$, $C(t) = (C_{ij}(t))$ are bounded linear operators on $L^p(\Omega;\mathbb{R}^N)$ and f is a nonlinear operator and, in keeping with standard notation used in delay equations theory, u_t denotes the function with values $u(t+\theta)$ with $\theta \in (-\infty, 0]$ (but recall that, since we work with the equations in an abstract form, each $u(t+\theta)$ is in fact a function of $x \in \Omega$). Yamada [98] established local existence and uniqueness of (strong) solutions, together with general results on the asymptotic stability properties. As a particular example, Yamada studied the predator prey system (also considered earlier by Pozio [62]):

$$\begin{aligned}
\frac{\partial u_1(t,x)}{\partial t} &= \mu_1 \Delta u_1(t,x) + u_1(t,x)[a_1 - b_1 u_1(t,x) - c_1 g_1(x, u_{2,t})],\\
\frac{\partial u_2(t,x)}{\partial t} &= \mu_2 \Delta u_2(t,x) + u_2(t,x)[-a_2 - b_2 u_2(t,x) + c_2 g_2(x, u_{1,t})],
\end{aligned} \tag{4.1}$$

on a finite domain with homogeneous Neumann boundary conditions, where

$$g_i(x,v) = \int_{-\infty}^{0} \int_{\Omega} G_i(x,y,\theta) v(\theta,y)\,dy\,d\theta \tag{4.2}$$

which implies, once one appreciates the notation in use, that $g_1(x,u_{2,t})$ and $g_2(x,u_{1,t})$ are time delayed terms involving all values of u_1 and u_2 prior to time t. Yamada did not derive the model (4.1)-(4.2) but made assumptions on the G_i that are perfectly (and amazingly!) consistent with the assumptions that one would have to impose if the model were derived on a random walk or other basis, namely that

$$\int_{\Omega} G_i(x,y,\theta)\,dx = \int_{\Omega} G_i(x,y,\theta)\,dy = h_i(\theta), \qquad \theta \leq 0 \tag{4.3}$$

with

$$\int_{-\infty}^{0} h_i(\theta)\,d\theta = 1 \quad \text{ and } \quad \theta h_i(\theta) \in L^1((-\infty,0];\mathbb{R}).$$

Assumption (4.3), that integration of $G_i(x,y,\theta)$ with respect to either x or y removes *both* the x and the y dependence, might seem to be rather a strong assumption. However, it is very easily seen to be reasonable when one uses the Green's functions that arise from random walk or other derivations of the model. Each $G_i(x,y,\theta)$ will in practice be a product of the form $K_i(x,y,\theta)h_i(\theta)$, and if $\Omega = (0,\pi)$, each $K_i(x,y,\theta)$ will turn out to be an expression of the form

$$K_i(x,y,\theta) = \frac{1}{\pi} + \frac{2}{\pi}\sum_{n=1}^{\infty} e^{\mu n^2 \theta}\cos nx \cos ny, \qquad \theta \in (-\infty, 0], \quad x,y \in [0,\pi] \tag{4.4}$$

so that (4.3) is indeed satisfied. The issue is discussed in more detail in [28]. Indeed, even for more complicated domains properties of Green's functions assure us that conditions such as (4.3) can be satisfied.

Pozio [62] had earlier proved global stability theorems for (4.1)-(4.2). Yamada [98] tackled the system using another approach, establishing eradication of species u_2 (the predator) under one set of conditions, and coexistence of predator and prey in another. The approach by Yamada was subsequently taken up by Gourley and Ruan [28] to derive similar results for a competition model having the same types of nonlocal delay terms.

Another important earlier work on classes of equations that include the types of nonlocal delays that we now know arise naturally from structured population models or probabilistic random walk derivations, was the work by Pozio [63] who considered systems of the form

$$\frac{\partial u}{\partial t}(t,x) = D\Delta u(t,x) + f(u(t,x),(Ru)(t,x)), \qquad (t,x) \in (0,\infty)\times\Omega$$

where the nonlocal delay term is given by

$$(Ru)_i(t,x) = \int_0^\infty d\eta_i(s)\left(\int_\Omega G_i(s,x,y)u_i(t-s,y)\,dy\right)$$

with assumptions on the η_i and the G_i similar to those in [98]. Here, $u(t,x) \in \mathbb{R}^n$, the domain Ω is bounded and homogeneous Neumann boundary conditions are applied. Pozio [63] obtained asymptotic stability of equilibria and attractivity results for the rather general systems just described, and illustrated the ideas on several concrete problems. The approach was to introduce contracting rectangles techniques for delay reaction-diffusion systems.

Also in the early 1980s came the work of R. Redlinger which has been very influential. In [67] he studied very general systems that even permit nonlinear boundary conditions that may vary in time, or different components of the state variable vector to satisfy different types of boundary conditions. His systems were of the form

$$\begin{aligned} &L_k u_k = f_k(t,x,u(\cdot)), \qquad (t,x)\in(0,T]\times\Omega, \qquad k=1,2,\ldots,N,\\ &\partial u_k/\partial\nu_k = g_k(t,x,u(\cdot)) \quad \text{ on } \partial\Omega, \quad k=1,2,\ldots,M,\\ &u_k = 0 \quad \text{ on } \partial\Omega, \quad k=M+1,\ldots,N,\\ &u=\psi, \qquad (t,x)\in[-\tau,0]\times\overline{\Omega}, \end{aligned} \tag{4.5}$$

where $u=(u_1,u_2,\ldots,u_N)\in\mathbb{R}^N$, L is any linear uniformly parabolic diagonal operator and Ω is a bounded domain. Both f, and the nonlinearity g in the boundary condition, may contain functionals of u. Large classes of functionals are admitted including those just mentioned in the works of Yamada and Pozio. At the heart of Redlinger's work in [67] was the development of a comparison principle for delay equations that has proved very useful to many investigators in the subsequent years, and which has inspired others (for example, Kyrychko, Gourley & Bartuccelli [38] recently derived a similar principle for a model on a finite lattice).

The notation in Redlinger's [67] paper is quite complicated and difficult to describe here, so we shall summarise his comparison principle only for the case when no component of the solution satisfies $u_k = 0$ on $\partial\Omega$. The case when every component satisfies homogeneous Neumann boundary conditions is admitted by the result to be summarised here. Let $X_0 = C(D_0, \mathbb{R}^N)$ where $D_0 = [-\tau, T] \times \overline{\Omega}$, and let

$$B_{v,w} = \{u \in X_0 : \;\; v \leq u \leq w \;\; \text{in } D_0\}.$$

A pair of functions v and w with $v \leq w$ in $[-\tau, T] \times \overline{\Omega}$ are known as a pair of sub- and supersolutions for (4.5) in the case when $M = N$ if, for any $(t, x) \in (0, T] \times \Omega$, we have for each $k = 1, 2, \ldots, N$,

$$\begin{aligned} (L_k v_k)(t,x) &\leq f_k(t, x, \phi^k(\cdot)), \\ (L_k w_k)(t,x) &\geq f_k(t, x, \eta^k(\cdot)), \end{aligned} \tag{4.6}$$

and, for any $(t, x) \in (0, T] \times \partial\Omega$ and each $k = 1, 2, \ldots, N$,

$$\begin{aligned} (\partial v_k / \partial \nu_k)(t,x) &\geq g_k(t, x, \phi^k(\cdot)), \\ (\partial w_k / \partial \nu_k)(t,x) &\leq g_k(t, x, \eta^k(\cdot)), \end{aligned} \tag{4.7}$$

for all functions $\phi^k, \eta^k \in B_{v,w}$ such that $\phi^k_k = v_k$, $\eta^k_k = w_k$ at (t, x). Also, $v \leq \psi \leq w$ in $[-\tau, 0] \times \overline{\Omega}$. Under some additional technical conditions which are not restrictive, Redlinger [67] proved that if a pair of sub- and supersolutions v and w can be found then system (4.5) (but recall that we summarise here the case $M = N$ only) has exactly one regular solution $u \in B_{v,w}$.

The theory due to Redlinger is important because it can be used to prove both existence of a solution and to obtain upper and lower bounds on it. In the homogeneous Neumann case it is common to find sub- and supersolutions that depend only on time. Indeed, in problems without delay or nonlocal effects it is common to take them simply as solutions of the corresponding ordinary differential equations without diffusion. In delay equations one can still do this if the equation is scalar and the right hand side is increasing with respect to the delayed variable. If the latter does not hold then things become more complicated although sub-and supersolutions may still be found. Successive refinement of pairs of sub- and supersolutions is a common strategy for establishing global convergence to equilibria, and one of the first papers using this strategy and based on Redlinger's comparison principle was by Redlinger himself [68]. Laister [39] has utilised the technique to study whole classes of equations and systems with ecologically realistic nonlocal terms on finite domains, including some Lotka Volterra competitive and cooperative systems.

4.2 Single species on a bounded domain. To illustrate how to derive the time delay term on a finite spatial domain we shall follow the treatment in Al-Omari and Gourley [3] which is concerned with a single species. The original system that motivated these authors was the frequently cited system due to Aiello and Freedman [1], namely

$$\begin{aligned} u_i'(t) &= \alpha u_m(t) - \gamma u_i(t) - \alpha\, e^{-\gamma\tau} u_m(t - \tau), \\ u_m'(t) &= \alpha\, e^{-\gamma\tau} u_m(t - \tau) - \beta u_m^2(t), \end{aligned} \tag{4.8}$$

which constitutes a simple reasonable model for a stage structured population divided into immature individuals u_i and mature individuals u_m. All solutions of (4.8), other than the trivial solution, converge to the equilibrium solution

$$(u_i, u_m) \equiv \left(\frac{\alpha^2}{\beta\gamma}e^{-\gamma\tau}(1-e^{-\gamma\tau}),\ \frac{\alpha}{\beta}e^{-\gamma\tau}\right).$$

The delay τ is the time taken from birth to maturity and is assumed to be the same for every individual. Al-Omari and Gourley [3] generalised this model in three main respects: incorporating a distribution of possible maturation times, allowing more general birth and death rates, and incorporation of spatial effects on a finite domain with homogeneous Neumann boundary conditions. The first two of these generalisations result in the model

$$\begin{aligned} u_i'(t) &= b(u_m(t)) - \gamma u_i(t) - \int_0^\infty b(u_m(t-s))f(s)e^{-\gamma s}\,ds, \\ u_m'(t) &= \int_0^\infty b(u_m(t-s))f(s)e^{-\gamma s}\,ds - d(u_m(t)), \end{aligned} \tag{4.9}$$

where $\int_0^\infty f(s)\,ds = 1$ and $f \geq 0$, because f is a probability density function. More precisely, the quantity $f(s)\,ds$, with ds infinitesimal, is the probability that an individual member takes an amount of time between s and $s+ds$ to mature. The quantity $b(u_m(t-s))$ is the birth rate at time $t-s$.

On a finite spatial domain, an appropriate reaction diffusion extension of system (4.9) is

$$\begin{aligned} \frac{\partial u_i(t,x)}{\partial t} &= D_i\Delta u_i(t,x) + b(u_m(t,x)) - \gamma u_i(t,x) \\ &\quad - \int_0^\infty \int_\Omega G(x,y,s)f(s)e^{-\gamma s}b(u_m(t-s,y))\,dy\,ds, \\ \frac{\partial u_m(t,x)}{\partial t} &= D_m\Delta u_m(t,x) + \int_0^\infty \int_\Omega G(x,y,s)f(s)e^{-\gamma s}b(u_m(t-s,y))\,dy\,ds \\ &\quad - d(u_m(t,x)), \end{aligned} \tag{4.10}$$

for $x \in \Omega \subset \mathbb{R}^N$, $t > 0$. Though [3] were concerned only with the Neumann problem for (4.10), we shall in this section make some comments about the Dirichlet problem as well.

Let us consider the Neumann problem first. Then (4.10) is supplemented with homogeneous Neumann boundary conditions

$$\mathbf{n}\cdot\nabla u_i = \mathbf{n}\cdot\nabla u_m = 0 \qquad \text{on } \partial\Omega$$

where $\mathbf{n}$ is an outward pointing normal to $\partial\Omega$. The initial conditions for (4.10) are

$$u_i(0,x) = u_i^0(x), \qquad u_m(s,x) = u_m^0(s,x), \quad (x,s) \in \overline{\Omega}\times(-\infty,0] \tag{4.11}$$

where u_i^0 and u_m^0 are prescribed. For the Neumann problem, in system (4.10), $G(x,y,t)$ is the solution, subject to homogeneous Neumann boundary conditions, of

$$\frac{\partial G}{\partial t} = D_i\Delta_x G, \qquad G(x,y,0) = \delta(x-y)$$

where δ is the Dirac delta function, and where Δ_x means the Laplacian computed with respect to the first argument of $G(x,y,t)$ (recall that x is a vector in $\mathbb{R}^N$ here), and $y \in \mathbb{R}^N$ is for these purposes a parameter. For the Neumann problem we can show that

$$\int_\Omega G(x,y,t)\,dx = \int_\Omega G(x,y,t)\,dy = 1$$

for all $t \geq 0$.

Let us show how to derive the term with the double integral (the adult recruitment term) in the right hand side of the second equation of (4.10), assuming that the birth rate at a particular location and time is some function of the number of adults at that point in space at that time. Let $u(x,t,a)$ be the density of individuals of age a at location $x \in \Omega$ at time t. At (x,t) we want to calculate the rate of adult recruitment, i.e. the rate at which individuals are just reaching maturity. Of all the individuals just reaching maturity, some will have taken time $s > 0$ to mature (more precisely, an amount of time between s and $s+ds$ with ds infinitesimal). Let us look at these individuals first. They will have been born at various locations in Ω and will have drifted around, being at point x on becoming mature. During their juvenile phase let us assume they evolved according to the McKendrick von Foerster equation

$$\frac{\partial u}{\partial t} + \frac{\partial u}{\partial a} = D_i \Delta u - \gamma u, \qquad x \in \Omega, \qquad 0 < a < s$$

under homogeneous Neumann boundary conditions, and subject to

$$u(t,0,x) = b(u_m(t,x))$$

where b is the birth function. Our aim at this stage is to find an expression for $u(t,s,x)$ in terms of u_m. Let $v(x,r,a) = u(a+r,a,x)$. Then

$$\begin{aligned} \frac{\partial v}{\partial a} &= \left[\frac{\partial u}{\partial t} + \frac{\partial u}{\partial a}\right]_{t=a+r} = [D_i \Delta u - \gamma u]_{t=a+r} \\ &= D_i \Delta u(a+r,a,x) - \gamma u(a+r,a,x) \end{aligned}$$

so that

$$\frac{\partial v}{\partial a} = D_i \Delta v - \gamma v, \qquad x \in \Omega, \qquad a > 0 \tag{4.12}$$

with

$$v(r,0,x) = b(u_m(r,x)). \tag{4.13}$$

The solution of (4.12) subject to (4.13) is

$$v(r,a,x) = e^{-\gamma a} \int_\Omega G(x,y,a) b(u_m(r,y))\, dy$$

where the kernel G is the function described above. Hence

$$u(t,s,x) = e^{-\gamma s} \int_\Omega G(x,y,s) b(u_m(t-s,y))\, dy.$$

The above quantity gives us the rate, at (x,t), at which individuals are becoming mature having taken time s to mature. The total rate, at (x,t), at which individuals become mature is

$$\int_0^\infty u(t,s,x) f(s)\, ds$$

or

$$\int_0^\infty \int_\Omega G(x,y,s) f(s) e^{-\gamma s} b(u_m(t-s,y))\, dy\, ds.$$

Al-Omari and Gourley [3] studied the Neumann problem for system (4.10) in the cases of monotonically increasing and non-monotone birth functions. For the monotone case they utilised the powerful results in the frequently cited paper of Martin and Smith [49, 50], and for this case the following assumptions were made: $b(0) = 0$, $b(u_m)$ is strictly increasing for all $u_m > 0$, $d(0) = 0$, equation (4.14) below has a root $\hat{u}_m > 0$ with $b(u_m)\int_0^\infty f(s)e^{-\gamma s}\,ds > d(u_m)$ when $0 < u_m < \hat{u}_m$ and $b(u_m)\int_0^\infty f(s)e^{-\gamma s}\,ds < d(u_m)$ when $u_m > \hat{u}_m$. Under these assumptions, if additionally the initial data for system (4.10) satisfies $u_m^0(s,x) \geq 0$ for all $(x,s) \in \overline{\Omega} \times (-\infty, 0]$ with $u_m^0(0,x) \not\equiv 0$, and homogeneous Neumann boundary conditions apply, then the solution of (4.10)-(4.11) satisfies

$$(u_i(t,x), u_m(t,x)) \to (\hat{u}_i, \hat{u}_m)$$

as $t \to \infty$, uniformly for $x \in \overline{\Omega}$, where $\hat{u}_m$ and $\hat{u}_i$ are given by equations (4.14) and (4.15) below:

$$b(\hat{u}_m)\int_0^\infty f(s)e^{-\gamma s}\,ds = d(\hat{u}_m) \tag{4.14}$$

and

$$\hat{u}_i = \frac{b(\hat{u}_m)}{\gamma}\left(1 - \int_0^\infty f(s)e^{-\gamma s}\,ds\right) > 0. \tag{4.15}$$

Actually, [3] confined attention to the case when $f(s)$ has compact support but their results hold without this restriction. The case when $b(u_m)$ qualitatively resembles $Pu_m e^{-Au_m}$ was also considered; this is particularly relevant in modelling populations which cannot reproduce if too many individuals are present. In this case if $\hat{u}_m$ exists and is in the interval of u_m for which $b(u_m)$ is increasing then global convergence to $\hat{u}_m$ follows. If $\hat{u}_m$ exceeds the value of u_m at which $b(u_m)$ attains its maximum then the situation is more delicate. This case was considered in [3] using linearised analysis. Formal linearised analysis of nonlocal equations on finite domains is perfectly tractable if the model has been correctly derived and the ecologically relevant kernel is being used as we demonstrate here. Let $(\mu_k, \phi_k(x))$, $k = 0, 1, 2, \ldots$, be the eigenvalues and (normalised) eigenfunctions of $-\Delta$ on Ω with homogeneous Neumann boundary conditions. Then it is easy to see that

$$G(x,y,t) = \sum_{k=0}^{\infty} e^{-D_i \mu_k t}\phi_k(x)\phi_k(y) \tag{4.16}$$

and that

$$\int_\Omega G(x,y,s)\phi_k(y)\,dy = e^{-D_i\mu_k s}\phi_k(x).$$

The linearised equation about the equilibrium $\hat{u}_m$ has solutions of the form $e^{\sigma t}\phi_k(x)$ whenever

$$\sigma = -D_m\mu_k + b'(\hat{u}_m)\int_0^\infty f(s)e^{-(\gamma+\sigma+D_i\mu_k)s}\,ds - d'(\hat{u}_m). \tag{4.17}$$

If $b(u_m)$ qualitatively resembles $Pu_m e^{-Au_m}$, and if the equilibrium $\hat{u}_m$ exceeds the value of u_m at which $b(u_m)$ attains its maximum (so that $b'(\hat{u}_m) < 0$) then it can be shown that a sufficient condition for all the roots σ of (4.17) to satisfy $\operatorname{Re}\sigma < 0$ for all $k = 0, 1, 2, \ldots$ (i.e. for the equilibrium $\hat{u}_m$ of the second equation of (4.10) to be linearly stable to arbitrary small perturbations), is

$$|b'(\hat{u}_m)|\int_0^\infty f(s)e^{-\gamma s}\,ds < d'(\hat{u}_m). \tag{4.18}$$

One situation in which (4.18) holds is if $\hat{u}_m$ is not much larger than the value at which $b(u_m)$ attains its maximum, since then $|b'(\hat{u}_m)|$ will be close to zero. Another situation in which (4.18) holds is if there is significant juvenile mortality (large γ), but one must remember that the feasibility of the equilibrium $\hat{u}_m$ (i.e. strict positivity of $\hat{u}_m$) depends on γ. We can show that $\hat{u}_m$ decreases as γ increases, and that $\hat{u}_m$ may either lose feasibility at a finite value of γ, or may remain feasible for all γ (eg. if the birth function is linear and the death function behaves quadratically at small densities). Naturally, the question arises as to whether $\hat{u}_m$ might lose stability while still feasible, leading to new dynamical behavior in the model, and the answer to this question is yes. It is not too hard to show that limit cycles can be found in the case when $f(s) = \delta(s - \tau)$ (the choice appropriate for situations in which all individuals take the same amount of time τ to become mature).

As is common in stage-structured population models, the initial conditions have to satisfy a compatibility condition in order to make sense ecologically. Positivity of solutions does not in general hold without this compatibility condition. For system (4.10) the compatibility condition takes the form

$$u_i^0(x) = \int_{-\infty}^{0} \int_{\Omega} \left\{ \int_{-s}^{\infty} f(\xi)\, d\xi \right\} b(u_m^0(s,y)) e^{\gamma s} G(x, y, -s)\, dy\, ds. \tag{4.19}$$

This condition is simply accounting for every immature individual that is present at time $t = 0$ at position x. Each such individual was born at some position y at some time $s < 0$ (hence the presence of the birth rate $b(u_m^0(s, y))$ in (4.19)) and has to have survived until time $t = 0$ (accounted for by $e^{\gamma s}$); also it has to have moved from its birth location y at time $s < 0$ to position x at time $t = 0$ (accounted for by $G(x, y, -s)$). Moreover, it has to have remained immature between its birth time $s < 0$ and time 0 when it will be of age $-s$. The probability of maturing before reaching age $-s$ is $\int_0^{-s} f(\xi)\, d\xi$. The probability of remaining immature is $1 - \int_0^{-s} f(\xi)\, d\xi$, which equals $\int_{-s}^{\infty} f(\xi)\, d\xi$, thereby explaining the presence of the latter quantity in (4.19).

We have seen that in system (4.10) on homogeneous Neumann boundary conditions the kernel $G(x, y, t)$ is given by (4.16). If $\Omega = (0, \pi)$ with homogeneous Neumann boundary conditions then $\mu_0 = 0$ with $\phi_0 = 1/\sqrt{\pi}$ and $\mu_k = k^2$ with $\phi_k(x) = \sqrt{(2/\pi)} \cos kx$ for $k = 1, 2, 3, \ldots$ so that

$$G(x, y, t) = \frac{1}{\pi} + \frac{2}{\pi} \sum_{k=1}^{\infty} e^{-D_i k^2 t} \cos kx \cos ky. \tag{4.20}$$

If instead both the adults and juveniles are subject to homogeneous Dirichlet boundary conditions on the domain $\Omega = (0, \pi)$ then $G(x, y, t)$ becomes

$$G(x, y, t) = \frac{2}{\pi} \sum_{k=1}^{\infty} e^{-D_i k^2 t} \sin kx \sin ky \tag{4.21}$$

and the problem becomes one of solving (4.10) on homogeneous Dirichlet boundary conditions with $G(x, y, t)$ now given by (4.21). The Dirichlet problem raises the interesting question of minimum domain size or maximum diffusivity for the population to persist and not become extinct. It is as easy to deal with this issue for general domains as for one-dimensional ones. Thus, let $(\lambda_k, \psi_k(x))$, $k = 1, 2, 3, \ldots$, be the eigenvalues and (normalised) eigenfunctions of $-\Delta$ on Ω with homogeneous Dirichlet boundary conditions. Then

$$G(x, y, t) = \sum_{k=1}^{\infty} e^{-D_i \lambda_k t} \psi_k(x) \psi_k(y) \tag{4.22}$$

so we now consider the Dirichlet problem for (4.10) with (4.22). Note that, with $G(x,y,t)$ given by (4.22),

$$\int_{\Omega} G(x,y,s)\psi_k(y)\,dy = e^{-D_i\lambda_k s}\psi_k(x).$$

The linearised equation for u_m near the extinction state $u_m = 0$ has solutions of the form $u_m(x,t) = e^{\sigma t}\psi_k(x)$ whenever σ satisfies

$$\sigma + D_m\lambda_k + d'(0) = b'(0)\int_0^\infty f(s)e^{-(\gamma+\sigma+D_i\lambda_k)s}\,ds. \tag{4.23}$$

It can be shown that if

$$b'(0)\int_0^\infty f(s)e^{-(\gamma+D_i\lambda_k)s}\,ds < D_m\lambda_k + d'(0) \tag{4.24}$$

then all the roots σ of (4.23) satisfy $\operatorname{Re}\sigma < 0$. If it is the case that (4.24) holds for all $k = 1,2,3,\dots$ then we can conclude that $u_m(x,t) \to 0$ as $t \to \infty$ in the linearised approximation for the Dirichlet problem. Now, in the Dirichlet problem, the eigenvalues λ_k satisfy $0 < \lambda_1 < \lambda_2 < \cdots$, so the right hand side of (4.24) is increasing in k while the left hand side is decreasing in k. Accordingly, (4.24) only has to hold for $k = 1$. We have proved the following theorem which gives a condition for extinction in the case of a general domain for the situation when both the immatures and matures are subject to homogeneous Dirichlet boundary conditions.

Theorem 4.1 *If*

$$b'(0)\int_0^\infty f(s)e^{-(\gamma+D_i\lambda_1)s}\,ds < D_m\lambda_1 + d'(0) \tag{4.25}$$

then the zero solution of the second equation of (4.10), subject to homogeneous Dirichlet boundary conditions and with $G(x,y,t)$ given by (4.22), is linearly asymptotically stable.

Note that (4.25) will hold if either the diffusivity D_i of the immatures, or the diffusivity D_m of the matures, is sufficiently large. The principal eigenvalue λ_1 of $-\Delta$ for the Dirichlet problem is, of course, strictly positive and depends on the size of the domain Ω, and gets larger as Ω is scaled down in size without change in shape. Thus, extinction is predicted for sufficiently small domains.

Finally, we remark that there is no reason why one should not study problems with mixed boundary conditions. For example, perhaps the immatures cannot escape from the domain, so that Neumann boundary conditions apply to them, but the matures are perfectly free to escape into the hostile exterior and therefore satisfy a Dirichlet problem. This particular scenario would result in a study of the second equation of (4.10) subject to homogeneous Dirichlet boundary conditions for the variable u_m, but with the kernel $G(x,y,t)$ being given by (4.16) rather than (4.22). Such a study would involve both the Dirichlet eigenfunctions $\psi_k(x)$ and the Neumann eigenfunctions $\phi_k(x)$. We shall not pursue this idea further here but it does seem to be a fruitful area requiring additional study. Indeed, the immature and mature members of a species often behave in very different ways. Consider, for example, the motion of the larvae and mature flies in a *lucilia cuprina* infestation at a farm. Larvae will be confined to a particular field as they live in a host sheep, but the adults can fly and need not recognise the boundaries of a particular field or farm.

4.3 A study in two dimensions. In a substantial numerical investigation, Liang, Wu and Zhang [41] looked at a single species model with two age classes, immature and mature, in the two-dimensional spatial domain $\Omega = [0, L_x] \times [0, L_y]$. The maturation period τ was taken as fixed and the same for every individual. Both Neumann and Dirichlet boundary conditions were considered. Starting with an age structure model for the density $u(t, a, x, y)$ of the form

$$\frac{\partial u}{\partial t} + \frac{\partial u}{\partial a} = D(a)\frac{\partial^2 u}{\partial x^2} + D(a)\frac{\partial^2 u}{\partial y^2} - d(a)u, \qquad t > 0,\ a > 0,\ (x, y) \in \Omega,$$

they formed an equation for the total number of mature adults

$$w(t, x, y) = \int_\tau^\infty u(t, a, x, y)\, da$$

of the form

$$\frac{\partial w}{\partial t} = u(t, \tau, x, y) + D_m \frac{\partial^2 w}{\partial x^2} + D_m \frac{\partial^2 w}{\partial y^2} - d_m w \tag{4.26}$$

where it is assumed that the diffusion and death rates are constant for the mature population, i.e. $D(a) = D_m$ and $d(a) = d_m$ for $a \in [\tau, \infty)$. As always, it is necessary to calculate $u(t, \tau, x, y)$, supposing a birth law of the form

$$u(t, 0, x, y) = b(w(t, x, y)). \tag{4.27}$$

This is done in a manner similar to the calculations described in Subsection 4.2. Since the domain is a rectangle the kernel can be computed explicitly and is basically a double Fourier series. The result is that the adult population $w(t, x, y)$ satisfies, for the situation when both immatures and matures satisfy homogeneous Neumann boundary conditions:

$$\begin{aligned}
&\frac{\partial w}{\partial t} = D_m \frac{\partial^2 w}{\partial x^2} + D_m \frac{\partial^2 w}{\partial y^2} - d_m w + F(x, y, w(t-\tau, \cdot)), \qquad (x, y) \in \Omega,\ t > 0,\\
&w(t, x, y) = w_0(t, x, y), \qquad (x, y) \in \Omega, \qquad t \in [-\tau, 0],\\
&\frac{\partial}{\partial x} w(t, 0, y) = 0, \qquad \frac{\partial}{\partial x} w(t, L_x, y) = 0, \qquad 0 \le y \le L_y,\\
&\frac{\partial}{\partial y} w(t, x, 0) = 0, \qquad \frac{\partial}{\partial y} w(t, x, L_y) = 0, \qquad 0 \le x \le L_x,
\end{aligned} \tag{4.28}$$

where

$$\begin{aligned}
&F(x,y,w(t-\tau,\cdot)) \\
&\quad = \frac{\epsilon}{L_x L_y}\int_0^{L_x}\int_0^{L_y} b(w(t-\tau,z_x,z_y))\cdot \\
&\qquad \left\{1+\sum_{n=1}^{\infty}\left[\cos\frac{n\pi(x-z_x)}{L_x}+\cos\frac{n\pi(x+z_x)}{L_x}\right]e^{-\alpha(n\pi/L_x)^2}\right. \\
&\qquad +\sum_{m=1}^{\infty}\left[\cos\frac{m\pi(y-z_y)}{L_y}+\cos\frac{m\pi(y+z_y)}{L_y}\right]e^{-\alpha(m\pi/L_y)^2} \\
&\qquad +\sum_{n=1}^{\infty}\sum_{m=1}^{\infty}\left[\cos\frac{n\pi(x-z_x)}{L_x}+\cos\frac{n\pi(x+z_x)}{L_x}\right]\cdot \\
&\qquad \left.\left[\cos\frac{m\pi(y-z_y)}{L_y}+\cos\frac{m\pi(y+z_y)}{L_y}\right]e^{-\alpha[(n\pi/L_x)^2+(m\pi/L_y)^2]}\right\}dz_x\,dz_y
\end{aligned}$$

in which the parameters ϵ and α are defined by

$$\epsilon = e^{-\int_0^\tau d(a)\,da}, \qquad \alpha = \int_0^\tau D(a)\,da.$$

It is possible also to write down systems for Dirichlet boundary conditions, mixed boundary conditions and periodic boundary conditions. If there is no immature dispersal or mortality (i.e. if $\alpha \to 0$ and $\epsilon \to 1$) then the above model reduces to the local delay model:

$$\frac{\partial w}{\partial t} = D_m\frac{\partial^2 w}{\partial x^2} + D_m\frac{\partial^2 w}{\partial y^2} - d_m w + b(w(t-\tau,x,y)), \qquad (x,y)\in\Omega,\ t>0$$

which has been studied by Yoshida [99], Memory [52], So and Yang [78] and So, Wu and Yang [79].

Liang, Wu and Zhang [41] conducted an extensive numerical investigation of their model (4.28) and its variants. The nonlocal term has to be dealt with carefully. As usual in delay systems, when time is discretised the variable $t-\tau$ might not be in the time grid and so some form of interpolation is required. Composite Simpson's method can be used to handle the spatial integrals and a finite difference scheme obtained for the two dimensional delayed model. For the Dirichlet problem they observed extinction of the mature species if its diffusivity D_m is large enough; this can be studied analytically as explained in Subsection 4.2. Other simulations, which were carried out for two widely used particular cases for the birth function $b(\cdot)$, showed solutions converging to stable steady state solutions or to periodic waves depending on parameter values and boundary conditions. Numerical simulation work on large domains and in circular geometry for these nonlocal delay problems would seem to be a worthwhile area for further study, especially since even the corresponding purely time dependent ODEs with delay can very easily generate periodic solutions, which in the reaction diffusion case in two spatial dimensions suggests the possibility of other types of interesting two dimensional behavior such as spiral waves and target patterns.

There have also been numerical studies of equations with nonlocal delay terms in one dimensional bounded domains. We refer to the paper by Liang, So, Zhang and Zou [40].

4.4 Derivation of a nonlocal predator prey system. Subsection 4.2 was concerned with the details of the model derivation for a single species. We hope also to have convinced the reader that nonlocal terms (whether on finite or infinite domains) do not necessarily make the formal linearised analysis intractable. They may make it more algebraically complicated, but the basic principles remain unchanged.

In the present subsection we wish to discuss the derivation of nonlocal terms in predator prey systems. There are not so many papers on predator prey systems with delay that include detailed derivations of the model equations. However, the paper by Thieme and Zhao [84] emphasizes and pays careful attention to the model derivation and it is this paper that we choose to summarise here.

In the predator-prey context, delay is often caused by the conversion of consumed prey biomass into predator biomass (either in the form of body size growth or reproduction). Let $u(t,x)$ be the prey biomass density and $v(t,x)$ the predator biomass density. It is necessary to take account of movements of a predator during the assimilation process. As in Subsection 4.2, this can be achieved using an age-structured model. Let $w(t,a,x)$ be the generalised predator biomass (i.e. predator biomass plus prey biomass ingested by the predator that has not yet been assimilated) of age a, where a is measured from the time of ingestion, and let $P(a)$ be the probability that generalised biomass of age a has been assimilated into predator biomass. Then the predator biomass $v(t,x)$ is given by

$$v(t,x) = \int_0^\infty w(t,a,x)P(a)\,da. \tag{4.29}$$

For the generalised predator biomass $w(t,a,x)$, assume it satisfies

$$\begin{aligned}&\left(\frac{\partial}{\partial t}+\frac{\partial}{\partial a}\right)w(t,a,x) = d\Delta_x w(t,a,x) - \mu(x)w(t,a,x),\\ &w(t,0,x) = f(x,u(t,x),v(t,x))\end{aligned} \tag{4.30}$$

in which $f(x,u(t,x),v(t,x))$ is the rate at which the predators ingest prey biomass at x. Solving (4.30) gives

$$w(t,a,x) = \int_\Omega \Gamma(x,y,a)f(y,u(t-a,y),v(t-a,y))\,dy \tag{4.31}$$

where Γ is the Greens function or fundamental solution associated with $d\Delta_x - \mu(x)$ and the boundary conditions if any (the actual model derivation in [84] is for either finite or infinite domains). Insertion of (4.31) into (4.29) gives

$$v(t,x) = \int_0^\infty \left(\int_\Omega \Gamma(x,y,a)f(y,u(t-a,y),v(t-a,y))\,dy\right)P(a)\,da.$$

Now suppose that the time taken to convert prey biomass into predator biomass is always the same, and denote this time by τ. Then $P(a) = H(a-\tau)$, a Heaviside function. This leads to a simpler expression for $v(t,x)$ which after a change of variables becomes

$$v(t,x) = \int_{-\infty}^{t-\tau}\left(\int_\Omega \Gamma(x,y,t-s)f(y,u(s,y),v(s,y))\,dy\right)ds.$$

Differentiating this expression then yields the following form for the predator equation:

$$\frac{\partial v(t,x)}{\partial t} = (d\Delta_x - \mu(x))v(t,x) + \int_\Omega \Gamma(x,y,\tau) f(y, u(t-\tau,y), v(t-\tau,y))\,dy. \tag{4.32}$$

Though their model derivation is for either unbounded or bounded domains, Thieme and Zhao [84] focused their mathematical analysis on the case of a bounded domain. They supplemented an equation having the form of (4.32) with another equation for the prey and studied the system

$$\begin{aligned} \frac{\partial u_1(t,x)}{\partial t} &= d_1\Delta u_1(t,x) + u_1(t,x)g(x,u_1(t,x)) - f_1(x,u_1(t,x),u_2(t,x)) \\ \frac{\partial u_2(t,x)}{\partial t} &= d_2\Delta u_2(t,x) - \mu(x)u_2(t,x) \\ &\quad + \int_\Omega \Gamma_2(x,y,\tau) f_2(y, u_1(t-\tau,y), u_2(t-\tau,y))\,dy, \end{aligned} \tag{4.33}$$

on a bounded domain with boundary conditions of the form $B_i u_i = 0$ on $\partial\Omega$, where $B_i w = \partial w/\partial\nu + \alpha_i(x)w$, $\partial/\partial\nu$ denoting derivative along the outward normal to $\partial\Omega$. They proved results on existence, uniqueness and boundedness. Given persistence of the prey, they derived sharp conditions for uniform persistence of the predator and existence of a coexistence steady state. They also presented conditions for both prey and predator to go extinct.

In some other works in the literature the delay appears in the prey equation rather than the predator equation, and sometimes the delay is attributed to depleted plant resources. If the prey is herbivorous then such delay terms may actually be local because although the prey moves, the plants are immobile. This highlights the importance of careful attention being given to the model derivation. Earlier in this article we pointed out another significant example of a situation in which delay does not lead to non-local effects, namely, a maturation delay in a species whose immatures members do not move (as is in fact the case for many insect species which go through immobile larval stages before becoming adults).

4.5 Diffusive monotonicity and threshold dynamics. Although non-local interaction often yields additional technical difficulties in the qualitative study of reaction diffusion equations, the delay induced non-locality may also suggest new methods to facilitate the development of the qualitative theory of abstract functional differential equations. In this subsection, we introduce the work [95], that gives an exponential ordering, suggested by the delay-induced non-locality, for abstract functional differential equations. We shall see how this induced monotonicity of the solution semiflows can be applied to describe the threshold dynamics (extinction or persistence/convergence to positive equilibria) for a nonlocal and delayed reaction-diffusion population model.

As discussed already, reaction diffusion equations with delayed reaction terms and, more generally, abstract functional differential equations, have been widely used to model the evolution of a physical system distributed over a spatial domain. In the celebrated work of Martin and Smith [49, 50], the monotonicity of the semiflow generated by an abstract functional differential equation is established and the powerful theory of monotone dynamical systems is applied to obtain some detailed descriptions of the generic dynamics of the semiflow. In order for the semiflow to be order-preserving with respect to the pointwise ordering of the phase space, it is required that the nonlinear reaction term satisfy a certain quasimonotonicity condition which, in the special case of a reaction diffusion equation with delay, requires the delayed reaction term to be monotone and thus limits the applications in some cases. It is therefore natural to ask whether the quasimonotonicity condition can be relaxed. This question is addressed in Smith and Thieme [72, 73] for the case of ordinary functional differential equations (that is, the spatial diffusion is absent), where they established the monotonicity of the semiflow in a restricted but sufficiently large subspace with a non-standard exponential ordering.

Extending exponential ordering and its induced monotonicity to abstract functional differential equations and delayed reaction diffusion equations is the main goal of [95]. Such an extension, however, seems to require some new ideas, as the interaction of spatial diffusion and temporal delay requires comparison of solutions at different locations. This will be illustrated by the example provided below.

We now state the abstract setting in the work [95]. Let (X, P) be an ordered Banach space with $int(P) \neq \emptyset$. For $u, v \in X$, we write $u \geq_X v$ if $u - v \in P$; $u >_X v$ if $u - v \in P \setminus \{0\}$; $u >>_X v$ if $u - v \in int(P)$. Since P is a closed subset of X, the topology and ordering on X are compatible in the sense that if $u_n \geq_X v_n$, $u_n \to u$, $v_n \to v$, then $u \geq_X v$.

Let $A : Dom(A) \to X$ be the infinitesimal generator of an analytic semigroup $T(t)$ satisfying $T(t)P \subset P, \forall t \geq 0$. For convenience, we denote $T(t)$ by e^{At}. Let $\tau \geq 0$ be fixed and let $C := C([-\tau, 0], X)$. For $\mu \geq 0$, we define

$$K_\mu = \{\phi \in C : \phi(s) \geq_X 0, \text{ and } \phi(t) \geq_X e^{(A-\mu I)(t-s)}\phi(s), \ \forall 0 \geq t \geq s \geq -\tau\}.$$

Then K_μ is a closed cone in C. Let $\geq_\mu$ be the partial ordering on C induced by K_μ. The meaning of $\leq_\mu$ and $\leq_X$ should be clear. It can be shown that if $\phi \in C$ is differentiable on $(-\tau, 0)$ and $\phi(t) \in Dom(A), \forall t \in (-\tau, 0)$. Then $\phi \geq_\mu 0$ if and only if

$$\phi(-\tau) \geq_X 0, \text{ and } \frac{d\phi(t)}{dt} - (A - \mu I)\phi(t) \geq_X 0, \ \forall t \in (-\tau, 0).$$

Let D be an open subset of C. Assume that $F : D \to X$ is continuous and satisfies a Lipschitz condition on each compact subset of D. We consider the abstract functional differential equation

$$\begin{cases} \dfrac{du(t)}{dt} = Au(t) + F(u_t), \quad t > 0, \\ u_0 = \phi \in D. \end{cases} \tag{4.34}$$

By the standard theory (see, e.g., [49, 91]), for each $\phi \in D$, equation (4.34) admits a unique mild solution $u(t, \phi)$ on its maximal interval $[0, \sigma_\phi)$. Moreover, if $\sigma_\phi > \tau$, then $u(t, \phi)$ is a classical solution to (4.34) for $t \in (r, \sigma_\phi)$. For simplicity, in the rest of this subsection we assume that for each $\phi \in C$, equation (4.34) admits a unique mild solution $u(t, \phi)$ defined on $[0, \infty)$. Then (4.34) generates a semiflow on C by $\Phi(t)(\phi) = u_t(\phi)$, $\phi \in C$.

In order to get a monotone solution semiflow of (4.34) with respect to $\geq_\mu$, we will impose the following monotonicity condition on F:

(M_μ): $\quad \mu(\psi(0) - \phi(0)) + F(\psi) - F(\phi) \geq_X 0 \quad$ for $\phi, \psi \in D$ with $\phi \leq_\mu \psi$.

It is shown in [95] that if (M_μ) holds, and if $\phi \leq_\mu \psi$, then $u_t(\phi) \leq_\mu u_t(\psi)$ for all $t \geq 0$. In other words, condition (M_μ) is sufficient for $\Phi(t) : C \to C$ to be monotone with respect to $\leq_\mu$ in the sense that $\Phi(t)(\phi) \leq_\mu \Phi(t)(\psi)$ whenever $\phi \leq_\mu \psi$ and $t \geq 0$. In some applications of monotone dynamical systems, however, we need a strong order-preserving property (see, e.g., [75]). The semiflow $\Phi(t) : C \to C$ is said to be strongly order-preserving with respect to $\leq_\mu$ if it is monotone and whenever $\phi <_\mu \psi$ there exist open subsets U, V of C with $\phi \in U$ and $\psi \in V$ and $t_0 > 0$ such that $\Phi(t_0)(U) \leq_\mu \Phi(t_0)(V)$. It is shown in [95] that if $T(t)(P\backslash\{0\}) \subset int(P)$, $\forall t > 0$, and (SM_μ) holds, then the following slightly stronger condition than (M_μ) is sufficient for $\Phi(t)$ to be strongly order-preserving:

(SM_μ): $\mu(\psi(0) - \phi(0)) + F(\psi) - F(\phi) >>_X 0$ for $\phi, \psi \in C$ with $\phi \leq_\mu \psi$ and $\phi(s) <<_X \psi(s)$, $\forall s \in [-\tau, 0]$.

We now illustrate the above results by considering the nonlocal delayed and diffusive predator-prey model of [84] described in last subsection:

$$\begin{cases} \dfrac{\partial v(t,x)}{\partial t} = d\Delta v(t,x) - k(x)v(t,x) \\ \qquad\qquad + \displaystyle\int_\Omega \Gamma(x,y,\tau)g(y, v(t-\tau, y))dy, \ x \in \Omega, \ t > 0, \\ Bv(t,x) = 0, \qquad x \in \partial\Omega, \ t > 0, \\ v(t,x) = \phi(t,x) \geq 0, \ \ x \in \Omega, \ t \in [-\tau, 0], \end{cases} \tag{4.35}$$

where Ω is a bounded and open subset of $\mathbb{R}^N$ with $\partial\Omega \in C^{2+\theta}$ for a real number $\theta > 0$, Δ denotes the Laplacian operator on $\mathbb{R}^N$, *either* $Bv = v$ *or* $Bv = \partial v/\partial\nu + \alpha v$ for some nonnegative function $\alpha \in C^{1+\theta}(\partial\Omega, \mathbb{R})$, Γ is the Green's function associated with $A := d\Delta - k(\cdot)I$ and boundary condition $Bv = 0$, and ϕ is a given function to be specified later.

Let $p \in (N, \infty)$ be fixed. For each $\beta \in (1/2 + N/(2p), 1)$, let X_β be the fractional power space of $L^p(\Omega)$ with respect to $(-A, B)$ (see, e.g., [35]). Then X_β is an ordered Banach space with the cone X_β^+ consisting of all nonnegative functions in X_β, and X_β^+ has nonempty interior $int(X_\beta^+)$. Moreover, $X_\beta \subset C^{1+\nu}(\overline{\Omega})$ with continuous inclusion for $\nu \in [0, 2\beta - 1 - N/p)$. We denote the norm in X_β by $\|\cdot\|_\beta$. It is well-known that A generates an analytic semigroup $T(t)$ on $L^p(\Omega)$. Moreover, the standard parabolic maximum principle implies that the semigroup $T(t) : X_\beta \to X_\beta$ is strongly positive, that is, $T(t)(X_\beta^+) \setminus \{0\}) \subset int(X_\beta^+)$, $\forall t > 0$. Let $C := C([-\tau, 0], X_\beta)$ and $C^+ := C([-\tau, 0], X_\beta^+)$. Then model (4.35) can be written as the following abstract functional differential equation

$$\begin{cases} \dfrac{dv(t)}{dt} = Av(t) + T(r)g(\cdot, v(t-\tau)), \ t > 0, \\ v_0 = \phi \in C^+. \end{cases} \tag{4.36}$$

We further assume that $k(\cdot)$ is a positive Hölder continuous function on $\overline{\Omega}$ and $g \in C^1(\overline{\Omega} \times \mathbb{R}^+, \mathbb{R}^+)$ satisfies the following condition:

(G): $g(\cdot, 0) \equiv 0$, $\partial_v g(x, 0) > 0, \forall x \in \Omega$, g is bounded on $\overline{\Omega} \times \mathbb{R}^+$, and for each $x \in \Omega$, $g(x, \cdot) : \mathbb{R}^+ \to \mathbb{R}^+$ is strictly sublinear in the sense that $g(x, \alpha v) > \alpha g(x, v)$, $\forall \alpha \in (0, 1)$, $v > 0$.

Using a standard argument, we can show that the nonlocal elliptic eigenvalue problem

$$\begin{cases} \lambda w(x) = d\Delta w - k(x)w(x) + \displaystyle\int_\Omega \Gamma(x,y,\tau)\partial_v g(y,0)w(y)dy, \quad x \in \Omega, \\ Bw = 0, \qquad x \in \partial\Omega \end{cases} \tag{4.37}$$

has a principal eigenvalue, which is denoted by $\lambda_0(d, \tau, \partial_v g(\cdot, 0))$.

For any $\phi \in C^+$, let $v(t, \phi)$ denote the solution of (4.35). Define $k_0 := \min\{k(x) : x \in \overline{\Omega}\}$, and

$$b(\tau) := \sup\left\{\int_\Omega \Gamma(x, y, \tau) g(y, \varphi(y)) dy : x \in \overline{\Omega}, \varphi \in X_\beta^+\right\}, \quad M(r) := \frac{b(\tau)}{k_0},$$
$$L(\tau) := \min\{\partial_v g(x, v) : x \in \overline{\Omega}, v \in [0, M(\tau)]\}.$$

Then we have the following threshold dynamics for model system (4.35): if the zero solution of (4.35) is linearly stable, then the species goes to extinction; if it is linearly unstable, then the species is uniformly persistent. More precisely, we have

Theorem 4.2 *Let $v^* \in int(X_\beta)$ be fixed and let (G) hold.*

(i): *If $\lambda_0(d, \tau, \partial_v g(\cdot, 0)) < 0$, then $\lim_{t\to\infty} \|v(t, \phi)\|_\beta = 0$ for every $\phi \in C^+$;*

(ii): *If $\lambda_0(d, \tau, \partial_v g(\cdot, 0)) > 0$, then (4.35) admits at least one steady state solution φ^* with $\varphi^*(x) > 0$, $\forall x \in \Omega$, and there exists a $\delta > 0$ such that for every $\phi \in C^+$ with $\phi(0, \cdot) \not\equiv 0$, there is $t_0 = t_0(\phi) > 0$ such that $v(t, \phi)(x) \geq \delta v^*(x)$, $\forall x \in \overline{\Omega}$, $t \geq t_0$.*

As an application of the order-preserving property, Wu and Zhao [95] also obtained the following sufficient conditions under which the species stabilizes eventually at positive steady states in case (ii):

Theorem 4.3 *Assume that (G) holds and $\lambda_0(d, \tau, \partial_v g(\cdot, 0)) > 0$.*

(i): *If $L(\tau) \geq 0$, then (4.35) admits a unique positive steady state φ^*, and $\lim_{t\to\infty} \|v(t, \phi) - \varphi^*\|_\beta = 0$ for every $\phi \in C^+$ with $\phi(0, \cdot) \not\equiv 0$.*

(ii): *If $L(\tau) < 0$ and $\tau|L(\tau)| < 1/e$, then there exists an open and dense subset S of C^+ with the property that for every $\phi \in S$ with $\phi(0, \cdot) \not\equiv 0$, there is a positive steady state ψ of (4.35) such that $\lim_{t\to\infty} \|v(t, \phi) - \psi\|_\beta = 0$.*

We briefly outline the proof here, in order to illustrate the application of the exponential ordering by considering the case (ii) where $L(\tau) < 0$ and $\tau|L(\tau)| < 1/e$. We define $f(\alpha) := \alpha + L(\tau)e^{\alpha\tau}$, $\forall \alpha \in [0, \infty)$. It then follows that $f(0) < 0$ and $f''(\alpha) \leq 0, \forall \alpha \in [0, \infty)$. If $\tau = 0$, then $f(\alpha) > 0$ for all $\alpha > |L(0)|$. If $0 < \tau|L(\tau)| < \frac{1}{e}$, then $f(\alpha)$ reaches its maximum value at $\alpha_0 = -\frac{1}{\tau}\ln(\tau|L(\tau)|) > 0$ and $f(\alpha_0) > 0$. Consequently, we can fix a real number $\mu > 0$ such that $f(\mu) = \mu + L(\tau)e^{\mu\tau} > 0$. Let $F : C^+ \to X_\beta$ be defined as the right hand side of the equation under consideration, and let K_μ be defined as above with $X = X_\beta$, $P = X_\beta^+$ and $A = d\Delta - k(\cdot)I$. By the definition of $L(\tau)$, there holds

$$g(x, v_2) - g(x, v_1) \geq L(\tau)(v_2 - v_1), \quad \forall x \in \overline{\Omega}, \quad 0 \leq v_1 \leq v_2 \leq M(\tau).$$

Assume that $\phi, \psi \in Y$ satisfy $\phi \leq_\mu \psi$ and $\phi(s) <<_{X_\beta} \psi(s)$, $\forall s \in [-\tau, 0]$. Clearly, $\psi - \phi \in K_\mu$ implies that

$$\psi(0) - \phi(0) \geq_{X_\beta} e^{(A-\mu I)\tau}(\psi(-\tau) - \phi(-\tau)) = T(\tau)e^{-\mu\tau}(\psi(-\tau) - \phi(-\tau)).$$

It then follows that

$$\begin{aligned}
&\mu(\psi(0) - \phi(0)) + F(\psi) - F(\phi) \\
&\geq_{X_\beta} \mu(\psi(0) - \phi(0)) + L(\tau)T(\tau)(\psi(-\tau) - \phi(-\tau)) \\
&\geq_{X_\beta} (\mu + L(\tau)e^{\mu\tau})\, e^{-\mu\tau}T(\tau)(\psi(-\tau) - \phi(-\tau)) >>_{X_\beta} 0. \qquad (4.38)
\end{aligned}$$

Thus condition (SM_μ) holds for $F: Y \to X_\beta$, and hence, $\Phi(t): Y \to Y$ is strongly order-preserving with respect to $\leq_\mu$. Let $\phi^* \geq_\mu 0$ be a given element so that $\phi^*(s) >>_{X_\beta} 0$, $\forall s \in [-r, 0]$.

Then for any $\psi \in Y$, either the sequence of points $\psi + \frac{1}{n}\phi^*$ or $\psi - \frac{1}{n}\phi^*$ is eventually contained in Y and approaches ψ as $n \to \infty$, and hence, each point of Y can be approximated either from above or from below in Y. Clearly, $\Phi(t): Y \to Y$ has a global compact attractor in Y. Note that the cone K_μ has empty interior in C. Fix a $\psi(\cdot) \in int(X_\beta^+)$ such that $d\Delta\psi - k(x)\psi \leq 0$, $\forall x \in \Omega$, and $B\psi = 0$, $\forall x \in \partial\Omega$ (e.g., taking $\psi(x)$ as a positive steady state of (4.35)). Then $\psi \in K_\mu$. Define

$$C_\psi = \{\phi \in C: \text{ there exists } \beta \geq 0 \text{ such that } -\beta\psi \leq_\mu \phi \leq_\mu \beta\psi\}$$

and

$$\|\phi\|_\psi = \inf\{\beta \geq 0: \ -\beta\psi \leq_\mu \phi \leq_\mu \beta\psi\}, \ \forall \phi \in C_\psi.$$

Then $(C_\psi, \|\cdot\|_\psi)$ is a Banach space and $C_\psi^+ := C_\psi \cap K_\mu$ is a closed cone in C_ψ with nonempty interior. Using the smoothing property of the semiflow $\Phi(t)$ on C^+ and the fundamental theory of abstract functional differential equations, we can show that for each $t > \tau$, $\Phi(t)Y \subset Y \cap C_\psi$, $\Phi(t): Y \to Y \cap C_\psi$ is continuous, $\Phi(t)\phi_2 - \Phi(t)\phi_1 \in int(C_\psi^+)$ for any $\phi_1, \phi_2 \in Y$ with $\phi_2 >_\mu \phi_1$, and for each nonnegative equilibrium φ of $\Phi(t)$, the Frechet derivative at φ of $\Phi(t): Y \cap C_\psi \to Y \cap C_\psi$ exists and is compact and strongly positive on C_ψ^+ (see, e.g., [73]). By [75, Theorem 2.4.7 and Remark 2.4.1], it then follows that there is an open and dense subset U of Y such that every orbit of $\Phi(t)$ starting from U converges to an equilibrium in Y.

Clearly, the condition that $L(\tau) < 0$ and $\tau|L(\tau)| < 1/e$ still holds under small perturbations of $b(\tau)$. It then follows that there is a small $\epsilon > 0$ such that the generic convergence also holds in

$$Y_\epsilon := \{\phi \in C^+: \ \phi(s, x) \leq M_\epsilon(\tau), \ \forall s \in [-\tau, 0], \ x \in \Omega\}$$

where $M_\epsilon(\tau) := b_\epsilon(\tau)/k_0 = M(\tau) + \epsilon/k_0$ and $b_\epsilon(\tau) := b(\tau) + \epsilon$. We can show that every orbit of $\Phi(t)$ in C^+ eventually enters into Y_ϵ, therefore, the conclusion (ii) follows from the generic convergence in Y_ϵ.

We refer also to [97] for some explicit description of the effects of time delay and dispersal rate on the global dynamics of structured population models in bounded domains.

5 Structured populations on pattices

Earlier in this article mention was made of the paper by So, Wu and Zou [80], in which a model was formulated for a single species living in two patches with two age classes. The persistence, convergence and Hopf bifurcation problem for the two-patch model was also discussed in [96].

There have been some works concentrating on species living in n patches with nonlocal effects due to delay and we summarise two of them here.

Smith and Thieme [73] considered the following model with n patches each of which offers a different quality of life:

$$\left(\frac{\partial}{\partial t} + \frac{\partial}{\partial a}\right) u_j(t, a) = \sum_{k=1}^{n} \gamma_{jk} u_k(t, a) - \left(\sum_{k=1}^{n} \gamma_{kj} + \mu_j\right) u_j(t, a) \tag{5.1}$$

with birth law

$$u_j(t,0) = g_j\left(\int_{\tau_j}^{\infty} u_j(t,a)\,da\right)$$

where each g_j is a bounded function. In this model the per capita reproduction rates, mortality rates and maturation delays can vary from patch to patch. Additionally the model permits individuals to migrate from any one patch directly to any other. The quantity γ_{jk} is the per capita migration rate from node k to node j, and the key assumption on the migration terms is that the matrix (γ_{jk}) be irreducible. Let $v_j(t)$ denote the number of reproductive adults in node j, so that

$$v_j(t) = \int_{\tau_j}^{\infty} u_j(t,a)\,da,$$

then under the standard and reasonable assumption that $u_j(t,\infty) = 0$ it follows that

$$v_j'(t) = \sum_{k=1}^{n} \gamma_{jk} v_k(t) - \zeta_j v_j(t) + u_j(t,\tau_j)$$

where

$$\zeta_j = \sum_{k=1}^{n} \gamma_{kj} + \mu_j.$$

Letting A denote the matrix $A = (\alpha_{jk})$ where $\alpha_{jk} = \gamma_{jk} - \zeta_j\delta_{jk}$, where δ_{jk} denotes the Kronecker delta, integration of (5.1) along characteristics gives

$$u_j(t,\tau_j) = (e^{\tau_j A}u(t-\tau_j,0))_j$$

so that the variables $v_j(t)$ satisfy the following delay equations with global interaction terms:

$$v_j'(t) = \sum_{k=1}^{n} \alpha_{jk} v_k(t) + \sum_{k=1}^{n} (e^{\tau_j A})_{jk}\, g_k(v_k(t-\tau_j)), \qquad t > 0. \tag{5.2}$$

Smith and Thieme [73] studied the convergence properties of (5.2) under the assumptions that the g_k are bounded (but not necessarily monotone) and the delays are sufficiently small.

Kyrychko, Gourley and Bartuccelli [38] derived a similar model for a single species living in N identical patches. Their model allows migration only on a nearest neighbor basis, though still has a global interaction term due to the delay. Let $u_j(t,a)$ denote the density of the population at the j-th patch at time $t \geq 0$ and age $a \geq 0$. Let $D(a)$ and $d(a)$ denote the diffusion and death rates of the population at age a. Assume that the patches are located at the integer nodes $j = 1, 2, \ldots, N$ of a one-dimensional lattice. A simple model with diffusion on a nearest neighbor basis is then

$$\left(\frac{\partial}{\partial t} + \frac{\partial}{\partial a}\right)\mathbf{u}(t,a) = D(a)A\mathbf{u}(t,a) - d(a)\mathbf{u}(t,a) \tag{5.3}$$

for $t > 0$, where

$$\mathbf{u}(t,a) = (u_1(t,a), \ldots, u_N(t,a))^T$$

and

$$A = \begin{pmatrix} -1 & 1 & 0 & \cdots & 0 & 0 & 0 \\ 1 & -2 & 1 & \cdots & 0 & 0 & 0 \\ \vdots & \vdots & \vdots & & \vdots & \vdots & \vdots \\ 0 & 0 & 0 & \cdots & 1 & -2 & 1 \\ 0 & 0 & 0 & \cdots & 0 & 1 & -1 \end{pmatrix}, \tag{5.4}$$

subject to

$$u_j(t,0) = b(w_j(t)) \tag{5.5}$$

where $w_j(t)$ is the total mature population at the j-th patch, given by $w_j(t) = \int_\tau^\infty u_j(t,a)\,da$ and $b(\cdot)$ is the birth function, which always satisfies $b(0) = 0$. Equation (5.3), with A given by (5.4), constitutes a discrete analogue of the homogeneous Neumann problem. From (5.3)-(5.5) a system of equations satisfied by the total matured population $w_j(t)$, $j = 1, 2, \ldots, N$ can be derived, and turns out to involve the kernel $\beta(t,k,j)$ defined by

$$\beta(t,k,j) = \frac{1}{N} + \frac{2}{N}\sum_{l=1}^{N} e^{-4\sin^2\left(\frac{l\pi}{2N}\right)t} \cos\left[(2j-1)\frac{l\pi}{2N}\right] \cos\left[(2k-1)\frac{l\pi}{2N}\right] \tag{5.6}$$

which is basically a discrete version of (4.20). It can be shown that $\beta(t,k,j) > 0$ for all $t > 0$ and all $1 \le k, j \le N$. If the diffusion and death rates of the mature population are age-independent, i.e.

$$D(a) = D_m, \qquad d(a) - d_m \quad \text{for } a \in [\tau, \infty),$$

where $D_m > 0$ and $d_m > 0$ are constants, if $\beta(t,k,j)$ is defined by (5.6) and if

$$\mu = e^{-\int_0^\tau d(z)\,dz}, \qquad \alpha = \int_0^\tau D(z)\,dz, \tag{5.7}$$

then for $t \ge \tau$ the total matured population $w_j(t)$ obeys

$$\frac{d}{dt}\begin{pmatrix} w_1(t) \\ w_2(t) \\ \vdots \\ w_{N-1}(t) \\ w_N(t) \end{pmatrix} = \mu \sum_{k=1}^{N} b(w_k(t-\tau)) \begin{pmatrix} \beta(\alpha,k,1) \\ \beta(\alpha,k,2) \\ \vdots \\ \beta(\alpha,k,N-1) \\ \beta(\alpha,k,N) \end{pmatrix}$$

$$+D_m \begin{pmatrix} -1 & 1 & 0 & \cdots & 0 & 0 & 0 \\ 1 & -2 & 1 & \cdots & 0 & 0 & 0 \\ \vdots & \vdots & \vdots & & \vdots & \vdots & \vdots \\ 0 & 0 & 0 & \cdots & 1 & -2 & 1 \\ 0 & 0 & 0 & \cdots & 0 & 1 & -1 \end{pmatrix} \begin{pmatrix} w_1(t) \\ w_2(t) \\ \vdots \\ w_{N-1}(t) \\ w_N(t) \end{pmatrix} - d_m \begin{pmatrix} w_1(t) \\ w_2(t) \\ \vdots \\ w_{N-1}(t) \\ w_N(t) \end{pmatrix}. \tag{5.8}$$

Ignoring the transient phase $t \in (0, \tau)$ (when some different nonautonomous equations apply), [38] studied (5.8) for $t > 0$ subject to

$$w_j(s) = w_j^0(s) \ge 0, \qquad j = 1, 2, \ldots, N, \qquad s \in [-\tau, 0] \tag{5.9}$$

with $w_j^0(s)$ prescribed. The central result in [38] is the following comparison theorem for the case of a non-monotone birth function $b(w)$. This comparison theorem can be compared to the one for continuous space proved by Redlinger [67] which was summarised in Subsection 4.1 of the present paper.

Theorem 5.1 *Let the birth function $b(w)$ be a differentiable function for all $w \geq 0$ and satisfy $b(0) = 0$, $b(w) > 0$ when $w > 0$. Let $\hat{w}$ and $\bar{w}$ be a pair of sub- and supersolutions for (5.8)-(5.9), i.e. a pair of functions satisfying*

(i) $\hat{w}_j(t) \leq \bar{w}_j(t)$ *for all* $t \in [-\tau, \infty)$, $j = 1, 2, \ldots, N$;

(ii) *letting* $\bar{\mathbf{w}}(t) = (\bar{w}_1(t), \ldots, \bar{w}_N(t))^T$ *and* $\hat{\mathbf{w}}(t) = (\hat{w}_1(t), \ldots, \hat{w}_N(t))^T$, *for* $t > 0$ *and* $j = 1, 2, \ldots, N$,

$$\frac{d\hat{\mathbf{w}}(t)}{dt} \leq D_m A\hat{\mathbf{w}}(t) - d_m\hat{\mathbf{w}}(t) + \mu \sum_{k=1}^{N} b(\varphi_k(t-\tau)) \begin{pmatrix} \beta(\alpha, k, 1) \\ \beta(\alpha, k, 2) \\ \vdots \\ \beta(\alpha, k, N-1) \\ \beta(\alpha, k, N) \end{pmatrix}, \tag{5.10}$$

and

$$\frac{d\bar{\mathbf{w}}(t)}{dt} \geq D_m A\bar{\mathbf{w}}(t) - d_m\bar{\mathbf{w}}(t) + \mu \sum_{k=1}^{N} b(\varphi_k(t-\tau)) \begin{pmatrix} \beta(\alpha, k, 1) \\ \beta(\alpha, k, 2) \\ \vdots \\ \beta(\alpha, k, N-1) \\ \beta(\alpha, k, N) \end{pmatrix}, \tag{5.11}$$

for all functions $\varphi_j(t)$ *such that* $\hat{w}_j(t) \leq \varphi_j(t) \leq \bar{w}_j(t)$, $t \in [-\tau, \infty)$, $j = 1, 2, \ldots, N$;

(iii) $\hat{w}_j(s) \leq w_j^0(s) \leq \bar{w}_j(s)$, $s \in [-\tau, 0]$, $j = 1, 2, \ldots, N$, *where* $w_j^0(s)$ *is the initial data for (5.8).*

Then the solution $w_j(t)$ *of (5.8)-(5.9) satisfies*

$$\hat{w}_j(t) \leq w_j(t) \leq \bar{w}_j(t) \quad \textit{for all } t > 0,\ j = 1, 2, \ldots, N.$$

Kyrychko, Gourley and Bartuccelli [38] used Theorem 5.1 to establish convergence results for (5.8), concentrating particularly on the case of a non-monotone birth function $b(w)$ having the qualitative properties of Pwe^{-Aw}. If a positive equilibrium exists then, as might be expected, whether it lies in the 'increasing' or 'decreasing' part of $b(w)$ is very important.

As mentioned before, in Weng, Huang and Wu [90], the following system of delay differential equations for a single species population with two age classes distributed over a patchy environment consisting of the integer nodes $j \in \mathbf{Z} := \{0, \pm 1, \pm 2, \ldots\}$ of a one-dimensional lattice is derived

$$\begin{aligned} \frac{dw_j(t)}{dt} &= \frac{\mu}{2\pi} \sum_{k=-\infty}^{\infty} \beta_\alpha(j-k) b(w_k(t-\tau)) + D_m [w_{j+1}(t) + w_{j-1}(t) - 2w_j(t)] \\ &\quad - d_m w_j(t), \qquad t > 0, \qquad j \in \mathbf{Z}. \end{aligned} \tag{5.12}$$

Here $w_j(t)$ denotes the total number of adults (i.e., the total number of age at least τ) in the jth patch, and $\tau > 0$ is the length of the juvenile phase. The function b denotes the birth function and satisfies $b(0) = 0$. The constants D_m and d_m are respectively the diffusion coefficient and death rate for the mature population, and the τ-dependent parameters μ and α are given by

$$\mu = \exp\left(-\int_0^\tau d(a)\, da\right), \qquad \alpha = \int_0^\tau D(a)\, da, \tag{5.13}$$

where $D(a)$ and $d(a)$ are the diffusion coefficient and death rate for the population at age a.

The coefficients $\beta_\alpha(l)$ in (5.12) are given by

$$\beta_\alpha(l) = 2e^{-2\alpha}\int_0^\pi \cos(l\omega)e^{2\alpha\cos\omega}\,d\omega. \tag{5.14}$$

Spatially uniform equilibria (i.e. equilibria independent of j) of (5.12) satisfy $\mu b(w) = d_m w$. Of course, zero is an equilibrium. Weng et al. [90] were concerned with the situation when there is one other equilibrium $w^+ > 0$, and with the possibility of traveling wave-front solutions connecting 0 to the other equilibrium w^+. In this discrete-space setting such a solution is a solution of the single variable $j + ct$ where $c > 0$ is the wave speed. They showed that such a wave-front exists for all c exceeding some minimum value c^*, and they also proved that c^* is the asymptotic speed of wave propagation if τ is not too large and the initial data satisfies certain biologically realistic conditions. More general lattice delay differential systems with non-local interactions are studied in [45, 46, 47], and analogous results for the continuous version of (5.12) will be discussed in a later section.

In [42], a very general and abstract framework was developed to deal with asymptotic speeds of monotone semiflows. These abstract results were also applied to (5.12) to show the existence of monotone traveling wave for $c = c^*$, the nonexistence of any traveling wave for $c < c^*$, and the coincidence of c^* with the asymptotic speed of propagation for any nonnegative time delay.

Let ℓ^∞ be the Banach space

$$\ell^\infty = \left\{ c = (c_j)_{j\in\mathbf{Z}}; \quad \|c\|_{\ell^\infty} := \sup_{j\in\mathbf{Z}} |c_j| < \infty \right\}.$$

It can be shown that if $b : \mathbb{R} \to \mathbb{R}$ is a C^1-smooth bounded function, then for any continuous $\phi : [-\tau, 0] \to \ell^\infty$, equation (5.12) has a unique solution $w^\phi : [-\tau,\infty) \to \ell^\infty$ with $w^\phi(s) = \phi(s)$ on $[-\tau, 0]$. Moreover, if $\phi_j(s) \geq 0$ for all $j \in \mathbf{Z}$ and $s \in [-\tau, 0]$, then $w_j^\phi(t) \geq 0$ for all $j \in \mathbf{Z}$ and $t \geq 0$.

The issue of extinction is considered in Gourley and Wu [26]. Recall that the condition $\mu b(w) < d_m w$ for all $w > 0$ ensures that there is no spatially homogeneous equilibrium other than 0 and is the weakest possible condition that ensures extinction for biologically sensible birth functions $b(w)$. The theorem below confirms that this is also a sufficient condition for extinction. More precisely, the theorem essentially states that extinction will occur if there is no positive equilibrium and if the initial data decays to zero sufficiently fast as $|j| \to \infty$. To state the theorem, we shall let ℓ^2 denote the Hilbert space of sequences $\{\xi_j\}_{j=-\infty}^\infty$ such that $\sum_{j=-\infty}^\infty \xi_j^2 < \infty$, with the norm

$$\|\xi\|_{\ell^2} = \left(\sum_{j=-\infty}^\infty \xi_j^2 \right)^{1/2}.$$

Theorem 5.2 *Let the initial data $\phi : [-\tau, 0] \to \ell^2$ be continuous and $\phi_j(s) \geq 0$ for each $s \in [-\tau, 0]$. Assume also that $\mu b(w) < d_m w$ for all $w > 0$ and that $\sup_{w\geq 0} |b'(w)| < \infty$. Then*

$$\sup_{j\in\mathbf{Z}} w_j^\phi(t) \to 0 \qquad \text{as } t \to \infty. \tag{5.15}$$

[26] also considered the case when (5.12) has a positive uniform equilibrium state, and they proved the existence of a family of periodic traveling waves, arising via a Hopf bifurcation from this uniform equilibrium state. Their approach is based on the Hopf bifurcation theory for functional differential equations of mixed type, due to A. Rustichini [69].

They considered traveling waves of the type $w_j(t) = \phi(t + cj)$ with $\phi : \mathbb{R} \to \mathbb{R}$. The profile ϕ satisfies, for $s = t + cj$, that

$$\begin{aligned} \dot{\phi}(s) = \quad & D_m[\phi(s+c) + \phi(s-c) - 2\phi(s)] - d_m\phi(s) \\ & + \frac{\mu}{2\pi} \sum_{l=-\infty}^{\infty} \beta_\alpha(l)\, b((\phi(s - r + lc)). \end{aligned} \tag{5.16}$$

They were interested in a periodic wave of period $2\pi/\omega$ (with $\omega > 0$), and so

$$\phi(s + 2\pi/\omega) = \phi(s), \quad s \in \mathbb{R}.$$

Their approach is to seek a Hopf bifurcation of (5.16) from the steady state K, where $K > 0$ satisfies

$$d_m K = \mu b(K), \quad K > 0. \tag{5.17}$$

Linearizing (5.16) at the constant solution K and, with

$$b'(K) = B, \tag{5.18}$$

they obtained

$$\begin{aligned} \dot{\phi}(s) = \quad & D_m[\phi(s+c) + \phi(s-c) - 2\phi(s)] - d_m\phi(s) \\ & + \frac{\mu}{2\pi} B \sum_{l=-\infty}^{\infty} \beta_\alpha(l)\phi(s - r + lc). \end{aligned} \tag{5.19}$$

This, with $\phi(s) = e^{\lambda s}$, gives the characteristic equation

$$\lambda = D_m[e^{\lambda c} + e^{-\lambda c} - 2] - d_m + \frac{\mu}{2\pi} B \sum_{l=-\infty}^{\infty} \beta_\alpha(l) e^{-\lambda r + l\lambda c}. \tag{5.20}$$

Letting $\lambda = iz$ in (5.20), with $z \geq 0$, they obtained

$$4D_m \sin^2(cz/2) + d_m = e^{-4\alpha \sin^2(cz/2)} \mu B \cos(z\tau) \tag{5.21}$$

and

$$-z = e^{-4\alpha \sin^2(cz/2)} \mu B \sin(z\tau). \tag{5.22}$$

Applying the Hopf bifurcation theorem in [69], they obtained

Theorem 5.3 *Assume that* $\mu|B| > d_m$. *Let*

$$\begin{aligned} x^* &= \min\{x \in [0,1]; \mu|B|e^{-4\alpha x} > 4D_m x + d_m\}; \\ z_0 &= \sqrt{(\mu B)^2 - d_m^2}; \\ z_1 &= \sqrt{(\mu B e^{-4\alpha x^*})^2 - (4D_m x^* + d_m)^2}; \\ \tau_0 &= \tfrac{1}{z_0}[\pi - \arctan \tfrac{z_0}{d_m}]; \\ \tau_1 &= \tfrac{1}{z_1}[\pi - \arctan \tfrac{z_1}{4D_m x^* + d_m}]. \end{aligned}$$

Then for every fixed $\tau \in (\tau_0, \tau_1)$ *there exists a unique pair* (c, z) *with* $c = c(\tau)$ *and* $\omega = \omega(\tau) > 0$ *such that* (5.21) *and* (5.22) *are satisfied and* $z\tau \in (\pi/2, \pi)$, $\omega(\tau) = 2\pi/z(\tau)$. *Moreover, for every* $\tau \in (\tau_0, \tau_1)$, (5.12) *has a family of periodic traveling waves* $w_j(t) = \phi(t + cj)$, $j \in \mathbf{Z}$, *of period* $2\pi/\omega(\tau)$, *for* c *near* $c(\tau)$, *where* ω *is close to* $\omega(\tau)$.

Similar results can be found in [48] for a lattice delay differential equation with local interaction.

6 Classification of dynamics in unbounded domains

In this section, we focus on the model equation (2.7), where the nonlinearity arises from the birth process. There are two typical birth functions used in most of the literature, these are

$$b(w) = pwe^{-aw} \tag{6.1}$$

and

$$b(w) = pw^2e^{-aw} \tag{6.2}$$

with two positive constants p and a. The case of a logistic birth function, $b(w) = rw(1-\beta w)$, was considered by Zou [101].

The first is basically of logistic type and has been used in the well-studied Nicholson's blowflies equation (see [30]). The general feature exhibited by this function is that the population growth rate $b'(0) = p$ is positive when the population size is small, and then the birth rate function increases until it reaches the so-called *carrying capacity* $b(\frac{1}{a}) = \frac{p}{ae}$ and then decreases due to the crowding effect. The second function has the feature that $b'(0) = 0$, and this is motivated by the fact that in some biological systems the population growth is very small when the population density is small due to the lack of group defense and due to the low mating probability. Like the first birth function, this function also changes the monotonicity once.

In the model equation (2.7), the spatially homogeneous equilibria are given by the algebraic equation

$$d_m w = b(w). \tag{6.3}$$

The dynamics of the model (2.7) in the case where equation (6.3) has only one trivial solution ($w = 0$) seems to be quite simple: all biologically realistic solutions tend to zero as $t \to \infty$ and the species goes to extinction. As discussed before, this also holds for the discrete analogue (5.12).

The classification of the nonlinearity of b, provided below, is related to relative location of the solutions to equation (6.3). As shown in Fig. 1 there will be four typical cases, though the corresponding reaction diffusion equation without delay and non-local interaction has only two cases: monostable and bistable cases.

Monostability: (6.3) has only two nonnegative solutions $0 = w_1 < w_2 = K$ and b remains to be monotonically increasing on $[w_1, w_2]$.

Crossing Monostability: (6.3) has only two nonnegative solutions $0 = w_1 < w_2 = K$ but b changes from being monotonically increasing to monotonically decreasing, once and only once on $[w_1, w_2]$. In this case, the derivatives of b at w_1, w_2 have opposite signs.

Bistability: (6.3) has three nonnegative solutions $0 = w_1 < w_3 < w_2 = K$ and b remains to be monotonically increasing on $[w_1, w_2]$.

Crossing Bistability: (6.3) has three nonnegative solutions $0 = w_1 < w_3 < w_2 = K$ and b changes the monotonicity from being monotonically increasing to monotonically decreasing, once and only once on $[w_1, w_2]$. Again, in this case, the derivatives of b at w_2, w_3 have opposite signs.

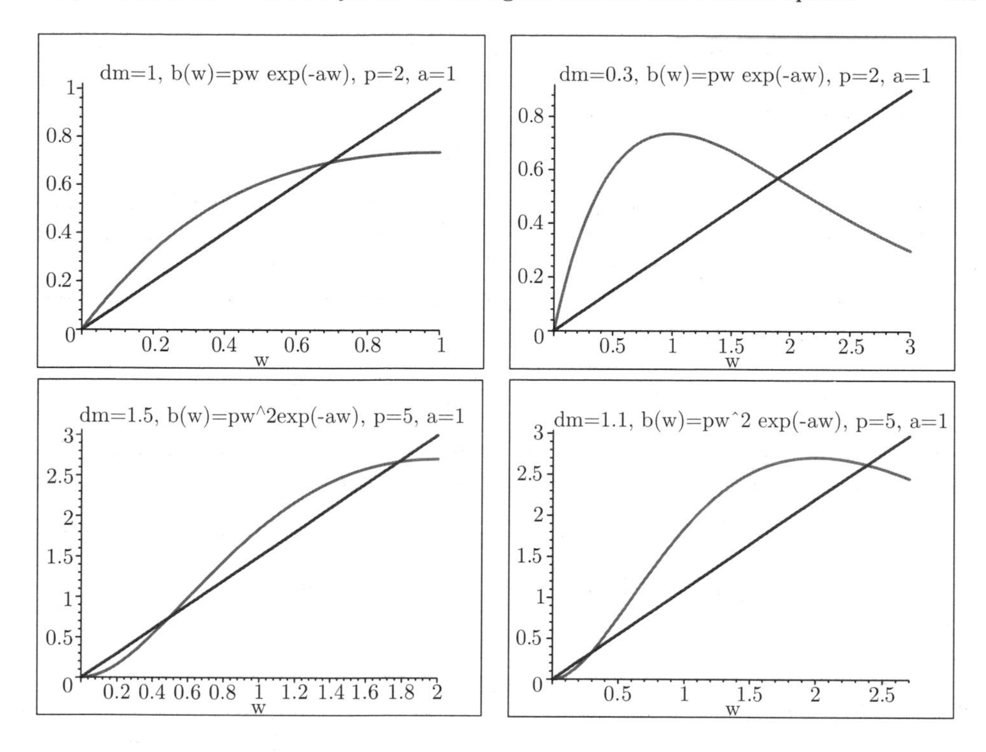

Figure 1 Classification of nonlinearity: left above–monostability with $b(w) = pwe^{-aw}$; right above–Crossing Monostability with $b(w) = pwe^{-aw}$; left below–Bistability with $b(w) = pw^2e^{-aw}$; right below: Crossing Bistability with $b(w) = pw^2e^{-aw}$

In the above classification, K is called the carrying capacity.

As shall be seen, we can give a relatively complete picture of the global dynamics in the monostability and bistability cases. The crossing monostability case has inspired some new methods and we have only partial information about the nonlinear dynamics. There seems to be no progress for the crossing-bistability case to our best knowledge.

6.1 Monostability case. For the delay equation (2.7) with nonlocal interaction on the whole spatial domain $\mathbb{R}$, to determine a solution for all future time $t > 0$, we need to specify an initial condition

$$w(s, x) = \psi(s, x), \quad x \in \mathbb{R}, s \in [-\tau, 0], \tag{6.4}$$

for a certain initial function to be defined later.

The initial value problem (2.7) subject to (6.4) can be solved using the method of steps, that is, we first solve (2.7) subject to (6.4) on $(0, \tau] \times \mathbb{R}$ and then on $(\tau, 2\tau] \times \mathbb{R}$, etc. To obtain more qualitative properties, we will need some abstract settings.

Let $X = BUC(\mathbb{R}, \mathbb{R})$ be the Banach space of all bounded and uniformly continuous functions from $\mathbb{R}$ to $\mathbb{R}$ with the usual supremum norm $|\cdot|_X$ and let $X^+ = \{\psi \in X : \psi(x) \geq 0, \text{ for all } x \in \mathbb{R}\}$. X^+ is a closed cone of X and X is a Banach lattice under the partial ordering induced by X^+. Let $T(t)$ be the analytic semigroup generated by $D_m\Delta$, and we note that $T(t)X^+ \subset X^+$ for $t \geq 0$. Let $C = C([-\tau, 0], X)$ be the Banach space of continuous functions from $[-\tau, 0]$ into X with the supremum norm $||\cdot||$ and let

$$C^+ = \{\psi \in C : \psi(s) \in X^+ \text{ for all } s \in [-\tau, 0]\}.$$

Then C^+ is a closed cone of C. For convenience, we will also identify an element $\psi \in C$ as a function from $[-\tau, 0] \times \mathbb{R}$ into $\mathbb{R}$ defined by $\psi(s, x) = \psi(s)(x)$. For any continuous function $y(\cdot) : [-\tau, b) \to X$, where $b > 0$, we define $y_t \in C$, $t \in [0, b)$, by $y_t(s) = y(t + s)$, $s \in [-\tau, 0]$. The right hand side of (2.7) induces a nonlinear functional $F : C^+ \to X$ by

$$F(\psi)(x) = -d_m\psi(0, x) + \varepsilon \int_{\infty}^{\infty} b(\psi(-\tau, y)) f_\alpha(x - y)\, dy, \ x \in \mathbb{R}, \ \psi \in C^+$$

if the birth function $b : \mathbb{R} \to \mathbb{R}$ is continuous. Moreover, if the restriction of b to $[0, \infty)$ is positive and locally Lipschitz continuous, then

$$\lim_{h\to 0} d(\psi(0)+hF(\psi), X^+) := \lim_{h\to 0} \inf\{|\psi(0)+hF(\psi)-\xi|_X : \xi \in X^+\} = 0 \text{ for } \psi \in C^+.$$

Furthermore, for any $R > 0$, there exists $L_R > 0$ such that

$$|F(\psi) - F(\tilde{\psi})|_X \leq L_R||\psi - \tilde{\psi}|| \quad \text{if} \quad \psi, \tilde{\psi} \in C^+ \quad \text{and} \quad ||\psi||, ||\tilde{\psi}|| \leq R.$$

Consequently, from the work of Martin and Smith [49] (see also Corollary 1.3 of [91], p. 270), we know that for each $\psi \in C^+$ there exists a unique non-continuable solution u on $[0, t_\psi)$, for some $t_\psi > 0$, of the following initial value problem of abstract integral equation

$$u(t) = T(t)u(0) + \int_0^t T(t - s)F(u_s)\, ds,$$
$$u_0 = \psi \in C^+$$

which satisfies $u(t) \in X^+$ for all $t \in [0, t_\psi)$. Such a solution is called a mild solution of (2.7) subject to (6.4). Since the semigroup $T(t)$ is analytic, such a mild solution of (2.7) must also be a classical solution of (2.7) for $t > \tau$ (see Corollary 2.5 of [91], p. 50).

We now introduce the result of [81] on the existence of traveling wave fronts in the monostable case. We will state the result for the particular birth function $b(w) = pwe^{-aw}$, though the results remain true in general in this case.

Here and in what follows, a traveling wave solution of (2.7) is a solution of the form

$$w(t, x) = \phi(x + ct),$$

where $c > 0$ is the wave speed, and ϕ is the profile. In the event that (6.4) has two equilibria w_1 and w_2 with $w_1 < w_2$ and the profile ϕ of the wave satisfies $\lim_{s\to-\infty} \phi(s) = w_1$ and $\lim_{s\to\infty} \phi(s) = w_2$, the travelling wave solution is called a travelling wave front.

It is easily seen that when $\frac{\varepsilon p}{d_m} > 1$, then (2.7) has two spatially homogeneous equilibria $w_1 = 0$ and $w_2 = \frac{1}{a} \ln \frac{\varepsilon p}{d_m} > 0$. To look for a traveling wave front $w(t, x) = \phi(x + ct)$ with ϕ saturating at w_1 and w_2, we need to find a monotone function $\phi(\xi)$, where $\xi = x + ct$, to the following equation

$$c\phi'(t) = D_m\phi''(t) - d_m\phi(t) + \varepsilon p \int_{-\infty}^{\infty} \phi(t - c r - y)e^{-a\phi(t-cr-y)} f_\alpha(y)\, dy, \quad (6.5)$$

subject to the boundary conditions

$$\phi(-\infty) = w_1, \quad \phi(\infty) = w_2.$$

Notice that, without causing unnecessary confusion, we are using t in place of ξ in (6.5).

Equation (6.5) is a second order functional differential equation of *mixed type* (namely, with both advanced and delayed arguments) whereas the equation from a similar local problem gives rise to a second order *delayed* differential equation only. This is the major difference between a local and a non-local problem in terms of traveling waves, and there is no systematic qualitative theory for functional differential equations of mixed type.

We concentrate, in this subsection, on the monostable case. Therefore, b is monotone on $[0, w_2]$. Equivalently, we require

$$1 < \frac{\varepsilon p}{d_m} \le e.$$

This enables [81] to construct a suitable pair of upper and lower solutions and to use the monotone iteration for delayed reaction diffusion equations ([92, 93, 94]) to obtain a family of traveling wave fronts parametrized by the wave speed.

To be more precise, for $\lambda \in \mathbb{R}$, define the function

$$\Delta_c(\lambda) = \varepsilon p e^{\alpha\lambda^2 - \lambda c r} - [c\lambda + d_m - D_m\lambda^2]. \tag{6.6}$$

We note that there exist $c^* > 0$ and $\lambda^* > 0$ such that

(i) $\Delta_{c^*}(\lambda^*) = 0$ and $\frac{\partial}{\partial\lambda}\Delta_{c^*}(\lambda)\Big|_{\lambda=\lambda^*} = 0$;
(ii) for $0 < c < c^*$ and $\lambda > 0$, we have $\Delta_c(\lambda) > 0$; and
(iii) for $c > c^*$ the equation $\Delta_c(\lambda) = 0$ has two positive real roots λ_1, λ_2 such that $0 < \lambda_1 < \lambda_2$ and

$$\Delta_c(\lambda) = \begin{cases} > 0 & \text{for} \quad \lambda < \lambda_1, \\ < 0 & \text{for} \quad \lambda \in (\lambda_1, \lambda_2), \\ > 0 & \text{for} \quad \lambda > \lambda_2. \end{cases}$$

Theorem 6.1 *For every $c > c^*$, (2.7) has a traveling wave front solution, which connects the trivial equilibrium $w_1 = 0$ and the positive equilibrium $w_2 = \frac{1}{a}\ln\frac{\varepsilon p}{d_m}$.*

The existence of the traveling wave with the minimal wave speed $c = c^*$ can be obtained using some limiting arguments, see [85].

To offer a heuristic explanation that c^* is the "minimal wave speed" in the sense that (2.7) has no travelling wave front with wave speed $c < c^*$, we observe that the formal linearization of (6.5) at the zero solution is given by

$$c\phi'(t) = D_m\phi''(t) - d_m\phi(t) + \varepsilon p \int_{-\infty}^{\infty} \phi(t - cr - y) f_\alpha(y)\, dy$$

and the function $\Delta_c(\lambda)$ is obtained by substituting $e^{\lambda t}$ for $\phi(t)$ to the above linearization. Therefore, using the properties of c^* ((ii) above), (6.5) should not have a solution (ϕ, c) with $c < c^*$ and $\phi(-\infty) = 0$.

We remark that the graph of $\lambda \mapsto \Delta_c(\lambda)$ moves upwards as α increases. Therefore, the minimal wave speed c^* is an increasing function of α. Hence waves with speeds near the minimal wave speed are going faster as the mobility of the immature population increases.

To describe the result of stability of the traveling wave due to Mei et al [51] and Gander et al [19], we let, for a given interval $I \subset \mathbb{R}$, $L^2(I)$ be the space of square integrable functions on an interval I, and $H^k(I)$ $(k \geq 0)$ the Sobolev space of L^2-functions f defined on the interval I whose derivatives $\partial_x^i f$, $i = 1, \cdots, k$, also belong to $L^2(I)$. $L^2_\omega(I)$ represents the weighted L^2-space with weight $\omega > 0$. Its norm is defined by

$$\|f\|_{L^2_\omega} = \Big(\int_I \omega(x) f(x)^2 dx \Big)^{1/2}.$$

$H^k_\omega(I)$ is the weighted Sobolev space with the norm

$$\|f\|_{H^k_\omega} = \Big(\sum_{j=0}^{k} \int_I \omega(x) |\partial_x^j f(x)|^2 dx \Big)^{1/2}.$$

Let $T > 0$ and let $\mathcal{B}$ be a Banach space. We denote by $C^0([0,T]; \mathcal{B})$ the space of $\mathcal{B}$-valued continuous functions on $[0,T]$, and $L^2([0,T]; \mathcal{B})$ as the space of $\mathcal{B}$-valued L^2-functions on $[0,T]$. The corresponding spaces of $\mathcal{B}$-valued functions on $[0,\infty)$ are defined similarly.

Let $\phi(\xi)$ be the traveling wave front obtained in Theorem 6.1 with $c > c^*$, connecting $w_- = w_1 = 0$ to $w_+ = w_2 > 0$, and satisfying $\phi'(\xi) > 0$ and $w_- < \phi(\xi) < w_+$, for all $\xi \in \mathbb{R}$. Then

Theorem 6.2 *Suppose that $\psi \in C^+$ is given so that $\psi(s,x) \to w_\pm$ for all $s \in [-\tau, 0]$ as $x \to \pm\infty$, and that $\psi(s,\cdot) - \phi(\cdot + cs) \in H^1(\mathbb{R})$ for each $s \in [-\tau, 0]$. Then there exists a unique solution $w(t,x)$ of the Cauchy problem (2.7) and (6.4) such that $w(t,x) - \phi(x+ct) \in C^0((0,+\infty); H^1(\mathbb{R}))$ and $w(t,x) \geq 0$ in $(0,\infty) \times \mathbb{R}$.*

To state the stability result of [19], let us define

$$\begin{aligned} I(z) := &\int_{\mathbb{R}} [\psi(0,x) - \phi(x+z)]dx - d_m \int_0^\infty \int_{\mathbb{R}} [w(s,x) - \phi(x+cs+z)]\, dx\, ds \\ &+ \int_0^\infty \int_{\mathbb{R}} [f_\alpha(w(s-\tau,x)) - f_\alpha(\phi(x+c(s-\tau)+z))]\, dx\, ds \end{aligned}$$

and

$$x_0 = I(0)/(w_+ - w_-).$$

Next, let

$$\begin{aligned} C_1 &:= \tfrac{\tau}{2}\Big(1 - \tfrac{\alpha}{2\tau D_m}\Big), \\ C_2 &:= a(d_m + D_m) + \lambda d_m \Big(\tfrac{1}{2} e^{-C_1 c^2} + 1\Big), \\ C_3 &:= \tfrac{1}{2C_2}\Big\{(1 - \lambda + 2\ln\lambda)d_m - \tfrac{1}{2}(1 - \ln\lambda) d_m e^{-C_1 c^2}\Big\}. \end{aligned}$$

In [19], it was shown that for a given traveling wave $w(t,x) = \phi(x+ct)$, if the following condition on the wave speed c:

$$(1 - \lambda + 2\ln\lambda)d_m - \frac{1}{2}(1 - \ln\lambda)d_m e^{-C_1 c^2} > 0$$

holds, then there exists a number x_* such that

$$\phi(\xi) > \frac{1}{a}\ln\lambda - C_3, \; |\phi''(\xi)| < C_3, \; \frac{1}{p} b'(\phi(\xi)) < \frac{1 - \ln\lambda}{\lambda} + C_3$$

holds for $\xi > x_*$. For this x_*, one can then define a weight function $\omega(\xi)$ as follows:

$$\omega(\xi) = \begin{cases} e^{-\beta(\xi - x_*)}, & \xi < x_*, \\ 1, & \xi \geq x_*, \end{cases}$$

where

$$\beta = \frac{c}{2D_m}.$$

This weight function plays an essential role in obtaining the following stability result.

Theorem 6.3 *For the shifted traveling wave front $\phi(x + ct + x_0)$, where the speed c satisfies the aforementioned condition and*

$$c > 2\sqrt{D_m d_m (3\lambda - 2)},$$

if $\psi(\cdot, x) - \phi(x + c \cdot + x_0) \in C^0([-\tau, 0]; H^1_w(\mathbb{R}))$, then there exist positive constants δ_0 and μ, which are dependent only on the coefficients D_m, d_m, ϵ, p, a, τ and the wave speed c, such that when $\|\psi(s, \cdot) - \phi(\cdot + cs + x_0)\|_{H^1_w} \leq \delta_0$ for $s \in [-\tau, 0]$, the solution $w(t, x)$ of (2.7) satisfies

$$w(t, x) - \phi(x + ct + x_0) \in C^0([0, \infty); H^1_\omega) \cap L^2([0, \infty); H^2_\omega)$$

and

$$\sup_{x \in \mathbb{R}} |w(t, x) - \phi(x + ct + x_0)| \leq Ce^{-\mu t}, \quad 0 \leq t \leq \infty.$$

The uniqueness and stability of traveling wave fronts of equation (2.7) is recently addressed in [89]. They considered the following general scalar equation

$$\frac{\partial}{\partial t} w(t, x) = D \frac{\partial^2}{\partial x^2} w(t, x) - dw(t, x) + \int_{-\infty}^{+\infty} f(x - y) b(w(t - \tau, y))\, dy, \tag{6.7}$$

under the following set of assumptions:

(W1) $b(0) = d - b(1) = 0 < -du + b(u)$ for $u \in (0, 1)$;

(W2) b is C^1 smooth on some open interval containing $[0, 1]$, $-d + b'(1) < 0 < -d + b'(0)$ and $0 < b(u) < b'(0)u$ for $u \in (0, 1)$;

(W3) There exist some constants $A > 0$ and $\nu \in (0, 1]$ so that $b'(0) - b(u) \leq Au^{1+\nu}$ for $u \in (0, 1)$;

(W4) $f(x) = f(-x)$ for all $x \in \mathbb{R}$, $\int_{-\infty}^{+\infty} f(x)dx = 1$ and $\int_{-\infty}^{+\infty} |x| f(x) dx < \infty$;

(W5) $\int_{-\infty}^{+\infty} f(y)e^{\lambda y} dy$ and $\int_{-\infty}^{+\infty} y f(y) e^{\lambda y} dy$ are uniformly convergent for $\lambda \in [0, \infty)$

In the theorem below, c^* is defined in a similar way as in Theorem 6.1.

Theorem 6.4 *For each $c > c^*$ there exists a unique (up to a translation) monotone traveling wave front of (6.7), $w(t, x) = \phi(x + ct)$. Furthermore, if w is a solution of (6.7) such that its initial condition satisfies*

$$\liminf_{x \to -\infty} w(0, x) > 0, \qquad \lim_{x \to \infty} \max_{s \in [-\tau, 0]} |w(s, x)e^{\lambda_1 x} - \rho_0 e^{\lambda_1 cs}| = 0$$

for some $c > c^*$ and some $\rho_0 > 0$, then

$$\lim_{t \to +\infty} \sup_{x \in \mathbb{R}} \left| \frac{w(t, x)}{\phi(x - \chi_0 - ct)} - 1 \right| = 0,$$

where $\chi_0 = \frac{1}{\lambda_1} \ln \rho_0$.

The above existence result is obtained using the monotone iteration techniques [92, 93, 94] and [81], but with a different lower solution. To obtain the uniqueness (up to translation), they first obtained the strict monotonicity and the exact information of the decay rate of the wave profile near zero. This approach for uniqueness is recently developed by [11] for the non-local reaction diffusion equation

$$u'(t) = J * u - u + f(u) \tag{6.8}$$

and the work of [88] shows that the Chmaj-Carr technique can be applied to delayed non-local problem with modifications: this success shows that some methods developed for non-local problems without delay can be adapted for delay induced non-local problems. A comprehensive review of non-local problems arising from different angles and comparison of methods and ideas should be extremely helpful for the development of the qualitative theory of the delayed non-local diffusive systems.

The global attractivity of the traveling wave front is obtained via the standard squeezing technique, previously used for other problems [12, 5, 76, 48].

As in the classical traveling wave theory of scalar reaction diffusion equations, in the monostability case the minimal wave speed coincides with the rate of propagation. This is shown in [85] for a more general integrodifferential equation, and in [90] for a discrete analogue of (2.7). In fact the paper by Thieme and Zhao [85] considers a general class of scalar nonlinear integral equation of the form

$$u(t,x) = u_0(t,x) + \int_0^t \int_{\mathbb{R}^n} F(u(t-s, x-y), s, y)\, dy\, ds \tag{6.9}$$

which covers many delayed reaction-diffusion models since they can be recast into the form of (6.9). In [85] the *asymptotic speed of spread* $c^* > 0$ for a function $u : \mathbb{R}_+ \times \mathbb{R}^n \to \mathbb{R}_+$ is defined by the criteria that $\lim_{t\to\infty, |x|\geq ct} u(t,x) = 0$ for every $c > c^*$ and that there exists some $\varepsilon > 0$ such that $\liminf_{t\to\infty,|x|\leq ct} u(t,x) \geq \varepsilon$ for every $c \in (0, c^*)$. A formula is obtained for the asymptotic speed of spread involving a Laplace-like transform of integral kernels which allows a calculus similar to the Laplace transform. Concerning (6.9) it is assumed that there exists a function $k : \mathbb{R}_+ \times \mathbb{R}^n \to \mathbb{R}_+$, integrable over $\mathbb{R}_+ \times \mathbb{R}^n$ with a number of technical properties that will not be repeated here in full but which include the requirement that $0 \leq F(u, s, x) \leq uk(s, x)$ for $u, s \geq 0$, $x \in \mathbb{R}^n$. The full list of properties imply that k is something like the derivative of F at $u = 0$ (though it is not actually required that F be differentiable at $u = 0$). A crucial role in the study of the asymptotic behavior of solutions of (6.9) is played by the function

$$\mathcal{K}(c,\lambda) = \int_0^\infty \int_{\mathbb{R}^n} e^{-\lambda(cs - z\cdot y)} k(s,y)\, dy\, ds, \tag{6.10}$$

where $z \in \mathbb{R}^n$ is fixed with $|z| = 1$, which is a transform of k that has certain properties comparable to those of the Laplace transform. Under some technical conditions (which are general enough to include whole classes of scalar delayed reaction-diffusion population models) it is shown that c^* is found by solving for c and λ the equations

$$\mathcal{K}(c,\lambda) = 1, \qquad \frac{d}{d\lambda}\mathcal{K}(c,\lambda) = 0. \tag{6.11}$$

[85] apply their results to a scalar equation that contains as particular cases the models of Gourley and Kuang [27] and So, Wu and Zou [81]. Application to a delayed and diffusive epidemic model is also considered.

6.2 Crossing monostability. For the aforementioned birth function, we have the crossing monostability when $\epsilon p/d_m > e$. In this case, a traveling wave front, if it exists, may not be monotone. In this subsection, we shall present some existence results for the crossing monostability case, but we should first remark that non-monotone traveling wave fronts have been observed by Liang and Wu [37] for system (2.7) (even with additional convection term), and by [22] in a slightly different reaction diffusion equation with delayed non-local effect.

A main reason why the dynamics of (2.7) in the crossing monostability case is so complicated is because the feedback at the positive equilibrium w_2 is a negative one. This, coupled with the existence of time delay, naturally gives rise to the possibility of a stable spatially homogeneous periodic solution. Therefore, a necessary condition for a traveling wave front connecting two equilibria w_1 and w_2 that possesses a certain stability property is the non-existence of periodic solutions, and this normally requires some smallness condition on the delay τ.

When the delay τ is zero, we have the classical monostable case where (2.7) has a traveling wave front. It is natural to ask if the existence of such a wavefront persists with a small delay. Also, if the answer to the above question is positive, then it is natural to ask the upper bound for this delay to ensure the persistence of wavefronts. In [14] this issue was addressed in a much more general setting. Their approach is based on an abstract formulation of the wave profile as a solution of an operational equation in a certain Banach space, coupled with an index formula of the associated Fredholm operator and some careful estimation of the nonlinear perturbation. The general result relates the existence of traveling wave solutions to the existence of heteroclinic connecting orbits of a corresponding ordinary delay differential equation.

More precisely, they considered the existence of traveling wave solutions for the following delayed reaction-diffusion equation with nonlocal interaction:

$$\frac{\partial u(t,x)}{\partial t} = D\Delta u(t,x) + F\Big(u(t,x), \int_{-\tau}^{0}\int_{\Omega} d\eta(\theta)d\mu(y)g\big(u(t+\theta, x+y)\big)\Big), \quad (6.12)$$

where $x \in \mathbb{R}^m$ is the spatial variable, $t \geq 0$ is the time, $u(t,x) \in \mathbb{R}^n$, $D = \operatorname{diag}(d_1, \cdots, d_n)$ with positive constants d_i, $i = 1, \cdots, n$, Δ is the Laplacian operator, τ is a positive constant, $\eta : [-\tau, 0] \to \mathbb{R}^{n\times n}$ is of bounded variation, μ is a bounded measure on $\Omega \subset \mathbb{R}^m$ with values in $\mathbb{R}^{n\times n}$, $F : \mathbb{R}^n \times \mathbb{R}^n \to \mathbb{R}^n$ and $g : \mathbb{R}^n \to \mathbb{R}^n$ are given mappings.

The associated ordinary delay differential equation on $\mathbb{R}^n$ is

$$\dot{u}(t) = F\Big(u(t), \int_{-r}^{0} d\eta(\theta)\mu_\Omega g\big(u(t+\theta)\big)\Big), \quad (6.13)$$

where $\mu_\Omega = \int_\Omega d\mu$.

We now formulate some assumptions about the nonlinearities F and g. We suppose that F and g are C^k-smooth functions, $k \geq 2$, and we let $F_u(u,v)$, $F_v(u,v)$ denote the partial derivatives of F with respect to the variables $u \in \mathbb{R}^n$ and $v \in \mathbb{R}^n$, respectively, and let $g_u(u)$ be the derivative of g with respect to the variable $u \in \mathbb{R}^n$. In addition, we suppose that (6.13) has two equilibria E_i, $i = 1, 2$, and we define

$$A_i = F_u\big(E_i, \int_{-r}^{0} d\eta(\theta)\mu_\Omega g(E_i)\big), \quad B_i = F_v\big(E_i, \int_{-r}^{0} d\eta(\theta)\mu_\Omega g(E_i)\big).$$

For a complex number λ we let

$$\Lambda_i(\lambda) = \det \left[\lambda I - A_i - B_i \int_{-r}^{0} d\eta(\theta) \mu_\Omega g_u(E_i) e^{\lambda\theta} \right].$$

We assume that the following hypotheses hold:

(H1) All eigenvalues corresponding to the equilibrium E_2 have negative real parts, that is, $\sup\{\Re\lambda : \Lambda_2(\lambda) = 0\} < 0$.

(H2) E_1 is hyperbolic and the unstable manifold at the equilibrium E_1 is M ($M \geq 1$) dimensional. In other words, $\Lambda_1(iv) \neq 0$ for all $v \in \mathbb{R}$ and $\Lambda_1(\lambda) = 0$ has exactly M roots with positive real parts, where the multiplicities are taken into account.

(H3) Equation (6.13) has a heteroclinic solution $u^* : \mathbb{R} \to \mathbb{R}^n$ from E_1 to E_2, i.e. a solution $u^*(t)$ defined for all $t \in \mathbb{R}$ such that

$$u^*(-\infty) := \lim_{t\to-\infty} u^*(t) = E_1, \quad u^*(\infty) := \lim_{t\to\infty} u^*(t) = E_2.$$

(H4) $\left\| \int_\Omega d|\mu|(y) \|y\|_{\mathbb{R}^m} \right\|_{\mathbb{R}^{n\times n}} < \infty$, where $|\mu| = \mu^+ - \mu^-$ with μ^+ and μ^- the positive and negative parts of μ, respectively.

The main result of [14] is as follows:

Theorem 6.5 *Under assumptions* **(H1)**–**(H4)**, *there is a* $c^* > 0$ *such that*

(i) *for each fixed unit vector* $\nu \in \mathbb{R}^m$ *and* $c > c^*$, *equation (6.12) has a traveling wave solution* $u(x,t) = U(\nu \cdot x + ct)$ *connecting* E_1 *to* E_2 *(that is,* $U(-\infty) = E_1$ *and* $U(\infty) = E_2$*);*

(ii) *if restricted to a small neighborhood of the heteroclinic solution* $u^* : \mathbb{R} \to \mathbb{R}^n$ *in the space* $C(\mathbb{R}, \mathbb{R}^n)$ *of bounded continuous functions equipped with the sup-norm, then for each fixed* $c > c^*$ *and* $\nu \in \mathbb{R}^m$, *the set of all traveling wave solutions connecting* E_1 *to* E_2 *in this neighborhood forms a* M*-dimensional manifold* $\mathcal{M}_\nu(c)$*;*

(iii) $\mathcal{M}_\nu(c)$ *is a* C^{k-1}*-smooth manifold which is also* C^{k-1}*-smooth with respect to* c. *More precisely, there is a* C^{k-1}*-function* $h : U \times (c^*, \infty) \to C(\mathbb{R}, \mathbb{R}^n)$, *where* U *is an open set in* $\mathbb{R}^M$, *such that* $\mathcal{M}_\nu(c)$ *has the form*

$$\mathcal{M}_\nu(c) = \{\psi : \psi = h(z, c),\ z \in U\}.$$

The idea of the proof is as follows. Let $\nu \cdot x + ct = s \in \mathbb{R}$ and $u(t,x) = U(\nu \cdot x + ct)$. Then, upon a straightforward substitution, a traveling wave $U(s)$ satisfies the second order equation

$$c\dot{U}(s) = D\ddot{U}(s) + F\Big(U(s), \int_{-r}^{0}\int_\Omega d\eta(\theta) d\mu(y) g\big(U(s + \nu\cdot y + c\theta)\big)\Big), \quad s \in \mathbb{R}. \tag{6.14}$$

Writing $V(s) = U(cs)$ and $\epsilon = 1/c^2$, then (6.14) leads to

$$\dot{V}(s) = \epsilon D\ddot{V}(s) + F\Big(V(s), \int_{-r}^{0}\int_\Omega d\eta(\theta) d\mu(y) g\big(V(s + \sqrt{\epsilon}\nu\cdot y + \theta)\big)\Big), \quad s \in \mathbb{R}. \tag{6.15}$$

In the case where c is sufficiently large, ϵ is small and hence (6.15) is a singularly perturbed equation. Such an equation has been extensively investigated via both geometric and analytic methods where the main idea is to study the corresponding slow motion and fast motion. The geometrical approach makes the connection of slow and fast motions by studying the intersection of the relevant invariant manifolds, while the analytic approach matches the slow and fast motion by using the asymptotic expansion of inner and outer layers. For both methods, to make a connection between slow and fast motions is far from being trivial.

In [14], a different approach is taken to avoid this difficulty. The central idea of the approach is to use a certain type of transformation to convert the singularly perturbed differential equation (6.15) into a regularly perturbed operational equation in a Banach space so that the Banach fixed point theorem and some existing results regarding the index of an associated Fredholm operator can be applied to prove the existence of traveling wave solutions. This approach also allows Faria, Huang and Wu to determine the number of traveling wave solutions as well as smooth dependence of traveling wave solutions on the wave speed c.

The aforementioned theorem relates the existence of traveling wave fronts for the reaction diffusion equation (6.12) with delay and non-local interaction to the existence of a connecting orbit between two hyperbolic equilibria of the associated ordinary delay differential equation (6.13), so that some existing results for invariant curves of semiflows generated by ordinary delay differential equations can be applied to derive systematically sharp sufficient conditions for the existence of traveling wave fronts of delayed reaction diffusion equations that, in turn, include most of the existing results in the literature as special cases.

In particular, for system (2.7), the associated ordinary differential equation is

$$\frac{dw}{dt} = -d_m w(t) + \epsilon b(w(t-\tau)) \tag{6.16}$$

with $b(w) = pwe^{-aw}$. If $\frac{\epsilon p}{d_m} > 1$, then (6.16) has exactly two nonnegative equilibria

$$E_1 = 0, \qquad E_2 = \frac{1}{a} \ln \frac{\epsilon p}{d_m}.$$

The corresponding characteristic equations are

$$\Lambda_1(\lambda) := \lambda + d_m - \epsilon p e^{-\lambda\tau} = 0$$

and

$$\Lambda_2(\lambda) := \lambda + d_m - \epsilon b'(E_2) e^{-\lambda\tau} = 0,$$

where

$$b'(E_2) = \frac{d_m}{\epsilon}(1 - \ln \frac{\epsilon p}{d_m}).$$

As $\epsilon p > d_m$, we can easily show that the unstable manifold for E_1 is at least one-dimensional. Furthermore, E_1 is hyperbolic for $\tau \neq \tau_n$, $n \in \mathbb{N}_0$, where

$$\tau_n = \frac{2\pi - \arccos(\frac{d_m}{\epsilon p})}{\sqrt{\epsilon^2 p^2 - d_m^2}} + 2n\pi.$$

Note that if $e < \frac{\epsilon p}{b d_m} \leq e^2$, then E_2 is asymptotically stable. In fact, in this case,

$$|\epsilon b'(E_2)| = |d_m(1 - \ln \frac{\epsilon p}{d_m})| \leq d_m$$

and hence all zeros of $\Lambda_2(\lambda)$ have negative real parts.

In the case where $\frac{\epsilon p}{d_m} > e^2$, the asymptotic stability of E_2 holds only when the delay τ is sufficiently small. Namely, in $\Lambda_2(\lambda) = 0$, we let $\lambda = i\omega$ to get

$$i\omega = -d_m + d_m(1 - \ln \frac{\epsilon p}{d_m})[\cos(\omega\tau) - i\sin(\omega\tau)], \tag{6.17}$$

from which we can find the minimal $\hat{\tau} > 0$ so that (6.17) has a solution $\omega > 0$. This is given by

$$\hat{\tau} = \frac{\pi - \arccos \frac{1}{\ln \frac{\epsilon p}{d_m} - 1}}{d_m \sqrt{(\ln \frac{\epsilon p}{d_m} - 1)^2 - 1}}. \tag{6.18}$$

It then follows that if $\frac{\epsilon p}{d_m} > e^2$ and $0 \le \tau < \hat{\tau}$ then E_2 is asymptotically stable.

It can be shown that if

$$\tau e^{\tau d_m} e\epsilon |b'_{min}| < 1,$$

then the semiflow of (6.16) is order-preserving with respect to a certain exponential ordering. This condition is equivalent to

$$0 < \tau < \tilde{\tau}, \text{ where } \tilde{\tau} \text{ is the unique solution of } \tau e^{\tau d_m} e\epsilon |b'_{min}| = 1.$$

Therefore, Faria, Huang and Wu [14] obtained

Theorem 6.6 *If $\frac{\epsilon p}{d_m} > e$, then there exist $\tau^* > 0$ and $c^* > 0$ such that if $\tau \in [0, \tau^*)$ then for every $c > c^*$, (2.7) has a traveling wave, which connects the trivial equilibrium $w_1 = 0$ to the positive equilibrium $w_2 = \frac{1}{a} \ln \frac{\varepsilon p}{d_m}$, where*

$$\tau^* = \begin{cases} \min\{\hat{\tau}, \tilde{\tau}, \tau_0\}, & \frac{\epsilon p}{d_m} > e^2, \\ \min\{\tilde{\tau}, \tau_0\}, & \frac{\epsilon p}{d_m} \le e^2. \end{cases}$$

To conclude this subsection, we note that the existence of traveling waves as described above is only guaranteed for large wave speeds $c > c^*$. We have no specific information about the precise value of c^* and we do not know how it is related to the linearized equation of the wave profiles at the two equilibria involved. It would be very nice to push the above general result further to obtain information on the minimal wave speed.

6.3 Bistability case. The situation when b remains to be monotonically increasing on the interval up to the carrying capacity is discussed in the work of Ma and Wu [44], where they obtained a complete description of the nonlinear dynamics for (2.7), including the existence, uniqueness and global asymptotic stability of traveling wave fronts. In this situation, for the prototype nonlinearity (6.2), we always assume that the birth function $b \in C^1(\mathbb{R}, \mathbb{R})$ and there exists a constant $w_2 > 0$ such that $b(0) = d_m K - \varepsilon b(w_2) = 0$. Let

$$u^+ := \sup\{u \in [0, w_2); d_m u = \varepsilon b(u)\}, \quad u^- := \inf\{u \in (0, w_2]; d_m u = \varepsilon b(u)\}.$$

The main result of [44] can be stated as follows:

Theorem 6.7 *Assume that*

(M1) $b'(\eta) \ge 0$, for $\eta \in [0, w_2]$;
(M2) $d_m > \varepsilon \max\{b'(0), b'(w_2)\} \ge \varepsilon \min\{b'(0), b'(w_2)\} > 0$;
(M3) $u^* := u^+ = u^-$ *and* $\varepsilon b'(u^*) > d$.

Then (2.7) has exactly one traveling wave solution $\phi(x + ct)$ with $0 \le \phi \le w_2$. The unique traveling wave solution $\phi(x + ct)$ is strictly increasing and globally asymptotically stable with phase shift in the sense that there exists $\gamma > 0$ such that for every bounded and uniformly continuous $\psi : [-\tau, 0] \times \mathbb{R} \to [0, w_2]$ with

$$\liminf_{x \to +\infty} \min_{s \in [-\tau, 0]} \psi(s, x) > u^*, \quad \limsup_{x \to -\infty} \max_{s \in [-\tau, 0]} \psi(s, x) < u^*,$$

the solution w of (2.7) with $w|_{[-\tau,0]\times\mathbb{R}} = \psi$ satisfies

$$|w(t,x) - \phi(x+ct+\xi)| \leq Ke^{-\gamma t}, \qquad t \geq 0,\ x \in \mathbb{R},$$

for some $K = K(\psi) > 0$ and $\xi = \xi(\psi) \in \mathbb{R}$.

The crossing bistability case is much more complicated and, as mentioned above, there seems to be no progress as yet. The potential difficulties arise because the delay increases to a certain value, there may be a Hopf bifurcation of periodic solutions bifurcating from the unstable equilibrium w_3 and so a traveling wave connecting w_1 and w_2, if exists, will have to go through a transient oscillation mode around w_3.

6.4 Food-limited population model. In [58, 60], a new method is developed to establish the existence of traveling wavefronts for a reaction diffusion equation with non-monotone delayed non-local effects. The approach is based on a combination of perturbation analysis, the Fredholm theory and some fixed point theorems. The method should apply to general situations, although they concentrate on the following food-limited reaction-diffusion equation

$$\frac{\partial u}{\partial t}(t,x) = \frac{\partial^2 u}{\partial x^2}(t,x) + u(t,x)\frac{1-(f*u)(t,x)}{1+\gamma(f*u)(t,x)}, \tag{6.19}$$

where the parameter $\gamma > 0$ and the spatio-temporal convolution $f*u$ is defined by

$$f*u = \int_{-\infty}^{t}\int_{-\infty}^{\infty} f(t,s,x,y)u(s,y)\,dy\,ds \tag{6.20}$$

with the kernel $f(t,s,x,y)$ satisfying the normalization condition

$$\int_{-\infty}^{t}\int_{-\infty}^{\infty} f(t,s,x,y)\,dy\,ds = 1.$$

The model on bounded domain was considered in [24]. The simplest version of equation (6.19) without diffusion is the following ODE

$$\frac{du}{dt} = ru(t)\frac{K-u(t)}{K+\gamma u(t)}, \tag{6.21}$$

where r, K and γ are positive constants. This equation was first proposed by Smith [71] as a mathematical model for populations of *Daphnia magna* (water flea) and a derivation of this equation is given in [61]. The equation can also be used to study the effects of environmental toxicants on a population [33]. The delayed food-limited model

$$\frac{du}{dt} = ru(t)\frac{K-u(t-\tau)}{K+\gamma u(t-\tau)}, \ \tau > 0 \tag{6.22}$$

has been studied recently by several authors; see [20] and [77]. It seems that the best result for the global stability of the positive equilibrium $u = K$ is given in [77].

Equation (6.21) incorporating spatial dispersal was investigated by Feng and Lu [15]. They considered both the reaction-diffusion equation without time delay

$$\frac{\partial u}{\partial t} - Au(t,x) = r(x)u(t,x)\frac{K(x)-u(t,x)}{K(x)+\gamma(x)u(t,x)}, \tag{6.23}$$

and the corresponding time-delay model

$$\frac{\partial u}{\partial t} - Au(t,x) = r(x)u(t,x)\frac{K(x)-au(t,x)-bu(t-\tau,x)}{K(x)+a\gamma(x)u(t,x)+b\gamma(x)u(t-\tau,x)}, \tag{6.24}$$

where $x = (x_1, x_2, \cdots, x_n) \in \Omega \subseteq \mathbb{R}^n$, with Ω bounded and the operator A, given by

$$A = \sum_{i,j=1}^{n} a_{ij}(x) \frac{\partial^2}{\partial x_i \partial x_j} + \sum_{j=1}^{n} \beta_j(x) \frac{\partial}{\partial x_j},$$

is uniformly strongly elliptic and has coefficient functions that are uniformly Hölder continuous in $\bar{\Omega}$. Feng and Lu studied the above problems subject to general boundary conditions that include both the zero-Dirichlet and zero-Neumann cases, and they established a global convergence result for a non-zero steady-state.

In [58] the general case (6.19) is considered, which includes various types of special cases by choosing the kernel function f as discussed below.

Reaction-Diffusion Model without Delay: If the kernel f is taken to be

$$f(t, s, x, y) = \delta(y - x)\delta(s - t),$$

equation (6.19) becomes the reaction-diffusion equation without delay

$$\frac{\partial u}{\partial t}(t, x) = \frac{\partial^2 u}{\partial x^2}(t, x) + u(x, t) \frac{1 - u(t, x)}{1 + \gamma u(t, x)}, \tag{6.25}$$

which is a special case of equation (6.23).

Non-local Equation with Discrete Delay: If the kernel function f has a discrete time lag τ and spatial averaging, that is,

$$f(t, s, x, y) = \frac{1}{\sqrt{4\pi(t-s)}} e^{-(x-y)^2/4(t-s)} \delta(t - s - \tau),$$

then equation (6.19) becomes

$$\frac{\partial u}{\partial t} = \frac{\partial^2 u}{\partial x^2} + u \left(\frac{1 - \int_{-\infty}^{\infty} \frac{e^{-(x-y)^2/4\tau}}{\sqrt{4\pi\tau}} u(t - \tau, y)\, dy}{1 + \gamma \int_{-\infty}^{\infty} \frac{e^{-(x-y)^2/4\tau}}{\sqrt{4\pi\tau}} u(t - \tau, y)\, dy} \right). \tag{6.26}$$

A derivation of this type of model, using probabilistic arguments, is given in [8]. In this model, the movement of individuals to their present position from where they have been at previous times is accounted for by a spatial convolution with a kernel that spreads normally with a dependence on the delay.

Local Model with Distributed Delay: If $f(t, s, x, y) = \delta(x - y)G(t - s)$, where

$$G(t) = \frac{1}{\tau} e^{-t/\tau} \quad \text{or} \quad G(t) = \frac{t}{\tau^2} e^{-t/\tau}, \tag{6.27}$$

equation (6.19) becomes a model of reaction diffusion equation with distributed delay:

$$\frac{\partial u}{\partial t} = \frac{\partial^2 u}{\partial x^2} + u \left(\frac{1 - \int_{-\infty}^{t} G(t - \eta) u(\eta, x)\, d\eta}{1 + \gamma \int_{-\infty}^{t} G(t - \eta) u(\eta, x)\, d\eta} \right). \tag{6.28}$$

Here, the parameter τ measures time delay and is comparable to the discrete delay τ in (6.26). The two kernel functions G in (6.27) are used frequently in the literature on delay differential equations. The first of the two functions G is sometimes called the "weak" generic kernel because it implies that the importance of events in the past decreases exponentially. The second kernel (the "strong" generic case) is different because it implies that a particular time in the past, namely τ time units ago, is more important than any other since this kernel achieves its unique maximum when $t = \tau$. This kernel can be viewed as a smoothed out version of the case $G(t) = \delta(t - \tau)$ which gives rise to the discrete delay model.

Non-local Model with Distributed Delay: If the kernel f is taken to be

$$f(t,s,x,y) = \frac{1}{\sqrt{4\pi(t-s)}} e^{-(x-y)^2/4(t-s)} G(t-s),$$

then equation (6.19) is a reaction diffusion equation with both distributed delay and spatial averaging. In the distributed delay case with $G(t) = \frac{1}{\tau}e^{-t/\tau}$, a formal asymptotic expansion of traveling wave front to (6.19) when τ is small was found recently by Gourley and Chaplain by using the so-called linear chain techniques; see [23]. The central idea of this trick is to recast the traveling wave equation into a higher dimensional system of ODEs without delay. When τ is small, Fenichel's geometrical singular perturbation theory (see [16] or part two of [4]) is applicable.

As mentioned in [23], traveling wave solutions to equation (6.19) in the discrete case are much more difficult to study than in the distributed case with specific kernels, because we are no longer able to recast the wave profile equation of (6.26) into a non-delay equation and thus Fenichel's geometrical singular perturbation theory cannot be directly used to find a heteroclinic connection in a finite dimensional manifold.

In [58] a new approach is developed which is suitable for all the aforementioned cases for the existence of traveling wave fronts. We describe their principal result for equation (6.26).

Theorem 6.8 *There exists a constant $\delta > 0$ so that for any $\tau \in [0, \delta]$ and $c \geq 2$, equation (6.26) possesses a traveling wave front $u(t,x) = \phi(x - ct)$ satisfying $\phi(-\infty) = 1$ and $\phi(\infty) = 0$.*

We briefly illustrate their argument here. Letting $\tau \to 0^+$, we arrive at

$$\frac{\partial u}{\partial t} = \frac{\partial^2 u}{\partial x^2} + u\left(\frac{1-u}{1+\gamma u}\right) \tag{6.29}$$

which is actually a modified version of the well-known Fisher equation. Obviously, equation (6.29) has two uniform steady-state solutions $u = 0$ and $u = 1$. Considering a traveling wavefront by setting $u(t,x) = U_0(z) = U_0(x - ct)$ in (6.29), we obtain the following second order ODE for $U_0(z)$:

$$U_0'' + cU_0' + U_0\left(\frac{1-U_0}{1+\gamma U_0}\right) = 0, \tag{6.30}$$

or equivalently the following first-order coupled system

$$\begin{cases} U_0' = V_0, \\ V_0' = -cV_0 - U_0\left(\dfrac{1-U_0}{1+\gamma U_0}\right). \end{cases} \tag{6.31}$$

The existence of a connecting orbit for system (6.31) can be established by using standard phase-plane techniques. In particular, we know that if $c \geq 2$, then in the (U_0, V_0) phase plane, a heteroclinic connection exists between the critical points $(U_0, V_0) = (1, 0)$ and $(0, 0)$. Furthermore, the traveling front $U_0(z)$ is strictly monotonically decreasing.

The idea in [58] is to obtain traveling fronts to (6.26) as approximation of the corresponding wavefronts $U_0(z)$ of (6.30) when τ is small. To describe their approach, we need the following notations. Let $C(\mathbb{R}, \mathbb{R})$ be the Banach space of continuous and bounded functions from $\mathbb{R}$ to $\mathbb{R}$ equipped with the standard norm

$||\phi|| = \sup\{|\phi(t)|, t \in \mathbb{R}\}$. Let $C^1 = C^1(\mathbb{R}, \mathbb{R}) = \{\phi \in C : \phi' \in C\}$, $C^2 = \{\phi \in C : \phi'' \in C\}$, $C_0 = \{\phi \in C : \lim_{t\to\pm\infty} \phi = 0\}$ and $C_0^1 = \{\phi \in C_0 : \phi' \in C_0\}$.

Set $u(t,x) = U(z) = U(x-ct)$ in (6.26). Then $U(z)$ satisfies the profile equation

$$-cU' = U'' + U\frac{1 - H(U)(z)}{1 + \gamma H(U)(z)} \tag{6.32}$$

where

$$H(U)(z) = \int_{-\infty}^{\infty} \frac{1}{\sqrt{4\pi\tau}} e^{-y^2/(4\tau)} U(z - y + c\tau)\, dy.$$

We suppose that U can be approximated by U_0 and hence assume that $U = U_0 + W$. Then an equation for W is given by

$$-cW' = W'' + (U_0 + W)\frac{1 - H(U_0 + W)(z)}{1 + \gamma H(U_0 + W)(z)} - U_0\frac{1 - U_0(z)}{1 + \gamma U_0(z)}. \tag{6.33}$$

Using Taylor expansions and letting $g(x) = x\frac{1-x}{1+\gamma x}$, we get

$$\begin{aligned} -cW' &= W'' + g'(U_0(z))W(z) \\ &\quad + R_1(z,\tau,W) + R_2(z,\tau) + R_3(z,\tau,W), \end{aligned} \tag{6.34}$$

where

$$\begin{aligned} R_1(z,\tau,W) &= (U_0 + W)\frac{1 - H(U_0 + W)(z)}{1 + \gamma H(U_0 + W)(z)} - U_0\frac{1 - H(U_0)}{1 + \gamma H(U_0)} \\ &\quad - W\frac{1 - H(U_0)}{1 + \gamma H(U_0)} + \frac{(1+\gamma)U_0}{(1 + \gamma H(U_0))^2} H(W), \\ R_2(z,\tau) &= U_0\frac{1 - H(U_0)}{1 + \gamma H(U_0)} - g(U_0), \\ R_3(z,\tau,W) &= W\frac{1 - H(U_0)}{1 + \gamma H(U_0)} - \frac{(1+\gamma)U_0}{(1 + \gamma H(U_0))^2} H(W) - g'(U_0(z))W(z). \end{aligned}$$

The next step is to rewrite (6.34) as an integral equation:

$$W = \frac{1}{\lambda_2 - \lambda_1}\left(\begin{array}{l} \int_{-\infty}^{z} e^{\lambda_1(z-s)}\left[(1 + g'(U_0(s)))W(s) + R_1 + R_2 + R_3\right] ds \\ + \int_{z}^{\infty} e^{\lambda_2(z-s)}\left[(1 + g'(U_0(s)))W(s) + R_1 + R_2 + R_3\right] ds \end{array}\right) \tag{6.35}$$

where

$$\lambda_1 = \frac{-c - \sqrt{c^2 + 4}}{2} < 0, \quad \lambda_2 = \frac{-c + \sqrt{c^2 + 4}}{2} > 0.$$

The linear part of equation (6.35) is related to the linear operator $L : C_0 \to C_0$ given by

$$\begin{aligned} L(W)(z) &= W - \frac{\int_{-\infty}^{z} e^{\lambda_1(z-s)}(1 + g'(U_0(s)))W(s)\, ds}{\lambda_2 - \lambda_1} \\ &\quad - \frac{\int_{z}^{\infty} e^{\lambda_2(z-s)}(1 + g'(U_0(s)))W(s)\, ds}{\lambda_2 - \lambda_1}. \end{aligned}$$

It is obvious that $L(W) \in C_0$ if $W \in C_0$.

We next define an operator $T : \Psi \in C^2 \to C$ from the homogeneous part of (6.34) as follows:

$$T\Psi(z) = c\Psi'(z) + \Psi''(z) + g'(U_0(z))\Psi(z). \tag{6.36}$$

The formal adjoint equation of $T\Psi = 0$ is given by

$$-c\Phi'(z) + \Phi''(z) + g'(U_0(z))\Phi(z) = 0, \quad z \in \mathbb{R}. \tag{6.37}$$

The existence of a traveling wavefront to equation (6.26) is then obtained using a long argument divided into the following steps:
Step 1. We show, using the classical Fredholm theory for asymptotic autonomous ODEs, that if $\Phi \in C$ is a solution of (6.37) and Φ is C^2-smooth, then $\Phi = 0$. Moreover, we have $\Re(T) = C$, where $\Re(T)$ is the range space of T.
Step 2. Let $\Theta \in C_0$ be given. We conclude, using the convergence result for asymptotically autonomous systems in [54], that if Ψ is a bounded solution of $T\Psi = \Theta$, then we have $\lim_{z\to\pm\infty} \Psi(z) = 0$.
Step 3. For the linear operator $L : C_0 \to C_0$ defined by

$$L(W)(z) = W - \frac{1}{\lambda_2 - \lambda_1}\left(\begin{array}{c} \int_{-\infty}^{z} e^{\lambda_1(z-s)}(1+g'(U_0(s)))W(s)\,ds \\ +\int_{z}^{\infty} e^{\lambda_2(z-s)}(1+g'(U_0(s)))W(s)\,ds \end{array}\right),$$

we prove, using results in Steps 1 and 2, that $\Re(L) = C_0$.
Step 4. Let $N(L)$ be the null space of the operator L. Define the substraction of $N(L)$ from C_0 as $N^{\perp}(L) = C_0/N(L)$. It is clear that $N^{\perp}(L)$ is a Banach space. If we let $S = L|_{N^{\perp}(L)}$ be the restriction of L to $N^{\perp}(L)$, then $S : N^{\perp}(L) \to C_0$ is one-to-one and onto. By the well known Banach inverse operator theorem, we have that $S^{-1} : C_0 \to C_0/N(L)$ is a linear bound operator.
Step 5. When L is restricted to $N^{\perp}(L)$, equation (6.35) can be written as

$$S(W)(z) = \frac{1}{\lambda_2 - \lambda_1}\left(\begin{array}{c} \int_{-\infty}^{z} e^{\lambda_1(z-s)}\,[R_1 + R_2 + R_3]\,ds \\ +\int_{z}^{\infty} e^{\lambda_2(z-s)}\,[R_1 + R_2 + R_3]\,ds \end{array}\right).$$

Since the norm $||S^{-1}||$ is independent of τ, careful estimates of the higher order terms R_i (i=1,2,3) yield that there exist $\sigma > 0, \delta > 0$, and $0 < \rho < 1$ such that for all $\tau \in (0, \delta]$ and $\varphi, \psi \in B(\sigma) \subset X_0$,

$$||F(z, W)\,|| \le \frac{1}{3}\,(||W|| + \sigma)$$

and

$$||F(z,\varphi) - F(z,\psi)||\ \le \rho||\varphi - \psi||,$$

where

$$F(z, W) = \frac{1}{\lambda_2 - \lambda_1} S^{-1}\left(\begin{array}{c} \int_{-\infty}^{z} e^{\lambda_1(z-s)}\,[R_1 + R_2 + R_3(\tau, s, W)]\,ds \\ +\int_{z}^{\infty} e^{\lambda_2(z-s)}\,[R_1 + R_2 + R_3(\tau, s, W)]\,ds \end{array}\right).$$

Hence $F(z,\varphi)$ is a uniform contractive mapping for $W \in C_0 \cap B(\sigma)$. By using the classical fixed point theorem, it follows that for $\tau \in [0, \delta]$, (6.35) has a unique solution $W \in C_0/N(L)$. Returning to the original variable, $W + U_0$ is a heteroclinic connection between the two equilibria 1 and 0. This completes the proof.

Unfortunately, the above result does not give an explicit bound on the smallness of the delay in order to ensure the existence of a traveling wavefront. However, as shown in [58], the argument outlined above can be adopted, together with the idea of Canosa, to yield the existence of traveling wavefront for (6.26) with an explicit bound on the delay τ. This method of Canosa [9] is originally a formal asymptotic analysis as the front speed approaches infinity. However, it is known that for Fisher's equation the method generates a solution that is accurate within a few percent of the true solution, even at the minimum speed. The method has also been applied to other reaction-diffusion equations, including coupled systems, with a very good accuracy; see [55] and [70]. The work of [58] described below provides a theoretical justification of the method for the food-limited model based on a fixed point theorem and the Fredholm theory for delay differential equations.

To be more precise, linearizing (6.32) for U far ahead of the front, where $U \to 0$, gives

$$-cU'(z) = U''(z) + U(z).$$

To ensure that we are studying ecologically realistic fronts that are positive for all values of z, we assume, as in Fisher's equation, that the wave speed $c \geq 2$. Following Canosa's approach, we introduce the small parameter

$$\varepsilon = 1/c^2 \leq \frac{1}{4}$$

and seek a solution of the form

$$U(z) = G(\zeta), \quad \zeta = \sqrt{\varepsilon} z.$$

Equation (6.32) becomes

$$\varepsilon G'' + G' + G\left(\frac{1 - \int_{-\infty}^{\infty} \frac{1}{\sqrt{4\pi\tau}} e^{-y^2/4\tau} G(\zeta - \sqrt{\varepsilon} y + \tau)\, dy}{1 + \gamma \int_{-\infty}^{\infty} \frac{1}{\sqrt{4\pi\tau}} e^{-y^2/4\tau} G(\zeta - \sqrt{\varepsilon} y + \tau)\, dy}\right) = 0. \tag{6.38}$$

When $\varepsilon = 0$, equation (6.38) reduces to

$$G' + G\left(\frac{1 - G(\zeta + \tau)}{1 + \gamma G(\zeta + \tau)}\right) = 0. \tag{6.39}$$

For equation (6.39), we can use the global convergence result of [77] to show that if

$$\tau/(1+\gamma) < \frac{3}{2},$$

then equation (6.39) has a heteroclinic orbit $g_0(\zeta)$ connecting the two equilibria $G = 1$ and $G = 0$.

For equation (6.38), set $\bar{G}(\zeta) = G(-\zeta)$. Then $\bar{G}$ satisfies the equation

$$\varepsilon \bar{G}'' - \bar{G}' + \bar{G}\left(\frac{1 - \int_{-\infty}^{\infty} \frac{1}{\sqrt{4\pi\tau}} e^{-y^2/4\tau} \bar{G}(\zeta + \sqrt{\varepsilon} y - \tau)\, dy}{1 + \gamma \int_{-\infty}^{\infty} \frac{1}{\sqrt{4\pi\tau}} e^{-y^2/4\tau} \bar{G}(\zeta + \sqrt{\varepsilon} y - \tau)\, dy}\right) = 0.$$

Now when ε is small, we use g_0 to approximate the wavefront $\bar{G}(\zeta)$ in (6.38). Let $\bar{G} = g_0 + W$. Then we have an equation for W:

$$W' = \varepsilon W'' + \varepsilon g_0'' + (g_0 + W)\frac{1 - h_1(g_0 + W)}{1 + \gamma h_1(g_0 + W)} - g_0\left(\frac{1 - g_0(\zeta - \tau)}{1 + \gamma g_0(\zeta - \tau)}\right) \tag{6.40}$$

where the functional h_1 is given by

$$h_1[U](\zeta) = \int_{-\infty}^{\infty} \frac{1}{\sqrt{4\pi\tau}} e^{-\eta^2/4\tau} U(\zeta + \sqrt{\varepsilon}\eta - \tau)\, d\eta, \tag{6.41}$$

which, using Taylor's expansion, can be written as

$$W' = \varepsilon W'' + P^0 W(z) + R_1(\zeta, \tau, W) + R_2(\zeta, \tau) + R_3(\zeta, \tau, W), \tag{6.42}$$

where the linear operator $P^0 : C \to C$ is defined by

$$P^0 W(\zeta) = \frac{1 - g_0(\zeta - \tau)}{1 + \gamma g_0(\zeta - \tau)} W(\zeta) - g_0 \frac{(1+\gamma)}{(1 + \gamma g_0(\zeta - \tau))^2} W(\zeta - \tau).$$

Equation (6.42), with

$$\lambda_1 = \frac{1 - \sqrt{1 + 4\varepsilon}}{2\varepsilon} < 0, \quad \lambda_2 = \frac{1 + \sqrt{1 + 4\varepsilon}}{2\varepsilon} > 0,$$

can be transformed into the following integral equation

$$\begin{aligned} W(\zeta) - \int_{-\infty}^{\zeta} e^{-(\zeta-t)}[W(t)+P^0W(t)]\,dt \\ = \int_{-\infty}^{\zeta}[\frac{e^{\lambda_1(\zeta-t)}}{\sqrt{1+4\varepsilon}} - e^{-(\zeta-t)}][W(t)+P^0W(t)]\,dt \\ +\frac{1}{\sqrt{1+4\varepsilon}}\int_{\zeta}^{\infty} e^{\lambda_2(\zeta-t)}[W(t)+P^0W(t)]\,dt \\ +\frac{1}{\sqrt{1+4\varepsilon}}\int_{-\infty}^{\zeta} e^{\lambda_1(\zeta-t)}[R_1+R_2+R_3\ ]\,dt \\ +\frac{1}{\sqrt{1+4\varepsilon}}\int_{\zeta}^{\infty} e^{\lambda_2(\zeta-t)}[R_1+R_2+R_3]\,dt. \end{aligned} \tag{6.43}$$

Let L be the linear operator defined by the left hand side of equation (6.43), namely

$$[LW](\zeta) = W(\zeta) - \int_{-\infty}^{\zeta} e^{-(\zeta-t)}[W(t)+P^0W(t)]\,dt.$$

It is obvious that if $W \in C_0$, then $LW \in C_0$. It remains to show that $\Re(L) = C_0$ where $\Re(L)$ is the range space of L, that is, for each $u \in C_0$, we need to show that equation $LW = u$, or equivalently,

$$W(\zeta) - \int_{-\infty}^{\zeta} e^{-(\zeta-t)}[W(t)+P^0W(t)]\,dt = u(\zeta), \qquad \zeta \in (-\infty,\infty)$$

has a solution in C_0. For this purpose, we set $w = W - u$. Upon substitution, we have an equation for w :

$$w' = P^0w(\zeta) + u(\zeta) + P^0u(\zeta). \tag{6.44}$$

Define an operator $T : C_0^1 \to C_0$ by

$$[Tw](\zeta) = w'(\zeta) - P^0w(\zeta)$$

and the formal adjoint equation of $Tw = 0$ by

$$\phi'(t) = -\frac{1-g_0(t-\tau)}{1+\gamma g_0(t-\tau)}\phi(t) + \frac{g_0(1+\gamma)}{(1+\gamma g_0(t-\tau))^2}\phi\,(t+\tau)\,,\ t \in (-\infty,\infty). \tag{6.45}$$

When $t \to \infty$, equation (6.45) tends asymptotically to

$$\phi'(t) = \frac{1}{1+\gamma}\phi(t+\tau).$$

When $\tau/(1+\gamma) < \frac{\pi}{2}$, it is easy to see that if ϕ is a bounded solution to equation (6.45), then $\phi = 0$. From [13], we see that T is Fredholm and $\Re(T) = C_0$. Therefore, equation (6.44) has a solution $w \in C_0$. We can then use the same argument as for Theorem 6.8 to verify that equation (6.43) has a solution $W \in C_0$.

7 Models in epidemiology

In [57], the impact of the spatial dispersal of juvenile foxes on the spread rate of rabies in continental Europe during the period 1945-1985 was examined using reaction-diffusion equations with non-local delayed terms which are implicitly given via a hyperbolic equation.

The focus of the work is on the front of the epizootic wave of rabies, starting on the edge of the German/Polish boarder and moved westward at an average speed of about 30-60 km a year. This traveling wavefront has been investigated quite successfully (see [55] and references therein), where the minimal wave speed was calculated from basic epidemiological and ecological parameters, and compared well with field observation data. It was also noted that juvenile foxes leave their home territory in the autumn traveling distances that typically may be 10 times a territory size in search of a new territory. If a fox happened to have contracted rabies around the time of such long-distance movement, it could certainly increase the spreading of the disease into uninfected areas. This observation has not been considered in the existing models. It turns out that incorporating the differential spatial movement behaviors of adult and juvenile foxes into a deterministic model yields a much more complicated system of reaction diffusion equations with delayed nonlinear non-local interactions.

To describe the model in Ou and Wu [57], we divide the fox population into the group of infectives and the group of susceptibles, with the former consisting of both rabid foxes and those in the incubation stage. The deterministic model is based on the following assumptions:

(H1) The rabies virus is contained in the saliva of the rabid fox and is normally transmitted by bite. Therefore, a contact between a rabid and a susceptible fox is necessary for the transmission of the disease.

(H2) Rabies is invariably fatal in foxes.

(H3) Adult susceptible foxes are territorial and seem to divide the countryside into non-overlapping home ranges which are marked out by scent. They do occasionally travel considerable distances but always return to their home territory. However, for young susceptible foxes, their behaviors are different, because they prefer to leave their home territories in search of new territories of their own.

(H4) The rabies virus enters the central nervous system and induces behavioral changes of foxes. If the spinal cord is involved, it often takes the form of paralysis. However, if the virus enters the limbic system, the foxes become aggressive, lose their sense of direction and territorial behavior, and wander about in a more or less random way.

Therefore, we need to consider the fox population with two ago classes: the immature and the mature. Let $I(t,a,x)$ and $S(t,a,x)$ denote the population densities at time t, age $a \geq 0$ and spatial location $x \in \mathbb{R}$ for infective and susceptible foxes respectively, and τ be the maturation time which is assumed to be a constant. Then the integrals

$$J(t,x) = \int_0^\infty I(t,a,x)\,da, \quad M(t,x) = \int_\tau^\infty S(t,a,x)\,da$$

give the total population of infective foxes and the density of the adult susceptible foxes.

Using Fick's diffusive law and the mass action incidence, we have

$$\left(\frac{\partial}{\partial t} + \frac{\partial}{\partial a}\right) I(t,a,x) = D_I \frac{\partial^2}{\partial x^2} I(t,a,x) + \beta S(t,a,x) J(t,x) - d_I I(t,a,x), \quad (7.1)$$

where D_I is the diffusive coefficient, d_I is the death rate for infective foxes and β is the transmission rate.

Using $I(t,\infty,x)=0$ and $I(t,0,x)=0$, we obtain that

$$\frac{\partial J(t,x)}{\partial t}=D_I\frac{\partial^2 J(t,x)}{\partial x^2}+\beta M(t,x)J(t,x)-d_I J(t,x)+\beta J(t,x)\int_0^\tau S(t,a,x)\,da. \tag{7.2}$$

For $S(t,a,x)$ with $a\geq\tau$, we have from the structured population model

$$\left(\frac{\partial}{\partial t}+\frac{\partial}{\partial a}\right)S(t,a,x)=-\beta S(t,a,x)J(t,x)-d_S S(t,a,x),$$

where the constant d_S is the death rate for susceptible foxes. Using $S(t,\infty,x)=0$, we get that

$$\frac{\partial M(t,x)}{\partial t}=-\beta M(t,x)J(t,x)-d_S M(t,x)+S(t,\tau,x). \tag{7.3}$$

To obtain a closed system for (J,M), we need to express $S(t,a,x)$ with $0\leq a\leq\tau$ in terms of (J,M). For this purpose, we need the following structured hyperbolic-parabolic equation

$$\left(\frac{\partial}{\partial t}+\frac{\partial}{\partial a}\right)S(t,a,x)=D_Y\frac{\partial^2}{\partial x^2}S(t,a,x)-\beta S(t,a,x)J(t,x)-d_S S(t,a,x), \tag{7.4}$$

subject to the boundary condition related to the birth process:

$$S(t,0,x)=b(M(t,x)), \tag{7.5}$$

where D_Y is the diffusive coefficient for the immatured susceptible foxes and $b(\cdot)$ is birth function of the susceptible foxes, and we will use the following function for illustration:

$$b(M)=(d_S+b_0)M\left(1-\frac{M}{S_0e^{-d_Y\tau}(1+\frac{d_S}{b_0})}\right).$$

Here, the term $b_0M(1-\frac{M}{S_0e^{-d_Y\tau}})$ is the net growth rate for mature susceptible foxes and $S_0e^{-d_Y\tau}$ is the maximum density of adult population.

Despite the fact that (7.4) is a linear equation in S, an explicit formula for $S(t,a,x)$ in terms of M at previous times cannot be obtained due to the time dependent nature of the coefficient $J(t,x)$. Therefore, equations (7.2)-(7.3) give a closed system in which the term $S(t,a,x)$ is implicitly determined by the hyperbolic equation (7.4) subject to the boundary condition (7.5). As will be shown later, this fact does not affect stability analysis of spatially homogeneous equilibria, since the corresponding $S(t,a,x)$ term of the linearization of (7.2)-(7.3) at a spatially homogeneous equilibrium can be solved explicitly from $M(s,y)$ at previous times $s\leq t$ and other spatial locations $y\in\mathbb{R}$.

The structure of the set of equilibria of biological interest is described in [57]. Namely, any equilibrium (M_0,J_0) is given by the following algebraic equations

$$\begin{cases}\beta M_0J_0-d_IJ_0+\beta J_0\dfrac{b(M_0)}{(d_Y+\beta J_0)}(1-e^{-(d_Y+\beta J_0)\tau})=0,\\ -\beta M_0J_0-d_SM_0+b(M_0)e^{-(d_Y+\beta J_0)\tau}=0.\end{cases} \tag{7.6}$$

When $J_0=0$, M_0 can take on two different values: $M_0=0$ or $M_0=M^\tau_{\max}$, where

$$M^\tau_{\max}=\left(1-\frac{d_S(e^{d_Y\tau}-1)}{b_0}\right)S_0e^{-d_Y\tau}. \tag{7.7}$$

To be biologically meaningful, $M_{\max}^{\tau} > 0$ must hold. This is equivalent to

$$d_S < b'(0)e^{-d_Y\tau} = (b_0 + d_S)e^{-d_Y\tau}. \tag{7.8}$$

Let

$$\tau_{\max} = \frac{1}{d_Y} \ln \frac{b_0 + d_S}{b_0}.$$

Then it is easy to see that

$$M_{\max}^{\tau} = \left(1 - \frac{d_S(e^{d_Y\tau} - 1)}{b_0}\right) S_0 e^{-d_Y\tau} > 0 \Leftrightarrow \tau < \tau_{\max}.$$

In what follows, we shall assume $\tau < \tau_{\max}$ and hence model system (7.2)-(7.4) admits two disease free equilibria $(0, 0)$ and $(0, M_{\max}^{\tau})$.

When $J_0 \neq 0$, (M_0, J_0) is implicitly determined by (7.6). Careful analysis shows that if

$$\frac{(d_S + b_0)}{b_0} > \frac{d_I}{\beta S_0 e^{-d_Y\tau}}, \tag{7.9}$$

then there is a disease endemic equilibrium (J_*^{τ}, M_*^{τ}) if and only if

$$C_0(\tau) := \frac{d_I}{\beta M_{\max}^{\tau}} - \frac{b(M_{\max}^{\tau})(1 - e^{-d_Y\tau})}{M_{\max}^{\tau} d_Y} < 1. \tag{7.10}$$

We now use the standard stability analysis to determine an explicit formula for the minimal wave speed. As usual, we linearize the wave equation of (7.2)-(7.4) near their equilibria and find all the eigenvalues and eigenvectors. Sketching this information in the system's phase plane yields useful information about possible heteroclinic connection between these equilibria. The key to the success of the work [57] is that in the linearization of (7.2)-(7.4) the maturation term $S(t, a, x)$ can now be explicitly given in terms of the state variables at a previous time. More precisely, we linearize equations (7.2)-(7.4) around its equilibrium (J_0, M_0) to obtain the following linearized system for ΔS :

$$\begin{cases} \dfrac{\partial \Delta S}{\partial t} = D_Y \dfrac{\partial^2}{\partial x^2} \Delta S - d_Y \Delta S - \beta J_0 \Delta S - \beta F(t - s, M_0, J_0)\Delta J, \\ \Delta S|_{t=s} = b'(M_0)\Delta M. \end{cases} \tag{7.11}$$

We can use the Fourier transform to solve this equation, obtaining

$$\begin{aligned} \Delta S(t, a, x) \;=\; & \frac{b'(M_0)}{\sqrt{4\pi D_Y a}} e^{-(d_Y + \beta J_0)a} \int_{-\infty}^{\infty} \Delta M(t - a, y) e^{-(x-y)^2/(4D_Y a)} dy \\ & -\beta \int_{-\infty}^{\infty} dy \int_0^a F(a - v, M_0, J_0)\Delta J(t - v, y) \\ & \times e^{-(d_Y + \beta J_0)v} \frac{e^{-(x-y)^2/(4D_Y v)}}{\sqrt{4\pi D_Y v}} dv. \end{aligned}$$

Next, we obtain from (7.2) and (7.3) the following linear system:

$$\begin{cases} \dfrac{\partial \Delta J}{\partial t} = D_I \dfrac{\partial^2 \Delta J}{\partial x^2} + \beta M_0 \Delta J + \beta J_0 \Delta M - d_I \Delta J + \beta \Delta J \displaystyle\int_0^{\tau} F(a, M_0, J_0)\, da \\ \qquad + \beta J_0 \displaystyle\int_0^{\tau} \Delta S(t, a, x)\, da, \\ \dfrac{\partial \Delta M}{\partial t} = -\beta M_0 \Delta J - \beta J_0 \Delta M - d_S \Delta M + \Delta S(t, \tau, x). \end{cases} \tag{7.12}$$

Similarly, near the equilibrium $(J, M) = (0, M^{\tau}_{\max})$, we have

$$\begin{cases} \dfrac{\partial \Delta J}{\partial t} = D_I \dfrac{\partial^2 \Delta J}{\partial x^2} + \beta M^{\tau}_{\max} \Delta J - d_I \Delta J + \beta \Delta J b(M^{\tau}_{\max}) \dfrac{1 - e^{d_Y \tau}}{d_Y}, \\ \quad \dfrac{\partial \Delta M}{\partial t} = -\beta M^{\tau}_{\max} \Delta J - d_S \Delta M + \Delta S(t, \tau, x), \end{cases} \tag{7.13}$$

where

$$\begin{aligned} &\Delta S(t, \tau, x) \\ &= \frac{b'(M^{\tau}_{\max})}{\sqrt{4\pi D_Y \tau}} e^{-d_Y \tau} \int_{-\infty}^{\infty} \Delta M(t - \tau, y) e^{-(x-y)^2/(4D_Y \tau)}\, dy \\ &\quad -\beta \int_{-\infty}^{\infty} dy \int_0^{\tau} F(\tau - v, M_0, J_0) \Delta J(t - v, y) e^{-d_Y v} \frac{e^{-(x-y)^2/(4D_Y v)}}{\sqrt{4\pi D_Y v}} dv. \end{aligned}$$

Looking for a traveling wavefront $\Delta J = f_1(x + ct)$, $\Delta M = g(x + ct)$, we have from (7.13) that

$$\begin{cases} c f_1' = D_I f_1'' + f_1 (\beta M^{\tau}_{\max} - d_I + \dfrac{\beta b(M^{\tau}_{\max})}{d_Y} (1 - e^{-d_Y \tau})) \\ c g' = -\beta M^{\tau}_{\max} f_1 - d_S g + \dfrac{\beta b'(M^{\tau}_{\max})}{\sqrt{4\pi D_Y \tau}} e^{-d_Y \tau} \displaystyle\int_{-\infty}^{\infty} g(y - c\tau) e^{-(\xi - y)^2/(4D_Y \tau)}\, dy \\ \quad -\beta \displaystyle\int_{-\infty}^{\infty} dy \int_0^{\tau} S_0(\tau - v) f_1(y - cv) e^{-d_Y v} \sqrt{\frac{1}{4\pi D_Y v}} e^{-(\xi - y)^2/(4D_Y v)}\, dv. \end{cases} \tag{7.14}$$

This is a linear system of functional differential equations with mixed arguments. The corresponding eigenvalues are given by either

$$\lambda^2 - \frac{c}{D_I} \lambda + \frac{k_1}{D_I} = 0, \tag{7.15}$$

or

$$-d_S + e^{-d_Y \tau} b'(M^{\tau}_{\max}) e^{\alpha \lambda^2 - \lambda c \tau} = c\, \lambda,$$

where

$$k_1 = (\beta M^{\tau}_{\max} - d_I + \frac{\beta b(M^{\tau}_{\max})}{d_y} (1 - e^{-d_y \tau})).$$

Solving (7.15) yields

$$\lambda_{1,2} = \frac{c \pm \sqrt{c^2 - 4k_1 D_I}}{2D_I}.$$

The corresponding eigenvectors to the following system, equivalent to (7.14) by letting $f_2 = f_1'$,

$$\begin{cases} f_1' = f_2, \\ D_I f_2' = c f_2 - f_1 (\beta M^{\tau}_{\max} - d_I + \dfrac{\beta b(M^{\tau}_{\max})}{d_y} (1 - e^{-d_y \tau})), \\ c g' = -\beta M^{\tau}_{\max} f_1 - d_S g + \dfrac{\beta b'(M^{\tau}_{\max})}{\sqrt{4\pi D_Y \tau}} e^{-d_Y \tau} \displaystyle\int_{-\infty}^{\infty} g(y - c\tau) e^{-(\xi - y)^2/(4D_Y \tau)}\, dy \\ \quad -\beta \displaystyle\int_{-\infty}^{\infty} dy \int_0^{\tau} S_0(\tau - v) f_1(y - cv) e^{-d_Y v} \sqrt{\frac{1}{4\pi D_Y v}} e^{-(\xi - y)^2/(4D_Y v)}\, dv \end{cases}$$

are

$$\vec{v}_1 = \begin{pmatrix} 1 \\ \lambda_1 \\ 0 \end{pmatrix}, \ \vec{v}_2 = \begin{pmatrix} 1 \\ \lambda_2 \\ 0 \end{pmatrix}.$$

When

$$0 < c < 2\sqrt{k_1 D_I},$$

the eigenvalues $\lambda_{1,2}$ are complex and the eigensolutions are oscillatory and can be negative. This is not biologically meaningful. Therefore, a natural condition for the existence of traveling wavefronts starting from $(0, M_{\max}^\tau)$ is

$$\begin{aligned} c \geq c_{\min}(\tau) &:= 2\sqrt{\beta M_{\max}^\tau D_I}\sqrt{(1 - \frac{d_I}{\beta M_{\max}^\tau} + \frac{b(M_{\max}^\tau)}{M_{\max}^\tau d_Y}(1 - e^{-d_Y \tau}))} \quad (7.16) \\ &= 2\sqrt{\beta M_{\max}^\tau D_I}\sqrt{1 - C_0(\tau)}. \end{aligned}$$

It can be shown that if τ is small, then it is impossible for a trajectory to go from $(0, M_{\max}^\tau)$ to $(0,0)$. So the solution starting from $(0, M_{\max}^\tau)$ should arrive at (J_*^τ, M_*^τ). Once this occurs, the condition (7.16) must be satisfied. The asymptotic behavior of traveling solutions approaching (J_*^τ, M_*^τ) depends on the eigenvalues of system (7.12) near the equilibrium (J_*^τ, M_*^τ). If all the eigenvalues with negative real parts are complex, then the traveling wave will tend to (J_*^τ, M_*^τ) with oscillatory damping. Otherwise it will approach (J_*^τ, M_*^τ) monotonically.

Some numerical evidences are given in the work [57] to show how the incorporation of the age-dependent diffusion affects the estimation of the minimal wave speed. The existence of traveling waves is also established rigorously under the condition $C_0(\tau) < 1$. The idea in the proof is to show the existence of a heteroclinic connection for a non-diffusive delayed system and then to show this is perturbed to a traveling wavefront with large wave speed using the general theory of [14] already discussed in Subsection 6.2.

8 Other developments and comments

We have discussed some delayed diffusive systems with non-local interactions where the time delay is a fixed constant. In many of the specific systems discussed so far, the time delay is the maturation period in a stage-structured population. It is conceivable that in many populations such a maturation period would vary from one individual to another and thus it would be feasible to use a distributed delay. The resulting model becomes a partial integro-differential diffusive system with non-local term, as has been discussed in previous sections. See also [87].

Time delay is not necessarily fixed once and for all, and in fact the maturation period for a population may evolve. The time delay can depend on the population density, see [34] for a comprehensive survey on state-dependent delays. In [66] a non-local PDE model is proposed for the evolution of a single species population that involves delayed feedback, where the delay such as the reproductive time in the delayed birth rate, is selective and the selection depends on the status of the system. More precisely, they consider the following non-local partial differential equation with state-dependent selective delay:

$$\frac{\partial}{\partial t}u(t,x) + Au(t,x) + du(t,x) = \int_{-\tau}^{0}\left\{\int_{\Omega} b(u(t+\theta,y))f(x-y)dy\right\}\xi(\theta,\|u(t)\|)\,d\theta, \qquad x\in\Omega, \tag{8.1}$$

where A is a densely-defined self-adjoint positive linear operator with domain $D(A)\subset L^2(\Omega)$ and with compact resolvent, so $A: D(A)\to L^2(\Omega)$ generates an analytic semigroup, Ω is a smooth bounded domain in $\mathbb{R}^n$, $f:\Omega\to\mathbb{R}$ is a bounded function to be specified later, $b:\mathbb{R}\to\mathbb{R}$ is a locally Lipschitz map and satisfies $|b(w)|\le C_1|w|+C_2$ with $C_1\ge 0$ and $C_2\ge 0$, d is a positive constant. Also, in the above equation, the function $u(\cdot,\cdot):[-\tau,+\infty)\times\Omega\to\mathbb{R}$, is given so that for any t the function $u(t)\equiv u(t,\cdot)\in L^2(\Omega)$, $\|\cdot\|$ is the $L^2(\Omega)$-norm. The function $\xi:[-\tau,0]\times\mathbb{R}\to\mathbb{R}$ represents the state-selective delay in the sense explained below. In the case where $\xi(\theta,s)=\xi(\theta)=e^{-\beta(c_1-\theta)^2}$ is independent of s, we get the term similar to $\int_{-\tau}^{0}u(t+\theta)\xi(\theta,\|u(t)\|)\,d\theta\equiv\int_{-\tau}^{0}u(t+\theta)e^{-\beta(c_1-\theta)^2}d\theta$ which is a "distributed delay analogue" to the term $u(t+c_1)=u(t-h)$ with the discrete delay $h=-c_1>0$. In the general case where

$$\xi(\theta,s)=e^{-\beta(g(s)-\theta)^2}, \quad \theta\in[-\tau,0], \quad s\in\mathbb{R}, \tag{8.2}$$

$g(s)$ gives the coordinate of the maximum of ξ. Thus the system selects the maximal historical impact on the current change rate according to the system's current state. This delay selection, in contrast with the usual state-dependent delay widely used in ordinary delay differential equations, ensures the Lipschitz continuity of the nonlinear functional in the classical phase space. The local theory, and the existence and upper semi-continuity of the global attractor with respect to parameters are developed in [66].

Motivated by whale and seal populations, [2] proposed a structured population model with a state dependent maturation delay. The mechanism here is that whales take longer to mature when there are many of them, as it is more difficult to find adequate food supplies. This leads to a delay that is a monotonically increasing function of total population size (immature plus mature whales). This leads to additional coupling in the model equations which would partially decouple if the delay were constant. The resulting model equations can even exhibit multiple positive equilibria where a constant time delay would allow at most one. It would be interesting to allow spatial dependence in such models with state dependent delays and to consider the impact this would have on the time delay terms.

We have discussed the issue of delayed diffusion which, together with the maturation time in a structured population leads to a hyperbolic parabolic partial differential equation with non-local delayed nonlinearity (2.13). The existence of traveling waves in the mono-stable case is established using a hybrid iterative scheme, and its stability is obtained using a Liapunov functional in a weighted Sobolev space, see [65]. The existence and uniqueness of the traveling wave in the bistability is discussed in the paper [59]. We must emphasize that caution must be exercised in modeling spatial diffusion with delay, as shown in the note [56] that a simple linear diffusion equation with delay (without reaction) can have very exotic spectral properties.

Many of the models we have had in mind in our review here have insect modelling in mind, so that the immatures are larvae. For many insect species it is correct to state that the larvae do not move much or perhaps not at all (for example, locust larvae attach themselves to tree roots and remain there for years). However, in crustaceans it may well be the larvae that do most of the moving. Part of the larval period of crustaceans is spent in the sea, swimming or simply drifting in the water, feeding on planktonic algae. A larva may spend over a year in the ocean, as with some types of lobsters, and could drift a long way from its birth location but might not move much as an adult. Drifting a long way from parents can help to ensure mixing of genes. However it might also mean drifting away from a favorable habitat.

Finally, the ecological and epidemiological considerations that have led to the types of models we have reviewed in this article are also, very frequently, situations where analysis at multiple scales or mixed boundary value problems might well be highly appropriate. Most of the models we have summarized here assume that interacting populations (or immature and mature members of a species) experience their spatial environment in the same way, whether as a single point, a continuum or a collection of patches. However, different species, or indeed immature and mature members of the same species, often experience space in vastly different ways. Cantrell and Cosner [10] carried out an interesting mathematical study of a predator prey interaction in which the predator operates on a large spatial scale and experiences its environment as patchy, while the prey operates on a very small spatial scale, each individual being confined to just one particular patch, but experiencing it as a continuum. These authors had in mind ladybirds preying on aphids inhabiting patches of host plants. Aphids disperse on the order of cm/day and ladybirds up to 1 km/day (and, of course, the ladybirds can move from patch to patch). Different time scales can be important too. Mathematical models of this kind should provide an interesting area for future research.

References

[1] Aiello, W. G. & Freedman, H. I. *A time-delay model of single species growth with stage structure.* Math. Biosci. **101** (1990), 139–153.

[2] Aiello, W. G., Freedman, H. I. & Wu, J. *Analysis of a model representing stage-structured population growth with state-dependent time delay.* SIAM J. Appl. Math. **52** (1992), 855–869.

[3] Al-Omari, J. F. M. & Gourley, S. A. *A nonlocal reaction-diffusion model for a single species with stage structure and distributed maturation delay.* European J. Appl. Math. **16** (2005), 37–51.

[4] Arnold, L., Jones, C., Mischaikow, K. & Raugel, G. *Dynamical systems.* Lectures given at the Second C.I.M.E. Session held in Montecatini Terme, June 13–22, 1994. Edited by R. Johnson. Lecture Notes in Mathematics, 1609. Springer-Verlag, Berlin, 1995.

[5] Bates, P. W., Fife, P. C., Ren, X. & Wang, X. *Traveling waves in a convolution model for phase transitions.* Arch. Rational Mech. Anal. **138** (1997), 105–136.

[6] Billingham, J. *Dynamics of a strongly nonlocal reaction-diffusion population model.* Nonlinearity **17** (2004), 313–346.

[7] Bocharov, G. & Hadeler, K. P. *Structured population models, conservation laws, and delay equations.* J. Diff. Eqns. **168** (2000), 212–237.

[8] Britton, N. F. *Spatial structures and periodic travelling waves in an integro–differential reaction–diffusion population model.* SIAM J. Appl. Math. **50** (1990), 1663-1688.

[9] Canosa, J. *On a nonlinear diffusion equation describing population growth.* IBM J. Res. Develop. **17** (1973), 307–313.

[10] Cantrell, R. S. & Cosner, C. *Models for predator-prey systems at multiple scales.* SIAM Rev. **38** (1996), 256–286.

[11] Carr, J. & Chmaj, A. *Uniqueness of traveling waves for nonlocal monostable equations.* Proc. Amer. Math. Soc. **132** (2004), 2433–2439.

[12] Chen, X. *Existence, uniqueness, and asymptotic stability of traveling waves in nonlocal evolution equations.* Adv. Differential Equations **2** (1997), 125–160.

[13] Chow, S. -N., Lin, X. -B. & Mallet-Paret, J. *Transition layers for singularly perturbed delay differential equations with monotone nonlinearities.* J. Dyna. Diff. Eqns. **1** (1989), 3–43.

[14] Faria, T., Huang, W. & Wu, J. *Traveling waves for delayed reaction-diffusion equations with global response.* Proc. Royal. Soc. London (A), in press.

[15] Feng, W. & Lu, X. *On diffusive population models with toxicants and time delays.* J. Math. Anal. Appl. **233** (1999), 373–386.

[16] Fenichel, N. *Geometric singular perturbation theory for ordinary differential equations.* J. Diff. Equns. **31** (1979), 5398.

[17] Fort, J. & Méndez, V. *Wavefronts in time-delayed reaction-diffusion systems. Theory and comparison to experiment.* Reports on Progress in Physics, **65** (2002), 895–954.

[18] Furter, J. & Grinfeld, M. *Local vs. non-local interactions in population dynamics.* J. Math. Biol. **27** (1989), 65–80.

[19] Gander, M., Mei, M., Schmidt, G. & So, J. M.-H. *Stability of traveling waves for a time-delayed reaction-diffusion equation with nonlocality*, preprint, 2004.

[20] Gopalsamy, K., Kulenovic, M. S. C. & Ladas, G. *Time lags in a "food-limited" population model.* Appl. Anal. **31** (1988), 225–237.

[21] Gourley, S. A. & Britton, N. F. *A predator prey reaction diffusion system with nonlocal effects.* J. Math. Biol. **34** (1996), 297–333.

[22] Gourley, S. A. *Travelling front solutions of a nonlocal Fisher equation.* J. Math. Biol. **41** (2000), 272–284.

[23] Gourley, S. A. & Chaplain, M. A. J. *Travelling fronts in a food-limited population model with time delay.* Proc. Roy. Soc. Edinburgh (A) **132** (2002), 75–89.

[24] Gourley, S. A. & So, J. W.-H. *Dynamics of a food-limited population model incorporating nonlocal delays on a finite domain.* J. Math. Biol. **44** (2002), 49–78.

[25] Gourley, S. A., So J. W. -H. & Wu J.: *Non-locality of reaction-diffusion equations induced by delay: biological modeling and nonlinear dynamics.* J. Math. Sci. **124** (2004), 5119-5153.

[26] Gourley, S. A. & Wu, J. *Extinction and periodic oscillations in an age-structured population model in a patchy environment.* J. Math. Anal. Appl. **289** (2004), 431-445.

[27] Gourley, S. A. & Kuang, Y. *Wavefronts and global stability in a time-delayed population model with stage structure.* Proc. Roy. Soc. Lond. Ser. A. **459** (2003), 1563–1579.

[28] Gourley, S. A. & Ruan, S. *Convergence and travelling fronts in functional differential equations with nonlocal terms: a competition model.* SIAM J. Math. Anal. **35** (2003), 806–822.

[29] Gurney, W. S. C. & Nisbet, R. M. *Fluctuation Periodicity, Generation Separation, and the Expression of Larval Competition.* Theoret. Population Biol. **28** (1985), 150–180.

[30] Gurney, W.S.C., Blythe, S.P. & Nisbet, R.M. *Nicholson's blowflies revisited. Nature* **287** (1980), 17–21.

[31] Gurney, W. S. C., Nisbet, R. M. & Blythe, S. P. *The systematic formulation of stage-structure models.* 1986. Pages 474–494 in J. A. J. Metz and O. Diekmann, editors. Lecture Notes in Biomathematics: The Dynamics of Physiologically Structured Populations. Springer-Verlag, W. Germany.

[32] Hale, J. K. *Asymptotic behavior of dissipative systems.* Amer. Math. Soc., Providence, 1988.

[33] Hallam, T. G., Lassiter, R. R. & Kooijman, S. A. L. M. *Effects of toxicants on aquatic populations. Applied mathematical ecology (Trieste, 1986).* Biomathematics, 18, 352–382, Springer, Berlin, 1989.

[34] Hartung, F., Krisztin, T., Walther, H.-O. & Wu, J. *Functional differential equations with state-dependent delays: theory and applications*, preprint, 2005.

[35] Henry, D. *Geometric theory of semilinear parabolic equations.* Lecture Notes in Math. 840, Springer-Verlag, Berlin, 1981.

[36] Huang, H., Jackie, L., Tracy, V. & Wu, J. *Aggregation and heterogentity from the nonlinear dynamic interaction of birth, maturation and spatial migration.* J. Nonlinear Analysis, **4:2**(B), (2003), 287–300.

[37] Liang, D. & Wu, J. *Travelling waves and numerical approximations in a reaction advection diffusion equation with nonlocal delayed effects.* J. Nonlinear Sci. **13:3**, (2003), 289–310.

[38] Kyrychko, Y., Gourley, S. A. & Bartuccelli, M. V. *Dynamics of a stage-structured population model on an isolated finite lattice.* SIAM. J. Math. Anal., in press.

[39] Laister, R. *Global asymptotic behaviour in some functional parabolic equations.* Nonlinear Anal. Ser. A: Theory Methods. **50** (2002), 347–361.
[40] Liang, D., So, J. W.-H., Zhang, F. & Zou, X. *Population dynamic models with nonlocal delay on bounded domains and their numerical computations.* Differential Equations Dynam. Systems **11** (2003), 117–139.
[41] Liang, D., Wu, J. & Zhang, F. *Modelling population growth with delayed nonlocal reaction in 2-dimensions.* Math. Biosci. Eng. **2** (2005), 111–132.
[42] Liang, X. & Zhao, X. *Asymptotic speeds of spread and traveling waves for monotone semiflows with applications.* Communications on Pure and Appl. Math., in press, 2005.
[43] Lika, K. & Hallam, T. G. *Traveling wave solutions of a nonlinear reaction-advection equation.* J. Math. Biol. **38** (1999), 346–358.
[44] Ma, S. & Wu, J. *Existence, uniqueness and asymptotic stability of traveling wavefronts in a non-local delayed reaction-diffusion equation*, Preprint, 2005.
[45] Ma, S & Zou, X. *Existence, uniqueness and stability of traveling waves in a discrete reaction-diffusion monostable equation with delay.* J. Diff. Eqns. **217** (2005), 54-87.
[46] Ma, S. & Zou, X. *Propagation and its failure in a lattice delay differential equation with global interaction.* J. Diff. Eqns. **212** (2005), 129–190.
[47] Ma, S., Weng, P. & Zou, X. *Asymptotic speed of propopagtion and traveling wavefronts in a non-local delayed lattice differential equation.* Nonlinear Analysis (A), in press.
[48] Madras, N., Wu, J. & Zou, J. *Local-nonlocal interaction and spatial-temporal patterns in single species population over a patchy environment.* Canadian Appl. Math. Quarterly, **4** (1996), 109–134.
[49] Martin, R. H. & Smith, H. L. *Abstract functional differential equations and reaction-diffusion systems.* Trans. Amer. Math. Soc. **321** (1990), 1–44.
[50] Martin, R. H. & Smith, H. L. *Reaction-diffusion systems with time delays: monotonicity, invariance, comparison and convergence.* J. reine angew. Math. **413** (1991), 1–35.
[51] Mei, M., So, J.W.-H., Li, M. & Shen S. *Stability of traveling waves for the Nicholson's blowflies equation with diffusion.* Proc. Roy. Soc. Edinburgh (A) **134** (2004), 579–594.
[52] Memory, M. C. *Bifurcation and asymptotic behavior of solutions of a delay-differential equation with diffusion.* SIAM J. Math. Anal. **20** (1989), 533–546.
[53] Metz, J. A. J. & Diekmann, O. *The dynamics of physiologically structured populations.* Springer–Verlag, New York, 1986.
[54] Mischaikow, K., Smith, K. & Thieme, H. R. *Asymptotically autonomous semiflows: chain recurrence and Lyapunov functions.* Trans. Amer. Math. Soc. **347** (1995), 1669–1685.
[55] Murray, J. D. *Mathematical biology.* Biomathematics 19. Springer-Verlag, Berlin, 1989.
[56] Ou, C., Rodrigues, H. M., & Wu, J. *Partial differential equations with delayed diffusion*, Preprint, 2005.
[57] Ou, C. & Wu, J. *Spatial spread of rabies revisited: role of age-dependent diffusion rate and non-local interaction*, preprint, 2005.
[58] Ou, C. & Wu, J. *Traveling wavefronts in a delayed food-limited population model* , preprint, 2005.
[59] Ou, C. & Wu, J. *Existence and uniqueness of a wavefront in a delayed hyperbolic parabolic model.* Nonlinear Analysis: Special Series on Hybrid Systems and Applications, **63** (2005), 364–387.
[60] Ou, C. & Wu, J. *Traveling wavefronts for delayed non-monotone reaction diffusion equations*, Preprint, 2005.
[61] Pielou, E. C. *Introduction to mathematical ecology,* Wiley, 1969.
[62] Pozio, M. A. *Behaviour of solutions of some abstract functional differential equations and application to predator-prey dynamics.* Nonlinear Anal. **4** (1980), 917–938.
[63] Pozio, M. A. *Some conditions for global asymptotic stability of equilibria of integro-differential equations.* J. Math. Anal. Appl. **95** (1983), 501–527.
[64] Raquepas, J. B. & Dockery, J. D. *Dynamics of a reaction-diffusion equation with nonlocal inhibition.* Phys. D. **134** (1999), 94–110.
[65] Raugel, G. & Wu, J. *Non-local delayed hyperbolic parabolic equations: model derivation, wavefronts and global attractors.* Preprint, 2005.
[66] Rezounenko, A & Wu, J. *A non-local PDE model for population dynamics with state-selective delay: local theory and global attractors.* J. Comp. Appl. Math., in press.

[67] Redlinger, R. *Existence theorems for semilinear parabolic systems with functionals.* Nonlinear Anal. **8** (1984), 667–682.
[68] Redlinger, R. *On Volterra's population equation with diffusion.* SIAM J. Math. Anal. **16** (1985), 135–142.
[69] Rustichini, A. *Hopf bifurcation for functional differential equations of mixed type.* J. Dyna. Diff. Eqns. **1** (1989), 145–177.
[70] Sherratt, J. A. *Wavefront propagation in a competition equation with a new motility term modelling contact inhibition between cell populations*, Proc. Roy. Soc. London (A) **456** (2000), no. 2002, 2365–2386.
[71] Smith, F. E. *Population dynamics in Daphnia magna*, Ecology **44** (1963), 651–663.
[72] Smith, H. L. & Thieme, H. R. *Monotone semiflows in scalar non-quasi-monotone functional differential equations.* J. Math. Anal. Appl. **150** (1990), 289–306.
[73] Smith, H. L. & Thieme, H. R. *Strongly order preserving semiflows generated by functional-differential equations.* J. Diff. Eqns. **93** (1991), 332–363.
[74] Smith, H. L. *A structured population model and a related functional-differential equation: global attractors and uniform persistence.* J. Dyn. Diff. Equ. **6**, 1994, 71–99.
[75] Smith, H. L. *Monotone Dynamical Systems, An Introduction to the Theory of Competitive and Cooperative Systems.* Mathematical Surveys and Monographs 41, Amer. Math. Soc., Providence, RI, 1995.
[76] Smith, H. L. & Zhao, X. -Q. *Global asymptotic stability of traveling waves in delayed reaction-diffusion equations.* SIAM J. Math. Anal. **21** (2000), 134–155.
[77] So, J. W. -H. & Yu, J. S. *On the uniform stability for a "food-limited" population model with time delay.* Proc. Roy. Soc. Edinburgh (A) **125** (1995), 991–1002.
[78] So, J. W. -H. & Yang, Y. *Dirichlet problem for the diffusive Nicholson's blowflies equation.* J. Diff. Eqns. **150** (1998), 317–348.
[79] So, J. W. -H., Wu, J. & Yang, Y. *Numerical steady state and Hopf bifurcation analysis on the diffusive Nicholson's blowflies equation.* Appl. Math. Comput. **111** (2000), 33–51.
[80] So, J. W. -H., Wu, J. & Zou, X. *Structured population on two patches: modeling dispersal and delay.* J. Math. Biol. **43** (2001), 37–51.
[81] So, J. W. -H., Wu, J. & Zou, X. *A reaction diffusion model for a single species with age structure, I. travelling wave fronts on unbounded domains.* Proc. Roy. Soc. Lond. Ser. A. **457** (2001), 1841–1853.
[82] Taylor, G. I. *Dispersion of soluble matter in solvent flowing slowly through a tube.* Proc. Roy. Soc. London (A) **219** (1953), 186–203.
[83] Thieme, H. R. *Persistence under relaxed point-dissipativity (with application to an endemic model).* SIAM J. Math. Anal. **24** (1993), 407–435.
[84] Thieme, H. R. & Zhao, X. -Q. *A nonlocal delayed and diffusive predator-prey model.* Nonlinear Analysis: Real World Applications **2** (2001), 145–160.
[85] Thieme, H. R. & Zhao, X. -Q.*Asymptotic speeds of spread and traveling waves for integral equations and delayed reaction-diffusion models.*, J. Diff. Eqns. **195** (2003), 430–470.
[86] Volpert, A. I. & Volpert, V. A. *Traveling wave solutions of parabolic systems.* Translated from the Russian manuscript by James F. Heyda. Translations of Mathematical Monographs, 140. American Mathematical Society, Providence, RI, 1994.
[87] Wang, Z., Li, W. & Ruan, S *Travelling wave fronts in reaction-diffusion systems with spatio-temporal delays.* J. Diff. Eqns., in press.
[88] Wen, X. & Wu, J. *Neutral equations arising from the interaction of maturation and delayed spatial dispersal in a structured population*, Preprint, 2005.
[89] Wen, X. & Wu, J. *The stability of traveling waves in a delayed reaction diffusion equation*, Preprint, 2005.
[90] Weng, P., Huang, H. & Wu, J. Asymptotic speed of propagation of wave fronts in a lattice delay differential equation with global interaction. IMA J. Appl. Math. **68** (2003), 409-439.
[91] Wu, J. *Theory and applications of partial functional-differential equations.* Applied Mathematical Sciences, 119, Springer-Verlag, New York, 1996.
[92] Wu, J. & Zou, X. *Asymptotic and periodic boundary value problems of mixed FDEs and wave solutions of lattice differential equations.* J. Diff. Eqns. **135** (1997), 315–357.
[93] Wu, J. & Zou, X. *Existence of traveling wave fronts in delayed reaction-diffusion systems via monotone iteration method.* Proc. Amer. Math. Soc. **125** (1997), 2589–2598.

[94] Wu, J. & Zou, X. *Traveling wave fronts of reaction-diffusion systems with delay.* J. Dynam. Diff. Eqns. **13** (2001), 651–687.

[95] Wu, J. & Zhao, X. *Diffusive monotonicity and threshold dynamics of delayed reaction diffusion equations.* J. Diff. Eqns. **186** (2002), 470–484.

[96] Xu, D. *Global dynamics and Hopf bifurcation of a structured population model.* Nonlinear Analysis: Real World Applications **6** (2005) 461–476.

[97] Xu, D. & Zhao, X. *A nonlocal reaction-diffusion population model with stage structure.* Canadian Appl. Math. Quarterly, **11** (2003), 303–320.

[98] Yamada, Y. *Asymptotic stability for some systems of semilinear Volterra diffusion equations.* J. Diff. Eqns. **52** (1984), 295–326.

[99] Yoshida, K. *The Hopf bifurcation and its stability for semilinear diffusion equations with time delay arising in ecology.* Hiroshima Math. J. **12** (1982), 321–348.

[100] Young, W. R. & Jones, S. *Shear dispersion.* Phys. Fluid, A **3** (1991), 1087-1101.

[101] Zou, X. *Delay induced traveling wave fronts in reaction diffusion equations of KPP-Fisher type.* J. Comput. Appl. Math. **146** (2002), 309–321.

Fields Institute Communications
Volume **48**, 2006

Asymptotic Behavior for Systems Comparable to Quasimonotone Systems

Jifa Jiang
Department of Applied Mathematics
Tongji University
Shanghai 200092, P. R. China
jiangjf@mail.tongji.edu.cn

Abstract. It is proved that every precompact orbit is convergent for those mappings that can be compared with strongly monotone mappings possessing totally ordered fixed point curve. Applications are made to various differential equations, such as ODEs, PDEs and FDEs, which can be compared with quasimonotone and irreducible systems having totally ordered equilibrium curve.

1 Introduction

Monotone methods and comparison arguments originate in the venerable subject of nonlinear elliptic and parabolic boundary value problems and have a very long history. The 1931 edition of Courant and Hilbert's famous book [12, Page 286] referred to a 1912 note of Bieberbach describing a monotone iteration scheme for solving a nonlinear elliptic equation. They developed this idea into "upper and lower solutions" technique and extended this method to a broad class of such problems. Around that period comparison principles were laid by M. Müller (1926) and E. Kamke (1932), and E. Hopf (1927) for ordinary and partial differential equations, respectively, which played very important roles in the fundamental theory of differential equations. Although during the development process of this theory some dynamics spirit was put into by many researchers (for example, see [4, 21, 22, 34], not until the path-breaking work of M. W. Hirsch [18] were monotone methods fully integrated with dynamical systems ideas. Now it is well-known that solutions are generically convergent to equilibrium (in autonomous case)or subharmonic solution (in periodic case) for the systems possessing strong comparison principle. If some additional conditions are imposed, then every forward orbit converges to an equilibrium or fixed point . These conditions could be *orbital stability*, proposed by Alikakos and his collaborators (see [1, 2, 3]) ; *fixed point stability* studied by Takáč

2000 *Mathematics Subject Classification.* Primary 34C12, 34D05, 34K25, 34K40, 35B40, 35B50, 35K57; Secondary 92B20, 92D30.

Supported by the NSF of China.

[31] and Dancer and Hess [13] ; possessing a *first integral* or *invariant function* with positive gradient , see, e.g., [19, 27, 28, 29, 33] for ordinary differential equations; [32] for partial differential equations; [5, 7, 23, 24, 36, 37] for functional differential equations or having *minimal equilibria* property, a notion introduced by Wu [35] and Haddock, Nkashama and Wu [17]. From their results, we observe that these systems have a common characteristic : the set of all equilibria (or fixed points) is a totally ordered curve.

Suppose a system is not quasimonotone but it can be compared with some quasimonotone one. An interesting problem to ask is what one can tell about asymptotic behavior of these systems. As far as the author knows only very few papers consider this case, most of which focus on so called *contracting rectangle* method, see Conway and Smoller [10], Gardner [15], Brown [9] and Smith [30]. Ding [14] and Yi and Huang[38, 39] consider the asymptotic behavior of the following retarded differential equation

$$x'(t) = -F(x(t)) + G(x(t-r)) \tag{1.1}$$

where $r > 0$ is a constant and F, G are continuous on $\mathbb{R}$, F is strictly increasing. This equation can be compared with the following one

$$x'(t) = -F(x(t)) + F(x(t-r)) \tag{1.2}$$

that is, either $G(x) \geq F(x)$ for all $x \in \mathbb{R}$ or $G(x) \leq F(x)$ for all $x \in \mathbb{R}$. They both showed that all bounded solutions of (1.1) are convergent. It is not difficult to see that (1.2) is eventually strongly monotone and possesses a first integral and the equilibrium set consists of all constants. Motivated by their study, we generally ask for given two systems, one is strongly monotone and its equilibrium set is a totally ordered curve, the other can be compared with the former, what the asymptotic behavior of the latter is ? This note will show that all precompact orbits are convergent for the latter. We will apply this abstract result to various differential equations which possess comparison principle.

2 Abstract result for comparable systems

Let (Y, Y_+) be ordered Banach space with $\mathrm{Int}(Y_+) \neq \emptyset$. For $x, y \in Y$, we define

$$x \leq y \quad \text{if} \quad y - x \in Y_+;$$

$$x < y \quad \text{if} \quad x \leq y \quad \text{and} \quad x \neq y;$$

$$x \ll y \quad \text{if} \quad y - x \in \mathrm{Int}Y_+.$$

For $x \ll y$, we define the *closed order interval* by

$$[x, y] = \{u \in Y : x \leq u \leq y\}.$$

Suppose that X is our state space which is either Y, or Y_+, or a closed order interval. A mapping $T : X \to X$ is called *monotone* if $Tu \leq Tv$ whenever $u, v \in X$ with $u \leq v$. Now we present our assumptions:

(AST) The mappings $S, T : X \to X$ are continuous, and either

(i) $Su \geq Tv$ whenever $u, v \in X$ with $u \geq v$; or

(ii) $Su \leq Tv$ whenever $u, v \in X$ with $u \leq v$;

(AT1) The mapping $T : X \to X$ is *eventually strongly monotone* in the sense that it is monotone and there is an integer m, which could depend on u and v, such that

$$T^n u \ll T^n v \quad \text{whenever} \quad u, v \in X \quad \text{with} \quad u < v \text{ and } n \geq m;$$

(AT2) The fixed point set $E(T)$ of eventually strongly monotone mapping T is a totally ordered curve (by *curve* we mean that it is homeomorphic to $\mathbb{R}$) and every compact subset $K \subset X$ has a lower bound in $E(T)$ in the case (i) and an upper bound in $E(T)$ in the case (ii) .

A mapping $S : X \to X$ is called *comparable* if there is a monotone mapping T such that (AST) is satisfied.

Theorem 2.1 *Suppose that the assumptions* (AST), (AT1) *and* (AT2) *hold. Then every precompact orbit for* S *is convergent, that is, its* $\omega-$*limit set is a singleton.*

Proof The idea here comes from [18] and [40, Theorem 2.2.4]. We only consider the case (i) in (AST). The other case (ii) being similar.

Suppose that $O_S(x) := \{S^n x : n = 0, 1, 2, \cdots\}$ is precompact. Then its $\omega-$limit set $\omega_S(x)$ is a nonempty compact invariant subset in X. By (AT2), $\omega_S(x)$ has a lower bound in $E(T)$. Since $E(T)$ is totally ordered, we can define the greatest lower bound in $E(T)$ as

$$u := \sup\{v \in E(T) : v \leq \omega_S(x)\}.$$

We claim that $\omega_S(x) = \{u\}$. Otherwise, there is a point $v \in \omega_S(x)$ with $u < v$. The assumptions (AT1) and (i) in (AST) deduce that $u = T^m u \ll T^m v \leq S^m v$. Thus, $u \ll S^m v \in \omega_S(x)$ by its invariance. Since $E(T)$ is a totally ordered curve, we can choose a $u_0 \in E(T)$ with $u \ll u_0 \ll S^m v$, which implies that we can find a sufficiently large integer n_0 such that $u_0 \ll S^{n_0} x$. Applying (i) in (AST) again, we get that $u_0 = T^n u_0 \leq S^{n+n_0} x$ for $n = 1, 2, \cdots$. Together this with the definition of limit point, we have $u_0 \leq \omega_S(x)$, contradicting the definition of u. This completes the proof. □

We note that the assumptions (AST) and (AT1) actually correspond to comparison principle, that is, for two given systems that can be compared, one of which is monotone, the corresponding orbits preserve the order of initial points. Theorem 2.1 says that if the strongly monotone system has some property, which is either orbit stability, or fixed point stability, or having a first integral, or minimal equilibria, then the comparable system has the same convergent property. So although our result is very concise, there are wide applications to various differential equations possessing comparison principles, which are given in the next section.

3 Applications

In this section, we will apply Theorem 2.1 to ODEs, FDEs, PDEs and nonlocal systems respectively. We will present them in examples rather than theorems.

3.1 ODE systems with comparison principle. Let us first state Kamke theorem for ODEs (see [11]). Consider

$$x' = F(t, x) \tag{3.1}$$

and

$$x' = G(t, x) \tag{3.2}$$

where $F(t,x)$ and $G(t,x)$ are continuous on $\mathbb{R} \times \mathbb{R}^n_+$ and $D_xF(t,x)$ exists and is continuous in $\mathbb{R} \times \mathbb{R}^n_+$. (3.1) is called *cooperative* if all off-diagonal elements for $D_xF(t,x)$ are nonnegative for each $(t,x) \in \mathbb{R} \times \mathbb{R}^n_+$; *irreducible* if $D_xF(t,x)$ are irreducible for each $(t,x) \in \mathbb{R} \times \mathbb{R}^n_+$.

Theorem 3.1 (KAMKE THEOREM) *Let $x(t)$ and $y(t)$ be solutions of (3.1) and (3.2) respectively, where both systems are assumed to have the uniqueness property for initial value problems. Assume both $x(t)$ and $y(t)$ belong to a domain $\mathbb{R}^n_+$ for $[t_0, t_1]$ in which one of the two systems is cooperative and*

$$F(t,z) \le G(t,z) \qquad (t,z) \in [t_0, t_1] \times \mathbb{R}^n_+. \tag{3.3}$$

If $x(t_0) \le y(t_0)$, then $x(t_1) \le y(t_1)$. If $F = G$ is cooperative and irreducible and $x(t_0) < y(t_0)$ then $x(t_1) \ll y(t_1)$.

Suppose that (3.1) and (3.2) are 2π-periodic with respect to t. We define their Poincaré mappings by

$$Tx = \phi(2\pi; x, F), \quad x \in \mathbb{R}^n_+$$

and

$$Sx = \phi(2\pi; x, G), \quad x \in \mathbb{R}^n_+$$

where $\phi(t;x,F)$ and $\phi(t;x,G)$ are the solutions of (3.1) and (3.2) passing through x at $t = 0$.

Assume that (3.1) is cooperative and irreducible and (3.3) holds . Then

$$T^n x \le S^n y \quad \text{whenever} \quad x \le y \quad \text{and} \quad n = 1, 2, \cdots,$$

which implies that (AST) and (AT1) hold. This shows (AST) and (AT1) just correspond to Kamke theorem or comparison principle in periodic ODEs.

Suppose that $F(t,0) \equiv 0$ and (3.1) possesses a first integral, that is, there is a C^1-function $H : \mathbb{R}^n_+ \to \mathbb{R}$ such that $\text{grad}H(u) \gg 0$ at each $u \in \mathbb{R}^n_+$, and $< \text{grad}H(u), F(t,u) >= 0$ for all $(t,u) \in \mathbb{R} \times \mathbb{R}^n_+$. Under these assumptions, $E(T)$ is a totally ordered curve from the origin (see [19, 27, 33, 29]. Furthermore, if (3.3) is satisfied, then every bounded solution for (3.2) is asymptotic to a 2π-periodic solution by Theorem 2.1.

3.2 FDE systems with comparison principle. There are a big amount of papers involving retarded FDEs and neutral type FDEs which satisfy (AT1) and (AT2), see [5, 6, 7, 23, 24, 35, 36, 37] and references therein. Therefore, our Theorem 2.1 can be applied to all their comparable systems. We don't intend to present them all rather than to give two typical examples showing that how our theorem can be applied to those comparable systems. Before that, we introduce their comparison principle (see [30]).

Let r be a positive constant and $C \equiv C([-r,0], \mathbb{R}^n)$. Then C is a strongly ordered Banach space with the usual uniform norm $| \phi |= \sup\{| \phi(\theta) | : -r \le \theta \le 0\}$ and positive cone C_+ composed of all nonnegative functions.

Consider

$$x'(t) = F(t, x_t) \tag{3.4}$$

and

$$x'(t) = G(t, x_t) \tag{3.5}$$

where $F(t, \phi)$ and $G(t, \phi)$ are continuous on $\mathbb{R}_+ \times C_+$ and satisfy the uniqueness property for initial value problems. The system (3.4) is called to satisfy *quasimonotone condition* if $F_i(t, \phi) \leq F_i(t, \psi)$ whenever $t \geq 0$, $\phi \leq \psi$ with $\phi_i(0) = \psi_i(0)$ for some i.

Theorem 3.2 (COMPARISON PRINCIPLE FOR RETARDED FDES) *Assume that either F or G satisfies quasimonotone condition, and*

$$F(t, \phi) \leq G(t, \phi) \qquad \forall (t, \phi) \in \mathbb{R}_+ \times C_+. \tag{3.6}$$

If $\phi \leq \psi$, then

$$x(t; \phi, F) \leq x(t; \psi, G)$$

holds for all $t \geq 0$ for which both are defined.

If (3.4) and (3.5) satisfy (3.6) and one of them is quasimonotone, then the other is comparable. One still can give additional conditions to guarantee the solution semiflow for quasimonotone systems is eventually strongly monotone, we omit the detail here and refer to Smith [30].

As an illustrating example for application of Theorem 2.1 to retarded FDEs, we consider an $n-$compartmental system with pipes described by the following FDEs (see [16, 6, 35])

$$\begin{cases} \dfrac{du_i(t)}{dt} = \displaystyle\sum_{j=1}^{n} g_{ij}(t - \tau_{ij}, u_j(t - \tau_{ij})) - \sum_{j=1}^{n} g_{ji}(t, u_i(t)), \ t > 0, \ 1 \leq i \leq n \\ u(s) = \phi(s), \ s \in [-\tau, 0] \end{cases} \tag{3.7}$$

where $\tau_{ij} \geq 0$, $\tau = \max_{1 \leq i,j \leq n}\{\tau_{ij}\}$, and $\phi \in C := C([-\tau, 0], \mathbb{R}^n)$. Clearly, C is an ordered Banach space with positive cone $C_+ = C([-\tau, 0], \mathbb{R}^n_+)$. We assume that

(D1) Each $g_{ij} \in C(\mathbb{R}^2, \mathbb{R})$ is continuous and $2\pi-$periodic in t;
(D2) $g_{ij}(\cdot, 0) \equiv 0$, and $\frac{\partial g_{ij}}{\partial u}(t, u) > 0$, $\forall (t, u) \in \mathbb{R}^2$, $\ 1 \leq i, j \leq n$.

Define its Poincaré mapping by

$$T\phi := u_{2\pi}(\phi, g), \forall \phi \in C_+$$

where $u(t; \phi, g)$ is the solution of (3.7) passing through ϕ at $t = 0$ and $u_t \in C$ is defined by $u_t(\theta) = u(t + \theta; \phi, g)$ for any $\theta \in [-r, 0]$. Then T is eventually strongly monotone (see [6, 35]) and $T\,\hat{0} = \hat{0}$. Since the system (3.7) has an order-increasing first integral $J : C_+ \to \mathbb{R}_+$ defined by

$$J(\phi) = \sum_{i=1}^{n} \phi_i(0) + \sum_{i,j=1}^{n} \int_{-\tau_{ij}}^{0} g_{ij}(s, \phi_j(s))ds,$$

$E(T)$ is a totally ordered curve originating from the origin (see [6, 35]). Now we consider the comparable system for (3.7):

$$\begin{cases} \dfrac{du_i(t)}{dt} = \displaystyle\sum_{j=1}^{n} G_{ij}(t - \tau_{ij}, u_j(t - \tau_{ij})) - \sum_{j=1}^{n} g_{ji}(t, u_i(t)), \ t > 0, \ 1 \leq i \leq n \\ u(s) = \phi(s), \ s \in [-\tau, 0] \end{cases} \tag{3.8}$$

where either $G_{i,j}(t, u) \geq g_{i,j}(t, u)$, $\forall (t, u) \in \mathbb{R} \times \mathbb{R}^n_+$ or $G_{i,j}(t, u) \leq g_{i,j}(t, u)$, $\forall (t, u) \in \mathbb{R} \times \mathbb{R}^n_+$. Let S denote its Poincaré mapping defined by

$$S\phi := u_{2\pi}(\phi, G), \forall \phi \in C_+.$$

Then either

$$T^n\phi \le S^n\psi,\ \forall \phi \le \psi, n = 1, 2, \ldots;$$

or

$$S^n\phi \le T^n\psi,\ \forall \phi \le \psi, n = 1, 2, \ldots.$$

All above assumptions imply that (AST), AT1) and (AT2) hold . Thus, we can conclude that every bounded solution of (3.8) is asymptotic to a $2\pi-$periodic solution. This result generalizes the one in [14] to systems.

For neutral FDEs, we still have comparison principle, which was deduced by Wu[37] and omitted here. The neutral type compartmental system has the following form:

$$\begin{cases} \dfrac{d}{dt}[x_i(t) - \displaystyle\sum_{j=1}^{n}\int_{-r_j}^{0} x_j(t+\theta)d\nu_{ij}(\theta)] \\ = -\displaystyle\sum_{j=0}^{n} g_{ji}(x_i(t))) + \sum_{j=1}^{n}\int_{-r_j}^{0} g_{ij}(x_j(t+\theta))d\eta_{ij}(\theta),\ t > 0,\ 1 \le i \le n \end{cases} \tag{3.9}$$

where $g_{ji}(x_i), j = 0, 1, \cdots, n$ and $i = 1, 2, \cdots, n$ are nondecreasing, continuously differentiable and $g_{ji}(0) = 0$. Assuming that $g_{0i}(x_i) \equiv 0, i = 1, 2, \cdots, n$ and a series of other conditions hold, Wu[37] proved that every solution of

$$\begin{cases} \dfrac{d}{dt}[x_i(t) - \displaystyle\sum_{j=1}^{n}\int_{-r_j}^{0} x_j(t+\theta)d\nu_{ij}(\theta)] \\ = -\displaystyle\sum_{j=1}^{n} g_{ji}(x_i(t))) + \sum_{j=1}^{n}\int_{-r_j}^{0} g_{ij}(x_j(t+\theta))d\eta_{ij}(\theta),\ t > 0,\ 1 \le i \le n \end{cases} \tag{3.10}$$

is convergent to an equilibrium and the equilibrium set of (3.10) is a totally ordered curve originating from the origin.

If $g_{0i}(x_i) \not\equiv 0$ for some i, then (3.9) can be regarded as the comparable system for (3.10). Applying Theorem 2.1 to this case , we obtain that either the origin is globally attractive in the case that $g_{0i}(x_i) > 0$ for $x_i > 0$, for some i or the equilibrium set is a segment with the origin as an endpoint and every solution of (3.9) is convergent to an equilibrium. It seems to us this result is new. Of course, we should note that all assumptions for (3.10) in Wu [37] should be made here.

We point out that there are also some monotone infinite delay FDEs possessing a first integral (see Wu [36]). But it is impossible to make the solution semiflow eventually strongly monotone, so our result cannot be applied to their comparable systems.

3.3 PDE systems with comparison principle. Consider the initial value problem

$$\begin{cases} \dfrac{\partial u}{\partial t} = \mathcal{A}(t)u + F(x,t,u), & \text{in } \Omega \times (0,\infty) \\ \mathcal{B}u = 0, & \text{on } \partial\Omega \times (0,\infty) \\ u(\cdot,0) = u_0 & \text{in } \Omega. \end{cases} \tag{3.11}$$

Here

$$\mathcal{A}(t) := \sum_{j,k=1}^{N} a_{jk}(x,t)\frac{\partial^2}{\partial x_j \partial x_k} + \sum_{j=1}^{N} a_j(x,t)\frac{\partial}{\partial x_j} + a_0(x,t)$$

is linear and uniformly elliptic on the bounded smooth domain $\Omega \subset \mathbb{R}^N$, with Hölder-continuous and $2\pi-$periodic coefficient functions. $\mathcal{B}$ is a linear boundary operator (Dirichlet, Neumann, or regular oblique derivative type), and F a smooth function on $\bar{\Omega} \times \mathbb{R} \times \mathbb{R}$. More precisely we take the coefficients of linear operator in $C^{\theta,\,\theta/2}(\overline{\Omega} \times \mathbb{R}), 0 < \theta < 1$, and $F : (x,t,\xi) \to F(x,t,\xi)$ continuous, $F(.,.,\xi) \in C^{\theta,\,\theta/2}(\overline{\Omega} \times \mathbb{R})$ uniformly for ξ in bounded intervals and identical hypotheses for $\frac{\partial F}{\partial \xi}$, and $\partial\Omega \in C^{2+\theta}$.

In the following, we first introduce the sufficient condition for orbit stability, which was proposed by Alikakos etal in[1, 2, 3].

We first assume F to be $2\pi-$*periodic* in t. The smooth function $\underline{u}(\in C^{2,1}(\bar{\Omega} \times (0,T]) \cap C^{1,0}(\bar{\Omega} \times [0,T]))$ is called a *subsolution* if

$$\begin{cases} \dfrac{\partial u}{\partial t} \leq \mathcal{A}(t)u + F(x,t,u), & \text{in } \Omega \times (0, 2\pi] \\ \mathcal{B}u \leq 0, & \text{on } \partial\Omega \times (0, 2\pi]; \end{cases} \tag{3.12}$$

a *supersolution* is defined similarly. We assume the existence of an ordered pair $\underline{u} < \bar{u}$ of periodic sub- and supersolutions. Set $u_0 := \underline{u}(0)$, $\bar{u}_0 := \bar{u}(0)$, as well as $Y := L^p(\Omega)(N < p < \infty)$ and $X := [\underline{u}_0, \bar{u}_0]$. Define T by $u_0 \in X \to T(u_0) := u(2\pi) \in Y$, where $u(t)$ is the solution of (3.11). By the parabolic maximum principle, T is well defined and maps X into itself. The smoothing property of analytic semigroups implies that $T(X) \subset Z$, where Z is either $C_0^1(\bar{\Omega})$(Dirichlet) or $C^1(\bar{\Omega})$. By the strong maximum principle, T is strongly order-preserving.

Define the $2\pi-$periodic function $\bar{F}$ by

$$\bar{F}(x,t) := \max\{\frac{\partial F}{\partial \xi}(x,t,\xi) |\ \underline{u}(x,t) \leq \xi \leq \bar{u}(x,t)\}.$$

It is shown in [8] that the periodic eigenvalue problem

$$\begin{cases} \dfrac{\partial w}{\partial t} - \mathcal{A}(t)w - \bar{F}(t)w = \lambda w, & \text{in } \Omega \times \mathbb{R} \\ \mathcal{B}w = 0, & \text{on } \partial\Omega \times \mathbb{R} \\ w(t) = w(t + 2\pi) & \text{in } \Omega, \forall t \in \mathbb{R} \end{cases} \tag{3.13}$$

has a real principal eigenvalue λ_1, characterized as the unique eigenvalue having an eigenfunction which is positive in $\Omega \times \mathbb{R}$. Alikakos etal. [3] proved that if

$$\lambda_1 \geq 0, \tag{3.14}$$

then every solution is stable in X and convergent to a $2\pi-$periodic solution, and $E(T)$ is totally ordered in X.

Now we consider the comparable systems for (3.11):

$$\begin{cases} \dfrac{\partial u}{\partial t} = \mathcal{A}(t)u + G(x,t,u), & \text{in } \Omega \times (0, \infty) \\ \mathcal{B}u = 0, & \text{on } \partial\Omega \times (0, \infty) \\ u(\cdot, 0) = u_0 & \text{in } \Omega, \end{cases} \tag{3.15}$$

where

$$0 \le G(x,t,u) \le F(x,t,u),\ \forall (x,t,u) \in \Omega \times \mathbb{R} \times \mathbb{R}_+. \tag{3.16}$$

Suppose that (3.14) and (3.16) are true, and $F(x,t,0) \equiv 0$. Then every solution of (3.15) passing through an initial value function in X is asymptotic to a $2\pi-$solution. In fact, denote the Poincaré mapping for (3.15) by S. In the case $E(T) = \{0\}$. Then the maximum principle implies that

$$0 \le S^n u \le T^n u,\ n = 1, 2, \cdots. \tag{3.17}$$

From [3], $T^n u \to 0$ as $n \to \infty$ for any $u \in X$. Thus $S^n u \to 0$ as $n \to \infty$ for any $u \in X$. Otherwise $E(T)$ is a segment with the origin as an endpoint. Let u^* be the other endpoint for $E(T)$. From (3.17) it follows that every orbit for S will asymptotic to the closed order interval $[0, u^*]$. Thus it suffices to focus our attention on $[0, u^*]$. When S is restricted to $[0, u^*]$, (AST), (AT1) and (AT2) hold obviously. The conclusion follows immediately from Theorem 2.1.

Our result can also be applied to semilinear parabolic initial-boundary value problem with Neumann boundary condition and spatial gradient term:

$$\begin{cases} \dfrac{\partial u}{\partial t} = \mathcal{A}u + F(x, u, \nabla u), & \text{in } \Omega \times (0, \infty) \\ \dfrac{\partial u}{\partial n} = 0, & \text{on } \partial\Omega \times (0, \infty) \\ u(\cdot, 0) = \phi & \text{in } \Omega. \end{cases} \tag{3.18}$$

We assume that

$$F(x, c, 0) \equiv 0 \qquad \text{on } \Omega. \tag{3.19}$$

Therefore, the equilibrium set for (3.18) is

$$E(T) = \{c : c \in \mathbb{R}\}.$$

Assume that F satisfies the so-called Nagumo type growth condition in spatial gradient. Then Haddock etal[17] proved that every solution of (3.18) approaches a constant. Applying our Theorem 2.1, we get that every bounded solution of any given comparable system of (3.18) converges to a constant if it has the same type as (3.18) and satisfies Nagumo type growth condition in spatial gradient.

We note that the same results hold if one adds diffusion term into (3.1) or (3.4) with every diffusive coefficients $2\pi-$periodic and positive. Therefore, their corresponding comparable systems have convergence property.

3.4 Nonlocal systems with comparison principle. In this subsection, we give a delayed nonlocal reaction diffusion equation , which is artificial, to show our result can also be applied to such kind of systems.

$$\begin{cases} \dfrac{\partial u(x,t)}{\partial t} = d\Delta u - F(u(x,t)) \\ \qquad + \int_\Omega \Gamma(x,y) F(u(y, t-r))dy & \text{in } \Omega \times (0, \infty) \\ \mathcal{B}u = 0, & \text{on } \partial\Omega \times (0, \infty) \\ u(x, \theta) = \phi(x, \theta) & x \in \Omega,\ \theta \in [-r, 0] \end{cases} \tag{3.20}$$

where Ω is a bounded and open subset of $\mathbb{R}^{\mathrm{N}}$ with $\partial\Omega \in C^{2+\theta}$, Δ is Laplacian operator on Ω, $d > 0$ is diffusion rate, $r > 0$ is delay, $\mathcal{B}$ is Neumann boundary operator, and Γ is Green's function associated with Δ and Neumann boundary condition.

Now set

$$C_\Omega = C(\bar{\Omega}, \mathbb{R}), \text{and } \mathrm{C} = \mathrm{C}([-\tau, 0], \mathrm{C}_\Omega).$$

Then using the abstract theory in [25, 26], we can define a solution semiflow on C. By the definition of Green's function and the characteristic for Neumann boundary condition, we get that any solution for (1.2) is a solution of (3.20).

Suppose that $F : \mathbb{R} \to \mathbb{R}$ is continuously differentiable and $F'(\xi) > 0$ for all $\xi \in \mathbb{R}$. Then from [25, 26] it follows that the solution semiflow is is eventually compact and strongly monotone. The above arguments show that all constants are included in the equilibrium set. Checking the proof of Theorem 2.1, we know that each solution of (3.20) is convergent to a constant. Furthermore, assume that $G : \mathbb{R} \to \mathbb{R}$ is continuously differentiable and either $G(\xi) \geq F(\xi)$ for all $\xi \in \mathbb{R}$ or $G(\xi) \leq F(\xi)$ for all $\xi \in \mathbb{R}$. Then

$$\begin{cases} \dfrac{\partial u(x,t)}{\partial t} = d\triangle u - F(u(x,t)) \\ \qquad + \int_\Omega \Gamma(x,y) G(u(y, t-r))dy & \text{in } \Omega \times (0, \infty) \\ \mathcal{B}u = 0, & \text{on } \partial\Omega \times (0, \infty) \\ u(x, \theta) = \phi(x, \theta) & x \in \Omega, \ \theta \in [-r, 0] \end{cases} \tag{3.21}$$

is a comparable system for (3.20). Applying Theorem 2.1 to (3.21), we conclude that every bounded solution of (3.21) approaches an equilibrium.

Finally, we note that in [20] we have established that every omega limit set for monotone and uniformly stable skew-product semiflows is conjugate to its base flow which is minimal and distal. This result is applied to study the asymptotic almost periodicity of solutions to almost periodic reaction-diffusion equations and differential systems with time delays to get that every solution is asymptotic to almost periodic solution if every solution is uniformly stable . In autonomous or periodic case, we have proved in this paper every precompact orbit for any comparable system is convergent if this comparable system can be compared with an eventually strongly monotone system every orbit of which is stable. An interesting problem is whether this result holds in almost periodic systems, that is, if every solution of an almost periodic system is uniformly stable, is each precompact solution of its comparable systems asymptotic to an almost periodic solution?

References

[1] N. D. Alikakos, P. Hess, On stabilization of discrete monotone dynamical systems, *Israel J. Math.*, 59(1987), 185-194.

[2] N. D. Alikakos, P. Bates, Stabilization of solutions for a class of degenerate equations of divergence for in one space dimension, *J. Differential Equations*, 73(1988), 363-393.

[3] N. D. Alikakos, P. Hess and H. Matano, Discrete order preserving semigroups and stability for periodic parabolic differential equations, *J. Differential Equations*, 82(1989), 322-341.

[4] H. Amann, On the number of solutions of equations in ordered Banach spaces, *J. Functional Analysis* 11(1972), 346-384.

[5] O. Arino, Monotone semi-flows which have a monotone first integral, in "Delay Differential Equations and Dynamical Systems", Lecture Notes in Math., Vol. 1475, Springer-Verlag, Berlin/Heidelberg, 1991, pp. 64-75.

[6] O. Arino and E. Haourigui, On the asymptotic behavior of solutions of some delay differential systems which have a first integral, *J. Math. Anal. Appl.*, 122(1987), 36-46.

[7] O. Arino and F. Bourad, On the asymptotic behavior of the solutions of a class of sclar neutral equations generating a monotone semi-flow, *J. Differential Equations* 87(1990), 84-95.

[8] A. Beltramo and P. Hess, On the principal eigenvalue of a periodic-parabolic operator, *Comm. Partial Differential Equations* 9(1984), 919-941.

[9] P. Brown, Decay to uniform states in ecology interation, *SIAM J Appl. Math.* 38(1980), 22-37.

[10] E. Conway and J. Smoller, A comparison theorem for systems of reaction-diffusion equations, *Comm. Part. Diff. Eqns.* 2(1977), 679-697.

[11] W. A. Coppel, Stability and Asymptotic Behavior of Differential Equations, Heath, Boston (1965).

[12] IR. Courant and D. Hilbert, "Methoden der Mathematischen Physik", Vol.2, Springer-Verlag, Berlin, 1937.

[13] E. N. Dancer and P. Hess, Stability of fixed points for order-preserving discrete-time dynamical systems, *J. Reine Angew. Math.*, 419(1991), 125-139.

[14] T. Ding, Asymptotic behavior of solutions of some retarded differential equations, *Science in China (Series A)*, 25(1982), 263-371.

[15] IR. Gardner, Comparison and stability theorems for reaction-diffusion systems, *SIAM J. Math. Anal.* 12(1980), 60-69.

[16] I. Gyori, Connection between compartmental systems with pipes and intergrodifferential equations, *Math. Modelling* 7(1986), 1215-1238.

[17] J. IR. Haddock, M.N.Nkashama and J. Wu, Asymptotic constancy for pseudo monotone dynamical systems on function spaces, *J. Differential Equations*, 100(1992), 292-311.

[18] M. W. Hirsch, Stability and convergence in strongly monotone dynamical systems, *J. Reine Angew. Math.*, 383(1988), 1-53.

[19] J. Jiang, Periodic monotone systems with an invariant function, *SIAM J. Math. Anal.*, 27(1996), 1738-1744.

[20] J. Jiang and X.-Q. Zhao, Convergence in Monotone and Uniformly Stable Skew-Product Semiflows with Applications, *J. Reine Angew. Math.*, in press, 2005.

[21] M. A. Krasnoselskii, "Positive Solutions of Operator Equations", Noordhoff, Groningen, 1964.

[22] M. A. Krasnoselskii and L. Ladyshenskaya, The structure of the spectrum of positive nonhomogeneous operators, Trudy Moskov. Mat. Odšč. 3(1954), 321-346[Russian].

[23] T. Krisztin and J. Wu, Monotone semiflows generated by neutral equations with different delays in neutral and retarded parts, *Acta math. Univ. Comeniane*, 63(1994), 207-220.

[24] T. Krisztin and J. Wu, Asymptotic periodicity, monotonicity, and oscillation of solutions of sclar neutral functional differential equations, *J. Math. Math. Appl.*, 199(1996), 502-525.

[25] R. H.Martin, H. L. Smith, Abstract functional differential equations and reaction-diffusion systems, *Trans. Amer. Math. Soc.* 321(1990),1-44.

[26] R. H.Martin, H. L. Smith, Reaction-diffusion systems with time delays: monotonicity, invariance, comparison and convergence, *J. Reine Angew. Math.* 413(1991), 1-35.

[27] J. Miecynski, Strictly cooperative systems with a first integral, *SIAM J. Math. Anal.*, 18(1987), 642-646.

[28] F. Nakajima, Periodic time-dependent groos-substitute systems, *SIAM J. Appl. Math.*, 36(1979), 421-427.

[29] W. Shen and X.-Q. Zhao, Convergence in almost periodic cooperative systems with a first integral, *Proc. Amer. Math. Soc.*, 133(2004), 203-212.

[30] H. L. Smith, "Monotone Dynamical Systems, An Introduction to the Theory of Competitive and Cooperative Systems", Mathematical Surveys and Monographs 41, Amer. Math. Soc., Providence, RI, 1995.

[31] P. Takáč, Convergence to equilibrium on invariant $d-$hypersurfaces for strongly increasing discrete-time semigroups, *J. Math. Anal. Appl.*, 148(1990), 223-244.

[32] P. Takáč, Domains of attraction of generic ω-limit sets for strongly monotone discrete-time semigroups, *J. Reine Angew. Math.*, 432(1992), 101-173.

[33] B. Tang, Y. Kuang and H. Smith, Strictly nonautonomous cooperative system with a first integral, *SIAM J. Math. Anal.*, 24(1993), 1331-1339.

[34] R. S. Uryson, Mittlehre Breite une Volumen der konvexen Körper im $n-$dimensionalen Räume, *Mat. Sb.* 31(1924), 477-486.

[35] J. Wu, Convergence of monotone dynamical systems with minimal equilibria, *Proc. Amer. Math. Soc.*, 106(1989), 907-911.

[36] J. Wu, Convergence in neutral equations with infinite delay arising from active compartmental systems, "World Congress of Nonlinear Analysts '92", Vol. I–IV (Tampa, FL, 1992), 1361-1369, de Gruyter, Berlin, 1996.

[37] J. Wu and H. I. Freedman, Monotone semiflows generated by neutral functional differential equations with application to compartmental systems, *Can. J. Math.*, 43(1991), 1098-1120.

[38] T. Yi and L. Huang, On the generalized Bernfeld-Haddock conjecture and its proof, to appear in *Science in China.*

[39] T. Yi and L. Huang, Convergence for psedo monotone semiflows on product ordered topological spaces, *J. Differential Equations*, 214(2005), 429-456.

[40] X.-Q. Zhao, *Dynamical Systems in Population Biology*, Springer-Verlag, New York, 2003.

Fields Institute Communications
Volume **48**, 2006

C^1-Smoothness of Center Manifolds for Differential Equations with State-dependent Delay

Tibor Krisztin
Bolyai Institute
University of Szeged
Aradi vértanúk tere 1, H-6720 Szeged, Hungary
krisztin@math.u-szeged.hu

Abstract. In the survey paper [4] a Lipschitz continuous local center manifold is constructed at a stationary point for a class of differential equations with state-dependent delay. Here we show that the obtained center manifold is continuously differentiable.

1 Introduction

Recently, for a class of functional differential equations including equations with state-dependent delays, Walther [9, 10] developed the fundamental theory. This class of equations has in general less smoothness properties than those representing equations with constant delay, and the classical theory (see, e.g., [2, 3]) is not applicable. Walther [9, 10] introduced the so called solution manifold, a smooth submanifold of finite codimension of a function space, and proved under mild smoothness hypotheses that the initial value problem is well-posed on the solution manifold, and the solutions define a semiflow of continuously differentiable solution operators. In addition, Walther resolved the problem of linearization for equations with state-dependent delay by demonstrating that the earlier heuristic linearization technique (see, e.g., [1]) is the true linearization in his framework.

Under the same hypotheses, at stationary points the continuously differentiable solution operators have local stable, center and unstable manifolds. It is shown in the survey paper [4] that these stable and unstable manifolds of maps yield local stable and unstable manifolds also for the semiflow. The same approach does not immediately work for center manifolds since the local center manifolds of the solution operators (or time-t maps) are not necessarily invariant under the semiflow, see [6].

The survey paper [4] constructed Lipschitz continuous local center manifolds for the semiflow at stationary points generated by a class of functional differential equations representing equations with state-dependent delays. The aim of this work

---

2000 *Mathematics Subject Classification.* Primary 34K19; Secondary 37D10.

is to prove that these center manifolds are continuously differentiable. In Section 2 we recall certain steps from the construction of [4] and the necessary technical tools. The proof of the smoothness is contained in Section 3. The proof applies the Lyapunov–Perron approach and closely follows that of [2]. However, as the right hand side of the equation has smoothness properties only on the space of continuously differentiable functions instead of the space of continuous functions as in the classical case ([2, 3]), the space, where the fixed point problem is formulated, is different, and some technical parts are also different. The smoothness proof is based on a slight modification (proved in [7]) of a result of Vanderbauwhede and van Gils [8] on contractions on embedded Banach spaces.

We remark that only C^1-smoothness of the semiflow is shown by Walther [9, 10], and as far as we know, more smoothness is a problem. It is also an open problem to construct center manifolds with more smoothness properties. This would be important in local bifurcation theory through the center manifold reduction. It is interesting that the paper [5] contains a result about C^k-smooth local unstable manifolds with $k \geq 1$ without using any smoothness property of the semiflow.

For motivation, applications and several additional results for functional differential equations with state-dependent delays we refer to the survey paper [4].

2 Preliminaries

Let $|\cdot|$ be a norm in $\mathbb{R}^n$. Fix an $h > 0$. The Banach spaces of continuous and continuously differentiable maps $\phi : [-h, 0] \to \mathbb{R}^n$ are denoted by

$$C = C([-h, 0], \mathbb{R}^n) \quad \text{and} \quad C^1([-h, 0], \mathbb{R}^n),$$

respectively. The norms are given by

$$|\phi|_C = \max_{-h \leq s \leq 0} |\phi(s)| \quad \text{and} \quad |\phi|_{C^1} = |\phi|_C + |\phi'|_C,$$

respectively. If I is an interval, $x : I \to \mathbb{R}^n$ is a function, and $t \in I$ with $t - h \in I$, then the segment $x_t : [-h, 0] \to \mathbb{R}^n$ is defined by

$$x_t(s) = x(t + s), \qquad -h \leq s \leq 0.$$

Let an open subset $U \subset C^1$ and a map $f : U \to \mathbb{R}^n$ be given. Throughout this paper, it is assumed that

(S1): f is continuously differentiable,

(S2): each derivative $Df(\phi)$, $\phi \in U$, extends to a linear continuous map $D_e f(\phi) : C \to \mathbb{R}^n$, and

(S3): the map

$$U \times C \ni (\phi, \chi) \mapsto D_e f(\phi)\chi \in \mathbb{R}^n$$

is continuous.

Consider the functional differential equation

$$x'(t) = f(x_t) \tag{2.1}$$

with the initial condition

$$x_0 = \phi \in U. \tag{2.2}$$

We refer to [9] to see that, for example, if $g \in C^1(\mathbb{R}, \mathbb{R}^n)$, and for $\tau : U \to [0, h]$ hypotheses (S1)-(S3) hold with $n = 1$, then

$$f(\phi) = g(\phi(-\tau(\phi))), \qquad \phi \in U,$$

satisfies (S1)-(S3). Thus, Equation (2.1) contains state-dependent delay differential equations of the form

$$x'(t) = g(x(t - \tau(x_t))).$$

By a solution of (2.1)-(2.2) we understand a continuously differentiable function $x : [-h, t_*) \to \mathbb{R}^n$, $0 < t_* \leq \infty$, satisfying $x_t \in U$, $0 \leq t < t_*$, $x_0 = \phi$, and (2.1) for $0 < t < t_*$.

The closed subset

$$X_f = \{\phi \in U : \phi'(0) = f(\phi)\}$$

of U is called the solution manifold of (2.1). In the sequel we assume $X_f \neq \emptyset$. The papers [9, 10] contain the following basic results. X_f is a C^1-smooth submanifold of U with codimension n. Each $\phi \in X_f$ uniquely defines a noncontinuable solution $x^\phi : [-h, t_+(\phi)) \to \mathbb{R}^n$ of (2.1)-(2.2). All segments x_t^ϕ, $0 \leq t < t_+(\phi)$, belong to X_f, and the relations

$$F(t, \phi) = x_t^\phi, \ \phi \in X_f, \ 0 \leq t < t_+(\phi),$$

define a continuous semiflow $F : \Omega \to X_f$, where $\Omega = \{(t, \phi) : \phi \in X_f, \ 0 \leq t < t_+(\phi)\}$. Each map

$$F(t, \cdot) : \{\phi \in X_f : (t, \phi) \in \Omega\} \to X_f$$

is continuously differentiable, and for all $(t, \phi) \in \Omega$ and $\chi \in T_\phi X_f$ we have

$$D_2 F(t, \phi)\chi = v_t^{\phi,\chi}$$

with the solution $v^{\phi,\chi} : [-h, t_+(\phi)) \to \mathbb{R}^n$ of the linear initial value problem

$$v'(t) = Df(F(t, \phi))v_t, \qquad v_0 = \chi.$$

The tangent spaces of the manifold X_f are

$$T_\phi X_f = \left\{\chi \in C^1 : \chi'(0) = Df(\phi)\chi\right\}.$$

Assume that $0 \in U$ and 0 is a stationary point of F, i.e., $f(0) = 0$. The linearization of F at 0 is the strongly continuous semigroup $T(t) = D_2 F(t, 0)$, $t \geq 0$, on the Banach space

$$T_0 X_f = \left\{\chi \in C^1 : \chi'(0) = Df(0)\chi\right\}$$

equipped with the norm $|\cdot|_{C^1}$.

The solutions of the linear initial value problem

$$y'(t) = D_e f(0) y_t, \ y_0 = \phi \in C$$

define the strongly continuous semigroup $T_e(t)$, $t \geq 0$, on C. The generator of T_e is $G_e : \mathrm{dom}(G_e) \to C$ with $\mathrm{dom}(G_e) = T_0 X_f$ and $G_e \phi = \phi'$.

Let G denote the generator of $T(t)$, $t \geq 0$. By [4] the domain of G is

$$\left\{\chi \in C^2 : \chi \in \mathrm{dom}(G_e), \ \chi' \in \mathrm{dom}(G_e)\right\},$$

and $G\chi = \chi'$. For the spectra of G_e and G,

$$\sigma(G_e) = \sigma(G)$$

holds. All points of $\sigma(G_e)$ are eigenvalues with finite dimensional generalized eigenspaces. The eigenvalues coincide with the zeros of the characteristic function

$$\mathbb{C} \ni z \mapsto \det(zI - D_e f(0) e^{z\cdot}) \in \mathbb{C}.$$

The realified generalized eigenspaces of G_e given by the eigenvalues with negative, zero and positive real part are the stable C_s, center C_c and unstable C_u spaces, respectively. We have the decomposition

$$C = C_s \oplus C_c \oplus C_u,$$

C_s is infinite dimensional, C_c and C_u are finite dimensional, $C_c \subset \mathrm{dom}(G_e)$, $C_u \subset \mathrm{dom}(G_e)$. The set $C_s^1 = C_s \cap C^1$ is a closed subset of C^1. Hence the decomposition

$$C^1 = C_s^1 \oplus C_c \oplus C_u \tag{2.3}$$

of C^1 holds as well.

The stable, center and unstable subspaces of G are $C_s \cap \mathrm{dom}(G_e) = C_s \cap \mathrm{dom}(G_e)$, C_c and C_u, respectively, and

$$T_0 X_f = \mathrm{dom}(G_e) = (C_s \cap \mathrm{dom}(G_e)) \oplus C_c \oplus C_u.$$

In the sequel we assume

$$\sigma(G_e) \cap i\mathbb{R} \neq \emptyset,$$

that is, $\dim C_c \geq 1$.

In [4] we constructed a Lipschitz continuous (local) center manifold of F at the stationary point 0.

Theorem A *There exist open neighbourhoods $C_{c,0}$ of 0 in C_c and $C^1_{su,0}$ of 0 in $C_s^1 \oplus C_u$ with $N = C_{c,0} + C^1_{su,0} \subset U$, a Lipschitz continuous map $w_c : C_{c,0} \to C^1_{su,0}$ such that $w_c(0) = 0$ and for the graph*

$$W_c = \{\phi + w_c(\phi) : \phi \in C_{c,0}\}$$

of w_c the following hold.

(i) *$W_c \subset X_f$, and W_c is a $\dim C_c$-dimensional Lipschitz smooth submanifold of X_f.*

(ii) *If $x : \mathbb{R} \to \mathbb{R}^n$ is a continuously differentiable solution of (2.1) on $\mathbb{R}$ with $x_t \in N$ for all $t \in \mathbb{R}$, then $x_t \in W_c$ for all $t \in \mathbb{R}$.*

(iii) *W_c is locally positively invariant with respect to the semiflow F, i.e., if $\phi \in W_c$ and $\alpha > 0$ such that $F(t,\phi)$ is defined for all $t \in [0,\alpha)$, and $F(t,\phi) \in N$ for all $t \in [0,\alpha)$, then $F(t,\phi) \in W_c$ for all $t \in [0,\alpha)$.*

The aim of this paper is to show the following

Theorem 2.1 *The map $w_c : C_{c,0} \to C^1_{su,0}$ is continuously differentiable, and $Dw_c(0) = 0$.*

The proof of Theorem A in [4] applies the Lyapunov–Perron approach which is based on a variation-of-constants formula of [2]. We recall some basic facts from [2], and also some steps from the proof of Theorem A.

We denote dual spaces and adjoint operators by an asterisk $*$ in the sequel. The elements $\phi^{\odot}$ of C^* for which the curve

$$[0,\infty) \ni t \mapsto T_e^*(t)\phi^{\odot} \in C^*$$

is continuous form a closed subspace $C^{\odot}$ (of C^*) which is positively invariant under $T_e^*(t)$, $t \geq 0$. The operators

$$T_e^{\odot}(t) : C^{\odot} \ni \phi^{\odot} \mapsto T_e^*(t)\phi^{\odot} \in C^{\odot},\ t \geq 0,$$

constitute a strongly continuous semigroup on $C^{\odot}$. Similarly, we can introduce the dual space $C^{\odot *}$ and the semigroup of adjoint operators $T_e^{\odot *}(t)$, $t \geq 0$, which is strongly continuous on $C^{\odot\odot}$. There is an isometric isomorphism between $\mathbb{R}^n \times L^\infty(-h,0;\mathbb{R}^n)$ equipped with the norm $|(\alpha,\phi)| = \max\{|\alpha|, |\phi|_\infty\}$ and $C^{\odot *}$, where $L^\infty(-h,0;\mathbb{R}^n)$ denotes the Banach space of measurable and essentially bounded functions from $[-h,0]$ into $\mathbb{R}^n$ equipped with the L^∞-norm $|\cdot|_\infty$. We will identify $C^{\odot *}$ with $\mathbb{R}^n \times L^\infty(-h,0;\mathbb{R}^n)$ and omit the isomorphism. The original state space is sun-reflexive in the sense that, for the norm-preserving linear map $\iota : C \to C^{\odot *}$ given by $\iota(\phi) = (\phi(0), \phi)$, we have $\iota(C) = C^{\odot\odot}$. We also omit the embedding operator ι and identify C and $C^{\odot\odot}$. All of these results as well as the decomposition of $C^{\odot *}$ and the variation-of-constants formula can be found in [2].

Let $Y^{\odot *}$ denote the subspace $\mathbb{R}^n \times \{0\}$ of $C^{\odot *}$. For the k-th unit vector e_k in $\mathbb{R}^n$ set $r_k^{\odot *} = (e_k, 0) \in Y^{\odot *}$. Let $l : \mathbb{R}^n \to Y^{\odot *}$ be the linear map given by $l(e_k) = r_k^{\odot *}$, $k \in \{1,2,\ldots,n\}$. Then l has an inverse l^{-1}, and $|l| = |l^{-1}| = 1$.

Let $G_e^{\odot *}$ denote the generator of $T_e^{\odot *}$. For the spectra $\sigma(G_e)$ and $\sigma(G_e^{\odot *})$ we have $\sigma(G_e) = \sigma(G_e^{\odot *})$. Recall that we assumed $\sigma(G_e) \cap i\mathbb{R} \neq \emptyset$. Then $C^{\odot *}$ can be decomposed as

$$C^{\odot *} = C_s^{\odot *} \oplus C_c \oplus C_u, \tag{2.4}$$

where $C_s^{\odot *}$, C_c, C_u are closed subspaces of $C^{\odot *}$, C_c and C_u are contained in C^1, $1 \le \dim C_c < \infty$, $\dim C_u < \infty$. The subspaces $C_s^{\odot *}$, C_c and C_u are invariant under $T_e^{\odot *}(t)$, $t \ge 0$, and $T_e(t)$ can be extended to a one-parameter group on both C_c and C_u. There exist real numbers $K \ge 1$, $a < 0$, $b > 0$ and $\epsilon > 0$ with $\epsilon < \min\{-a, b\}$ such that

$$\begin{aligned} |T_e(t)\phi| &\le K e^{bt}|\phi|, && t \le 0,\ \phi \in C_u, \\ |T_e(t)\phi| &\le K e^{\epsilon|t|}|\phi|, && t \in \mathbb{R},\ \phi \in C_c, \\ |T_e^{\odot *}(t)\phi| &\le K e^{at}|\phi|, && t \ge 0,\ \phi \in C_s^{\odot *}. \end{aligned} \tag{2.5}$$

Using the identification of C and $C^{\odot\odot}$, we obtain $C_s^1 = C^1 \cap C_s^{\odot *}$. The decompositions (2.3) and (2.4) define the projection operators P_s, P_c, P_u and $P_s^{\odot *}, P_c^{\odot *}, P_u^{\odot *}$ with ranges C_s^1, C_c, C_u and $C_s^{\odot *}, C_c, C_u$, respectively. P_{su}^1 denotes the projection of C^1 onto $C_s^1 \oplus C_u$ along C_c.

We need a variation-of-constants formula for solutions of

$$x'(t) = D_e f(0) x_t + q(t) \tag{2.6}$$

with a continuous function $q : \mathbb{R} \to \mathbb{R}^n$.

If c, d are reals with $c \le d$, and $w : [c,d] \to C^{\odot *}$ is continuous, then the weak-star integral

$$\int_c^d T_e^{\odot *}(d-\tau) w(\tau)\, d\tau \in C^{\odot *}$$

is defined by

$$\left(\int_c^d T_e^{\odot *}(d-\tau) w(\tau)\, d\tau\right)(\phi^\odot) = \int_c^d T_e^{\odot *}(d-\tau) w(\tau)(\phi^\odot)\, d\tau$$

for all $\phi^\odot \in C^\odot$.

If $I \subset \mathbb{R}$ is an interval, $q : I \to \mathbb{R}^n$ is continuous and $x : I + [-h,0] \to \mathbb{R}^n$ is a solution of (2.6) on I, that is, x is continuous on $I + [-h,0]$, continuously differentiable on I, and (2.6) holds for all $t \in I$, then the curve $u : I \ni t \mapsto x_t \in C$ satisfies the integral equation

$$u(t) = T_e(t-s)u(s) + \int_s^t T_e^{\odot *}(t-\tau) Q(\tau)\, d\tau, \qquad t, s \in I,\ s \le t, \tag{2.7}$$

with $Q(t)=l(q(t))$, $t\in I$. Moreover, if $Q:I\to Y^{\odot *}$ is continuous, and a continuous $u:I\to C$ satisfies (2.7), then there is a continuous function $x:I+[-h,0]\to\mathbb{R}^n$ such that $x_t=u(t)$ for all $t\in I$, x is continuously differentiable on I, and x satisfies (2.6) with $q(t)=l^{-1}(Q(t))$, $t\in I$. So, there is a one-to-one correspondence between the solutions of (2.6) and (2.7).

For a Banach space B with norm $|\cdot|$ and a real $\eta\ge 0$, define the Banach space

$$C_\eta(\mathbb{R},B)=\left\{b\in C(\mathbb{R},B):\sup_{t\in\mathbb{R}}e^{-\eta|t|}|b(t)|<\infty\right\}$$

with norm

$$|b|_{C_\eta(\mathbb{R},B)}=\sup_{t\in\mathbb{R}}e^{-\eta|t|}|b(t)|.$$

For $\eta\ge 0$, we introduce the notation

$$Y_\eta=C_\eta(\mathbb{R},Y^{\odot *}),\ C_\eta^0=C_\eta(\mathbb{R},C),\ C_\eta^1=C_\eta(\mathbb{R},C^1).$$

For a given $Q:\mathbb{R}\to Y^{\odot *}$ we (formally) define

$$(\mathcal{K}Q)(t)=\int_0^t T_e^{\odot *}(t-\tau)P_c^{\odot *}Q(\tau)\,d\tau+\int_\infty^t T_e^{\odot *}(t-\tau)P_u^{\odot *}Q(\tau)\,d\tau$$
$$+\int_{-\infty}^t T_e^{\odot *}(t-\tau)P_s^{\odot *}Q(\tau)\,d\tau.$$

The weak-star integrals over unbounded intervals are defined as for finite intervals above, see the details in [2].

The results in the remaining part of this section are either shown in [4] or can be obtained in a straightforward manner.

For any $\eta\in(\epsilon,\min\{-a,b\})$,

$$\mathcal{K}(Y_\eta)\subset C_\eta^1,$$

and the induced linear map $\mathcal{K}_\eta:Y_\eta\to C_\eta^1$ is linear bounded with

$$|\mathcal{K}_\eta|\le c(\eta),$$

where

$$c(\eta)=K\left(1+e^{\eta h}|D_ef(0)|\right)\left(\frac{1}{\eta-\epsilon}+\frac{1}{-a-\eta}+\frac{1}{b-\eta}\right)+e^{\eta h}.$$

Moreover, if $Q\in Y_\eta$ then $u=\mathcal{K}Q$ is the unique solution of

$$u(t)=T_e(t-s)u(s)+\int_s^t T_e^{\odot *}(t-\tau)Q(\tau)\,d\tau,\qquad -\infty<s\le t<\infty,\tag{2.8}$$

in C_η^1 with $P_c^{\odot *}u(0)=0$.

As $\dim C_c<\infty$, there is a norm $|\cdot|_c$ on C_c which is C^∞-smooth on $C_c\setminus\{0\}$. Then

$$|\phi|_1=\max\{|P_c\phi|_c,|P_{su}^1\phi|_{C^1}\},\qquad \phi\in C^1,$$

defines the new norm $|\cdot|_1$ on C^1 which is equivalent to $|\cdot|_{C^1}$.

Let $\rho:\mathbb{R}\to\mathbb{R}$ be a C^∞-smooth function so that $\rho(t)=1$ for $t\le 1$, $\rho(t)=0$ for $t\ge 2$, and $\rho(t)\in(0,1)$ for $t\in(1,2)$.

Define $r:U\ni\phi\mapsto f(\phi)-D_ef(0)\phi\in\mathbb{R}^n$ and

$$\hat{r}(\phi)=\begin{cases}r(\phi), & \text{if } \phi\in U;\\ 0, & \text{if } \phi\notin U.\end{cases}$$

For any $\delta > 0$, let

$$r_\delta(\phi) = \hat{r}(\phi)\rho\left(\frac{|P_c\phi|_c}{\delta}\right)\rho\left(\frac{|P^1_{su}\phi|_1}{\delta}\right), \qquad \phi \in C^1.$$

Clearly, $r_\delta : C^1 \to \mathbb{R}^n$ is continuous. For $\gamma > 0$ set $B_\gamma(C^1) = \{\phi \in C^1 : |\phi|_1 < \gamma\}$.

Choose $\delta_0 > 0$ so that

$$B_{2\delta_0}(C^1) \subset U,$$

and $r|_{B_{2\delta_0}(C^1)}$, $Dr|_{B_{2\delta_0}(C^1)}$ are bounded. Then, for any $\delta \in (0, \delta_0)$,

$$r_\delta|_{\{\phi \in C^1 : |P^1_{su}\phi|_1 < \delta\}}(\phi) = \hat{r}(\phi)\rho\left(\frac{|P_c\phi|_c}{\delta}\right), \qquad \phi \in C^1,$$

and $r_\delta|_{\{\phi \in C^1 : |P^1_{su}\phi|_1 < \delta\}}$ is a bounded and C^1-smooth function with bounded derivative.

There exist $\delta_1 \in (0, \delta_0)$ and a nondecreasing function $\mu : [0, \delta_1] \to [0, 1]$ such that μ is continuous at 0, $\mu(0) = 0$, and for all $\delta \in (0, \delta_1]$ and for all $\phi, \psi \in C^1$

$$\begin{aligned} |r_\delta(\phi)| &\le \delta\mu(\delta), \\ |r_\delta(\phi) - r_\delta(\psi)| &\le \mu(\delta)|\phi - \psi|_{C^1}. \end{aligned} \tag{2.9}$$

For $\delta \in (0, \delta_1]$ we consider the modified equations

$$x'(t) = D_e f(0)x_t + r_\delta(x_t), \qquad t \in \mathbb{R}, \tag{2.10}$$

and

$$u(t) = T_e(t-s)u(s) + \int_s^t T_e^{\odot *}(t-\tau)l(r_\delta(u(\tau)))\,d\tau, \qquad -\infty < s \le t < \infty. \tag{2.11}$$

These equations are equivalent in the following sense: If $x : \mathbb{R} \to \mathbb{R}^n$ is C^1-smooth and is a solution of Equation (2.10), then $u : \mathbb{R} \ni t \mapsto x_t \in C^1$ is a solution of Equation (2.11), and conversely, a continuous $u : \mathbb{R} \to C^1$ satisfying (2.11) defines a C^1-smooth solution of (2.10) by $x(t) = u(t)(0)$, $t \in \mathbb{R}$.

As in [4] we fix an $\eta_0 \in (\epsilon, \min\{-a, b\})$ and a $\delta \in (0, \delta_1)$ such that

$$c(\eta_0)\mu(\delta) < \frac{1}{2}.$$

By the continuity of $c(\eta)$ for $\epsilon < \eta < \min\{-a, b\}$, there is $\eta_1 \in (\eta_0, \min\{-a, b\})$ such that

$$c(\eta)\mu(\delta) < \frac{1}{2} \qquad \text{for all } \eta \in [\eta_0, \eta_1]$$

also holds.

Define the substitution operator

$$R : (C^1)^{\mathbb{R}} \to (Y^{\odot *})^{\mathbb{R}}$$

by

$$R(u)(t) = l(r_\delta(u(t))).$$

Then, for each $\eta > 0$, we have

$$R(C^1_\eta) \subset Y_\eta.$$

By using inequalities (2.9) and $|l| = 1$, it follows that for the induced maps

$$R_\eta : C^1_\eta \to Y_\eta$$

the inequalities

$$|R_\eta(u)|_{Y_\eta} \le \delta\mu(\delta),$$
$$|R_\eta(u) - R_\eta(v)|_{Y_\eta} \le \mu(\delta)|u - v|_{C^1_\eta}$$

hold for all $u \in C^1_\eta$, $v \in C^1_\eta$ and $\eta \in [\eta_0, \eta_1]$.

Define the mapping $S : C_c \to C^1_\eta$ by

$$(S\phi)(t) = T_e(t)\phi.$$

If $\eta > \epsilon$ then $S(C_c) \subset C^1_\eta$, and the induced map $S_\eta : C_c \to C^1_\eta$ is bounded linear with $|S_\eta| \le K$.

For each $\eta \in [\eta_0, \eta_1]$ a mapping

$$G_\eta : C^1_\eta \times C_c \to C^1_\eta$$

can be defined by

$$G_\eta(u, \phi) = S_\eta(\phi) + \mathcal{K}_\eta \circ R_\eta(u).$$

For all $u, v \in C^1_\eta$, $\phi \in C_c$ and $\eta \in [\eta_0, \eta_1]$, this mapping satisfies

$$\begin{aligned} |G_\eta(u,\phi) - G_\eta(v,\phi)|_{C^1_\eta} &\le |\mathcal{K}_\eta|\,|R_\eta(u) - R_\eta(v)|_{Y_\eta} \\ &\le c(\eta)\mu(\delta)|u - v|_{C^1_\eta} \\ &\le \frac{1}{2}|u - v|_{C^1_\eta}. \end{aligned}$$

So for every $\phi \in C_c$ there is a unique fixed point $u_\eta(\phi) \in C^1_\eta$ of the contractions $G_\eta(\cdot, \phi) : C^1_\eta \to C^1_\eta$, $\eta \in [\eta_0, \eta_1]$. We have

$$P_c u_\eta(\phi)(0) = (S_\eta(\phi))(0) = \phi.$$

Therefore $u \in C^1_\eta$ is a solution of Equation (2.11) with $P_c u(0) = \phi \in C_c$ if and only if $u = u_\eta(\phi)$. The maps $u_\eta : C_c \to C^1_\eta$, $\eta_0 \le \eta \le \eta_1$, are Lipschitz continuous, and $u_\eta(0) = 0$.

Introduce the maps

$$w_\eta : C_c \ni \phi \mapsto P^1_{su} u_\eta(\phi)(0) \in C^1_s \oplus C_u$$

for $\eta_0 \le \eta \le \eta_1$. The sets

$$W_\eta = \{u_\eta(\phi)(0) : \phi \in C_c\} = \{\phi + w_\eta(\phi) : \phi \in C_c\}$$

are called the global center manifolds of the modified equation (2.11) at the stationary point 0, $\eta_0 \le \eta \le \eta_1$.

An important observation of [4] was that, for each $\phi \in C_c$ and $\eta \in [\eta_0, \eta_1]$,

$$\begin{aligned} |w_\eta(\phi)|_1 &= |P^1_{su} u_\eta(\phi)(0)|_{C^1} \\ &= |\mathcal{K}_\eta(R_\eta(u_\eta(\phi)))(0)|_{C^1} \le |\mathcal{K}_\eta(R_\eta(u_\eta(\phi)))|_{C^1_\eta} \\ &\le |\mathcal{K}_\eta|\,|R_\eta(u_\eta(\phi))|_{Y_\eta} \\ &\le c(\eta)\delta\mu(\delta) < \delta. \end{aligned}$$

It is also observed in [4] that for all $t \in \mathbb{R}$,

$$u_\eta(\phi)(t) = u_\eta(P_c u_\eta(\phi)(t))(0), \qquad \phi \in C_c,\ \eta \in [\eta_0, \eta_1].$$

Therefore,

$$P^1_{su} u_\eta(\phi)(t) = w_\eta(P_c u_\eta(\phi)(t))$$

and

$$\left|P^1_{su} u_\eta(\phi)(t)\right|_1 < \delta$$

hold for all $\eta \in [\eta_0, \eta_1]$, $\phi \in C_c$ and $t \in \mathbb{R}$. Setting

$$O_\delta = \{\phi \in C^1 : |(\mathrm{id}_{C^1} - P_c)\phi|_1 < \delta\}$$

we obtain

$$\{u_\eta(\phi)(t) : \phi \in C_c,\ t \in \mathbb{R}\} \subset O_\delta, \qquad \eta \in [\eta_0, \eta_1]. \tag{2.12}$$

[4] shows that Theorem A holds by setting

$$C_{c,0} = \{\phi \in C_c : |\phi|_1 < \delta\},\ C^1_{su,0} = \{\phi \in C^1_s \oplus C_u : |\phi|_1 < \delta\},$$
$$N = C_{c,0} + C^1_{su,0} = \{\phi \in C^1 : |\phi|_1 < \delta\},$$
$$w_c = w_{\eta_0}|_{C_{c,0}},\ W_c = \{\phi + w_c(\phi) : \phi \in C_{c,0}\}.$$

It remains to prove that w_c is C^1-smooth, and $Dw_c(0) = 0$.

3 Proof of the C^1-smoothness

Let $j_{\eta_0\eta_1}$ denote the linear continuous inclusion map

$$C^1_{\eta_0} \ni u \mapsto u \in C^1_{\eta_1}.$$

For the fixed point $u_{\eta_0}(\phi)$ of $G_{\eta_0}(\cdot, \phi)$, $\phi \in C_c$, we have

$$\begin{aligned} G_{\eta_1}(j_{\eta_0\eta_1}u_{\eta_0}(\phi), \phi) &= T_e(\cdot)(\phi) + \mathcal{K}(R(u_{\eta_0}(\phi))) \\ &= j_{\eta_0\eta_1}G_{\eta_0}(u_{\eta_0}(\phi), \phi) \\ &= j_{\eta_0\eta_1}u_{\eta_0}(\phi). \end{aligned}$$

So $j_{\eta_0\eta_1}u_{\eta_0}(\phi)$ is a fixed point of $G_{\eta_1}(\cdot, \phi) : C^1_{\eta_1} \to C^1_{\eta_1}$, and by the uniqueness of the fixed point,

$$u_{\eta_1}(\phi) = j_{\eta_0\eta_1}u_{\eta_0}(\phi), \qquad \phi \in C_c.$$

Then $u_{\eta_0}(\phi)(0) = u_{\eta_1}(\phi)(0)$, $\phi \in C_c$, follows, and consequently

$$w_{\eta_0}(\phi) = w_{\eta_1}(\phi), \qquad \phi \in C_c.$$

Let $ev : C^1_{\eta_1} \to C^1$ be given by $ev(u) = u(0)$. Then

$$w_{\eta_1} = P^1_{su} \circ ev \circ u_{\eta_1}.$$

As P^1_{su} and ev are linear bounded maps, in order to conclude the C^1-smoothness of w_{η_1}, it is sufficient to show that $u_{\eta_1} : C_c \to C^1_{\eta_1}$ is C^1-smooth. At the end of the proof we shall conclude $Dw_{\eta_1}(0) = 0$ as well.

We need two abstract results. The first one is a slightly modified version of Lemma II.6 in [7] about the smoothness of substitution operators. Although its proof follows very closely that of Lemma II.6 in [7], we show it here.

Lemma 3.1 *Let E and F be real Banach spaces. For $\eta \geq 0$ set $E_\eta = C_\eta(\mathbb{R}, E)$ and $F_\eta = C_\eta(\mathbb{R}, F)$. Let a subset $O \subset E$ and a continuous and bounded map $q : O \to F$ be given. Consider the substitution operator*

$$Q : O^{\mathbb{R}} \to F^{\mathbb{R}}, \quad Q(u)(t) = q(u(t)) \quad \text{for } u \in O^{\mathbb{R}},\ t \in \mathbb{R}.$$

(i) *If $\eta \geq 0$ and $\tilde{\eta} \geq 0$, then $Q(O^{\mathbb{R}} \cap E_\eta) \subset F_{\tilde{\eta}}$.*
(ii) *Assume that O is open, q is C^1-smooth with Dq bounded, and $0 \leq \eta \leq \tilde{\eta}$. Then, for every $u \in C(\mathbb{R}, O)$, the linear map*

$$A(u) : E^{\mathbb{R}} \to F^{\mathbb{R}}$$

given by

$$A(u)(v)(t) = Dq(u(t))v(t) \qquad \text{for } v \in E^{\mathbb{R}},\ t \in \mathbb{R},$$

satisfies

$$A(u)E_\eta \subset F_{\tilde\eta}, \qquad \sup_{v\in E_\eta,\ |v|_{E_\eta}\le 1} |A(u)v|_{F_{\tilde\eta}} \le \sup_{x\in O} |Dq(x)|,$$

the induced linear maps

$$A_{\eta\tilde\eta}(u) : E_\eta \to F_{\tilde\eta}$$

are continuous, and in case $\eta < \tilde\eta$ *the map*

$$A_{\eta\tilde\eta} : C(\mathbb{R}, O) \cap E_\eta \ni u \mapsto A_{\eta\tilde\eta}(u) \in L(E_\eta, F_{\tilde\eta})$$

is also continuous.

(iii) *If, in addition,* $\eta < \tilde\eta$ *and* O *is convex, then for every* $\tilde\epsilon > 0$ *and* $u \in C(\mathbb{R}, O) \cap E_\eta$ *there exists* $\tilde\delta > 0$ *so that for every* $v \in C(\mathbb{R}, O) \cap E_\eta$ *with* $|v-u|_{E_\eta} < \tilde\delta$, *we have*

$$|Q(v) - Q(u) - A_{\eta\tilde\eta}(u)[v-u]|_{F_{\tilde\eta}} \le \tilde\epsilon |v-u|_{E_\eta}.$$

Proof 1. Proof of (i). The map $\mathbb{R} \ni t \mapsto q(u(t)) \in F$ is continuous provided $u \in O^{\mathbb{R}} \cap E_\eta$, and

$$\sup_{t\in\mathbb{R}} e^{-\tilde\eta|t|} |q(u(t))| \le \sup_{x\in O} |q(x)|,$$

yielding the inclusion.

2. Proof of (ii). For $u \in C(\mathbb{R}, O)$ and $v \in E_\eta$, the map $\mathbb{R} \ni t \mapsto Dq(u(t))v(t) \in F$ is continuous. In addition

$$e^{-\tilde\eta|t|} |Dq(u(t))v(t)| \le e^{-(\tilde\eta-\eta)|t|} e^{-\eta|t|} |v(t)| \sup_{x\in O} |Dq(x)|,$$

yielding $A(u)E_\eta \subset F_{\tilde\eta}$, and

$$\sup_{v\in E_\eta,\ |v|_{E_\eta}\le 1} |A(u)v|_{F_{\tilde\eta}} \le \sup_{x\in O} |Dq(x)|.$$

In order to show the continuity of $A_{\eta\tilde\eta}$, let $u \in C(\mathbb{R}, O) \cap E_\eta$ be given and let $\tilde\epsilon > 0$. Choose $t_0 > 0$ such that

$$2e^{-(\tilde\eta-\eta)|t|} \sup_{x\in O} |Dq(x)| < \tilde\epsilon \qquad \text{for } |t| \ge t_0.$$

There is such a $t_0 > 0$ because of $\tilde\eta > \eta$. Find $\tilde\delta > 0$ so that for all $t \in [-t_0, t_0]$ the ball

$$B_t = \{y \in E : |y - u(t)| < \tilde\delta e^{\eta t_0}\}$$

is contained in O, and for all $y \in B_t$ we have

$$|Dq(y) - Dq(u(t))| < \tilde\epsilon.$$

Consider $\hat u \in C(\mathbb{R}, O) \cap E_\eta$ with $|\hat u - u|_{E_\eta} < \tilde\delta$, and $v \in E_\eta$ with $|v|_{E_\eta} \le 1$. Then, for $|t| \ge t_0$,

$$\begin{aligned} e^{-\tilde\eta|t|} |[Dq(\hat u(t)) - Dq(u(t))]v(t)| &\le 2e^{-(\tilde\eta-\eta)|t|} e^{-\eta|t|} |v(t)| \sup_{x\in O} |Dq(x)| \\ &\le 2e^{-(\tilde\eta-\eta)|t|} \sup_{x\in O} |Dq(x)| \\ &< \tilde\epsilon, \end{aligned}$$

and, for $|t| \le t_0$, by applying $|\hat{u}(t) - u(t)| < \delta e^{\eta|t|} \le \delta e^{\eta t_0}$, we find

$$e^{-\tilde{\eta}|t|}|[Dq(\hat{u}(t)) - Dq(u(t))]v(t)| \le e^{-(\tilde{\eta}-\eta)|t|}e^{-\eta|t|}|v(t)|\,|Dq(\hat{u}(t)) - Dq(u(t))|$$
$$\le |Dq(\hat{u}(t)) - Dq(u(t))|$$
$$< \tilde{\epsilon},$$

It follows that

$$|A_{\eta\tilde{\eta}}(\hat{u}) - A_{\eta\tilde{\eta}}(u)|_{L(E_\eta, F_{\tilde{\eta}})} \le \tilde{\epsilon}.$$

3. Proof of (iii). For all u, v in $C(\mathbb{R}, O) \cap E_\eta$ and $t \in \mathbb{R}$, by using the convexity of O, we obtain

$$\begin{aligned}
&e^{-\tilde{\eta}|t|}|q(v(t)) - q(u(t)) - Dq(u(t))[v(t) - u(t)]| \\
&\quad = e^{-\tilde{\eta}|t|}\left|\int_0^1 [Dq(sv(t) + (1-s)u(t)) - Dq(u(t))][v(t) - u(t)]\,ds\right| \\
&\quad \le e^{-(\tilde{\eta}-\eta)|t|}e^{-\eta|t|}|v(t) - u(t)| \max_{s\in[0,1]} |Dq(sv(t) + (1-s)u(t)) - Dq(u(t))| \\
&\quad \le e^{-(\tilde{\eta}-\eta)|t|}|v - u|_{E_\eta} \max_{s\in[0,1]} |Dq(sv(t) + (1-s)u(t)) - Dq(u(t))|.
\end{aligned} \tag{3.1}$$

Let $\tilde{\epsilon} > 0$ and $u \in C(\mathbb{R}, O) \cap E_\eta$ be given. Choose $t_0 > 0$ and $\tilde{\delta} > 0$ as in part 2. Let $v \in C(\mathbb{R}, O) \cap E_\eta$ be given with $|v - u|_{E_\eta} < \tilde{\delta}$. For $|t| \ge t_0$, combining (3.1) and the choice of t_0, it follows that

$$\begin{aligned}
&e^{-\tilde{\eta}|t|}|q(v(t)) - q(u(t)) - Dq(u(t))[v(t) - u(t)]| \\
&\le 2e^{-(\tilde{\eta}-\eta)|t|} \max_{x\in O}|Dq(x)|\,|v - u|_{E_\eta} \le \tilde{\epsilon}|v - u|_{E_\eta}.
\end{aligned}$$

For $|t| \le t_0$, we get $|v(t) - u(t)| \le \tilde{\delta}e^{\eta t_0}$, $|t| \le t_0$. Combining this inequality, the asumption $\eta < \tilde{\eta}$, estimation (3.1), and the choice of $\tilde{\delta}$, one concludes

$$e^{-\tilde{\eta}|t|}|q(v(t)) - q(u(t)) - Dq(u(t))[v(t) - u(t)]| \le \tilde{\epsilon}|v - u|_{E_\eta}.$$

This completes the proof of Lemma 3.1. □

The following C^1-smoothness result is contained in [7]. It is a slightly modified version of that of [8].

Lemma 3.2 *Let X, Λ be Banach spaces over $\mathbb{R}$, and let an open set $P \subset \Lambda$, a map $h : X \times P \to X$ and a constant $\kappa \in [0, 1)$ be given with*

$$|h(x, p) - h(\tilde{x}, p)| \le \kappa|x - \tilde{x}|$$

for all $x, \tilde{x}$ in X and all $p \in P$. Consider a convex subset M of X and a map $\Phi : P \to M$ so that for every $p \in P$, $\Phi(p)$ is the unique fixed point of $h(\cdot, p) : X \to X$. Suppose that the following hold.

(i) *The restriction $h_0 = h|_{M\times P}$ has a partial derivative $D_2h_0 : M \times P \to L(\Lambda, X)$, and D_2h_0 is continuous.*

(ii) *There are a Banach space X_1 over $\mathbb{R}$ and a continuous injective map $j : X \to X_1$ so that the map $k = j \circ h_0$ is continuously differentiable with respect to M in the sense that there is a continuous map*

$$B : M \times P \to L(X, X_1)$$

so that for every $(x, p) \in M \times P$ and every $\epsilon^* > 0$ there exists $\delta^* > 0$ with

$$|k(\tilde{x}, p) - k(x, p) - B(x, p)[\tilde{x} - x]| \le \epsilon^*|\tilde{x} - x|$$

for all $\tilde{x} \in M$ *with* $|\tilde{x} - x| \le \delta^*$.

(iii) *There exist maps*

$$h^{(1)} : M \times P \to L(X, X), \quad h_1^{(1)} : M \times P \to L(X_1, X_1)$$

such that

$$B(x,p)\hat{x} = jh^{(1)}(x,p)\hat{x} = h_1^{(1)}(x,p)j\hat{x} \qquad \text{for all } (x,p,\hat{x}) \in M \times P \times X$$

and

$$|h^{(1)}(x,p)| \le \kappa, \ |h_1^{(1)}(x,p)| \le \kappa \qquad \text{for all } (x,p) \in M \times P.$$

(iv) *The map*

$$M \times P \ni (x,p) \mapsto j \circ h^{(1)}(x,p) \in L(X, X_1)$$

is continuous.

Then the map $j \circ \Phi : P \to X_1$ *is* C^1*-smooth, and*

$$D(j \circ \Phi)(p) = h_1^{(1)}(\Phi(p), p) \circ D(j \circ \Phi)(p) + j \circ D_2 h_0(\Phi(p), p) \quad \text{for all } p \in P.$$

Now we employ Lemma 3.2 to prove that the map u_{η_1} is C^1-smooth.

Recall that r_δ restricted to the convex open set O_δ of C^1 is C^1-smooth, and, by inequality (2.9),

$$\sup_{\phi \in O_\delta} |Dr_\delta(\phi)| \le \mu(\delta).$$

An application of Lemma 3.1 with $E = C^1$, $F = Y^{\odot *}$, $O = O_\delta$, $q = l \circ r_\delta$ and $\eta = \eta_0$, $\tilde{\eta} = \eta_1$ shows that the linear maps

$$A(u) : (C^1)^{\mathbb{R}} \to (Y^{\odot *})^{\mathbb{R}}$$

induce a continuous map $A_{\eta_0 \eta_1}$ from the convex set

$$M = \{u \in C^1_{\eta_0} : u(t) \in O_\delta \text{ for all } t \in \mathbb{R}\} \subset C^1_{\eta_0}$$

into $L(C^1_{\eta_0}, Y_{\eta_1})$ so that for every $u \in M$ and for every $\tilde{\epsilon} > 0$ there exists $\tilde{\delta} > 0$ such that for every $v \in M$ with $|v - u|_{C^1_{\eta_0}} \le \tilde{\delta}$, we have

$$R(u) \in Y_{\eta_1}, \ R(v) \in Y_{\eta_1}, \ |R(u) - R(v) - A_{\eta_0\eta_1}(u)[v - u]|_{Y_{\eta_1}} \le \tilde{\epsilon}|v - u|_{C^1_{\eta_0}}. \quad (3.2)$$

Set $X = C^1_{\eta_0}$, $\Lambda = P = C_c$, $h = G_{\eta_0}$, and $\kappa = \frac{1}{2}$. By (2.12), $u_{\eta_0}(P) \subset M$ holds. So, the unique fixed point of $h(\cdot, \phi) : X \to X$ is the map $\Phi : P \to M$ given by $\Phi(\phi) = u_{\eta_0}(\phi)$. In the next steps we verify the hypotheses of Lemma 3.2.

The map $h_0 = h|_{M \times P}$ satisfies

$$h_0(u, \phi) = G_{\eta_0}(u, \phi) = S_{\eta_0}(\phi) + \mathcal{K}_{\eta_0} \circ R_{\eta_0}(u).$$

So, the partial derivative $D_2 h_0 : M \times P \to L(\Lambda, X)$ exists, and for every $(u, \phi) \in M \times P$ it is given by

$$D_2 h_0(u, \phi)\psi = S_{\eta_0}(\psi) \in C^1_{\eta_0}, \qquad \psi \in C_c.$$

It is clearly continuous since it is a constant map.

Setting $X_1 = C^1_{\eta_1}$ and $j = j_{\eta_0\eta_1}$, the map $k = j \circ h_0$ is given by

$$k(u, \phi) = S_{\eta_1}(\phi) + \mathcal{K}_{\eta_1} \circ R_{\eta_1} \circ j(u),$$

and

$$B : M \times P \ni (u, \phi) \mapsto \mathcal{K}_{\eta_1} \circ A_{\eta_0\eta_1}(u) \in L(X, X_1)$$

is continuous.

Let $\epsilon^* > 0$ and $(u, \phi) \in M \times P$ be given. Define

$$\delta^* = \tilde{\delta}\left(\frac{\epsilon^*}{1 + |\mathcal{K}_{\eta_1}|}\right),$$

where $\tilde{\delta}(\tilde{\epsilon})$ is chosen so that (3.2) holds for all $v \in M$ with $|v - u|_{C^1_{\eta_0}} \le \delta^*$ and $\tilde{\epsilon} = \epsilon^*/(1 + |\mathcal{K}_{\eta_1}|)$. Then for all such $v \in M$ we find

$$\begin{aligned}
|k(v, \phi) - &k(u, \phi) - B(u, \phi)[v - u]|_{X_1} \\
&= |\mathcal{K}_{\eta_1}(R(v)) - \mathcal{K}_{\eta_1}(R(u)) - \mathcal{K}_{\eta_1}(A_{\eta_0\eta_1}(u)[v - u])|_{C^1_{\eta_1}} \\
&\le |\mathcal{K}_{\eta_1}| \frac{\epsilon^*}{1 + |\mathcal{K}_{\eta_1}|} |v - u|_{C^1_{\eta_0}} \\
&\le \epsilon^* |v - u|_{C^1_{\eta_0}}.
\end{aligned}$$

From the facts $\sup_{\phi \in O_\delta} |Dr_\delta(\phi)| \le \mu(\delta)$ and $|l| = 1$, it is clear that, for every $u \in M$, $A(u)$ defines the elements

$$\mathcal{K}_{\eta_0} \circ A_{\eta_0\eta_0}(u) \in L(X, X) \quad \text{with } |\mathcal{K}_{\eta_0} \circ A_{\eta_0\eta_0}(u)| \le \mu(\delta)$$

and

$$\mathcal{K}_{\eta_1} \circ A_{\eta_1\eta_1}(u) \in L(X_1, X_1) \quad \text{with } |\mathcal{K}_{\eta_1} \circ A_{\eta_1\eta_1}(u)| \le \mu(\delta).$$

Now we can define

$$h^{(1)} : M \times P \ni (u, \phi) \mapsto \mathcal{K}_{\eta_0} \circ A_{\eta_0\eta_0}(u) \in L(X, X)$$

and

$$h_1^{(1)} : M \times P \ni (u, \phi) \mapsto \mathcal{K}_{\eta_1} \circ A_{\eta_1\eta_1}(u) \in L(X_1, X_1).$$

For all $(u, \phi, v) \in M \times P \times X$,

$$B(u, \phi)v = \mathcal{K}(A(u)v) = jh^{(1)}(u, \phi)v = h_1^{(1)}(u, \phi)jv$$

holds. Using $|A_{\eta_0\eta_0}(u)| \le \mu(\delta)$, $|A_{\eta_1\eta_1}(u)| \le \mu(\delta)$, and the choice of η_0, η_1, δ, it is obvious that

$$|h^{(1)}(u, \phi)| \le \kappa, \; |h_1^{(1)}(u, \phi)| \le \kappa \qquad \text{for all } (u, \phi) \in M \times P.$$

For the map $M \times P \ni (x, p) \mapsto j \circ h^{(1)}(x, p) \in L(X, X_1)$ we have

$$j \circ h^{(1)}(u, \phi)v = (j \circ \mathcal{K}_{\eta_0} \circ A_{\eta_0\eta_0}(u))v = \mathcal{K}(A(u)v) = B(u, \phi)v$$

for all $(u, \phi, v) \in M \times P \times X$, and the continuity of $M \times P \ni (x, p) \mapsto j \circ h^{(1)}(x, p) \in L(X, X_1)$ follows from that of B.

Therefore all hypotheses of Lemma 3.2 are fulfilled, and $j \circ \Phi = u_{\eta_1} : C_c \to C^1_{\eta_1}$ is C^1-smooth. Moreover,

$$Du_{\eta_1}(\phi) = h_1^{(1)}(u_{\eta_0}(\phi), \phi) \circ Du_{\eta_1}(\phi) + j \circ D_2h_0(u_{\eta_0}(\phi), \phi)$$

for all $\phi \in C_c$. From $Dr_\delta(0) = 0$, $A(0) = 0$ and $h_1^{(1)}(0, 0) = 0$ follows. Then, using also $u_{\eta_0}(0) = 0$, we obtain

$$Du_{\eta_1}(0)\psi = S_{\eta_1}(\psi) \qquad \text{for all } \psi \in C_c.$$

Hence

$$\begin{aligned}
Dw_{\eta_1}(0)\psi &= (P^1_{su} \circ ev \circ Du_{\eta_1}(0))\psi \\
&= P^1_{su}\psi = 0
\end{aligned}$$

follows for all $\psi \in C_c$. Therefore

$$Dw_{\eta_1}(0) = 0.$$

This completes the proof of Theorem 2.1.

Acknowledgement. This work was supported by the Hungarian Scientific Research Fund, grant no. T 049516.

References

[1] Cooke, K., and W. Huang, *On the problem of linearization for state-dependent delay differential equations.* Proceedings of the A.M.S. 124 (1996), 1417–1426.

[2] Diekmann, O., van Gils, S. A., Verduyn Lunel, S. M., and H. O. Walther, *Delay Equations: Functional-, Complex- and Nonlinear Analysis.* Springer, New York, 1995.

[3] Hale, J. K., and S. Verduyn Lunel, *Introduction to Functional Differential Equations.* Springer, New York 1993.

[4] Hartung, F., T. Krisztin, H.-O. Walther and J. Wu, *Functional differential equations with state-dependent delays: theory and applications.* Preprint.

[5] Krisztin, T., *A local unstable manifold for differential equations with state-dependent delay.* Discrete and Continuous Dynamical Systems 9 (2003), 993–1028.

[6] Krisztin, T., *Invariance and noninvariance of center manifolds of time-t maps with respect to the semiflow.* SIAM J. Math. Anal. 36 (2004), 717–739.

[7] Krisztin, T., Walther, H. O., and J. Wu, *Shape, Smoothness, and Invariant Stratification of an Attracting Set for Delayed Monotone Positive Feedback.* Fields Institute Monograph series, vol. 11, Amer. Math. Soc., Providence, 1999.

[8] Vanderbauwhede, A. and S.A. van Gils, *Center manifolds and contractions on a scale of Banach spaces.* J. Functional Anal. 71 (1987), 209–224.

[9] Walther, H. O., *The solution manifold and C^1-smoothness of solution operators for differential equations with state dependent delay.* J. Differential Equations 195 (2003), 46–65.

[10] Walther, H. O., *Smoothness properties of semiflows for differential equations with state dependent delay.* Russian, in *Proceedings of the International Conference on Differential and Functional Differential Equations, Moscow, 2002*, vol. 1, pp. 40-55, Moscow State Aviation Institute (MAI), Moscow 2003. English version: Journal of the Mathematical Sciences 124 (2004), 5193–5207.

Fields Institute Communications
Volume **48**, 2006

Normal Forms for Germs of Analytic Families of Planar Vector Fields Unfolding a Generic Saddle-node or Resonant Saddle

Christiane Rousseau
Département de mathématiques et de statistique and CRM
Université de Montréal
C.P. 6128, Succursale Centre-ville
Montréal (Qué.), H3C 3J7, Canada
rousseac@dms.umontreal.ca

Abstract. Normal form theory provides an algorithmic way to decide if two germs of planar vector fields with a saddle-node or a resonant saddle are equivalent under a C^N-change of coordinates, in which case, the normal forms are polynomial. However, in the analytic case, the formal change of coordinates to normal form generically diverges. An explanation of this is found by considering unfoldings of the vector fields and explaining the divergence in the limit process. We consider the orbital equivalence problem for germs of families of vector fields unfolding a generic saddle-node or resonant saddle and give a complete modulus of analytic classification for such families.

1 Introduction

The paper is a contribution to the general question of the equivalence problem for analytic vector fields on $\mathbb{R}^n$, namely

1. When are two germs of vector fields v and w locally orbitally equivalent?
2. When are two germs of families of vector fields v_ϵ and w_ϵ locally orbitally equivalent? (ϵ can be a multi-parameter.)

We can of course always suppose that the germs of vector fields and parameters are all localized at the origin.

Question 1 (resp. 2) has the answer "always" when $v(0), w(0) \neq 0$ (resp. $v_0(0), w_0(0) \neq 0$). This is just a consequence of the blow-box theorem which is valid for analytic families and implies that there is a unique equivalence class under the hypothesis that the vector field does not vanish.

2000 *Mathematics Subject Classification.* Primary 34C20, 34M25, 37G05; Secondary 35G10.
This work was supported by NSERC in Canada.

A strategy to try to solve the equivalence problem and identify the different equivalence classes is to use normal form theory. Indeed the normal form can be seen as a canonical element of the equivalence class.

If we look to the case of a nonzero family of vector fields the flow-box theorem states that the family is locally orbitally equivalent to

$$\begin{array}{rcl} \dot{x}_1 & = & 1 \\ \dot{x}_2 & = & 0 \\ \vdots & & \\ \dot{x}_n & = & 0 \end{array} \tag{1.1}$$

which can be seen as a normal form for the family.

So the first nontrivial case is when the vector field has a singular point at the origin $v(0) = 0$. In dimension 1 the problem has been completely solved by Kostov [7] (except for Question 2 in the case where $v_0 \equiv 0$). Indeed if $v'(0) \neq 0$, then for any family v_ϵ unfolding v there exists an analytic linearizing change of coordinates $h_\epsilon(x) = h(\epsilon, x)$ defined in a neighborhood of the origin in (x, ϵ) space. When $v'(0) = 0$ and $v^{(k+1)}(0) \neq 0$ (this is the codimension k case), then for any family v_ϵ unfolding v there exists an analytic change of coordinates $h_\epsilon(x) = h(\epsilon, x)$ defined in a neighborhood of the origin in (x, ϵ) and an analytic scaling of time bringing the family to the normal form

$$\dot{x} = (\epsilon_0 + \epsilon_1 x + \cdots + \epsilon_{k-1} x^{k-1} + x^{k+1})(1 + a(\epsilon) x^k). \tag{1.2}$$

For the rest of the paper we will limit ourselves to the two-dimensional case $n = 2$.

The node case. We consider a vector field v_0 which has a node at the origin with eigenvalues λ_1 and λ_2 such that $\lambda_1/\lambda_2 = \mu_0 \in \mathbb{R}^+$. Let v_ϵ be an unfolding of v_0 and μ_ϵ be the quotient of the eigenvalues at the singular point. It goes back to Poincaré that the family v_ϵ is locally orbitally equivalent to:

1. If $\mu_0 \notin \mathbb{N} \cup 1/\mathbb{N}$

$$\begin{array}{rcl} \dot{x} & = & x \\ \dot{y} & = & \mu_\epsilon y. \end{array} \tag{1.3}$$

2. If $\mu_0 \in \mathbb{N}$

$$\begin{array}{rcl} \dot{x} & = & x \\ \dot{y} & = & \mu_\epsilon y + a(\epsilon) x^{\mu_0}. \end{array} \tag{1.4}$$

3. If $\mu_0 \in 1/\mathbb{N}$

$$\begin{array}{rcl} \dot{x} & = & x + b(\epsilon) y^{1/\mu_0} \\ \dot{y} & = & \mu_\epsilon y. \end{array} \tag{1.5}$$

The saddle case. We consider a vector field v_0 which has a saddle at the origin with eigenvalues λ_1 and λ_2 such that $\lambda_1/\lambda_2 = -\mu_0 \in \mathbb{R}^-$.

1. If λ_1/λ_2 is irrational then there is a formal change of coordinates and a formal time scaling to the linear system

$$\begin{array}{rcl} \dot{x} & = & x \\ \dot{y} & = & -\frac{\mu_0}{y}. \end{array} \tag{1.6}$$

If μ_0 is diophantian (badly approximated by the rational numbers) then there exists an analytic change of coordinates and time scaling to the linear system. However if μ_0 is Liouvillian then *divergence is the rule and convergence is the exception* [6]. The question we ask is **"Why?"**

2. If $\lambda_2/\lambda_1 = -p/q$, then, in the generic (codimension 1) case, there exists a formal change of coordinates and a formal time scaling transforming the system to the polynomial normal form

$$\begin{aligned} \dot{x} &= x \\ \dot{y} &= y(-\tfrac{p}{q} + u + au^2), \end{aligned} \tag{1.7}$$

where $u = x^p y^q$. Again *divergence is the rule.* It is very exceptional that the change of coordinate to normal form converges. Here also we ask **"Why?"**

The Hopf bifurcation case. We consider a vector field v_0 which has a weak focus of order 1 at the origin with eigenvalues $\lambda_{1,2} = \pm i\omega$. Then there is a formal change of coordinates and a formal time scaling to the system

$$\begin{aligned} \dot{x} &= -\omega y + x(x^2+y^2) + ax(x^2+y^2)^2 \\ \dot{y} &= \omega x + y(x^2+y^2) + ay(x^2+y^2)^2. \end{aligned} \tag{1.8}$$

Again *divergence is the rule* and it is very exceptional that the change of coordinate to normal form converges. Again we ask **"Why?"**

The saddle-node case. We consider a vector field v_0 which has a saddle-node of codimension 1 at the origin. Then there exists a formal change of coordinates and a formal time scaling to the system

$$\begin{aligned} \dot{x} &= x^2 \\ \dot{y} &= y(1+ax). \end{aligned} \tag{1.9}$$

Here also *divergence is the rule* and only exceptionally the change of coordinate to normal form converges. Again we ask **"Why?"**

These questions are all related as non resonant (resp. resonant) saddles appear in the perturbation of a resonant (resp. non resonant) saddle. And a saddle-node is the coallescence of a saddle and a node. Hence, to answer these questions, it is natural to study the unfoldings of these situations. Moreover as we are considering convergence of power series it is necessary to enlarge the variables (x, y) to the complex domain, namely to a neighborhood of $(0,0)$ in $\mathbb{C}^2$. This point of view allows to unify the weak focus case with the saddle case as the ratio of eigenvalues of a weak focus is -1.

The spirit of the general answer is the following. The dynamics of the original system is very rich. It is much too rich to be encoded in the simple dynamics of the normal form which depends of at most one parameter. Hence the divergence of the normalizing series.

Strategy. We must learn to read the rich dynamics of the original system in order to solve the equivalence problem.

In Section 2 we discuss the example of the saddle-node. In Section 3 we discuss the equivalence problem for saddles and saddle-nodes via the holonomy map. In Section 4 we discuss analytic changes of coordinates to normal form. Finally in

Section 5 we discuss applications to problems of finite cyclicity of graphics. We end up with perspectives.

2 The example of the saddle-node

As mentioned above, if v_0 has a saddle-node of codimension 1 at the origin, then there exists a formal change of coordinates and a formal time scaling to the system

$$\begin{array}{rcl}\dot{x} & = & x^2 \\ \dot{y} & = & y(1+ax).\end{array} \tag{2.1}$$

If v_η is a generic family unfolding the saddle node then, for any $k \in \mathbb{N}$, there exists a C^k-change of coordinates and parameters and a C^k time scaling bringing the family to the normal form

$$\begin{array}{rcl}\dot{x} & = & x^2 - \epsilon \\ \dot{y} & = & y(1+a(\epsilon)x).\end{array} \tag{2.2}$$

We call (2.1) the *model* and (2.2) the *model family*. Their phase portrait appear in Figure 1.

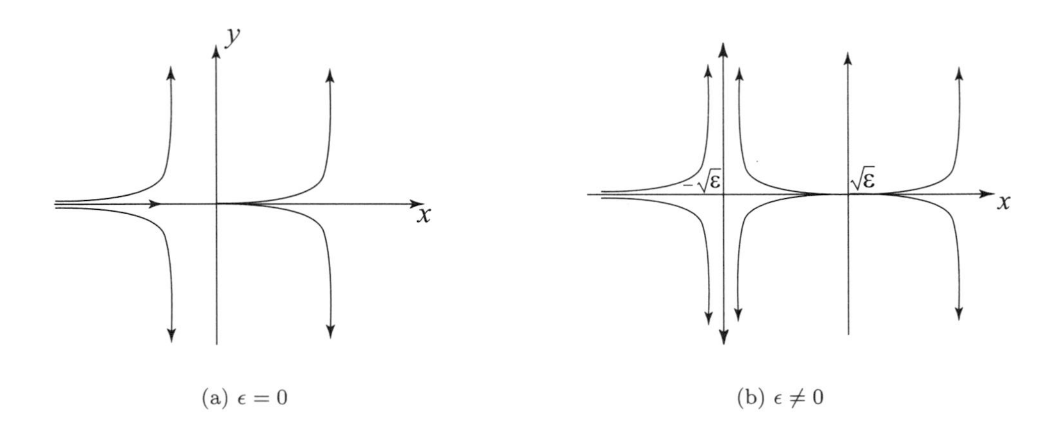

(a) $\epsilon = 0$ (b) $\epsilon \neq 0$

Figure 1 The "model"

Starting with a single analytic vector field v_0 with a saddle-node at the origin it is possible to find an analytic change of coordinates and analytic time scaling to bring the system to the form

$$\begin{array}{rcl}\dot{x} & = & x^2 \\ \dot{y} & = & f_0(x) + y(1+ax) + \sum_{j=2}^{\infty} f_j(x)y^j.\end{array} \tag{2.3}$$

The model has the analytic center manifold $y = 0$. One obstruction to bring (2.3) to normal form is the non-existence of an analytic center manifold. Indeed there exists a formal center manifold

$$y = \sum_{n=2}^{\infty} b_n x^n, \tag{2.4}$$

but generically the series is divergent. The generic divergence is this case is formulated in this way. Suppose that the system (2.3) depends analytically on a finite number of parameters $\eta \in \mathbb{R}^n$. As soon as there exists a single value η_0 for which the series (2.4) is divergent, then the set of values of η for which the series is convergent is an analytic set. In particular divergence occurs on a dense open set.

To understand why divergence is the rule in this case we complexify x and y so that (x, y) is now defined on a neighborhood of the origin in $\mathbb{C}^2$ and we unfold. The model and model family appear in Figure 2.

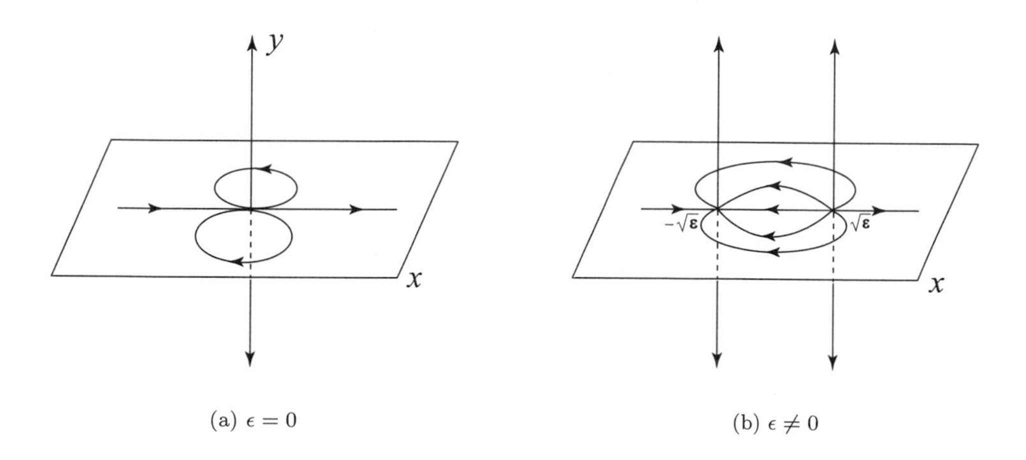

(a) $\epsilon = 0$ (b) $\epsilon \neq 0$

Figure 2 The "model"

Given a generic unfolding of v_0 it is possible to find an analytic change of coordinates and parameter and an analytic time scaling to bring the family to the form

$$\begin{aligned} \dot{x} &= x^2 - \epsilon \\ \dot{y} &= f_{0,\epsilon}(x) + y(1 + a(\epsilon)x) + \textstyle\sum_{j=2}^{\infty} f_{j,\epsilon}(x) y^j. \end{aligned} \tag{2.5}$$

Let us now discuss the case where there exists no analytic center manifold, i.e. the series given in (2.4) is divergent. However this series is 1-summable (or Borel-summable) [10] and yields a solution in a sectorial domain V of the universal covering of the x-space punctured at 0 of the form $V = \{\hat{x}; |\hat{x}| < r, arg(\hat{x}) \in (-\pi/2 + \delta, 5\pi/2 - \delta)\}$, where $r, \delta > 0$ are small. As we will not use the theory of summability we do not define it and refer the interested reader to [1]. This solution is ramified (see Figure 3a).

The Figure 3b) represents the case $\epsilon > 0$, i.e. the system has a saddle and a node. It is known that the saddle always has an analytic stable manifold with equation $y = g_\epsilon(x)$. Let us now discuss the local model at the node. We have two cases depending if the node is non resonant or resonant.

1. If the node is non resonant then the local model at the node is given by the linear system

$$\begin{aligned} \dot{x} &= \lambda_1 x \\ \dot{y} &= y \end{aligned} \tag{2.6}$$

with $\lambda_1 \notin 1/\mathbb{N}$. All solution curves (except $x = 0$) are of the form $y = Cx^{1/\lambda_1}$. They are all ramified but one! We get the following:

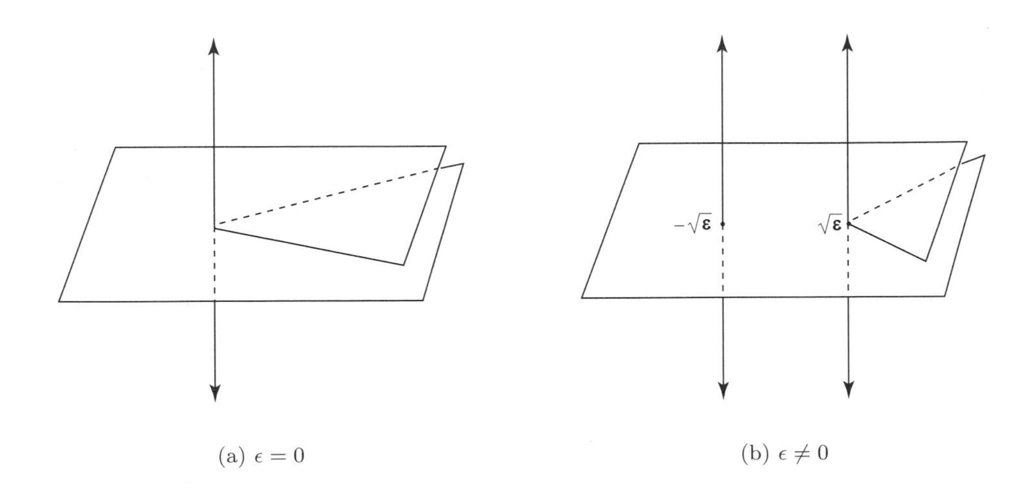

(a) $\epsilon = 0$ (b) $\epsilon \neq 0$

Figure 3 The invariant manifold

Conclusion 1: *When we unfold a system with no analytic center manifold, then the analytic separatrices of the saddle and of the node do not match.*

2. If the node is resonant then the local model at the node is the normal form

$$\begin{aligned} \dot{x} &= \frac{x}{n} \\ \dot{y} &= y + Ax^n. \end{aligned} \tag{2.7}$$

If $A = 0$ then all solution curves at the node (except $x = 0$) are analytic of the form $y = Cx^n$. This case is obviously impossible when unfolding a system as in Figure 3a) and we are forced to have $A \neq 0$, yielding that all solutions (except $x = 0$) are of the form $y = nAx^n \ln x + Cx^n$. Hence we get:

Conclusion 2: *When we unfold a system with no analytic center manifold then the node is non linearizable as soon as resonant. This is the "parametric resurgence phenomenon".*

Putting together Conclusions 1 and 2 we get:

Conclusion: *The divergence of the series giving the center manifold reflects an incompatibility between the two equilibrium points as their analytic separatrices do not match. For sequences of parameter values the incompatibility is carried by the singular point itself: this is the parametric resurgence phenomenon.*

The non analyticity of the center manifold is expressed by the divergence of the series (2.4) which is equivalent to the divergence of the change of coordinate $Y = y - \sum_{n=2}^{\infty} b_n x^n$ removing the term $f_0(x)$ in (2.3). This is the first step in transforming the system to the normal form which is our model.

The phenomenon we have described above is very general. If we were considering a system (2.3) in which $f_0 \equiv 0$ (the system has an analytic center manifold) we can look for a change of coordinates

$$Y = y + \sum_{k=0}^{\infty}\sum_{j=2}^{\infty} b_{k,n}x^k y^n \tag{2.8}$$

removing the terms $f_j(x)y^j$ in (2.3). This power series is generically divergent. Its sum is defined as a ramified function on a domain of the form $V \times W$, where W is a neighborhood of the origin in y-space and V is a sectorial domain of the universal covering of the x-space punctured at 0 of the form $V = \{\hat{x}; |\hat{x}| < r, arg(\hat{x}) \in (-3\pi/2 + \delta, 3\pi/2 - \delta)\}$, where $r, \delta > 0$ are small. If the unfolding is of the form (2.5) with $f_{0,\epsilon} \equiv 0$, then we can find an unfolded change of coordinates which is regular at the node and ramified at the saddle, the ramification reflecting an incompatibility between normalizing changes of coordinates at the saddle and at the node. Here again we observe parametric resurgence phenomena for sequences of parameter values for which the saddle is resonant: for these parameter values the saddle is not integrable.

This last step remains true when the system (2.3) has no analytic center manifold, but the calculations cannot be done in a simple way on the series and the use of geometric methods to handle the proofs is necessary.

Remark 2.1 Many papers in the literature describe the fact that, whenever the normalizing series may diverge, then divergence is the rule and convergence the exception (see for instance [6]). The above geometric explanation for the divergence of the series for the center manifold explains why this is the case. It is indeed the general situation that the analytic separatrices of the saddle and the node do not match. It is also the generic situation that a resonant node be nonlinearizable. When these generic behaviours persist till the limit case of the saddle-node there exists no analytic center manifold.

3 The equivalence problem for resonant saddles and saddle-nodes

The equivalence problem is a problem in two-variables for 2-dimensional vector fields. We will introduce the holonomy of each separatrix (resp. of the strong separatrix) in the case of a saddle (resp. saddle-node). The holonomy is a 1-dimensional map. We can make a parallel between its use and the use of the Poincaré return map which allows to reduce the search for periodic trajectories of a 2-dimensional vector field to the search of fixed points of a 1-dimensional map. As the separatrices of an analytic saddle are analytic and similarly for the strong separatrix of a saddle-node we can always use an analytic change of coordinates transforming these separatrices to the coordinate axes.

To define the holonomy we need to extend the system to a neighborhood of the origin in $\mathbb{C}^2$. We also allow the time to be complex. So the trajectories are parametrized by open sets in $\mathbb{C}$ and hence are complex curves in $\mathbb{C}^2$, which we can think of as real 2-dimensional surfaces in $\mathbb{R}^4$. The trajectories are usually called the *leaves of the foliation* given by the differential equation.

Definition 3.1 We consider a saddle point

$$\begin{aligned} \dot{x} &= \lambda_1 x(1 + h_1(x, y)) \\ \dot{y} &= \lambda_2 y(1 + h_2(x, y)) \end{aligned} \tag{3.1}$$

or a saddle-node

$$\begin{array}{rcl}\dot{x} & = & x^2(1+h_1(x,y)) \\ \dot{y} & = & \lambda_3 y + h_3(x,y)\end{array} \tag{3.2}$$

of a 2-dimensional vector field, where $\lambda_j \neq 0$, $\lambda_1\lambda_2 < 0$ and $h_{1,2}(x,y) = O(|x,y|)$, $h_3(x,y) = o(|x,y|)$. We consider a section $\Sigma = \{|x| < \delta, y = y_0\}$ where $y_0, \delta > 0$. The holonomy of the y-separatrix is a map $f : U \subset \Sigma \to \Sigma$, where U is a neighborhood of 0 in Σ. Let $(x, y_0) \in U$. We lift the curve $y = y_0 e^{i\theta}$, $\theta \in [0, 2\pi]$ into the leaf of the foliation passing through $(x, y_0) \in U$. The end point is the point $(f(x), y_0)$, yielding the definition of f (Figure 4).

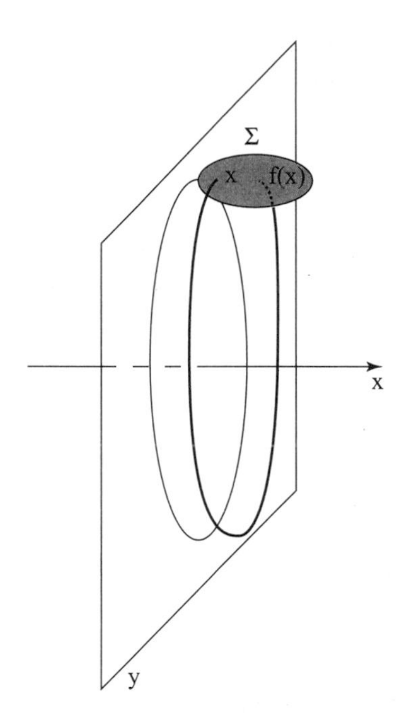

Figure 4 The holonomy map

The following proposition is classical.

Proposition 3.2 The holonomy f of the y-axis has the form

$$f(x) = \exp\left(2\pi i \frac{\lambda_2}{\lambda_1}\right) x + o(x) \tag{3.3}$$

for (3.1) and

$$f(x) = x + ax^2 + o(x^2), \tag{3.4}$$

with $a \neq 0$ for a saddle-node.

Remark 3.3 If we choose a different section $\Sigma_1 = \{y = y_1\}$ and if f_1 is the holonomy for the new section, then f and f_1 are conjugate, i.e. there exists an analytic diffeomorphism defined in a neighborhood of 0 such that $f_1 = h^{-1} \circ f \circ h$.

Theorem 3.4 1. *Two germs of vector fields* (3.1) *with a saddle at the origin and same formal normal form, are locally analytically orbitally equivalent if and only if the holonomies of their y-separatrices are conjugate.*

2. *Two germs of vector fields (3.2) with a saddle-node at the origin and same formal normal form, are locally analytically orbitally equivalent if and only if the holonomies of their strong separatrices are conjugate.*

Part 1 of the Theorem was proved by Mattei-Moussu [9] for resonant saddles and Pérez-Marco-Yoccoz [14] for non resonant saddles. Part 2 was proved by Martinet-Ramis [10].

The Theorem 3.4 shows that the equivalence problem for germs of vector fields with a saddle or saddle-node is reduced to the conjugacy problem for germs of diffeomorphisms with a fixed point at the origin and multiplier on the unit circle. The kind of results we will describe below applies to generic (codimension 1) resonant saddles ($\lambda_2/\lambda_1 \in \mathbb{Q}$) and saddle-nodes. But for the sake of simplicity we will limit ourselves to saddles with $\lambda_2/\lambda_1 = -1$ and saddle-nodes, which both have a holonomy map tangent to the identity. We limit ourselves to saddle points which are non integrable of order 1, i.e. orbitally analytically equivalent to the form

$$\begin{aligned} \dot{x} &= x \\ \dot{y} &= -y(1 + Axy + o(|xy|)). \end{aligned} \tag{3.5}$$

Then, under an adequate scaling for x, the holonomy of the y-axis has the form

$$f(x) = x + x^2 + o(x^2). \tag{3.6}$$

The conjugacy problem for germs of diffeomorphisms of the form (3.6) has been solved by Ecalle-Voronin ([4] and [16]). (Ecalle has also solved the conjugacy problem in the more general case of a resonant multiplier $\exp(2\pi i \frac{p}{q})$.)

Theorem 3.4 can be generalized to generic families unfolding a resonant saddle or a saddle-node.

Theorem 3.5 *([2] and [15]) Two germs of generic families of analytic vector fields unfolding a vector field with a resonant saddle (resp. saddle-node) at the origin are analytically orbitally equivalent if and only if the families of their unfolded holonomies are conjugate.*

Hence the equivalence problem for germs of families of vector fields is reduced to the conjugacy problem for germs of families of diffeomorphisms

$$f_\epsilon(x) = x + x(x - \epsilon) + o(x^2), \tag{3.7}$$

in the saddle case and

$$f_\epsilon(x) = x + x^2 - \epsilon + o(x^2), \tag{3.8}$$

in the saddle-node case. To solve the conjugacy problem we identify a complete modulus of analytic classification, so that two families are analytically conjugate if and only if they have the same modulus.

Before describing the modulus for the families we describe the Ecalle-Voronin modulus in the case $\epsilon = 0$. The principle is the following: Two germs of diffeomorphisms $f_1, f_2 : (\mathbb{C}, 0) \to (\mathbb{C}, 0)$ of the form

$$f(x) = x + x^2 + Ax^3 + o(x^3) \tag{3.9}$$

with same constant A (determined by the formal normal form) are conjugate if and only if they have the same orbit space. The Ecalle-Voronin modulus is one way to describe the orbit space. The orbit space is described in Figure 5: to explain its construction we first remark that the diffeomorphism is topologically like the time-one map of the vector field whose flow lines appear in Figure 5. We give ourselves a

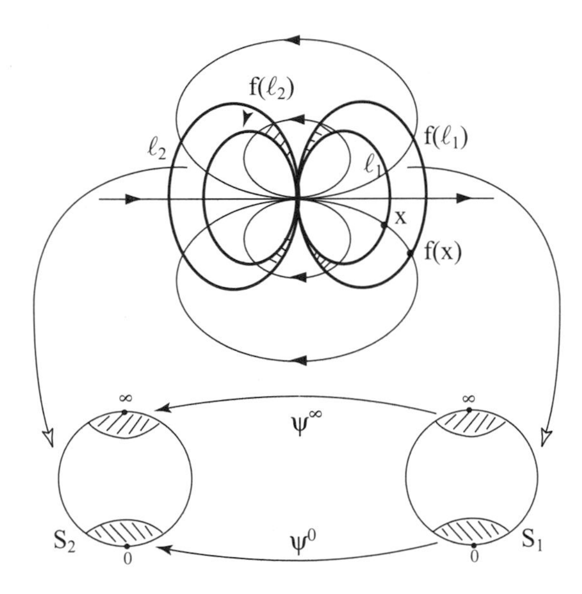

Figure 5 Orbit space of a generic diffeomorphim

first fundamental domain limited by the curve ℓ_1 and its image $f(\ell_1)$. If we identify $x \in \ell_1$ with its image $f(x)$ the fundamental domain is conformally equivalent to a sphere S_1. The ends of the crescent limited by ℓ_1 and $f(\ell_1)$ correspond to the points 0 and ∞ on the sphere. All orbits of f (except that of 0) are represented by a most one point of the sphere. However there exists points in the neighborhood of 0 whose orbits have no representative on the sphere. To cover the orbit space we therefore need to take a second fundamental neighborhood limited by a second curve ℓ_2 and its image $f(\ell_2)$. As before we identify $x \in \ell_2$ with its image $f(x)$ and this fundamental domain is also conformally equivalent to a sphere S_2. But there exists points in the neighborhood of 0 (resp. ∞) in S_1 and S_2 which belong to the same orbit. So we need to identify a neighborhood of 0 (resp. ∞) in S_1 with a neighborhood of 0 (resp. ∞) in S_2. This is done via an analytic diffeomorphism ψ^0 (resp. ψ^∞) sending 0 to 0 (resp. ∞ to ∞). The size of the neighborhoods of 0 and ∞ depend on the curves ℓ_i but what is intrinsic is the germs of analytic diffeomorphims:

$$\begin{cases} \psi^0 : (\mathbb{C}, 0) \to (\mathbb{C}, 0) \\ \psi^\infty : (\mathbb{C}, \infty) \to (\mathbb{C}, \infty). \end{cases} \tag{3.10}$$

The only analytic changes of coordinates on S_j which preserve 0 and ∞ are the linear maps. If we choose different coordinates on S_j we get different germs $\overline{\psi}^0$ and $\overline{\psi}^\infty$. The equivalence relation corresponding to changes of coordinates on S_j preserving 0 and ∞ is

$$(\psi^0, \psi^\infty) \sim (\overline{\psi}^0, \overline{\psi}^\infty) \iff \exists C_1, C_2 \begin{cases} \overline{\psi}^0(w) = C_2\psi^0(C_1 w) \\ \overline{\psi}^\infty(w) = C_2\psi^\infty(C_1 w). \end{cases} \tag{3.11}$$

Definition 3.6 The Ecalle-modulus of the diffeomorphism f is given by the tuple $(\psi^0, \psi^\infty)/\sim$.

All tuples (ψ^0, ψ^∞) are realizable as the Ecalle-Voronin modulus of a germ of diffeomorphism of the form (3.9).

Definition 3.7 The normal form of a germ of diffeomorphism of the form (3.9) is the time-one map of a vector field

$$\dot{x} = \frac{x^2}{1+ax} \tag{3.12}$$

where $a = A - 1$. a is called the *formal invariant.*

Theorem 3.8 *([4] and [16]) For a germ of diffeomorphism of the form (3.9) there always exist a formal change of coordinate to normal form. This change of coordinate converges if and only if ψ^0 and ψ^∞ are linear.*

This theorem remains true for generic families unfolding (3.9). Considering such a family it is always possible to "prepare" the family so that the parameter becomes canonical.

Theorem 3.9 *We consider a generic germ of analytic family $f_\eta(x)$ depending on the parameter η unfolding the germ $f(x)$ given in (3.9) (the family is generic if $\frac{\partial f_\eta}{\partial \eta} \neq 0$). There exists a germ of analytic diffeomorphism $(x,\eta) \mapsto (y,\epsilon)$ such that in the coordinate y the diffeomorphism (3.9) becomes*

$$\overline{f}_\epsilon(y) = y + (y^2 - \epsilon)(1 + B(\epsilon) + (y^2 - \epsilon)h_\epsilon(y)), \tag{3.13}$$

where

- *The fixed points of $\overline{f}$ are given by $y_\pm = \pm\sqrt{\epsilon}$.*
- *The multipliers $\lambda_\pm$ of $y_\pm$ satisfy*

$$\frac{1}{\ln \lambda_+} - \frac{1}{\ln \lambda_-} = \frac{1}{\sqrt{\epsilon}}. \tag{3.14}$$

Moreover

$$a(\epsilon) = \frac{1}{\ln \lambda_+} + \frac{1}{\ln \lambda_-}, \tag{3.15}$$

so $a(\epsilon)$ is a "shift" between the two singular points.

Definition 3.10 The family (3.13) of Theorem 3.9 is called prepared. Its parameter ϵ is called the canonical parameter.

We now consider the conjugacy problem for two prepared families. Then necessarily the canonical parameter must be preserved as it is an analytic invariant of the family.

Theorem 3.11 [11] *Two germs of generic analytic famlies of diffeomorphisms*

$$f_\epsilon(x) = x + (x^2 - \epsilon) + o(x^2) \tag{3.16}$$

are analytically conjugate if and only if they have the same unfolded Ecalle-Voronin modulus $((\psi^0_{\hat\epsilon}, \psi^\infty_{\hat\epsilon})/\sim)_{\hat\epsilon \in V}$, where

- *V is a sectorial neighborhood of the origin in the universal covering of ϵ-space punctured at the origin. The radius $r(\delta)$ of V depends on its opening defined with the help of an arbitrarily small $\delta > 0$:*

$$V = \{\hat\epsilon; |\hat\epsilon| < r(\delta), arg(\hat\epsilon) \in (-\pi + \delta, 3\pi - \delta)\}. \tag{3.17}$$

- *The fundamental neighborhoods unfold as in Figure 6. As before we glue*

Figure 6 Orbit space of a generic family of diffeomorphims

together the two curves ℓ_j and $f_\epsilon(\ell_j)$ limiting the fundamental neighborhoods, thus yielding domains which have the conformal structure of spheres $S_{j,\epsilon}$, $j = 1, 2$, the distinguished point 0 (resp. ∞) corresponding to the fixed point $-\sqrt{\epsilon}$ (resp. $\sqrt{\epsilon}$).

- $$\begin{cases} \psi^0_{\hat{\epsilon}} : (\mathbb{C}, 0) \to (\mathbb{C}, 0) \\ \psi^\infty_{\hat{\epsilon}} : (\mathbb{C}, \infty) \to (\mathbb{C}, \infty) \end{cases} \tag{3.18}$$

  are germs of analytic diffeomorphisms depending analytically on $\hat{\epsilon} \neq 0$ and continuously on $\hat{\epsilon}$ near $\hat{\epsilon} = 0$.

Theorem 3.12 [15] *A germ of generic analytic family unfolding a saddle-node is analytically orbitally equivalent to a prepared family*

$$\begin{aligned} \dot{x} &= x^2 - \epsilon \\ \dot{y} &= g_0(x)(x^2 - \epsilon) + y(1 + a(\epsilon)x) + O(y^2). \end{aligned} \tag{3.19}$$

The modulus $((\psi^0_{\hat{\epsilon}}, \psi^\infty_{\hat{\epsilon}})/\sim)_{\hat{\epsilon}\in V}$ for the unfolded holonomy of the y-separatrix with prepared form (3.13) is such that $\psi^\infty_{\hat{\epsilon}}$ is an affine transformation.

Open questions:

1. Identify precisely the modulus space of $((\psi^0_{\hat{\epsilon}}, \psi^\infty_{\hat{\epsilon}})/\sim)_{\hat{\epsilon}\in V}$ which are realizable as moduli of families of diffeomorphims. It is known that all (ψ^0, ψ^∞) are realizable for a single diffeomorphim ($\epsilon = 0$), see[4] and [16]. The difficulty is to identify precisely the dependence on $\hat{\epsilon}$ near $\hat{\epsilon} = 0$. The $(\psi^0_{\hat{\epsilon}}, \psi^\infty_{\hat{\epsilon}})$ surely depend more than continuously on $\hat{\epsilon}$.

2. Derive similar theorems for the higher codimension case. The existence of fundamental domains for all values of the parameters has been done by Oudkerk ([12] and later work). It remains to organize them nicely in the parameter space. As appearing in Figure 6 there can be different non equivalent choices of fundamental domains to describe the orbit space for a single value of ϵ.

4 Changes of coordinates to normal form

We discuss the case of the generic saddle-node family (3.19).

Theorem 4.1 [15] *There exists a change of coordinate*

$$Y = y + \sum_{j=0}^{\infty} y^j h_j(\hat{\epsilon}, x) \tag{4.1}$$

defined on a domain $U \times D \times V$ where

- *U is a domain in the universal covering of x-space ramified at $\pm\sqrt{\epsilon}$ as in Figure 7,*
- *D is a neighborhood of 0 in y-space,*
- *V is a sectorial neighborhood of $\hat{\epsilon} = 0$ as in (3.17),*

bringing the family (3.19) to the model family

$$\begin{aligned} \dot{x} &= x^2 - \epsilon \\ \dot{Y} &= Y(1 + a(\epsilon)x). \end{aligned} \tag{4.2}$$

The model family has a first integral

$$H(x, Y) = Y k_a(x, \epsilon), \tag{4.3}$$

with

$$k_a(x, \epsilon) = (x - \sqrt{\epsilon})^{-\frac{1+a\sqrt{\epsilon}}{2\sqrt{\epsilon}}} (x + \sqrt{\epsilon})^{\frac{1-a\sqrt{\epsilon}}{2\sqrt{\epsilon}}}. \tag{4.4}$$

This yields a first integral $\overline{H}_{\hat{\epsilon}}(x, y)$ for (3.19) which is ramified at $x = \pm\sqrt{\epsilon}$. If we

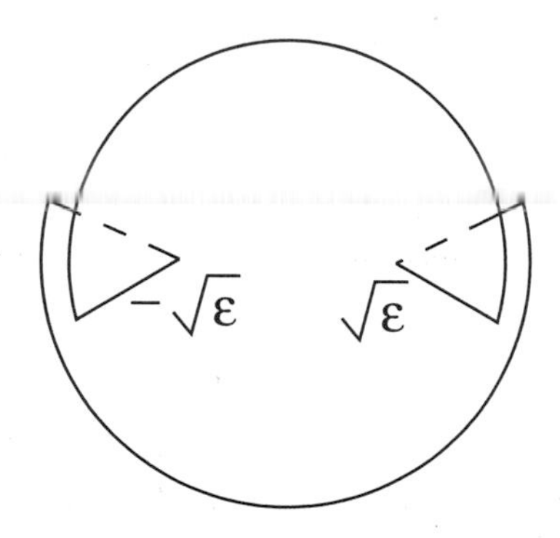

Figure 7 Domain U in x-space

turn around $\sqrt{\epsilon}$ we get two branches $\overline{H}_{1,\hat{\epsilon}}$ and $\overline{H}_{2,\hat{\epsilon}}$ related in the following way

$$\overline{H}_{2,\hat{\epsilon}} = L(a, \hat{\epsilon}) \circ \psi_{\hat{\epsilon}}^{\infty}(\overline{H}_{1,\hat{\epsilon}}) \tag{4.5}$$

where $L(a,\hat{\epsilon})$ is a linear map. Similarly if we turn around $-\sqrt{\epsilon}$ we get two branches $\tilde{H}_{1,\hat{\epsilon}}$ and $\tilde{H}_{2,\hat{\epsilon}}$ related in the following way

$$\tilde{H}_{2,\hat{\epsilon}} = L(a,\hat{\epsilon}) \circ \psi^0_{\hat{\epsilon}}(\tilde{H}_{1,\hat{\epsilon}}). \tag{4.6}$$

5 Applications to problems of finite cyclicity of graphics

Definition 5.1 A graphic of a vector field is a union of singular points and characteristic trajectories joining them which is likely to produce limit cycles or periodic trajectories under perturbation.

Definition 5.2 A graphic Γ of a vector field v_0 has *finite cyclicity inside a family* v_λ unfolding v_0 (where λ is a multi-parameter) if there exists $N \in \mathbb{N}$, there exists $\epsilon > 0, \delta > 0$ such that, for any λ with $|\lambda| < \delta$, the vector field v_λ has at most N limit cycles $\gamma_1, \dots, \gamma_n$, $n \leq N$ such that $dist_H(\gamma_i, \Gamma) < \epsilon$, where $dist_H$ is the Hausdorff distance between compact sets. The graphic Γ has *finite absolute cyclicity* (or simply *finite cyclicity*) if N can be chosen independent of the C^∞ unfolding v_λ.

5.1 The lips. The lips is a continuum of graphics as in Figure 8 with two saddle-nodes and central transition through them. Each graphic is likely to create periodic solutions or limit cycles in a perturbation where the singular points disappear. In the finite cyclicity questions we are interested to give a bound on the number of limit cycles which can be created from a single limit periodic set in the bifurcation process.

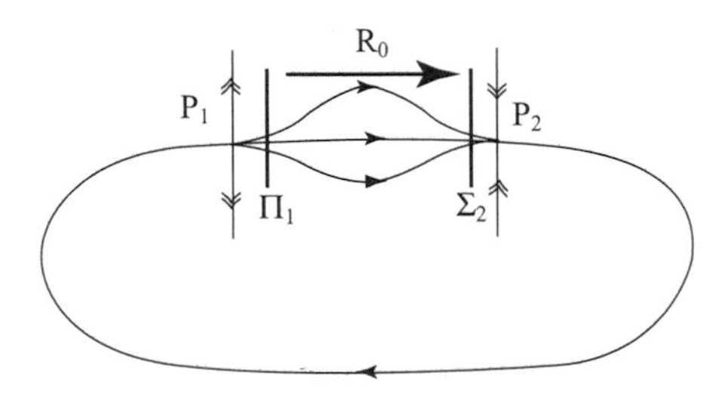

Figure 8 The lips

Theorem 5.3 [8] *We consider a C^∞ family of vector fields v_η having for v_0 lips as in Figure (8). We consider C^k-normalizing changes of coordinates bringing the vector field in the neighborhood of P_1 and P_2 to the respective normal forms*

$$\begin{cases} \dot{x}_1 = x_1^2 - \epsilon_1 \\ \dot{y}_1 = y_1(1 + a_1 x_1) \end{cases} \qquad \begin{cases} \dot{x}_2 = x_2^2 - \epsilon_2 \\ \dot{y}_2 = -y_2(1 + a_2 x_2), \end{cases} \tag{5.1}$$

and sections parallel to the y_j-axes in the normalizing charts. We consider a graphic intersecting Π_1 in $y_{1,0}$. If the regular transition $R_0 : \Pi_1 \to \Sigma_2$ satisfies $R_0^{(n)}(y_{1,0}) \neq 0$, where $1 < n < k$, then the corresponding graphic has cyclicity less than or equal to n. Moreover there exists a perturbation of v_0 with exactly n limit cycles.

Proof The proof is easy. We consider sections Σ_i and Π_i, $i = 1, 2$, as in Figure 9. Limit cycles are given by zeros of the displacement map

$$V_\eta(y_1) = R_\eta(y_1) - D_2^{-1} \circ S_\eta^{-1} \circ D_1^{-1}(y_1). \tag{5.2}$$

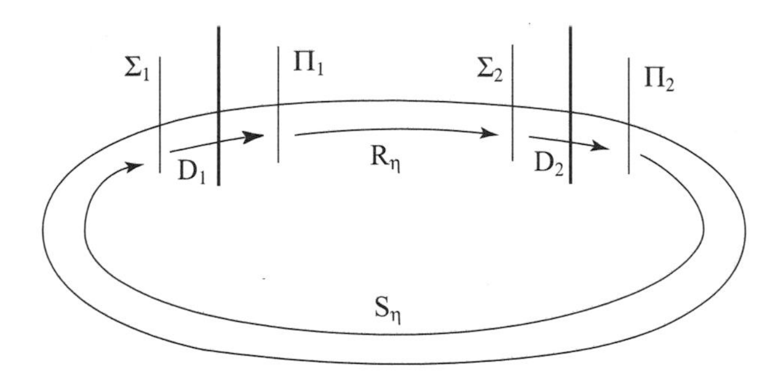

Figure 9 Transition maps for the lips

As the system is integrable in the neighborhood of P_1 and P_2 this allows to calculate the Dulac maps $D_j : \Sigma_j \to \Pi_j$. These are linear maps of the form $D_1(y_1) = M(\epsilon_1)y_1$ with $\lim_{\epsilon_1 \to 0} M(\epsilon_1) = +\infty$ and $D_2(y_2) = m(\epsilon_2)$ with $\lim_{\epsilon_2 \to 0} m(\epsilon_2) = 0$.

We exploit the freedom on the choice of normalizing coordinates to simplify the map S_η and bring it to a mere affine map

$$S_\eta(y_2) = A(\eta)y_2 + B(\eta), \tag{5.3}$$

with $A(\eta) > 0$ and $B(0) = 0$ [5]. Then $D_2^{-1} \circ S_\eta^{-1} \circ D_1^{-1}$ is affine. Hence

$$V_\eta^{(n)}(y_1) = R_\eta^{(n)}(y_1) \neq 0 \tag{5.4}$$

for (y_1, η) in a small neighborhood of $(0,0)$, which determines a neighborhood U_1 of the origin on the section Π_1. By Rolle's theorem there are at most n periodic solutions intersecting Π_1 on U_1. □

Remark 5.4

1. The proof illustrates the power of a good "preparation" to deal with problems of finite cyclicity. Indeed composing a Dulac map with $M(\epsilon_1)$ very large with one with $m(\epsilon_2)$ very small yields indeterminacy. The trick of transforming S_η to an affine map allows to see that the indeterminacy is limited to constant and linear terms.
2. The theorem may seem completely useless in practice for C^∞ vector fields because it is impossible in practice to check the hypothesis. Fortunately this is not so for analytic vector fields and we will see that there are a number of cases where the hypothesis can be checked without nearly any calculation.

Theorem 5.5 [3] *We consider an analytic family of vector fields with a saddle-node at the origin. It is possible to choose C^k normalizing changes of coordinates bringing the family to the normal form*

$$\begin{aligned} \dot{X} &= X^2 - \epsilon \\ \dot{Y} &= Y(1 + a(\epsilon)X) \end{aligned} \tag{5.5}$$

in such a way that there exists $X_0 > 0$ such that, for all values of the parameters, the sections $X = \pm X_0$ are analytic and parameterized by analytic coordinates.

Proof The proof uses the case $\epsilon = 0$ of Theorem 4.1. This particular case of Theorem 4.1 was proved in [3], with an additional step to prove that the change of coordinates can be taken real. □

Theorem 5.6 [3] *We consider an analytic family of vector fields* v_η *having for* v_0 *lips as in Figure 8.*

1. *If* R_0 *is nonlinear at one point* $y_{1,0}$ *it is nonlinear everywhere.*
2. *Each graphic of the lips has finite cyclicity if for instance one of the following conditions is satisfied*
   - *One of the graphics has an additional saddle point with ratio of eigenvalues different of* -1 *(Figure 10(a)).*

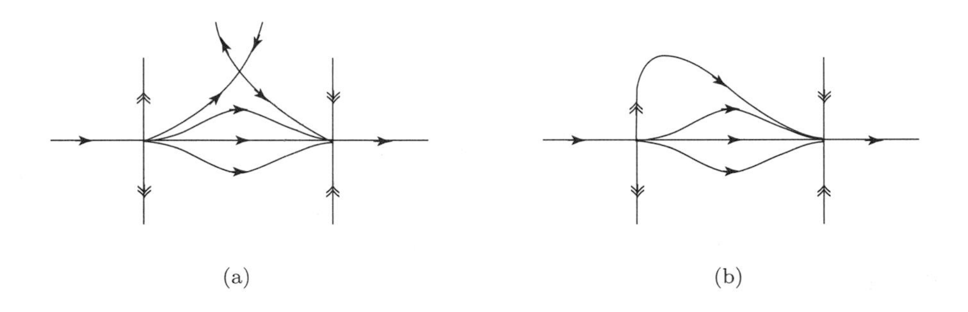

Figure 10 Two types of bordering graphics

   - *The family of graphics ends in a graphic entering one saddle-node through the strong manifold and the other saddle-node through a center manifold (Figure 10(b)).*

Proof 1. is just analytic extension principle.

2.
   - Let $\lambda_1 < 0$ and $\lambda_2 > 0$ be the eigenvalues of the saddle point and $r = -\lambda_1/\lambda_2$. Let $y = y_1 - y_{1,0}$, where $y_{1,0}$ corresponds to the graphic through the saddle point. The transition map has the form $R_0(y) = y^r(C + O(y))$ with $C > 0$ in the neighborhood of the graphic through the saddle. This map is obviously nonlinear as soon as $r \neq 1$.
   - No affine map can send a semi-infinite domain on a finite one. Hence the map R_0 is non affine.

□

Remark 5.7 1. These results only use the modulus of the vector field ($\epsilon = 0$) and not the modulus of the family. This is because of the genericity condition $R_0^{(n)}(0) \neq 0$ for $n > 1$.

2. We have seen that the method of Theorem 5.6 is extremely powerful, since nearly no calculations are needed. The price to pay however is that we just get results of finite cyclicity but cannot determine the exact cyclicity.

In the non generic case where R_0 is linear, then we can hope for a full result of finite cyclicity for analytic families depending on a finite number of parameters and unfolding a vector field with lips. The tool would be an improved Theorem 4.1 where we would have a better control on the dependence on the parameter in the neighborhood of $\hat{\epsilon} = 0$.

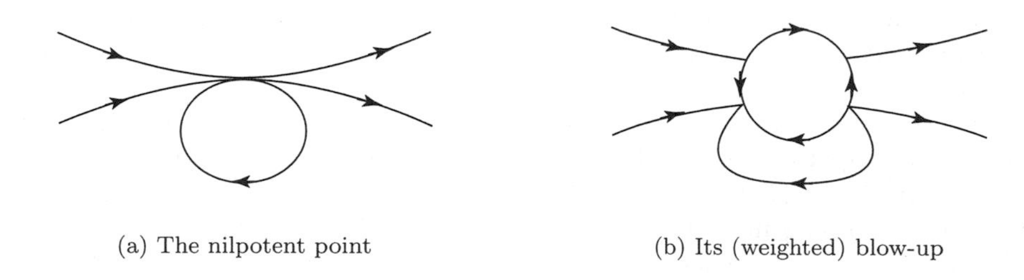

(a) The nilpotent point (b) Its (weighted) blow-up

Figure 11 A nilpotent elliptic point of multiplicity 3

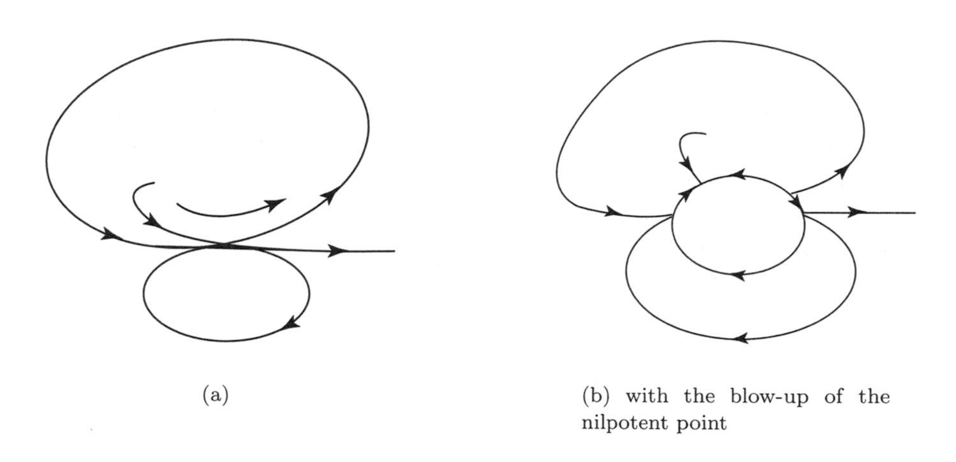

(a) (b) with the blow-up of the nilpotent point

Figure 12 An hp-graphic

5.2 The hp-graphic through a nilpotent elliptic point. We consider a nilpotent elliptic point of multiplicity 3 inside a C^∞ family of vector fields. Such a point has a 3-jet C^∞ orbitally equivalent to

$$\begin{aligned} \dot{x} &= y \\ \dot{y} &= -x^3 + y(bx + o(x)) + y^2 h(x, y) \end{aligned} \tag{5.6}$$

with $b > 2\sqrt{2}$. Such a point has a phase portrait as in Figure 11.

We consider an hp-graphic through such a point, i.e. a connection between a separatrix of a hyperbolic sector and a characteristic curve of a parabolic sector (Figure 12). The following theorem proved in [18] made an essential use of Theorem 5.5.

Theorem 5.8 *We consider an hp-graphic through a nilpotent elliptic point of multiplicity 3 in a C^∞ vector field v_0. There exists $N \in \mathbb{N}$ such that the graphic has finite cyclicty inside any C^∞ family of vector fields v_λ unfolding v_0.*

6 Perspectives

Sections 2 and 3 illustrate that the divergence of the normalizing series reflects that the dynamics of the system is much more complicated than that of the normal form. We have also seen that embedding a vector field in a family so as to unfold the situation and extending the phase variables to the complex domain allows to give a geometric explanation of the divergence and why the divergence is so often the rule.

Here we have only discussed the divergence of normalizing series in the resonant cases: the divergence can essentially be explained as the incompatibility between a finite number of objects: for instance two fixed points in the case of a generic diffeomorphism tangent to the identity, one fixed point and one periodic orbit of period q in the case of a generic diffeomorphism with multiplier $\exp(2\pi i \frac{p}{q})$, one fixed point and one periodic orbit in the case of a generic Hopf bifurcation, one saddle and one node in the case of a saddle-node. There is another source of divergence coming from the "small denominators". This occurs for instance in the case of a diffeomorphism

$$f(z) = \exp(2\pi i \alpha)z + o(z) \tag{6.1}$$

where α is irrational. It is also the case of a saddle point for which the quotient of eigenvalues is irrational. In both cases the system is formally linearizable but the linearizing change of coordinates is generically divergent when α (in the case of a diffeomorphism) or the quotient of eigenvalues (in the case of a saddle) is Liouvillian. Here again an explanation is suggested by unfolding. For instance if we embed the diffeomorphism (6.1) inside a family

$$f_\epsilon(z) = \exp(2\pi i(\alpha + \epsilon))z + o(z) \tag{6.2}$$

then we generically have the birth of a periodic orbit as soon as $\alpha + \epsilon = \frac{p}{q} \in \mathbb{Q}$. If α is Liouvillian (i.e. very well approximated by the rationals) then an infinite number of periodic orbits cannot escape from a neighborhood of the origin sufficiently rapidly and form an obstruction to linerizability. This was conjectured by Arnold and Yoccoz showed that it indeed occurs [17]. Pérez-Marco showed that there are also other kinds of obstructions to linearizability [13].

One interesting direction of research is to put together the two approaches: indeed small divisors phenomena occur in the unfolding of a resonant situation. The quadratic family

$$f(z) = \exp(2\pi i \alpha)z + z^2 \tag{6.3}$$

is a family for which the change of coordinate to normal form in the neighborhood of the origin diverges for all rational α and all Liouvillian irrational α.

A second direction of research is to make a systematic study of the meaning of the divergence of the normalizing series for more complicated singular points.

In Section 5 we have shown the power of good normal forms in some applications to finite cyclicity problems. The results on normal forms presented here are still partial as the dependence on the parameters is not enough precise. For instance in the conjugacy problem for a germ of analytic family unfolding a germ of diffeomorphism (3.6) we are still missing a realization theorem describing exactly the modulus space, i.e. identifying precisely which pairs of germs $(\psi^0_{\hat{\epsilon}}, \psi^\infty_{\hat{\epsilon}})_{\hat{\epsilon} \in V}$ can be realized as the modulus of a germ of family. This problem is probably very difficult. However its solution would open new perspectives in finite cyclicity problems.

References

[1] W. Balser, *From divergent series to analytic differential equations*, Springer Lecture Notes in Mathematics, 1582, Springer-Verlag, Berlin, (1994).

[2] C. Christopher and C. Rousseau, *Modulus of analytic classification for the generic unfolding of a codimension one resonant diffeomorphism or resonant saddle*, preprint CRM 2004.

[3] F. Dumortier, Y. Ilyashenko, and C. Rousseau, *Normal forms near a saddle-node and applications to the finite cyclicity of graphics*, Ergod. Theor. Dyn. Systems **22** (2002), 783–818.

[4] J. Ecalle, *Les fonctions résurgentes*, Publications mathématiques d'Orsay, 1985.

[5] A. Guzmán, and C. Rousseau, *Genericity conditions for finite cyclicity of elementary graphics*, J. Differential Equations **155** (1999), 44–72.

[6] Y. Ilyashenko, *In the theory of normal forms of analytic differential equations, divergence is the rule and convergence the exception when the Bryuno conditions are violated*, Moscow University Mathematics Bulletin **36** (1981) 11–18.

[7] V. Kostov, *Versal deformations of differential forms of degree α on the line*, Functional Anal. Appl. **18** (1984), 335–337.

[8] A. Kotova, and V. Stanzo, *On few-parameter generic families of vector fields on the two-dimensional sphere*, in *Concerning the Hilbert's problem*, Amer. Math. Soc. Trans. Ser. 2, **165**, AMS, Providence, RI, (1995), 155-201.

[9] J.-F. Mattei, and R. Moussu, *Holonomie et intégrales premières*, Ann. Scient. Éc. Norm. Sup., 4e série, **13** (1980), 469–523.

[10] J. Martinet, and J.-P. Ramis, *Problèmes de modules pour des équations différentielles non linéaires du premier ordre*, Publ. Math., Inst. Hautes Etud. Sci. **55** (1982), 63–164.

[11] P. Mardešić, R. Roussarie and C. Rousseau, *Modulus of analytic classification for unfoldings of generic parabolic diffeomorphisms*, Moscow Mathematical Journal **4** (2004), 455–502.

[12] R. Oudkerk, *The parabolic implosion for $f_0(z) = z + z^{\nu+1} + O\left(z^{\nu+2}\right)$*, thesis, University of Warwick, (1999).

[13] R. Pérez-Marco, *Solution complète au problème de Siegel de linéarisation d'une application holomorphe au voisinage d'un point fixe (D'après J.-C. Yoccoz)*, Astérisque **206** (1992), 273–310.

[14] R. Pérez-Marco and J.-C. Yoccoz, *Germes de feuilletages holomorphes à holonomie prescrite*, Astérisque **222** (1994), 345–371.

[15] C. Rousseau, *Modulus of orbital analytic classification for a family unfolding a saddle-node*, Moscow Mathematical Journal **5** (2005), 243–267.

[16] S. M. Voronin, *Analytic classification of germs of conformal maps $(\mathbb{C}, 0) \to (\mathbb{C}, 0)$ with identical linear part*, Funktsional. Anal. i Prilozhen **15** (1981), 1–17 (Russian), Funct. Anal. Appl. **15** (1981), 1–13.

[17] J.-C. Yoccoz, *Théorème de Siegel, nombres de Bruno et polynômes quadratiques*, Astérisque, **231** (1995), 3–88.

[18] H. Zhu, and C. Rousseau, *Finite cyclicity of graphics through a nilpotent singularity of elliptic or saddle type*, J. Differential Equations **178** (2002), 325-436.

Fields Institute Communications
Volume **48**, 2006

Generic Properties of Symplectic Diffeomorphisms

Radu Saghin
Department of Mathematics
University of Toronto
Toronto, Ontario, Canada M5S 2E4
rsaghin@fields.utoronto.ca

Zhihong Xia
Department of Mathematics
Northwestern University
Evanston, Illinois 60208 USA
xia@math.northwestern.edu

Abstract. In this short survey, we present some of the recent results on symplectic diffeomorphisms and Hamiltonian systems. We particularly focus on the generic properties of this class of dynamical systems.

1 Introduction

Let M be a compact symplectic manifold with a symplectic form ω. Let $\mathrm{Diff}^r_\mu(M)$ be the set of all C^r, $r = 1, 2, \ldots, \infty$ diffeomorphisms of M. A property for C^r symplectic diffeomorphisms of M is said to be *generic*, if there is a residual subset $R \subset \mathrm{Diff}^r_\mu(M)$ such that the property holds for all $f \in R$. The following are some examples of known generic properties:

- All periodic points are either elliptic, quasi-elliptic, or hyperbolic. This is proved by Robinson [31]. A periodic point is said to be *elliptic* if all of its eigenvalues are on unit circle and are away from ± 1. A periodic point is said to be *quasi-elliptic* if some pairs of eigenvalues are unit circle, away from ± 1 and other pairs of eigenvalues are away from unit circle. Note that if λ is an eigenvalue for a periodic point of a symplectic diffeomorphism, then so is λ^{-1}.
- For any two hyperbolic periodic points p, q, the intersections of the stable manifold $W^s(p)$ and the unstable manifold $W^u(q)$ are transversal. This is again proved by Robinson [31]. This and above property together are often referred as the Kupka-Smale condition for symplectic diffeomorphisms.
- All elliptic periodic points are Moser stable. Moser stable means that the normal form at the elliptic periodic points are non-degenerate in the sense

2000 *Mathematics Subject Classification.* Primary 37J10, 37J45, 37G20.

Research for both authors are supported in part by National Science Foundation.

of KAM theory (cf. Siegel & Moser [35]). This implies that there are invariant tori surrounding each elliptic periodic point. This condition requires that the map is sufficiently smooth, say $r \geq 16$ for area-preserving surface diffeomorphisms (cf. Douady [10]).

Poincaré pioneered the study of the general properties of Hamiltonian dynamics and area-preserving surface diffeomorphisms. He already noted the importance of the generic properties of symplectic diffeomorphisms in his study of the three-body problem. Poincaré [26] certainly believed that the following two fundamental conjectures are true.

Conjecture 1 *For generic C^r area-preserving diffeomorphisms on compact surface M, the set of all periodic points are dense.*

Conjecture 2 *There exist a residual set $R \in \mathrm{Diff}^r_\mu(M)$ such that if $f \in R$ and p is a hyperbolic periodic point of f then the homoclinic points of p is dense in both stable and unstable manifolds of p. In other words, let J be a segment in $W^s(p)$ (or $W^u(p)$), then $\overline{W^s(p) \cap W^u(p) \cap J} = J$.*

In C^1 topology, both of the above conjectures are proved to be true. The first one is a consequence of the so-called closing lemma. It is proved by Pugh [27] and later improved to various cases by Pugh & Robinson [28]. A different proof was given by Liao [18] and Mai [19] (also cf. Arnaud [2]). The second conjecture in C^1 topology is a result of Takens [37]. The high dimensional analog was proved by Xia [41]. It can also be regarded as a part of so-called C^1 connection lemma for general dynamical systems. The C^1 connecting lemma was first proved by Hayashi [15] and later simplified and generalized by Wen & Xia [38] [39]. See also Arnaud, Bonatti and Crovisier [4].

In C^r topology with $r > 1$, little progress has been made for these two conjectures and it is known to be a difficult problem (cf. Smale [36]). The local perturbation methods used in the C^1 case no longer seem to work and examples suggests that a more global approach has to be developed (Gutierrez [12]). We will describe some of progresses in this direction.

Our survey will mostly be centered around above two conjectures.

2 Closing lemmas

In this section, we describe various results on closing lemmas. The first one is the classical result of Pugh [27].

Theorem 2.1 *Let M be a compact manifold and f a C^1 diffeomorphism on M. Let $x \in M$ be a non-wandering point of f. Then for any neighborhood U of f in $\mathrm{Diff}^1(M)$, there is a map $g \in U$ such that x is a periodic point for g.*

A point x is said to be non-wandering if for neighborhood of U of x, there is an integer k such that $f^k(U) \cap U \neq \emptyset$. One might naively think that it is easy to make a small perturbation to turn x into a periodic point. Indeed, it is trivial if we only require that the perturbation be C^0 small. For C^1 small perturbations, it is very difficult. Pugh's proof is very long and technical. A different proof given by Liao [18] and Mai [19] is also long and difficult. As in similar types of problems in perturbation theory, the difficulty lies in the interferences of intermediate trajectories of the original point. In both proofs, one has to try to avoid, in the perturbations, touching the intermediate points of the intended periodic point.

The above theorem applies to symplectic and volume-preserving diffeomorphisms. i.e., the perturbations can be made symplectic or volume preserving. For volume-preserving and symplectic diffeomorphisms on compact manifold, every point is non-wandering. Together with the fact that non-degenerate periodic points survives small perturbations, the above theorem implies that generic C^1 volume-preserving and symplectic diffeomorphisms have dense periodic points.

To generalize the proof of the C^1 closing lemma to C^r case for $r > 1$ has been a challenge for almost half of a century. Gutierrez [12] gave an example of flow on two-torus $\mathbb{T}^2$ where local perturbation method as used in Pugh's proof fails to close a recurrent point. In that case, a global perturbation is necessary to produce a periodic point.

On the other hand, if the diffeomorphism is hyperbolic, then there are abundance of periodic orbits. This is the consequence of the so-called Anosov closing lemma [1].

Theorem 2.2 *Let $\Lambda \subset M$ be a hyperbolic invariant set for a C^r, $r \geq 1$ diffeomorphism $f : M \to M$. Assume that Λ is chain transitive and maximal, then the periodic points are dense in Λ.*

This theorem is a consequence of the so-called the shadowing property for hyperbolic invariant sets. It's proof can be found in standard textbooks in dynamical systems. We refer readers to Robinson [33].

The Anosov closing lemma can also be extended to non-uniformly hyperbolic systems. Let $f \in \text{Diff}^r(M)(M)$, $r > 1$, be a diffeomorphism that preserves a probability measure μ on M. The measure is said to be hyperbolic if μ-almost every point of M has non-zero Liapunov exponents. If, in addition, μ is ergodic, then the Liapunov exponents are constants on a full measure set. Pesin theory [24] studies the hyperbolic invariant probability measures. It turns out that one can generalize the shadowing lemma to this non-uniformly hyperbolic situation. Therefore we have the Anosov closing lemma on a full measure set. More precisely, we have the following theorem.

Theorem 2.3 *Let μ be a hyperbolic invariant measure for a C^r, $r > 1$ diffeomorphism $f : M \to M$. Then the set of periodic points is dense in the support of μ.*

It is interesting to note that in the Pesin theory, one requires a little stronger smoothness for the diffeomorphisms. Usually, it suffices to assume that f has a α-Hölder first derivative with some $\alpha > 0$.

Recently, there are some new results in C^r closing lemma, by using global perturbation techniques. Gutierrez [13] proved a C^r closing lemma for flows on orientable surfaces. A negative result was obtained for flows on non-orientable surfaces(Gutierrez and Pires [14]).

Another recent results on C^r closing lemma, obtained by Xia and Zhang [43] concerns partially hyperbolic symplectic diffeomorphisms. A diffeomorphism $f : M \to M$ is partially hyperbolic if the tangent bundle TM admits a Tf invariant splitting $TM = E^u \oplus E^c \oplus E^s$ and there is a Riemannian metric on M such that there exist real numbers $\lambda_1 > \lambda_2 > 1 > \mu_2 > \mu_1 > 0$ satisfying

$$m(Tf|_{E^u}) \geq \lambda_1 > \lambda_2 \geq \|Tf|_{E^c}\| \geq m(Tf|_{E_c}) \geq \mu_2 > \mu_1 \geq \|Tf|_{E^s}\| > 0.$$

Here the co-norm $m(A)$ of a linear operator A between two Banach spaces is defined by $m(A) := inf_{\|v\|=1}\|A(v)\| = \|A^{-1}\|^{-1}$.

To avoid triviality, we assume at least two of the sub-bundles are non-zero. Partial hyperbolicity is a C^1 open condition, as can be easily verified by it's associated invariant cone fields.

We remark that our definition of partial hyperbolicity here is not the most general one. One can allow the parameters $\lambda_1, \lambda_2, \mu_1, \mu_2$ to depend on each trajectory in general cases. The systems that we are considering satisfy the definition given here.

For symplectic cases, the stable distribution E^s and the unstable distribution E^u have the same dimension. Moreover, one can choose the parameters such that $\lambda_1 = \mu_1^{-1}$ and $\lambda_2 = \mu_2^{-1}$.

Theorem 2.4 *Let M_1 be a compact symplectic manifold and $f_1 \in \mathrm{Diff}^r_{\omega_1}(M_1)$ an Anosov diffeomorphism. Let M_2 be a compact symplectic surface (orientable surface) with an area form ω_2 and let $f_2 \in \mathrm{Diff}^r_{\omega_2}(M_2)$ be an area preserving diffeomorphism on M_2. Let $\omega = \omega_1 + \omega_2$ be the symplectic form defined on $M_1 \times M_2$. Assume that $f_1 \times f_2 : M_1 \times M_2 \to M_1 \times M_2$ is partially hyperbolic with TM_2 as its center splitting and satisfies the following bunching condition: there exist hyperbolic constants $\lambda > \tau > 1$, $m(Tf_1) \geq \lambda > \tau \geq \|Tf_2\|$, such that $\tau^r < \lambda$. Then there exists a neighborhood U of $f_1 \times f_2$ in $\mathrm{Diff}^r_\omega(M_1 \times M_2)$ and a residual subset $R \in U$ such that for any $g \in U$, the set of periodic points of g is dense in $M_1 \times M_2$.*

The proof took advantage of the partial hyperbolicity and a recent results of Franks & Le Calvez[11] and Xia [42] on surface diffeomorphisms.

The bunching condition is satisfied when f_2 is close to identity. This is a technical condition one hopes to remove.

M. Herman [16] has a counter example for the C^r closing lemma with r large for symplectic diffeomorphisms where the symplectic form is not exact.

3 Connecting lemmas

Like closing lemma, the connecting lemma is also one of the fundamental problems in dynamical systems. We start by considering a very general problem in connecting trajectories in a given dynamical system. Let $f \in \mathrm{Diff}^r(M)$ with a compact manifold M and let p and q be points in M. Suppose that the ω-limit set of p intersects with the α-limit set of q. Can we perturb the map f such that there is a trajectory going from p to q? It is believed and conjectured that such a perturbation is always possible. Like the closing lemmas, the problem is trivial in C^0 topology, the problem is mostly solved in C^1 topology and the problem remains open in C^r, $r > 1$, besides a few special cases.

First, we point out that if the manifold is not compact, there are counter examples to the above connecting using any small C^r, $r \geq 1$, perturbations.

Connecting stable and unstable manifolds of hyperbolic invariant sets is of special importance in global dynamics. In this context, one asks the following question: if the stable manifold of a hyperbolic invariant set accumulates on the unstable manifold of another hyperbolic invariant set, can one make a small perturbation to create a heteroclinic connection? The answer to this question is affirmative in C^1 topology. This is the result of Hayashi [15].

Theorem 3.1 (Hayashi) *Let $f : M \to M$ be diffeomorphism on a compact manifold M. Let p be a hyperbolic periodic point for f. Assume that $q \in \overline{W^u(p)} \cap W^s(p)$ and $q \neq p$. Then for any neighborhood U of f in $\mathrm{Diff}^1(M)$, there is $g \in U$ such that p is a hyperbolic periodic of g and q is a homoclinic point of p for g.*

The Hayashi's connecting lemma has been generalized to a few other situations with a simplified proof by Wen and Xia [39] [38]. Particularly, they proved the following theorem:

Theorem 3.2 *Let $f : M \to M$ be diffeomorphism on a compact manifold M. Let p and q be points in M. Let $\omega(p)$ be the ω-limit set of p and let $\alpha(q)$ be the α-limit set of q. Assume that the intersection $\omega(p) \cap \alpha(q) \neq \emptyset$ and the intersection is not a single non-hyperbolic periodic point, then for any neighborhood U of f in $\mathrm{Diff}^1(M)$, there is $g \in U$ such that $g^k(p) = q$ for some positive integer k.*

It is interesting to note that it is still an open problem if the intersection in the above theorem is a non-hyperbolic periodic orbit. It is rather peculiar that a technical difficulty in this case prevented us from obtaining a general result.

We remark that the proof of the connecting lemma is very much similar, in technical details, to that of closing lemma. The key idea of Hayashi [15] is to cut pieces of trajectories that interfere with perturbations.

In the context of symplectic and volume-preserving diffeomorphisms, the connecting lemmas take some stronger forms. This is due to the fact that every point is non-wandering point and C^r generically, stable manifold of every hyperbolic periodic point accumulates on its unstable manifold and vice versa.

Takens [37] was the first to show that for C^1 generic volume-preserving and symplectic diffeomorphisms, every hyperbolic periodic point, which is dense in the manifold, has a transversal homoclinic point. Moreover, if M is a compact surface, then the transversal homoclinic points of any hyperbolic periodic points are dense in both its stable and unstable manifolds. Xia [41] generalized this density result to higher dimensional manifolds. More precisely, we have the following theorem.

Theorem 3.3 (Takens, Xia) *Let M^n be a compact n-dimensional manifold with a symplectic or volume form ω. Then there is a residual subset $R \subset \mathrm{Diff}^1_\mu(M)$ such that if $\phi \in R$ and $p \in M^n$ are such that p is a hyperbolic periodic point of ϕ, then $W^s_\phi(p) \bigcap W^u_\phi(p)$ is dense in both $W^s_\phi(p)$ and $W^u_\phi(p)$.*

This implies that the Conjecture 2 is true in C^1 topology.

Using connecting lemma types of techniques, recently, Arnaud, Bonatti and Crovisier [4] showed that every C^1 generically volume-preserving and symplectic diffeomorphisms are topologically transitive. On compact surfaces, this is certainly not true for higher smoothness, say C^r, $r > 3$. Generic elliptic periodic points are surrounded by invariant curves by KAM theory. For higher dimensional volume-preserving maps with sufficient smoothness, the persistent existence of co-dimension one invariant tori also prevented transitivity (cf. Cheng & Sun [8], Xia [40] and Yoccoz [44]). However, the problem remains open for higher dimensional symplectic diffeomorphisms in C^r topology, $r > 1$. This is one of the fundamental problems in Hamiltonian dynamics.

We return to the problem of C^r connecting. On two-sphere S^2, Robinson [32] showed that if the stable manifold of a hyperbolic fixed point accumulates on a point in its unstable manifold, then an arbitrary small C^r perturbation can create a homoclinic point for the fixed point. Pixton [25] extended the result to hyperbolic periodic points. Oliveira [22] further extended the result to area-preserving diffeomorphisms on $\mathbb{T}^2$. We have the following.

Theorem 3.4 (Robinson, Pixton, Oliveira) *Let M be either S^2 or $\mathbb{T}^2$. There is a residual subset R of $\mathrm{Diff}^r_\mu(M)$ such that for any $f \in R$, if $p \in M$ is a hyperbolic periodic of f, then p has a transversal homoclinic point.*

Recently Oliveira [23] further generalized the above theorem to other compact orientable surfaces for certain homotopic types of area-preserving diffeomorphisms.

Recent results concerning generic C^r diffeomorphisms obtained by Franks & Le Calvez [11] and Xia [42] has shed some new light in C^r closing lemma and C^r connecting lemma on compact surfaces. Their results raise hope that one might be able to prove that the set of periodic points is dense for generic area-preserving C^r diffeomorphisms. We state their results as follows.

Theorem 3.5 *Let M be a compact orientable surface. There exists a residual subset $R \subset \mathrm{Diff}^r_\mu(M)$ such that if $f \in R$ and P is the set of all hyperbolic periodic points of f, then both the sets $\cup_{p\in P} W^s(p)$ and $\cup_{p\in P} W^u(p)$ are dense in M.*

Furthermore, let $U \subset M$ be an open connected subset such that it contains no periodic point for f and suppose that, for some hyperbolic periodic point p, $U \cap W^s(p) \neq \emptyset$, then $W^s(p)$ is dense in U.

The theorem was proved by Franks & Le Calvez [11] for S^2 and the general case is proved by Xia [42]. The main tool of the proof is the theory of prime ends [7] [20] and Lefschetz fixed point theorem. Also, Arnold's conjecture for symplectic fixed points, as proved by Conley & Zehnder [9] for the case of torus, turned out to be essential for the results on $\mathbb{T}^2$.

One corollary of the above theorem is that, on compact surfaces, Conjecture 2 implies Conjecture 1.

4 Elliptic and quasi-elliptic periodic points; zero Liapunov exponents

In this section we present some results regarding the generic existence of totally elliptic and quasi-elliptic periodic points, as well as zero Liapunov exponents. These will have some consequences on the global dynamics of the diffeomorphisms, like the structural stability, robust transitivity, stable ergodicity.

If the symplectic diffeomorphism is smooth enough and under some generic non-degeneracy conditions KAM theory implies that the elliptic and quasi-elliptic periodic points are accumulated by many invariant tori (possible a positive measure set for the totally elliptic case) on which the restriction of the map is conjugated to an irrational translation, as well as other quasi-elliptic and hyperbolic periodic points together with homoclinic intersections.

On the other hand the Anosov diffeomorphisms have a well understood global behavior: they have a Markov partition, they are robustly transitive, stably ergodic if they are smooth enough ($C^{1+\alpha}$), and structurally stable because of the shadowing property. The uniform hyperbolicity and the existence of quasi-elliptic periodic points are mutually exclusive, and C^1 generically these are the only two possibilities (see [21]):

Theorem 4.1 (Newhouse) *There exist a residual subset $\mathcal{R}$ in the set $\mathit{Diff}^1_\omega(M)$ of C^1 symplectic diffeomorphisms on M such that every map in $\mathcal{R}$ either is Anosov or it has dense quasi-elliptic periodic points.*

In this section M is considered to be a compact connected symplectic manifold. This result implies that a structurally stable symplectic diffeomorphism must be uniformly hyperbolic. For example if the topology of the manifold rules out the Anosov case then there are no structurally stable symplectic diffeomorphisms.

In dimension 2, i.e., M is a surface, the quasi-elliptic periodic points and the elliptic ones coincide. The natural way to extend this result for totally elliptic periodic points is to replace the Anosov maps with partially hyperbolic ones. Indeed, there is a C^1 generic dichotomy between partially hyperbolic symplectic diffeomorphisms and diffeomorphisms with totally elliptic periodic points (see Arnaud [3] for dimension 4, Saghin and Xia [34] for any dimension):

Theorem 4.2 *There exist a residual subset $\mathcal{R}$ in $Diff^1_\omega(M)$ such that every map in $\mathcal{R}$ either is partially hyperbolic or it has dense totally elliptic periodic points.*

Here we have to consider the general definition of partial hyperbolicity, i.e. we allow the parameters $\lambda_1, \lambda_2, \mu_1, \mu_2$ from the definition of partial hyperbolicity to depend on trajectories.

Transitivity is probably not a generic property for high differentiability, definitely not for surfaces because of the persistence of invariant circles around elliptic periodic points; but C^1 generically the symplectic diffeomorphisms are transitive. Actually M is C^1 generically the homoclinic class of a periodic hyperbolic point (see [4]). However a C^1 robustly transitive symplectic diffeomorphism must be partially hyperbolic, because the existence of a totally elliptic periodic point is an obstruction to robust transitivity, in the same way the existence of a quasi-elliptic periodic point is an obstruction to structural stability.

A C^2 diffeomorphism f is called stably ergodic if every C^2 diffeomorphism from a C^1 neighborhood of f is ergodic (with respect to the volume form corresponding to the symplectic form in our case). It is generally believed that most of the partially hyperbolic volume preserving or symplectic diffeomorphisms are stably ergodic(Pugh & Shub [29] [30]):

Conjecture 3 (Pugh, Shub) *The set of partially hyperbolic volume preserving or symplectic diffeomorphisms contain a C^1 open and C^r dense subset of stably ergodic ones.*

There are large classes of partially hyperbolic diffeomorphisms which are known to be stably ergodic. In fact it has been recently proved that accessibility and center bunching implies ergodicity (see Burns & Wilkinson [6]) and stable accessibility is C^1 dense in general, even C^r dense in the case when the center bundle is one dimensional. The center bunching condition is also satisfied for the one dimensional center bundle, so in this case the conjecture is true.

The converse of this conjecture is also true for symplectic diffeomorphisms, because again the existence of a totally elliptic periodic point destroys stable ergodicity (see Saghin & Xia [34], Horita & Tahzibi [17]):

Corollary 4.3 *A stably ergodic symplectic diffeomorphism must be partially hyperbolic.*

Other results of this type are concerned with the Liapunov exponents of almost every point of generic symplectic diffeomorphisms. There is a natural volume form preserved by every symplectic map, obtained by wedging the symplectic form ω with itself several times. Then, for a fixed diffeomorphism f preserving this volume, almost every point of the manifold (with respect to this volume) will have Liapunov exponents, measuring the exponential rate of expansion of tangent vectors under the derivative of f. In the case of symplectic diffeomorphisms the Liapunov exponents come in pairs, if λ is an exponent then $-\lambda$ is also an exponent with the same multiplicity, in particular if zero is an exponent then it must have even multiplicity.

Again, for symplectic diffeomorphisms one gets a dichotomy between uniform hyperbolicity and the existence of zero Liapunov exponents at almost every point of M (see Bochi & Viana [5]).

Theorem 4.4 (Mañé, Bochi, Viana) *There exist a residual subset $\mathcal{R}$ of $Diff^1_r(M)$ such that every map in $\mathcal{R}$ either is Anosov or it has at least two zero Liapunov exponents at almost every point.*

It is an open question weather this result could be extended in the same way the Newhouse theorem was extended for partially hyperbolic symplectic diffeomorphisms.

To conclude the section, we remark that C^1 generically we can have three types of symplectic diffeomorphisms:

-Anosov: they are structurally stable, stably ergodic if they are smooth enough, all the Liapunov exponents are nonzero (for all the points where they exist); may not exist for some manifolds;

-partially hyperbolic but not Anosov: they have dense quasi-elliptic periodic points and consequently they are not structurally stable; almost every point has zero as a Liapunov exponent with multiplicity at least two; they are believed to be stably ergodic if smooth enough;

-not partially hyperbolic: they have dense totally elliptic periodic points and consequently they are not structurally stable, robustly transitive or stably ergodic; almost every point has zero Liapunov exponent with multiplicity at least two (and it is believed that all the Liapunov exponents are zero for almost every point).

These are all C^1 generic properties. It is not known weather they can be extended to the C^r case.

References

[1] D.V. Anosov. Geodesic flows on closed Riemannian manifolds of negative curvature. *Proc. Steklov. Inst. Math.*, 90, 1967.

[2] M. C. Arnaud. Création de connexions en topologie C^1. *Ergodic Theory Dynamical Systems*, 21(2):339–381, 2001.

[3] M. C. Arnaud. The generic c^1 symplectic diffeomorphisms of symplectic 4-dimensional manifolds are hyperbolic, partially hyperbolic, or have a completely elliptic point. *Ergodic Theory Dynamical Systems*, 22:1621–1639, 2002.

[4] M. C. Arnaud, C. Bonatti, and Crovisier S. Dynamiques symplectiques génériques. *Preprint*, 2004.

[5] J. Bochi and M. Viana. A sharp dichotomy for conservative systems: zero Liapunov exponents or projective hyperbolicity. *Preprint*, 2002.

[6] K. Burns and A. Wilkinson. Bunching plus accessibility implies ergodicity. *Preprint*, 2005.

[7] C. Caratheodory. Über die begrenzung einfach zusammenhangender gebiete. *Math. Ann.*, 73:323–370, 1913.

[8] C-Q. Cheng and Y-S. Sun. Existence of invariant tori in three-dimensional measure-preserving mappings. *Celest. Mech. Dyn. Astr.*, 47:275–292, 1990.

[9] C. Conley and E. Zehnder. The Birkhoff-Lewis fixed point theorem and a conjecture of V. I. Arnold. *Invent. Math.*, 73(1):33–49, 1983.

[10] R. Douady. Applications du théorème des tores invariants. *Thèse de troisiéme cycle, Université de Paris 7*, 1992.

[11] J. Franks and P. Le Calvez. Regions of instability for non-twist maps. *Ergodic Theory Dynam. Systems*, 23(1):111–141, 2003.

[12] C. Gutierrez. A counter-example to a C^2 closing lemma. *Ergodic Theory Dynamical Systems*, 7(4):509–530, 1987.

[13] C. Gutierrez. On C^r-closing for flows on 2-manifolds. *Nonlinearity*, 13(6):1883–1888, 2000.

[14] C. Gutierrez and B. Pires. On peixoto's conjecture for flows on non-orientable 2-manifolds. *Proc. Amer. Math. Soc.*, 133(4):1063–1074, 2004.

[15] S. Hayashi. Connecting invariant manifolds and the solution of the C^1 stability and ω-stability conjectures for flows. *Ann. of Math.*, 145(1):81–137, 1997.

[16] M. Herman. Exemples de flots hamiltoniens dont aucune perturbation en topologie C^∞ n'a d'orbites périodiques sur un ouvert de surfaces d'énergie. *C.R. Acad. Sci. Paris, t.*, 312:989–994, 1991.

[17] V. Horita and A Tahzibi. Partial hyperbolicity for symplectic diffeomorphisms. *Preprint*, 2004.

[18] S.T. Liao. An extension of the C^1 closing lemma. *Acta Sci. Natur. Univ. Pekinensis*, 2:1–41, 1979.

[19] J. Mai. A simpler proof of C^1 closing lemma. *Scientia Sinica*, 10:1021–1031, 1986.

[20] J. Mather. Topological proofs of some purely topological consequences of Caratheodory's theory of prime ends. *in Selected Studies. Eds. Th. M. Rassias and G. M. Rassias*, pages 225–255, 1982.

[21] S. Newhouse. Quasi-elliptic periodic points in conservative dynamical systems. *Amer. J. Math.*, 99(2):1061–87, 1977.

[22] F. Oliveira. On the generic existence of homoclinic points. *Ergodic Theory Dynamical Systems.*, 7:567–595, 1987.

[23] F. Oliveira. On C^∞ genericity of homoclinic orbits. *Nonlinearity*, 13:653–662, 2000.

[24] Ya. Pesin. Characteristic Liapunov exponents and smooth ergodic theory. *Russian Math. Surveys*, 32(4):55–112, 1977.

[25] D. Pixton. Planar homoclinic points. *J. Differential Equations*, 44:1365–382, 1982.

[26] H. Poincaré. *Les méthodes nouvelles de la mécanique céleste.* Paris, 1892.

[27] C. Pugh. The closing lemma. *Amer. J. Math.*, 89:956–1021, 1967.

[28] C. Pugh and C. Robinson. The C^1 closing lemma, including Hamiltonians. *Ergodic Theory Dynamical Systems*, 3:261–313, 1983.

[29] C. Pugh and M. Shub. Stable ergodicity and partial hyperbolicity. *International Conference on Dynamical Systems (Montevideo, 1995)*, pages 182–187, 1995.

[30] C. Pugh and M. Shub. Stable ergodicity. *Bulletin Amer. Math. Soc.*, 41(1):1–41, 2003.

[31] C. Robinson. Generic properties of conservative systems, i, ii. *Amer. J. Math.*, 92:562–603, 897–906, 1970.

[32] C. Robinson. Closing stable and unstable manifolds on the two-sphere. *Proc. Amer. Math. Soc.*, 41:299–303, 1973.

[33] C. Robinson. *Dynamical systems, Stability, symbolic dynamics and chaos. Second edition.* CRC Press, Boca Raton, FL, 1999.

[34] R. Saghin and Z. Xia. Partial hyperbolicity or dense elliptic periodic points for C^1 generic symplectic diffeomorphisms. *Preprint, to appear in Trans. Amer. Math. Soc.*, 2004.

[35] C.L. Siegel and J.K. Moser. *Lectures on Celestial Mechanics.* Springer, 1971.

[36] S. Smale. Mathematical problems for the next century. *Math. Intelligencer*, 20(2):7–15, 1998.

[37] F. Takens. Homoclinic points in conservative systems. *Invent. Math.*, 18:267–292, 1972.

[38] L. Wen and Z. Xia. A basic C^1 perturbation theorem. *J. Differential Equations*, 154(2):267–283, 1999.

[39] L. Wen and Z. Xia. On C^1 connecting lemmas. *Trans. Amer. Math. Soc.*, 352(10), 2000.

[40] Z. Xia. Existence of invariant tori for certain non-symplectic diffeomorphism. *Hamiltonian dynamical systems (Cincinnati, OH, 1992), IMA Vol. Math. Appl*, 63:373–385, 1995.

[41] Z. Xia. Homoclinic points in symplectic and volume-preserving diffeomorphism. *Commun. Math. Phys.*, 177:435–449, 1996.

[42] Z. Xia. Area-preserving surface diffeomorphisms. *Preprint, Mathematics ArXiv: math.DS/0503223, to appear in Commun. Math. Physics*, 2004.

[43] Z. Xia and Hua Zhang. A C^r closing lemma for a class of symplectic diffeomorphisms. *Preprint, to appear in Nonlinearity*, 2005.

[44] Y. Yoccoz. Travaux de herman sur les tores invariants. *S'eminaire Bourbaki*, 754, 1992.

Fields Institute Communications
Volume **48**, 2006

Mathematical Aspects of Modelling Tumour Angiogenesis

B. D. Sleeman
School of Mathematics,
University of Leeds,
Leeds, LS2 9JT UK
bds@maths.leeds.ac.uk

Abstract. Tumour angiogenesis, the formation of new blood vessels is a key event in many types of cancer progression. In this paper we describe how mathematical models may be formulated on the basis of the complex biochemical processes involved. The models not only give results which are in good qualitative agreement with observations but also indicate possible approaches to anti-angiogenic therapeutic strategies.An additional outcome is that the models are mathematically challenging and present problems of independent interest.

We begin with a brief outline of tumour biology. The mathematical modelling covers two fundamental aspects of aniogenesis. First we describe the cell-based biochemistry in terms of cell-kinetics founded on the mass-action laws of chemistry. Secondly we model cell migration on the basis of the theory of reinforced random walks. Some simulations are carried out using ideas drawn from the theory of circular statistics.

At the continuum,i.e., cell density or macroscopic level, the models take the form of quasi-linear strongly couple partial differential equations. Local existence and uniqueness of solutions are discussed. It turns out that for some related problems analytical solutions can be constructed which provide insight into the underlying mechanisms of angiogenesis. We also discuss a related systems which have spike solutions. These systems are relevant to neo-vascular remodelling.

Finally we discuss some further aspects of tumour angiogenesis; the role of the angiopoietins; inclusion of blood flow in the modelling programme and also the problems of modelling the mechanics of cell locomotion.

1 Intoduction to tumour biology

The most common cause of primary tumours is the genetic mutation of one or more cells resulting in uncontrolled proliferation. The mutated cells have a proliferative advantage over neighbouring healthy cells and are able to form a growing

2000 *Mathematics Subject Classification.* Primary 92C50, 35A05; Secondary 92C17, 35A320, 34C60.

mass. The reason for this advantage is not necessarily an increase in the proliferation rate, but may be a decrease in the cell death rate. For example, one of the key functions of tumour suppressor genes such as p53 is to induce apoptosis (programmed cell death) in damaged cells. Loss of p53 function allows propagation of damaged DNA.

If the mutated cells remain contained within a single cluster, with a well defined boundary separating them from neighbouring normal cells, the tumour is said to be benign, and surgical removal will often provide a complete cure. However, if the tumour cells are inter-mixed with normal cells and attempt to invade surrounding tissue, the growth ceases to be contained and the tumour is described as malignant.

A tumour may persist in a diffusion limited state, usually not more than 2mm in diameter with cell proliferation balanced by cell death, for many months or years. It rarely causes significant damage in this dormant phase and often goes undetected. A tumour may however emerge from dormancy by inducing the growth of new blood vessels, a process termed angiogenesis, or neovascularisation. This process allows the tumour to progress from the avascular to the vascular state. There are a large number of pro-angiogenic and anti-angiogenic factors. It is a shifting of the balance from the anti-to the pro-angiogenic factors, the so-called angiogenic switch, that causes the transition from the dormant to the angiogenic phase. This switch is a highly complex process which is not fully understood, but oxygen deficiency in the tumour is thought to be an important factor stimulating the production of pro-angiogenic molecules by the tumour cells.

The most essential component of blood vessels is the endothelial cell (EC). Every vessel, from the aorta to the smallest capillaries, consists of a monolayer of EC (called the endothelium) arranged in a mosaic pattern around a central lumen, through which blood can flow. The endothelium controls the passage of nutrients, white blood cells and other materials between the bloodstream and the tissues. The healthy endothelium represents a highly stable population of cells; cell-cell connections are tight and the cell turnover period is measured in months or years.

Outside the endothelium is an extracellular lining called the basement membrane separating the EC from the surrounding connective tissue. This is composed of protein fibres, mainly laminin and collagen and also peri-endothelial support cells. The basement membrane serves as a scaffold on which the EC rest and helps to maintain the endothelium in its quiescent state. In quiescent endothelia, EC proliferation and migration arise, for example, in the case of tissue damage. In particular EC proliferation and migration are crucial for angiogenesis. Angiogenesis is essential for embryonic growth, tissue growth and repair. It can however be induced under certain pathological conditions such as rheumatoid arthritis, wound healing, diabetic retinopathy and solid tumour growth, which is the main concern in this paper.

When it was first suggested by Folkman [2]that the growth of a tumour beyond a diameter of about 2mm is dependent on its ability to recruit new blood vessels, it was not known how this process might take place, nor how the tumour might induce it. It was postulated that the tumour secretes some diffusible substance which would stimulate the growth of new capillaries.

Interest in angiogenesis subsequently increased and exerimental models, allowing the *in vivo* formation of new blood vessels to be observed directly, were developed. The first direct evidence for Folkman's hypothesis came when basic fibroblast growth factor was shown to be capable of inducing an angiogenic response *in vitro*. Subsequently many other angiogenic growth factors have been isolated including vascular endothelial growth factor (VEGF) which acts specifically on EC.

On receiving a net angiogenic stimulus, EC in capillaries near the tumour become activated; they loosen the normally tight contacts with adjacent cells and secrete proteolytic enzymes (or protease) whose collective behaviour is to degrade extracellular tissue. The first target of the proteases produced by the EC is the basement membrane. When this has been sufficiently degraded, the EC are able to move through the gap in the basement membrane and into the ECM. Neighbouring EC move in to fill the gap and subsequently follow the leading cells into the ECM. The first function of the angiogenic growth factors therefore is to stimulate the production of proteases by EC. This is a key step in the angiogenic cascade and is the starting point of the mathematical modelling.

Following extravasation, the EC continue to secrete proteolytic enzymes which also degrade the ECM. This is necessary to create a pathway along which the cells can move.They continue to move away from the parent vessel and towards the tumour, thus forming small sprouts. More EC are recruited from the parent vessel, elongating the new sprouts. These sprouts may initially take the form of solid strands of cells, but the EC subsequently form a central lumen, thereby creating the necessary structure for a new blood vessel.

EC migration is mainly governed by a chemotactic response to concentration gradients of diffusible growth factors (e.g. VEGF) produced by the tumour, which provide a potent directional stimulus. Thus the second key function of the angiogenic growth factors is to induce directed EC migration towards the tumour.

Haptotaxis, cell movement in response to an adhesive gradient, also plays a role. The effect of haptotaxis, however is more complicated, and not fully understood, because the EC are continually modifying the adhesive properties of their micro-environment via proteolysis and the synthesis of new ECM components.In our mathematical modelling we adopt the view, on the basis of a number of *in vivo* and *in vitro* reported experiments that EC move into areas where high concentrations of fibronectin are sufficiently degraded and at low concentration levels of undegraded fibronectin EC exhibit their natural tendency to migrate up a concentration gradient to a region of higher cell-matrix adhesion.

In quiescent endothelia, the turnover of EC is very slow and for a short period following extravasation,the low mitosis levels continue. After this initial period of migration, rapid EC proliferation begins a short distance behind the sprout tips, increasing the rate of sprout elongation. EC proliferation is therefore necessary for vascularisation and is the final key function of the angiogenic growth factors. Sprouts are observed to branch, adding to the number of migrating tips. They begin by growing approximately parallel to each other but, at a certain distance from the parent vessel, begin to incline towards other sprouts. This leads to the formation of closed loops (anastomoses),which is essential for blood circulation in the new vessels.

This whole orchestration of complex events leads to a micro-vascular structure that eventually reaches and penetrates the tumour, vastly improving its blood supply and allowing for rapid and unconstrained growth.

For an up-to-date account of the biochemistry of tumour angiogenesis we refer to Plank and Sleeman [15] and cited references.

2 Mathematical modelling the biochemistry

We begin with a number of EC in a capillary in the neighbouhood of a source of VEGF at the tumour. The ECM and the basement membrane are assumed to be composed of fibronectin, a principal extra-cellular component. The fibronectin concentration is viewed as representing the thickness of the basement membrane in the capillary and density of the ECM. VEGF diffuses through the ECM and into the capillary where it binds to receptors on the surface of the EC. This stimulates the EC to produce a proteolytic enzyme which in turn degrades the fibronectin levels in the capillary. When the fibronectin falls below a certain threshold level, the basement membrane has been sufficiently degraded to allow EC to migrate into the ECM. The EC continue to produce protease in the ECM and can degrade fibronectin levels there.

We take the view that ligand binding can be modelled as an enzymatic biochemical reaction [6]. The reaction of interest here is the binding of VEGF to receptors located on EC. A molecule of VEGF (denoted by V) binds to a receptor on the surface of an EC (denoted by R_V)forming a receptor-ligand complex (denoted by $R_V V$). The complex is internalized and triggers an intracellular sequence of transcription events, the result of which is the secretion of a number of molecules of protease (denoted by C). The complex decomposes into a modified receptor (denoted by R_V') which is subsequently recycled back to the cell surface to become R_V, where it may repeat the process. The transduction cascade triggered by the binding of VEGF is highly complex and not fully understood. Here we shall consider a simplified mechanism, which still retains the most important biochemical features. Using standard chemistry symbolism we write;

$$\begin{aligned} V + R_V &\rightleftharpoons R_V V, \text{ rates: } k_1,\ k_{-1} \\ R_V V &\rightarrow C + R_V, \text{ rate: } k_2. \end{aligned}$$

Denoting the concentration of a substance X by $[X]$ We apply the law of mass action, which states that the reaction rate is proportional to the product of the concentrations of the reactants. This process leads to a set of four coupled rate equations for $[R_V]$, $[V]$, $[R_V V]$ and $[C]$; viz:

$$\begin{aligned} \frac{\partial[R_V]}{\partial t} &= -k_1[R_V][V] + (k_{-1}+k_2)[R_V V] + \frac{\partial[R_T]}{\partial t}, \\ \frac{\partial[V]}{\partial t} &= -k_1[R_V][V] + k_{-1}[R_V V], \\ \frac{\partial[R_V V]}{\partial t} &= k_1[R_V][V] - (k_{-1}+k_2)[R_V V], \\ \frac{\partial[C]}{\partial t} &= k_2[R_V V]. \end{aligned} \tag{1}$$

The inclusion of the term $\frac{\partial[R_T]}{\partial t}$ is to allow for local crowding or dispersion of cells.

Degradation of fibronectin (F) by protease is also thought of as an enzymatic reaction. In addition angiostatin acts by inhibiting protease. We therefore assume that the total protease (C) consists of a proportion that is active (C_A), a proportion that is inactive (C_I), as well as a proportion that is in the intermediate complex ($C_A F$). Angiostatin (A) combines with the active protease to form inactive protease, which is inhibited from functioning in the degradation of fibronectin. Again using the symbolism of chemistry we write;

$$\begin{aligned} C_A + F &\rightleftharpoons C_A F, \text{ rates: } k_3,\ k_{-3} \\ C_A F &\rightarrow C_A + F', \text{ rate: } k_4, \\ A + C_A &\rightleftharpoons C_I, \text{ rates: } k_5,\ k_{-5}. \end{aligned}$$

where F' represents proteolytic fragments of fibronectin. The law of mass action yields a set of six rate equations for $[C_A]$, $[F]$ $[C_AF]$, $[F']$, $[A]$ and $[C_I]$.

These together with the above rate equations form the basis of a mathematical description of the biochemistry of chemical -cell interaction both in the capillary and the ECM. A full description together with the modelling arguments are given in [9, 10, 11, 15]

Let $v(x,t)$, $c(x,t)$, $f(x,t)$ and $a(x,t)$ respectively denote the concentrations of VEGF, protease, fibronectin and angiostatin in the capillary. Also let $p(x,t)$ denote the EC density. Combining the reaction kinetics with a source term, $v_r(x,t)$, allowing VEGF to pass into the capillary from the ECM; natural decay of protease; logistic production of fibronectin by the EC together with a source of angiostatin, $a_r(x,t)$ and natural decay of angiostatin, leads to the following system of differential equations:

$$\begin{aligned} \frac{\partial v}{\partial t} &= -\frac{\lambda_1 \delta_e v p}{1+\nu_1 v} + v_r(x,t), \\ \frac{\partial c}{\partial t} &= \frac{\lambda_1 \delta_e v p}{1+\nu_1 v} - \mu c, \\ \frac{\partial f}{\partial t} &= \frac{4}{T_f} f \left(1 - \frac{f}{f_0}\right) \frac{p}{p_0} - \lambda_3 c_a f, \\ \frac{\partial a}{\partial t} &= a_r(x,t) - \frac{a}{T_a}. \end{aligned} \tag{2}$$

Only the active protease, c_a takes part in the degradation of fibronectin, and we have

$$c_a = \frac{c}{1+\nu_e a + \nu_3 f}.$$

Diffusion is neglected here since it takes place on a much longer time scale than the kinetic reactions in the capillary. However, it is necessary to include diffusion terms for the VEGF, fibronectin and angiostatin in the ECM. Using capital letters for the various quantities in the ECM we have the following system of partial differential equations:

$$\begin{aligned} \frac{\partial V}{\partial t} &= D_V \Delta V - \frac{\lambda_1 \delta_e V P}{1+\nu_1 V}, \\ \frac{\partial C}{\partial t} &= \frac{\lambda_1 \delta_e V P}{1+\nu_1 V} - \mu C, \\ \frac{\partial F}{\partial t} &= D_F \Delta F + \frac{4}{T_F} F \left(1 - \frac{F}{f_0}\right) - \lambda_3 C_A F, \\ \frac{\partial A}{\partial t} &= D_A \Delta A + a_r(x,t) \left(1 - \frac{F}{f_0}\right) - \frac{A}{T_a}, \end{aligned} \tag{3}$$

where D_V, D_F, D_A and T_F are constants.

The active protease law also applies in the ECM:

$$C_A = \frac{C}{1 + \nu_e A + \nu_3 F}.$$

3 Cell movement and reinforced random walks

In the quiescent endothelium, connections between neighbouring EC are tight and there is little or no cell movement. However during angiogenesis cell-cell connections are loosened and EC are observed to migrate towards the tumour.

To account for EC movement we begin by assuming that the EC lining the capillary wall move on a regular one-dimensional lattice of mesh size h. Let $p_n(t)$ be the EC density at mesh point n at time t. Let $\hat{\tau}_n^{\pm}$ be the transition probability rate of EC moving from mesh point n to mesh point $n \pm 1$.

Consider the master equation for the EC:

$$\frac{\partial p_n}{\partial t} = \hat{\tau}_{n-1}^{+} p_{n-1} + \hat{\tau}_{n+1}^{-} p_{n+1} - (\hat{\tau}_n^{+} + \hat{\tau}_n^{-}) p_n. \tag{4}$$

Assume that the decision of when to move, is independent of the decision of where to move [14]. This means that the mean waiting time is constant.

Consequently we let

$$\hat{\tau}_n^{+} + \hat{\tau}_n^{-} = 2\lambda,$$

where λ is a constant.

We further assume that the transition probability rates depend only on the control substances at the nearest 1/2 neighbour mesh points and let

$$\hat{\tau}_n^{\pm}(\omega) = 2\lambda \frac{\tau(\omega_{n\pm 1/2}}{\tau(\omega_{n-1/2}) + \tau(\omega_{n+1/2})} \equiv 2\lambda N_n^{\pm}(\omega), \tag{5}$$

where ω is the vector of control substances.

$$\frac{1}{2\lambda} \frac{\partial p_n}{\partial t} = N_{n-1}^{+} p_{n-1} + N_{n+1}^{-} p_{n+1} - p_n. \tag{6}$$

It can be shown that the continuum limit, $h \to 0$, $\lambda \to \infty$, such that $\lambda h^2 = D_p$, of (6) is

$$\frac{\partial p}{\partial t} = D_p \frac{\partial}{\partial x} \left(p \frac{\partial}{\partial x} \left(ln \frac{p}{\tau} \right) \right). \tag{7}$$

This PDE is extremely important and will be discussed in some depth later.

VEGF is a potent chemoattractant for EC and so EC will migrate towards regions of high VEGF concentration. Similarly, protease will be expressed in response to VEGF, in areas of angiogenic activity, and so we assume that protease also has a chemotactic effect on the EC. The effect of ECM components, such as fibronectin, on the directional movement of cells is unclear and there is conflicting experimental evidence. Taking fibronectin concentration as a measure of the penetrability of the tissue, we therefore take the view that EC move into regions of low fibronectin. That is they will move towards areas where the ECM has been degraded, and where they can move more freely. Consequently we model $\tau(c_a, v, f)$ as

$$\tau(c_a, v, f) = \left(\frac{c_a + \alpha_1}{c_a + \alpha_2} \right)^{\gamma_1} \left(\frac{f + \beta_1}{f + \beta_2} \right)^{\gamma_2} \left(\frac{v + \delta_1}{v + \delta_2} \right)^{\gamma_3}$$

where $0 < \alpha_1, \delta_1 \ll 1 < \alpha_2, \delta_2$ $0 < \beta_2 \ll 1 < \beta_1$ and $\gamma_1, \gamma_2, \gamma_3 > 0$.

In the ECM a two-dimensional version of the master equation is used which in the continuum limit gives rise to the PDE

$$\frac{\partial P}{\partial t} = D_P \nabla \cdot \left(P \nabla \left(ln \frac{P}{T} \right) \right). \qquad (8)$$

In addition to migration, EC proliferation is an important component of angiogenesis. To accomodate this we add a proliferation term Γ to the master equation which is taken to be of the form

$$\Gamma = P \left[\left(Q + G(C_A) \frac{\partial C_A}{\partial t} \right) - \mu_1 \right],$$

where $Q, \mu_1 > 0$ are constants. The first term in square brackets represents EC proliferation: the second term represents natural death by apoptosis (programmed cell death). The QP term represents a constant low level of background proliferation, whilst the $G(C_A)$ represents a protease dependent contribution to cell proliferation.

To complete the modelling of cell-movement we require initial state and boundary conditions together with transmission conditions which are needed to model the coupling of the capillary and ECM equations. Full details are given in [16].

The setting for our mathematical model is as follows. The capillary is assumed to be of infinitesimal thickness and is located on the subset of the x-axis. The ECM separating the capillary and the tumour is viewed as a rectangular subset of the $x - y$ plane. We begin with a number of EC in the capillary and a source of VEGF at the tumour. The ECM and the basement membrane are assumed to be composed of fibronectin. Thus fibronectin concentration is viewed as representing the thickness of the basement membrane in the capillary and the density of the ECM.

Cell movement is governed by the master equations. For each cell at each time step, the probabilities of staying still, and moving in each direction are calculated. For example, in the capillary, the probabilities are given by

$$\begin{aligned} P(\text{stay still}) &= 1 - 2\lambda k, \\ P(\text{move left}) &= 2\lambda k \frac{\tau_{n-1/2}}{\tau_{n-1/2} + \tau_{n+1/2}}, \\ P(\text{move right}) &= 2\lambda k \frac{\tau_{n+1/2}}{\tau_{n-1/2} + \tau_{n+1/2}}, \end{aligned}$$

In the two-dimensional ECM similar probabilities are calculated.

Using the method of Sleeman and Wallace [23], the interval $[0, 1)$ is divided into three (or in the ECM, five) intervals corresponding to the above probabilities. A random number is generated in $[0, 1)$ and depending into which probability interval the random number falls, the cell either stays still or moves accordingly.

In the capillary, cell aggregation is allowed as this may be one mechanism by which the cells move together to degrade the capillary wall. In the ECM however, each cell is viewed as a 'leader' cell, tracing the tip of a newly formed capillary tip. Other EC (not included in the model) are assumed to follow the path of this leader to form the capillary lumen. When a cell divides, a new cell is created and the capillary branches. When a cell dies, the capillary tip can move no further. When a cell collides with another cell, we have tip-to-tip anastomosis. When a cell collides with the trail of another, we have tip-to branch anastomosis and the colliding capillary tip ceases to move. See [16] for details of these simulations together with a discussion of therapeutic strategies.

4 Non-lattice models

In the previous section cells are constrained to move on a fixed square lattice. Here we propose a different method of simulating cell movement based on an idea of Hill and Häder [3] in their study of the trajectories of swimming mico-organisms. Here we characterise each EC by its speed, $s(t)$, and its direction of motion, $\theta(t)$. Treating the direction independently of the speed, an EC can be thought of as performing a random walk on the unit circle. At each time step, k, the EC is assumed to have a probability, $a(\theta(t))$, of moving clockwise through an angle δ and a probability, $b(\theta(t))$, of moving anti-clockwise through an angle δ and a probability $1 - a(\theta(t)) - b(\theta(t))$ of continuing in the same direction. This may be expressed as

$$\begin{aligned} P(\Theta(t+k) - \Theta(t) = \delta) &= a(\theta(t)), \\ P(\Theta(t+k) - \Theta(t) = -\delta) &= b(\theta(t)), \\ P(\Theta(t+k) - \Theta(t) = 0) &= 1 - a(\theta(t)) - b(\theta(t)). \end{aligned}$$

The probability density function,f, for $\Theta(t)$ is defined as usual by

$$f(\theta, t) = P(\theta \leq \Theta(t) < \theta + d\theta).$$

In the limit as $k \to 0, \delta \to 0$ such that $\delta^2/k < \infty$, it can be shown that f satisfies the Fokker-Planck equation

$$\frac{\partial}{\partial t} f(\theta, t) = -\frac{\partial}{\partial \theta}(\mu(\theta) f(\theta, t)) + \frac{1}{2}\frac{\partial^2}{\partial \theta^2}(\sigma^2(\theta) f(\theta, t)),$$

where

$$\begin{aligned} \mu(\theta) &= lim_{k \to 0, \delta \to 0} \left(\frac{1}{k} E(\Theta(t+k) - \Theta(t)) \right), \\ \sigma^2(\theta) &= lim_{k \to 0, \delta \to 0} \left(\frac{1}{k} Var(\Theta(t+k) - \Theta(t)) \right) \end{aligned}$$

and $E(X)$ and $Var(X)$ respectively denote the expectation and variance of X. Writing the Fokker-Planck equation in the chemotaxis form

$$\frac{\partial}{\partial t} f(\theta, t) = \frac{\partial}{\partial \theta} \left(D_r(\theta) f(\theta, t) \frac{\partial}{\partial \theta} \left(ln \frac{f(\theta, t)}{\tau(\theta)} \right) \right), \tag{9}$$

we find

$$D_r(\theta) = \frac{\sigma^2(\theta)}{2} \tag{10}$$

$$\tau(\theta) = \frac{1}{\sigma^2(\theta)} exp \left(2 \int \frac{\mu(\theta)}{\sigma^2(\theta)} d\theta \right). \tag{11}$$

Now equation (9) is the continuum limit of the reinforced random walk master equation

$$\frac{\partial f_n}{\partial t} = \hat{\tau}_{n-1}^{+} f_{n-1} + \hat{\tau}_{n+1}^{-} f_{n+1} - (\hat{\tau}_n^{+} + \hat{\tau}_n^{-}) f_n. \tag{12}$$

where $f_n(t) = f(n\delta, t)$.

With $\tau(\theta)$ defined by (11) we construct the probability transition rates $\hat{\tau}_n^{\pm}$ as above by

$$\hat{\tau}_n^{\pm} = 2\lambda \frac{\tau((n \pm 1/2)\delta)}{\tau((n-1/2)\delta) + \tau((n+1/2)\delta)}, \tag{13}$$

where $\lambda\delta^2 = D_r$ is the diffusion constant.

Simulations are carried out, as in section 4, by generating a random number between 0 and 1 and then constructing confidence intervals such that if r lies in the interval $0 \leq r < \hat{\tau}_n^+ k$ turn anti-clockwise through an angle δ; if r lies in the interval $\hat{\tau}_n^+ k \leq r < 2\lambda k$ turn clockwise through angle δ; if r lies in the interval $2\lambda k \leq < 1$ continue in the current direction.

The movement of an EC in the x, y-plane is governed by the rate equations

$$\frac{dx}{dt} = S\cos\theta(t), \ \frac{dy}{dt} = S\sin\theta(t),$$

where S denotes the speed.

Ideally functional forms for the mean μ and the variance σ^2 of the turning rate would be obtained via a statistical analysis of *in vitro* or *in vivo* capillary networks. However such statistical data is currently unavailable. We therefore adopt a sinusoidal reorientation model which seems to agree well with experimental observations. That is we assume the mean and variance of the turning rate are given by;

$$\mu(\theta) = -d\sin(\theta - \theta_0), \ \sigma^2(\theta) = \sigma_0^2,$$

where θ_0 is the preferred direction of motion and σ_0^2 and d are positive constants. The turning rate variance is constant and the mean turning rate is at its greatest when the direction of motion is perpendicular to the preferred direction. We choose the preferred direction to be in the direction of increasing VEGF concentration. Specifically if the VEGF concentration at a point (x, y) is given by $v(x, y)$ then we take θ_0 to be in the direction of the gradient of $v(x, y)$.

The quantity $D_r = \sigma_0^2/2$ is the rotational diffusivity. A high value of D_r means a high variance in the turning rate and a large degree of randomness in the direction of motion. Conversely a low value of D_r means a low variance and a probability of turning. That is cells exhibit a tendency to continue in their current direction and the random walk is said to be highly correlated.

The turning coefficient,d, is a measure of a cells ability to re-orient itself: the higher d is, the greater the mean turning rate and the more quickly a cell will be able to turn towards the preferred direction.

In addition to chemotaxis we need to encorporate haptotaxis. Consequently there now two preferred directions of motion; θ_1 in the direction of the gradient of VEGF and θ_2 in the direction of the gradient of fibronectin. We take a simple addition form for the mean turning rate;

$$\mu(\theta) = -d_1\sin(\theta - \theta_1) - d_2\sin(\theta - \theta_2)$$

For a full discussion of these constructions see [17]. Specifically the transition rate $\tau(\theta)$ may be given by;

$$\tau(\theta) = exp\left(\frac{d_v \mid \nabla v \mid}{D_r(1+\gamma v)^p}\cos(\theta - \theta_1)\right) exp\left(\frac{d_f \mid \nabla f \mid}{D_r}\cos(\theta - \theta_2)\right)$$

where

$$\theta_1 = \frac{v_y}{\mid v_y \mid} cos^{-1}\left(\frac{v_x}{\sqrt{v_x^2+v_y^2}}\right),$$

$$\theta_2 = \frac{f_y}{\mid f_y \mid} cos^{-1}\left(\frac{f_x}{\sqrt{f_x^2+f_y^2}}\right).$$

In figure 1 we show the results of some typical non-lattice simulations.

5 Existence and uniqueness

In this section we study the local existence and uniqueness of solutions to a wide class of systems of quasi-linear strongly coupled partial differential equations. These systems include not only the tumour angiogenesis models presented here but also many other chemotaxis models. The systems we wish to consider have the general form:

$$\begin{aligned}
\frac{\partial p}{\partial t} &= D\nabla\cdot\left(p\nabla\left(ln\left(\frac{p}{\Phi(\omega)}\right)\right)\right), \text{ for } (x,t)\in\Omega\times(0,T),\\
\frac{\partial \omega}{\partial t} &= F(p,\omega),\\
0 &= p\nabla\left(ln\left(\frac{p}{\Phi(\omega)}\right)\right)\cdot\mathbf{n}, \text{ for } (x,t)\in\partial\Omega\times(0,T),\\
p(x,0) &= p_0(x)>0,\\
\Phi(\omega(x,0)) &= \Phi(\omega_0))>0, \text{ for } \in\bar{\Omega}.
\end{aligned} \tag{14}$$

We begin with some basic notation and terminology. Let Ω be a bounded open subset of $\mathbb{R}^n$ with boundary $\partial\Omega$, $Q_T\subset\mathbb{R}^{n+1}$ the cylinder $\Omega\times(0,T), \Gamma_T$ the boundary of Q_T and $S_T=\Gamma_T\cup\{(x,t)\ :\ x\in\Omega,\ t=0\}$. Set $x=(x_1,\ldots,x_n)$, $p=(p_1,\ldots,p_n)$, $u_x=(u_{x_1}\ldots,u_{x_n})$ and set

$$\begin{aligned}
a(x,t,u,p) &= a(x_1,\ldots,x_n,t,u,p_1,\ldots,p_n),\\
a(x,t,u,u_x) &= a(x_1,\ldots,x_n,t,u,_{x_1},\ldots,u_{x_n}).
\end{aligned}$$

The Euclidean norm is denoted by $\mid x\mid=\left(\sum_{i=1}^n x_i^2\right)^{1/2}$. The notation p^2 is understood to mean $\mid p\mid^2$.

We shall need the Sobolev space $H^l(\bar{\Omega})$ in which elements are continuous functions in Ω with continuous derivatives in $\bar{\Omega}$ up to order $[l]$ inclusively. Also we shall need the Sobolev space $H^{l,l/2}(\bar{Q}_T)$ which is the Banach space of functions that are continuous in $\bar{Q}_T$, together with all derivatives of the form $D_t^r D_x^s$ for $2r+s<l$.

Let Λ_T be a subset of $\mathbb{R}$ and U_T be a subset of $\mathbb{R}^m$. Functions $\Phi(x,t,q)$: $\bar{Q}_T\times U_T\to\mathbb{R}$ and $F(x,t,v,q):\bar{Q}_T\times\Lambda_T\times U_T\to\mathbb{R}^m$ are continuous and possess continuous derivatives up to third order. Furthermore we assume $\Phi(x,t,q)\neq 0$ for $(x,t,q)\in\bar{Q}_T\times U_T$.

Let $X(t)=H^{2+\beta,1+\beta/2}(\bar{Q}_T)$ and $Y(t)=X(t)\times\cdots\times X(t)=X^m(t)$ and

$$\begin{aligned}
\Sigma_T &= p\in X(T)\ :\ p(x,t)\in\Lambda_T,\ (x,t)\in\bar{Q}_T,\\
\Theta_T &= \omega\in Y(T)\ :\ \omega(x,t)\in U_T,\ (x,t)\in\bar{Q}_T.
\end{aligned}$$

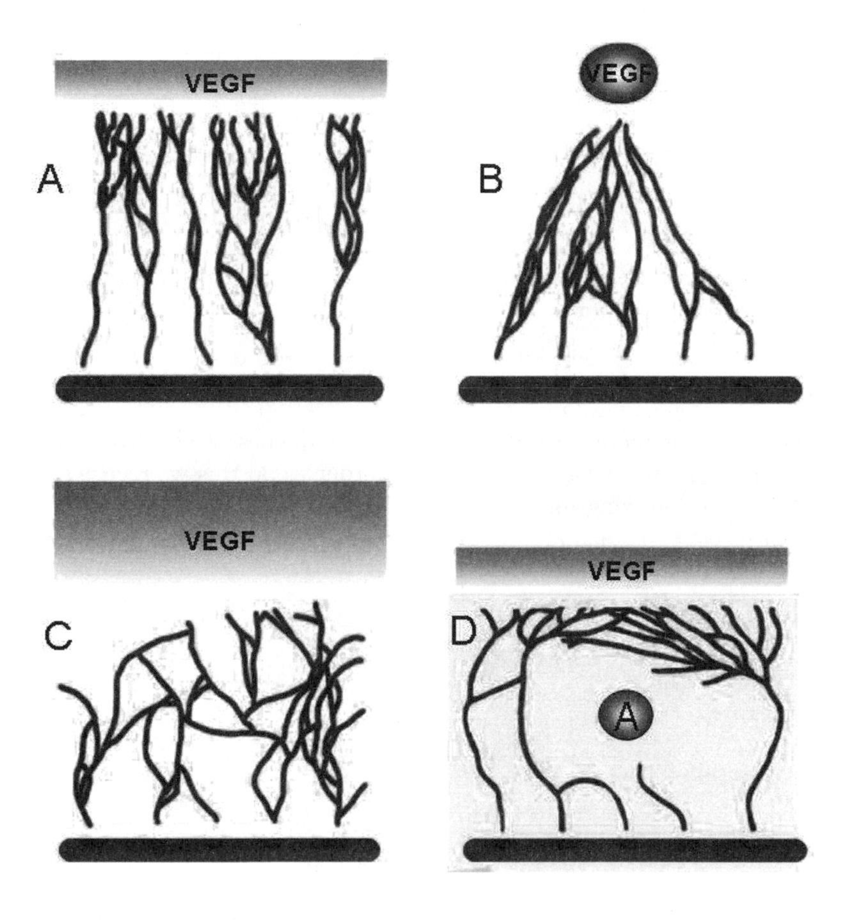

Figure 1 Capillary networks formed in the non-lattice model: (A) with a source of VEGF representing an extended tumour colony;(B) with a point source of VEGF representing a small tumour; (C) Capillary network formed with desensitisation. Capillaries lose the ability to reorient as they approach the tumour. (D) Capillary network formed with control substance dynamics. The shaded circle represents a source of diffusible angiostatin in the ECM. The angiostatin inhibits angiogenesis in its immediate vicinity, but capillaries can still grow around it to achieve tumour vascularisation.

Our model under consideration can now be written as :

$$\begin{aligned}
\frac{\partial p}{\partial t} &= D\nabla\cdot\left(p\nabla\left(ln\left(\frac{p(x,t)}{\Phi(x,t,\omega(x,t))}\right)\right)\right), \text{ for } (x,t)\in\Omega\times(0,T),\\
\frac{\partial \omega}{\partial t} &= F(x,t,p,\omega),\\
0 &= p\nabla\left(ln\left(\frac{p(x,t)}{\Phi(x,t,\omega(x,t))}\right)\right)\cdot\mathbf{n}, \text{ for } (x,t)\in\partial\Omega\times(0,T),\\
p(x,0) &= p_0(x)>0,\\
\omega(x,0) &= \omega_0(x),\\
\Phi(x,0,\omega(x,0)) &= \Phi(x,0,\omega_0))>0, \text{ for } \in\bar{\Omega},
\end{aligned} \tag{15}$$

where $p_0(x)\in\Sigma_T$ and $\omega_0(x)\in\Theta_T$.

To address the problems of existence and uniqueness we base our investigations on the work of Ladyzenskaya et al. [7] . In order to do this we recast problem (15) in the following following form.

Let $u=\frac{p}{\Phi(x,t,\omega)}-u_0(x)$, where $u_0(x)=\frac{p_0(x)}{\Phi(x,0,\omega_0(x))}$. Then $\frac{\partial u_0}{\partial \mathbf{n}}=0$,for $x\in\Gamma$, and we have;

$$\begin{aligned}
\frac{\partial u}{\partial t} &- D\Delta u - D\frac{1}{\Phi}\sum_{i=1}^{n}\left(\frac{\partial \Phi}{\partial x_i}+\sum_{j=1}^{m}\frac{\partial \Phi}{\partial \omega_j}\frac{\partial \omega_j}{\partial x_j}\right)\frac{\partial u}{\partial x_i}\\
&+ \frac{u+u_0}{\Phi}\left[\frac{\partial \Phi}{\partial t}+\langle\Phi_\omega, F(x,t,(u+u_0)\Phi,\omega)\rangle\right]\\
&+ K(\omega)=0, \text{ for } (x,t)\in Q_T, \\
\frac{\partial \omega}{\partial t} &= F(x,t,(u+u_0)\Phi(x,t,\omega),\omega)\\
\frac{\partial u}{\partial \mathbf{n}} &= 0, \text{ for } (x,t)\in\Gamma_T,\\
u(x,0) &= 0\\
\omega(x,) &= \omega_0(x), \text{ for } x\in\bar{\Omega},
\end{aligned} \tag{16}$$

where

$$K(\omega)=-D\Delta u_0 - D\frac{1}{\Phi}\sum_{i=1}^{n}\left(\frac{\partial \Phi}{\partial x_i}+\sum_{j=1}^{m}\frac{\partial \Phi}{\partial \omega_j}\frac{\partial \omega_j}{\partial x_j}\right)\frac{\partial u_0}{\partial x_i}.$$

This reformulation allows us to consider the problem in essentially two parts; one governing u and one governing ω.

We then prove, under fairly mild conditions, see Yang et al. [27];

Theorem 5.1 *There exists $\tau>0$ such that the system (16) has a unique solution $(u(x,t),\omega(x,t))$ in the class $H^{2+\beta,1+\beta/2}(\bar{Q}_\tau)\times\cdots\times H^{2+\beta,1+\beta/2}(\bar{Q}_\tau)$. This implies that the system (15) has a unique solution $(p(x,t),\omega(x,t))$ in the above class.*

In [27] we develop this result to prove under some further conditions the existence of a unique global solution.

5.1 Examples.

Example 5.1 Consider the problem

$$\begin{aligned}
\frac{\partial p}{\partial t} &= D\nabla\cdot\left(p\nabla\left(ln\left(\frac{p}{\omega}\right)\right)\right), \text{ for } (x,t)\in\Omega\times(0,T),\\
\frac{\partial \omega}{\partial t} &= \beta p-\mu\omega,\\
0 &= p\nabla\left(ln\left(\frac{p}{\omega}\right)\right)\cdot\mathbf{n}, \text{ for } (x,t)\in\partial\Omega\times(0,T),\\
p(x,0) &= p_0(x)>0,\\
\omega(x,0) &= \omega_0>0, \text{ for } \in\bar{\Omega}.
\end{aligned}\tag{17}$$

From Theorem 5.1 we have local existence of a unique solution. Furthermore we have, for any finite time $T>0$

$$\min_{(x,t)\in\bar{Q}_T} u(x,t)>0, \quad \min_{(x,t)\in\bar{Q}_T}\omega(x,t)>0.$$

Also we have the estimate

$$u(x,t)\le e^{\mu t}\left[c_0+\frac{\beta}{\mu}(e^{\mu t}-1\right]^{-1}.$$

and so from [27] theorem 3.3 there exists a unique global solution.

Example 5.2 For the following system

$$\begin{aligned}
\frac{\partial p}{\partial t} &= D\nabla\cdot\left(p\nabla\left(ln\left(\frac{p}{\omega}\right)\right)\right), \text{ for } (x,t)\in\Omega\times(0,T),\\
\frac{\partial \omega}{\partial t} &= (\beta p-\mu)\omega,\\
0 &= p\nabla\left(ln\left(\frac{p}{\omega}\right)\right)\cdot\mathbf{n}, \text{ for } (x,t)\in\partial\Omega\times(0,T),\\
p(x,0) &= p_0(x)>0,\\
\omega(x,0) &= \omega_0>0, \text{ for } \in\bar{\Omega}.
\end{aligned}\tag{18}$$

we obtain local in time existence of solutions. However, as we shall see in the next section section, there exist both global and non-global solutions.

Example 5.3 Here we consider the system;

$$\begin{aligned}
\frac{\partial p}{\partial t} &= D\nabla\cdot\left(p\nabla\left(ln\left(\frac{p}{\omega}\right)\right)\right), \text{ for } (x,t)\in\Omega\times(0,T),\\
\frac{\partial \omega}{\partial t} &= F(p,\omega),\\
0 &= p\nabla\left(ln\left(\frac{p}{\omega}\right)\right)\cdot\mathbf{n}, \text{ for } (x,t)\in\partial\Omega\times(0,T),\\
p(x,0) &= p_0(x)>0,\\
\omega(x,0) &= \omega_0>0, \text{ for } \in\bar{\Omega}.
\end{aligned}\tag{19}$$

where

$$(i)\;\; F(p,\omega)=\frac{\lambda p\omega}{k_1+\omega}+\frac{\gamma p}{k_2+p}-\mu\omega,$$

with λ, k_1, k_2, γ, $\mu\ge 0$ are positive constants., or

$$(ii)\;\; F(p,\omega)=\frac{\alpha_1 p}{\alpha_2+\alpha_3 p\omega}+\frac{\alpha_4\omega}{\alpha_5+\alpha_6\omega}-\mu\omega,$$

where α_i $(i = 1, \ldots, 6), \mu \geq 0$ are constants.

For this system we have unique local in time existence of a solution.

Also for any finite time T

$$\min_{(x,t)\in\bar{Q}_T} u(x,t) > 0, \quad \min_{(x,t)\in\bar{Q}_T} \omega(x,t) > 0.$$

By a simple maximum principle argument we can find a constant $M > 0$ such that

$$p(x,t) < M(1 + e^{\mu t}), \;\; \omega(x,t) < M(1 + e^{\mu t}).$$

Again from [27] theorem 3.3 we see that this system has a global solution (p, ω) in the class $H^{2+\beta,1+\beta/2}(Q_\infty) \times H^{2+\beta,1+\beta/2}(Q_\infty)$.

As a final example we consider a problem studied by Rascle [20] which was the inspiration for the ideas used here.

Example 5.4

$$\begin{aligned}
\frac{\partial u}{\partial t} &= \mu\Delta u + \operatorname{div}(u \text{ grad } U), \;\; \text{for } (x,t) \in Q_T, \\
\frac{\partial u}{\partial \mathbf{n}} &= 0, \;\; \text{for } (x,t) \in \Gamma_T, \\
u(x,0) &= u_0(x) > 0, \;\; \text{for } x \in \bar{\Omega}, \\
U(x,t) &= \int_0^t u(x,s)ds, \;\; \text{for } x \in Q_T.
\end{aligned}$$

If we let $\omega = -\frac{1}{\mu}U$, $p = u$, $\Phi(\omega) = e^\omega$ and $F(p,\omega) = -\frac{1}{\mu}p$ then we can write the system in the form;

$$\begin{aligned}
\frac{\partial p}{\partial t} &= \mu\nabla \cdot \left(p\nabla \left(ln \left(\frac{p(x,t)}{\Phi(\omega)}\right)\right)\right), \;\; \text{for } (x,t) \in Q_T, \\
\frac{\partial \omega}{\partial t} &= -\frac{1}{\mu}, \\
0 &= p\nabla \left(ln \left(\frac{p(x,t)}{\Phi(\omega)}\right)\right) \cdot \mathbf{n}, \;\; \text{for } (x,t) \in \Gamma_T, \\
p(x,0) &= p_0(x) > 0, \\
\omega(x,0) &= \omega_0(x) = 0, \\
\Phi(\omega(x,0)) &= 1 > 0, \;\; \text{for} x \in \bar{\Omega}.
\end{aligned} \tag{20}$$

Theorem 5.1 can now be used directly to establish local in time existence of a unique solution.

6 Qualitative behaviour

Although the full model of angiogenesis cannot be solved analytically it is possible to consider the qualitative properties of some of the related models whose existence and uniqueness of solutions was disussed in the previous section.

We begin by considering the following problem related to Example 5.2. That is we consider the problem;

$$\begin{aligned}
\frac{\partial p}{\partial t} &= D\frac{\partial}{\partial x}\left(p\frac{\partial}{\partial x}\left(ln\left(\frac{p}{\Phi(\omega)}\right)\right)\right), \text{ for } (x,t) \in (0,l) \times (0,T), \\
\frac{\partial \omega}{\partial t} &= \lambda p\omega - \mu\omega, \\
0 &= p\frac{\partial}{\partial x}\left(ln\left(\frac{p}{\Phi(\omega)}\right)\right), \text{ for } x = 0, l, \\
p(x,0) &= p_0(x) > 0, \\
\omega(x,0) &= \omega_0(x) \geq 0,
\end{aligned} \tag{21}$$

where

$$\Phi(\omega) = \omega^a, \tag{22}$$

with $a = \pm\, 1$.

This problem has been discussed in considerable depth in [8], [14] [22] and [26] Without loss of generality we can rescale and normalise the system and consider the problem;

$$\begin{aligned}
\frac{\partial p}{\partial t} &= D\frac{\partial}{\partial x}\left(p\frac{\partial}{\partial x}\left(ln\left(\frac{p}{\Phi(\omega)}\right)\right)\right), \text{ for } (x,t) \in (0,l) \times (0,T), \\
\frac{\partial \omega}{\partial t} &= p\omega, \\
0 &= \frac{p_x}{p} - a\frac{\omega_x}{\omega}, \text{ for } x = 0, \pi, \\
p(x,0) &= p_0(x) > 0, \\
\omega(x,0) &= \omega_0(x) \geq 0,.
\end{aligned} \tag{23}$$

If we set

$$\psi = \ln \omega, \tag{24}$$

then $p(x,t) = \psi_t(x,t)$ and

$$\psi_{tt} = \psi_{xxt} - a(\psi_x \psi_t)_x. \tag{25}$$

It turns out that when $a = 1$ equation (25) is of mixed type and be solved by the method of separation of variables resulting in a solution of the form

$$p(x,t) = \psi_t(x,t) = 1 + 2Nc\sum_{n=1}^{\infty} \epsilon^n exp(Nnct) \cos Nnx, \tag{26}$$

where ϵ is an arbitrary parameter and $N,\ c$ are related by the indicial equation;

$$c^2 + Nc - 1 = 0.$$

This solution exists as long as $t < T(\epsilon, N) = -\ln \mid \epsilon \mid /Nc$, for then the series converges absolutely and uniformly on compact subsets of $[0,\pi] \times [0,T]$. In fact on summing the series (26) we find

$$p(x,t) = 1 - 2Nc + 2Nc\left(\frac{1 - \epsilon e^{Nct}\cos Nx}{1 - 2\epsilon e^{Nct}\cos Nx + \epsilon^2 e^{2Nct}}\right), \tag{27}$$

which blows-up at the single point $(x,t) = (x_0, T)$ where $\cos Nx_0 = \pm 1,\ 0 < x_0 \leq \pi$.

In the case $a = -1$ separation of variables can again be used to find an exact solution $p(x,t)$ which decays exponentially to 1 as $t \to \infty$.

In a further development of the above ideas,with $a = 1$, Yang et al. [26] have obtained the class of solutions;

$$\begin{aligned} p(x,t) &= \alpha - \frac{Ak_1e^{k_1t} + Bk_2e^{k_2t} \pm 2c_1\sqrt{AB}e^{c_1t}\cos nx}{Ae^{k_1t} + Be^{k_2t} \pm 2\sqrt{AB}e^{c_1t}\cos nx}, \\ \omega(x,t) &= \frac{e^{\alpha t}}{Ae^{k_1t} + Be^{k_2t} \pm 2\sqrt{AB}e^{c_1t}\cos nx}, \end{aligned}$$

where α, c_1, A, B, are arbitrary constants and $k_{1,2} = c_1 \pm n\sqrt{n - c_1}$.

Theorem 6.1 [26]

(a) If $A > B > 0$ and $c_1 < \alpha - n^2\{\sqrt{A} + \sqrt{B}/\sqrt{A} - \sqrt{B}\}^2$, then $p(x,t)$, $\omega(x,t)$ exists globally.

(b) If $0 < A < B$ then for any $c_1 < \alpha$ there exists $T > 0$ such that $p(x,t)$, $\omega(x,t)$ exists on $0 < t < T$ and blows-up at finite time T at some point $x_0 \in [0,\pi]$.

Using a novel approach based on the theory of harmonic functions due to H.F.Weinberger (see [8])we obtain the class of solutions:

$$\begin{aligned} p(x,t) &= \frac{1}{a} + \frac{1}{a}\frac{N\sinh 2(Nt+\beta)}{\cos^2(Nx-\alpha) + \sinh^2(Nt+\beta)}, \\ \omega(x,t) &= e^{t/\alpha+\beta}\left[cos^2(Nx-\alpha) + \sinh^2(Nt+\beta)\right]^{1/\alpha}. \end{aligned} \quad (28)$$

The corresponding initial conditions being;

$$\begin{aligned} p(x,0) &= \frac{1}{a} + \frac{1}{a}\frac{N\sinh 2\beta}{\cos^2(Nx-\alpha) + \sinh^2\beta}, \\ \omega(x,0) &= e^{\beta}\left[cos^2(Nx-\alpha) + \sinh^2\beta\right]^{1/\alpha}. \end{aligned} \quad (29)$$

The consequences of these solutions are:

Theorem 6.2 [8]

(i) If $a = 1$, $\beta \le 0$ then $p(x,t)$ blows up in the finite time $T = -\beta/N$ at the point $x_0 = 2\alpha + (2m+1)\pi/2N$ where m is an integer such that $x_0 \in (0,\pi)$.Furthermore, at the blow-up point (x_0,T), $\omega(x_0,T) = 0$. If however $\beta > 0$ then $p(x,t)$, $\omega(x,t)$ exist globally. The consequences of this observation is that there is sensitive dependence on the initial data as β passes through $\beta = 0$.

(ii) If $a = -1$, the situation is similar in that if $\beta \le 0$ we have blow-up in both $p(x,t)$ and $\omega(x,t)$ and global existence of solution if $\beta > 0$.

In addition to the search for exact solutions we have investigated in [8] and [22] we have also investigated the structure of aggregating solutions (i.e those which converge to a nonconstant steady state in finite or infinite time) and travelling wave solutions in which Burgers equation plays a central role.

In work related to the system (21) Rascle and Ziti [21] considered the following system in $\mathbb{R}^n$;

$$\begin{aligned} \frac{\partial p}{\partial t} &= D_1\Delta p - \nabla \cdot \left[p\omega^{-\alpha}\nabla\omega\right], \\ \frac{\partial \omega}{\partial t} &= D_2\Delta\omega - k\omega^m p, \end{aligned}$$

in which all the parameters are positive. Rascle and Ziti constructed similarity solutions of the form

$$(p, \omega) = ((T-t)^a P(\xi), (T-t)^b W(\xi)),$$

where $\xi = (T-t)^{-1} \mid x \mid^2$ for $x \in \mathbb{R}^n$ in one , two and three space dimensions , when $0 < m < \alpha = 1$ and $D_2 = 0$. When $D_1 = 0$ as well, they construct such solutions which blow up in finite time in one and two space dimensions. For $D_1 > 0$ they are able to construct only global self-similar solutions.

7 A problem exhibiting spike solutions

Let us now consider a system of equations which may help with the understanding not only of angiogenesis and more generally to chemotaxis. This system has been investigated in [24] and takes the form

$$\begin{aligned}
\frac{\partial p}{\partial t} &= D_1 \nabla \cdot \left(p \nabla \left(\ln \frac{p}{\omega^\alpha} \right) \right), \\
\frac{\partial \omega}{\partial t} &= D_2 \Delta \omega - \mu \omega + \frac{p\omega}{1+\gamma\omega}, \quad \text{in } \Omega \times (0,\infty), \qquad (30) \\
\frac{\partial p}{\partial \nu} &= \frac{\partial \omega}{\partial \nu} = 0, \quad \text{on } \partial\Omega \times (0,\infty), \qquad (31) \\
p(x,0) &= p(x_0) \geq 0, \;\; \omega(x,0) = \omega_0(x) \geq 0,
\end{aligned}$$

where D_1, D_2 are positive constants, $\Omega \subset \mathbb{R}^n$, $(n \leq 3)$ is a smooth bounded domain, $\mu\ \gamma\ \alpha > 1$ are also positive constants and $\nu = \nu(x)$ is the unit normal at $x \in \partial\Omega$.

The object here is to show that when the diffusion coefficient D_2 is small then the system (30), (31) has stable *spike* patterns. The relevance of this to angiogenesis may come from the following experimental observations.

Recently Holash et al. [5] have demonstrated that once a tumour has become vascularized, the resulting capillary network may undergo periods of dramatic collapse and remodelling. Paradoxically, the coopted vasculature does not undergo angiogenesis to support the growing tumor, but instead regresses via a process that involves disruption of EC and smooth muscle cell interactions and EC apoptosis. This vessel regression in turn results in necrosis within the central part of the tumour. However vigorous angiogenesis is initiated at the boundary, rescuing the surviving tumour and supporting further growth. We suggest that this behaviour is indicative of the existence of point condensation solutions or spike-type patterns which could be modelled by systems of the type (30), (31).

Without loss of generality we take $\mu = 1$, $D_1 = 1$, $D_2 = \epsilon^2 \ll 1$ and consider the system;

$$\begin{aligned}
\frac{\partial p}{\partial t} &= \nabla \cdot \left(p \nabla \left(\ln \frac{p}{\omega^\alpha} \right) \right), \\
\frac{\partial \omega}{\partial t} &= \epsilon^2 \Delta \omega - \omega + \frac{p\omega}{1+\gamma\omega}, \quad \text{in } \Omega \times (0,\infty), \qquad (32) \\
\frac{\partial p}{\partial \nu} &= \frac{\partial \omega}{\partial \nu} = 0, \quad \text{on } \partial\Omega \times (0,\infty), \qquad (33) \\
p(x,0) &= p(x_0) \geq 0, \;\; \omega(x,0) = \omega_0(x) \geq 0,
\end{aligned}$$

Integrating the equation for p over Ω and using the divergence theorem we find

$$\int_\Omega p(x,t) = \int_\Omega p(x,0) = m.$$

To simplify the computations we assume $m = 1$.

The steady state problem associated with (32),(33) is;

$$\begin{aligned} \nabla \cdot \left(p\nabla \left(\ln \frac{p}{\omega^\alpha}\right)\right) &= 0, \\ \epsilon^2 \Delta\omega - \omega + \frac{p\omega}{1+\gamma\omega} &= 0, \quad \text{in } \Omega \times (0,\infty), \\ \frac{\partial p}{\partial \nu} = \frac{\partial \omega}{\partial \nu} &= 0, \quad \text{on } \partial\Omega \times (0,\infty), \\ \int_\Omega p(x) &= 1. \end{aligned} \tag{34}$$

From the equation for p we get

$$p(x) = \frac{1}{\int_\Omega \omega^\alpha} \omega^\alpha(x). \tag{35}$$

Now define the normalizing function $\hat{\omega}$ by

$$\omega(x) = \frac{1}{\gamma \int_\Omega \omega^\alpha} \hat{\omega(x)}. \tag{36}$$

Substituting in (34) results in the the problem

$$\begin{aligned} \epsilon^2 \Delta\hat{\omega} - \hat{\omega} + \frac{\hat{\omega}^{\alpha+1}}{\int_\Omega \hat{\omega}^\alpha + \hat{\omega}} &= 0, \quad \text{in } \Omega, \\ \hat{\omega} &> 0, \text{ in } \Omega, \\ \frac{\partial \hat{\omega}}{\partial \nu} &= 0 \text{ on } \partial\Omega. \end{aligned} \tag{37}$$

We begin with statements on the existence of steady state solutions.

Theorem 7.1 [24]
Assume that

$$1 < \alpha < +\infty, \ \textit{if}\, n = 1,2; \ \ 1 < \alpha < 5 \ \textit{if}\, n = 3.$$

Then, for $\epsilon \ll 1$, there exists a steady state solution of the form:

$$(p_\epsilon,\ \omega_\epsilon) = \left(\frac{\hat{\omega}_\epsilon^\alpha}{\int_\Omega \hat{\omega}_\epsilon^\alpha}, \frac{\hat{\omega}_\epsilon}{\gamma \int_\Omega \hat{\omega}_\epsilon^\alpha}\right), \ \textit{where}\ \hat{\omega}_\epsilon = \omega\left(\frac{x - Q_\epsilon}{\epsilon}\right) + O(\epsilon). \tag{38}$$

Here $\omega(y)$ is the unique solution of

$$\Delta\omega - \omega + \omega^\alpha = 0, \ \ \omega > 0 \ \textit{in}\ \mathbb{R}^n, \ \ \omega(0) = \max_{y\in\mathbb{R}^n} \omega(y), \ \ y \to 0 \ \textit{as}\ |y| \to \infty. \tag{39}$$

The point Q_ϵ is classified either by

(a) (single boundary spike) $Q_\epsilon \in \partial\Omega$, $H(Q_\epsilon) \to \max_{Q\in\partial\Omega} H(Q)$, where $H(Q)$ is the mean curvature at $Q \in \partial\Omega$, or

(b) (single interior spike) $Q_\epsilon \in \Omega$, $d(Q_\epsilon, \partial\Omega) \to \max_{Q\in\Omega} d(Q,\partial\Omega)$, where $d(Q,\partial\Omega)$ is the distance function at $Q \in \Omega$.

To study the linearized stability of the solution (38) we linearize (32), (33) about $(p_\epsilon,\ \omega_\epsilon)$, to obtain the eigenvalue problem:

$$\nabla \cdot \left(\psi_\epsilon \nabla \ln\left(\frac{p_\epsilon}{\omega_\epsilon^\alpha}\right)\right) + \nabla \cdot \left(p_\epsilon \nabla\left(\frac{\psi_\epsilon}{p_\epsilon} - \alpha\frac{\phi_\epsilon}{\omega_\epsilon}\right)\right) = \lambda_\epsilon \psi_\epsilon \ \text{ in } \Omega, \tag{40}$$

$$\epsilon^2 \Delta\phi_\epsilon - \phi_\epsilon + \frac{p_\epsilon}{(1+\gamma\omega_\epsilon)^2}\phi_\epsilon + \frac{\omega_\epsilon}{(1+\gamma\omega_\epsilon)}\psi_\epsilon = \lambda_\epsilon\phi_\epsilon \ \text{ in } \Omega. \tag{41}$$

$$\frac{\partial\phi_\epsilon}{\partial\nu} = \frac{\partial\psi_\epsilon}{\partial\nu} = 0 \ \text{ on } \partial\Omega, \tag{42}$$

where $\lambda_\epsilon \in \mathcal{C}$.

We say that $(p_\epsilon,\ \omega_\epsilon)$, is linearly stable if for all eigenvalues λ_ϵ we have $Re(\lambda_\epsilon) < 0$. $(p_\epsilon,\ \omega_\epsilon)$, is linearly unstable if there exists an eigenvalue λ_ϵ such that $Re(\lambda_\epsilon) > 0$. We say that $(p_\epsilon,\ \omega_\epsilon)$, is metastable if for all eigenvalues λ_ϵ we have $Re(\lambda_\epsilon) < 0$ or $\mid \lambda_\epsilon \mid = O(e^{-d/\epsilon})$ for some $d > 0$ independent of $\epsilon > 0$.

Theorem 7.2 [24]
Assume that

$$1 < \alpha < +\infty,\ if\, n = 1;\ \ 2 < \alpha \leq 5\ if\, n = 2;\ \ 2 < \alpha \leq 3\ if\, n = 3.$$

Let $(p_\epsilon,\ \omega_\epsilon)$ be the solution given in Theorem 7.2.

(a) (metastability) the single interior spike is metastable.

(b) (stability) If $n = 1$, then the single boundary spike is linearly stable.

(c) (stability) If $\Omega = B_R(0) = x| \mid x \mid < R$ and $(p(x,t),\ \omega(x,t)) = (p(\mid x \mid, t),\ \omega(\mid x \mid, t))$, then the single interior spike is linearly stable.

(d) (stability) If $n = 2, 3$ and Q_0 is a nondegenerate global maximum point of $H(Q)$, where $Q_\epsilon \to Q_0$, then the single boundary spike is linearly stable.

We note that in both Theorems 7.1 and 7.2 we have assumed that $D_1 = 1$. Theorem 7.1 holds for any $D_1 > 0$. While Theorem 7.2 holds provided $\frac{\epsilon^2}{D_1} \ll 1$.

Biologically this means that if the substrate ω diffuses more slowly than p, then p will move toward the boundary and form nontrivial stable spike patterns. This is the basis for our conjecture underlying the observations of Holash et al. [5] in angiogenesis.

8 Other aspects of angiogenesis and future directions

Two members of the recently discovered angiopoietin family, Ang-1 [1] and Ang-2 [13] have been found to be important regulators of angiogenesis. In particular, they are key players in the angiogenic balance between quiescence and activation of the endothelium.

The angiopoietins are ligands for the EC-specific receptor tyrosine kinase, Tie-2. Ang-1 is widely expressed throughout tissue and is thought to play a stabilizing role, maintaining cell-cell interactions, inhibiting apoptosis and mediating interactions between the EC and the basement membrane. In the vascular endothelium, Ang-2 is a natural antagonist for Ang-1: it binds to the Tie-2 receptor, but does not activate it, thus blocking the normal effects of Ang-1. In the presence of Ang-2 vessels become destabilised: cell-cell and cell-matrix connections are loosened and both the basement membrane and the peri-endothelial support cells become disassociated from the endothelium. In [18], [19] a mathematical model, based on the known biochemistry, of the description of the role of the angiopoietins in angiogenesis is developed. Model predictions are in qualitative agreement with experimental observations and may have implications for anti-angiogenic cancer therapies.

An aspect of tumour angiogenesis not considered in this paper is blood flow in the neo-vasclular network. This is important to the understanding of chemotherapy. In order to do this we must first understand the mechanisms underlying anastomoses (the formation of closed loops in the vascular network). Also in view of the size of the growing capillaries it is not clear as to whether blood should be considered as a non-newtonian fluid or as a particulate. In [12] and [25] an investigation has been initiated wherein blood flow in vascular networks has been simulated by adapting modelling techniques drawn from the field of petroleum engineering. In these works the authors have used this idea to investigate the conditons underwhich a chemotherapeutic drug can be delivered to a solid tumour in an optimal manner.

Another fundamental problem not discussed here is to understand the mechanisms underlying cellular,in our case EC, locomotion. Bacterial organisms may use appendages such as flagellae or cilia to facilitate motion. Amoeboid motion, exhibited by eucaryotic cells are seen to flatten onto surfaces and extend thin sheets of cytosol called lamellipodia. These in turn make attachments to the surface and by initiation of internal contractions within the cell, a forward motion is achieved. The processes which govern this behaviour are extremely complex. In [4] we have singled out osmotic/hydrostatic expansion and cellular contraction mediated by intracellular calcium as the fundamental factors involved and have modelled the motion of a non-muscle motile cell based on these factors. The mathematical model uses ideas from visco-elasticity theory and develops a dynamic finite element method to simulate motion of an EC subject to a chemoattractant.

It is the hope that the study and development of mathematical models of tumour angiogenesis and other aspects of tumour growth and spread will lead to suggesting workable and robust criteria of help to clinicians and surgeons in their search for the most efficacious treatment of patients.

Mathematically we have seen that the modelling processes used have led to highly non-linear coupled partial differential equations and dynamical sytems of interest in their own right. As a consequence there are a host of novel problems arising for the mathematical analyst as well as challenging computational and simulation problems.

Acknowledgements. The research reported in this paper was supported by grants from the EPSRC of Great Britain; The Royal Society of London; NSF of the United States of America and the NNSF of China. The author is indebted to Pamela.Jones (Molecular Medicine Unit, St. James's University Hospital, Leeds)for her collaboration and patience in answering my innumerable questions and also for enhancing Figure 1 computed from [17]. Thanks also to my colleague Carmen. Molina-Paris who has helped with the vagaries of LaTeX.

References

[1] Davis, S., Aldrich, T.H.,Jones, P.F., Acheson, A., Compton, D.L., Jain, V., Ryan, T.E., Bruno, J., Radziejewski, C., Maisonpierre, P.C.and Yancopoulos, G.D (1996), Isolation of angiopoietin-1, a ligand for the Tie-2 receptor, by secretion-trap expression cloning. *Cell*, **87**, 1161-1169.

[2] Folkman, J.(1971), Tumour angiogenesis: therapeutic implications. *N. Engl. J. Med.*, **285**, 1182-118.

[3] Hill, N.A and Häder, D.P. (1997), A biased random walk model for the trajectories of swimming micro-organisms. *J. Theor. Biol.*, **186**, 503-526.

[4] Holmes, M.J. and Sleeman, B.D.(2001), A Mathematical Model for Cellular Locomotion Exhibiting Chemotaxis. *J. Theor. Med.,* **3**, 101-123.

[5] Holash, J.,Maisonpierre, P.C., Compton, D., Boland, P.,Alexander, C.R., Zagzag, D., Yancopoulos, G.D. and Wiegand. S.J. (1999), Vessel cooption, regression and growth in tumors mediated by angiopoietins and VEGF. *Science*, **284**, 1994-1998.

[6] Kendall, R.L., Rutledge, R.Z., Mao, X., Tebben, A.J., Hungate, R.W and Thomas, K.A. (1999), Vascular endothelial growth factor receptor kdr tyrosine kinase activity is increased by autophosphorylation of two activation loop tyrosine residues. *J. Biol. Chem.*, **274**, 6453-6460.

[7] Ladyzenskaya, O.A., Solonnikov, V.A. and Ural'ceva, N.N.(1968), Linear and Quasi-linear Equations of Parabolic Type, American Mathematical Society Translations, **23**, American Mathematical Society, Providence, RI.

[8] Levine, H.A. and Sleeman., B.D. (1997), A system of reaction diffusion equations arising in the theory of reinforced random walks. *SIAM. J. Appl. Math.*, **57**, 683-730.

[9] Levine, H.A., Sleeman, B.D and Nilsen-Hamilton, M.(2000), A mathematical model of the roles of pericytes and macrophages in the initiation of angiogenesis: I, the role of protease inhibitors in preventing angiogenesis. *Math. Biosciences,* **168**, 77-115.

[10] Levine, H.A., Sleeman, B.D and Nilsen-Hamilton, M.(2001), Mathematical modelling of the onset of capillary formation initiating angiogenesis. *J. Math. Biol.,* **42**, 195-238.

[11] Levine, H.A., Pamuk, S., Sleeman, B.D and Nilsen-Hamilton, M.(2001), A mathematical model of capillary formation and development in tumour angiogenesis: penetration into the stroma. *Bull. Math. Biol.,* **63**, 801-863.

[12] McDougall, S.R., Anderson, A.R.A., Chaplain, M.A.J. and Sherratt, J.A. (2002), Mathematical modelling of flow through vascular networks: Implications for tumour-induced angiogenesis and chemotherapy stategies. *Bull. Math. Biol.,* **64**, 673-702.

[13] Maisonpierre, P.C., Suri, C., Jones, P.F., Bartunkova, S., Wiegand, S.J., Radziejewski, C., Compton,D., McClain, J., Aldrich, T.H., Papandopoulos, N., Daly, S. and Sato, T.N (1997), Angiopoietin-2, a natural antagonist for Tie-2 that disrupts *in vivo* angiogenesis. *Science,* **277**, 55-60.

[14] Othmer, H.G and Stevens, A. (1997), Aggregation, blow-up and collapse: the ABC's of taxis and reinforced random walks. *SIAM. J. Appl. Math.*, **57**, 1044-1081.

[15] Plank, M.J and Sleeman, B.D.(2003), Tumour-induced Angiogenesis: A Review. *J. Theor. Med.,* **5**, 137-153.

[16] Plank, M.J and Sleeman, B.D.(2003), A reinforced random walk model of tumour angiogenesis and anti-angiogenic strategies. *IMA. J. Math. Med and Biol.*, **20**, 135-181.

[17] Plank, M.J and Sleeman, B.D.(2004), Lattice and Non-Lattice Models of Tumour Angiogenesis. *Bull. Math. Biol.,* **66**, 1785-1819.

[18] Plank, M.J.,Sleeman, B.D. and Jones, P.F(2004), The Role of the Angiopoietins in Tumour Angiogenesis. *Growth Factors,* **22**, 1-11.

[19] Plank, M.J., Sleeman, B.D. and Jones, P.F.(2004), A mathematical model of tumour angiogenesis, regulated by vascular endothelial growth factor and the angiopoietins. *J. Theor. Biol.,* **229**, 435-454.

[20] Rascle, M. (1979), Sur une équation integro-differentielle non lineaire issue de la biologie. *J. Diff. Equations,* **32**, 420-453.

[21] Rascle, M and Ziti, C.(1995), Finite time blow-up in some models of chemotaxis. *J. Math. Biol.,* **33**, 388-414.

[22] Sleeman, B.D and Levine, H.A. (2001), Partial differential equations of chemotaxis and angiogenesis.*Math. Meth. Appl. Sci.,* **24**, 405-426.

[23] Sleeman, B.D and Wallis, I.P.(2002), Tumour induced angiogenesis as a reinforced random walk: modelling capillary network formation without endothelial cell proliferation. *J. Math. Comp. Modelling,* **36**, 339-358.

[24] Sleeman, B.D., Ward, M.J. and Wei, J.C.(2005), The existence and stability of spike patterns in a chemotaxis model. *SIAM. J. Appl. Math.*,**65**, 790-817.

[25] Stéphanou, A., McDougall, S.R., Anderson, A.R.A. and Chaplain, M.A.J. (2005), Mathmatical modelling of Flow in 2D and 3D Vascular networks: Applications to Anti-Angiogenic and Chemotherapeutic Drug Strategies. *Math. Comp. Modelling.,* **41**, 1137-1156.

[26] Yang, Y., Chen, H and Liu, W. (2001), On existence of global solutions and blow-up to a system of reaction-diffusion equations modeling chemotaxis. *SIAM. J. Math. Anal.*, **33**,736-785.

[27] Yang, Y., Chen, H., Liu, W. and Sleeman.B.D. (2005), The solvability of some chemotaxis systems. *J.Diff. Equations,* **212**, 432-451.

Fields Institute Communications
Volume **48**, 2006

Interpretation of the Generalized Asymmetric May-Leonard Model of Three Species Competition as a Food Web in a Chemostat

Gail S.K. Wolkowicz
Department of Mathematics and Statistics
McMaster University
1280 Main St West
Hamilton, Ontario, Canada L8S 4K1
wolkowic@mcmaster.ca

Abstract. Consider a simple model of a food web in a chemostat involving three species competing for a single, nonreproducing, growth-limiting nutrient in which one of the competitors also predates on one of the other competitors. If it is assumed that the response functions satisfy the law of mass action, it is shown that under certain assumptions on the parameters, this model is equivalent to a special case of the generalized asymmetric May-Leonard (Lotka-Volterra) model of three species competition or to a Lotka-Volterra model in which two of the species compete, and two are involved in predator-prey interaction. In both cases there is a repelling heteroclinic cycle connecting the three single species boundary equilibria, and a positive three species coexistence equilibrium that is globally asymptotically stable with respect to the interior of the positive cone.

1 Introduction

Consider the Gause-Lotka-Volterra model of three species competition:

$$\begin{aligned} x_1'(t) &= r_1x_1(t)(1-x_1(t)-\alpha_1x_2(t)-\beta_1x_3(t)), \\ x_2'(t) &= r_2x_2(t)(1-\beta_2x_1(t)-x_2(t)-\alpha_2x_3(t)), \\ x_3'(t) &= r_3x_3(t)(1-\alpha_3x_1(t)-\beta_3x_2(t)-x_3(t)), \\ &x_1(0)>0,\ x_2(0)>0,\ x_3(0)>0, \end{aligned} \tag{1.1}$$

2000 *Mathematics Subject Classification.* Primary 34D23; Secondary 92D25.
Key words and phrases. Lotka-Volterra, competition, predator-prey, chemostat, global asymptotic stability, heteroclinic cycle, rock-paper-scissors game, food web.
The work of this author was partially supported by the Natural Sciences and Engineering Research Council of Canada.

where r_i, α_i and β_i $i = 1, 2, 3$, are all positive constants. Under the additional assumption that

$$0 < \alpha_i < 1 < \beta_i, \quad i = 1, 2, 3, \tag{1.2}$$

we shall refer to this model as the *generalized* asymmetric May-Leonard model (GAML). In the case that $r_i = r$, $i = 1, 2, 3$, the model is referred to as the asymmetric May-Leonard model (AML) or the rock-paper-scissors game.

In model (1.1), t denotes time, and x_i, $i = 1, 2, 3$, denote some measure of the size of the ith competitor population at time t, e.g., density or concentration. The r_i, $i = 1, 2, 3$, denote the intrinsic growth rates of each population and the α_i and β_i, $i = 1, 2, 3$, denote the competition coefficients. Condition (1.2) implies that there exists a heterclinic cycle connecting the single species equilibria on the boundary. In particular, let e_i, $i = 1, 2, 3$, denote the single species equilibrium on each x_i axis, with only species x_i present. Therefore, $e_1 = (1, 0, 0)$, $e_2 = (0, 1, 0)$, and $e_3 = (0, 0, 1)$. From results on two species competition (see for example Waltman [10]), it follows that (1.2) ensures that there is a heteroclinic orbit $\mathcal{O}_3$ on the the $x_1 - x_2$ plane from e_2 to e_1, a heteroclinic orbit $\mathcal{O}_2$ on the $x_1 - x_3$ plane from e_1 to e_3, and a heteroclinic orbit $\mathcal{O}_1$ on the the $x_2 - x_3$ plane from e_3 to e_2. Define

$$\bar{\mathcal{O}} \triangleq \text{ closure of } \cup_{i=1}^{3} \mathcal{O}_i. \tag{1.3}$$

That model (1.1) admits the heteroclinc cycle, $\bar{\mathcal{O}}$, when (1.2) holds was first proved by May and Leonard [7] in the so-called symmetric case, i.e., $\alpha_i = \alpha$, $\beta_i = \beta$, $i = 1, 2, 3$. In fact, they argued that in the case that $\alpha + \beta > 2$ and $0 < \alpha < 1 < \beta$, $\bar{\mathcal{O}}$ attracts all solutions with positive initial conditions, except the unique interior equilibrium point $P = \frac{1}{1+\alpha+\beta}(1, 1, 1)$ and its one dimensional stable manifold. In particular, they argued that and provided collaborating numerical simulations to show that asymptotically, solutions move in population space from a neigbourhood of e_1, to a neighbourhood of e_3, to a neigbourhood of e_2, back toward e_1, and so on, that the time spent in the vicinity of any one point is proportional to the total time elapsed up to that state, and that the total time spent in completing one cycle is proportional to the total length of time the system has been running.

Schuster, Sigmund and Wolf [8] considered the (AML) model and proved that if in addition to (1.2), one assumes that

$$\beta_i - 1 > 1 - \alpha_j, \quad 1 \leq i, j \leq 3, \tag{1.4}$$

then there exists an open set of orbits in the interior of $\mathbb{R}^3_+$ having $\bar{\mathcal{O}}_i$ as ω limit set. Hofbauer and Sigmund [5] provided conditions that allow one to decide whether the (GAML) model is permanent or the heteroclinic cycle is an attractor. In particular, defining $A_i = 1 - \alpha_i$ and $B_i = \beta_i - 1$ for $i = 1, 2, 3$, they proved that the (GAML) model is permanent if $A_1A_2A_3 > B_1B_2B_3$, but that the heteroclinic cycle is an attractor if $A_1A_2A_3 < B_1B_2B_3$. In Hofbauer and Sigmund ([6], section 15.3), the Volterra-Liapunov Theorem provides sufficient conditions for the global asymptotic stability of the interior equilibrium. They also proved in [5] that if one restricts $r_i = r$, $i = 1, 2, 3$, to obtain the (AML) model, there are no periodic orbits except in the case that $A_1A_2A_3 = B_1B_2B_3$, and that in this case there is a center, i.e., all orbits are periodic. Therefore, for the (AML) model either $A_1A_2A_3 > B_1B_2B_3$ and the interior equilibrium is globally asymptotically stable with respect to the interior of the positive cone, or $A_1A_2A_3 < B_1B_2B_3$ and the heteroclinic cycle on the boundary attracts almost all orbits, or $A_1A_2A_3 = B_1B_2B_3$, and there is a center. Chi, Hsu and Wu [3] also proved this result for the (AML) model using a different method.

Zeeman and Zeeman [13] proved that for competitive Lotka-Volterra systems (1.1), if an interior equilibrium P exists, and the carrying simplex of the system lies to one side of its tangent hyperplane at P, then there is no nontrivial recurrence and so the global dynamics are known. They also gave algebraic criteria for verifying this geometric condition and provided a computational algorithm.

Next, consider the following model of a food web in a chemostat:

$$\begin{aligned}
S'(t) &= (S^0 - S(t))D - \frac{x_1(t)p_1(S(t))}{\eta_1} - \frac{x_2(t)p_2(S(t))}{\eta_2} - \frac{x_3(t)p_3(S(t))}{\eta_3}, \\
x_1'(t) &= x_1(t)\left(-D + p_1(S(t)) - x_3(t)\frac{q(x_1(t))}{z}\right), \\
x_2'(t) &= x_2(t)\left(-D + p_2(S(t))\right), \\
x_3'(t) &= x_3(t)\left(-D + p_3(S(t)) + q(x_1(t))\right), \\
&\quad S(0) \geq 0,\ x_1(0) > 0,\ x_2(0) > 0,\ x_3(0) > 0, \\
&\quad S^0 > 0, D > 0, \eta_i > 0,\ i = 1,2,3, \text{ and } z > 0.
\end{aligned} \tag{1.5}$$

As for model (1.1), in model (1.5), t denotes time, and x_i, $i = 1,2,3$, denote some measure of the size of the ith competitor population (in the culture vessel) at time t. However, in model (1.5), population $x_3(t)$ is also a predator, predating on population $x_1(t)$. Here $S(t)$ denotes the concentration of the nutrient in the culture vessel at time t. Parameter S^0 denotes the concentration of the nutrient in the feed vessel and D denotes the dilution rate. The species specific death rates are assumed to be insignificant compared to the dilution rate and are ignored. The culture vessel is assumed to be well-stirred and for convenience its volume is assumed to be one cubic unit. The functional response for each population $x_i(t)$ is assumed to satisfy the law of mass action and so we define $p_i(S) = m_i S$, $i = 1,2,3$, and $q(x_1) = nx_1$. In addition we assume that growth rate is proportional to the consumption rate and so the consumption rate of nutrient $S(t)$ by population i is given by $\frac{p_i(S(t))}{\eta_i}$ and the consumption rate of $x_1(t)$ by $x_3(t)$ is given by $\frac{q(x_1(t))}{z}$. The positive constants η_i, $i = 1,2,3$, and z are referred to as yield constants.

Model (1.5) is a special case of a more general model first studied in Daoussis [4], where a global analysis was given. In Wolkowicz, Ballyk, and Daoussis [11], this model was provided as an example of competitor-mediated competition, i.e. a scenario in which introduction of a population that exploits common resources promotes *greater* diversity.

In the next section we perform a series of substitutions and transformations on model (1.5) to show that under certain conditions it is equivalent to model (1.1)-(1.2). Hence, under these conditions, we provide a new interpretation of the May-Leonard three species competition model as a simple food web in a chemostat involving three species competing for a single, nonreproducing, growth-limiting nutrient in which one of the competitors also predates on one of the other competitors. The global dynamics of model (2.1) are completely understood. We summarize this in Section 3. However, the global dynamics of the (GAML) model are still not completely understood. Therefore, it is still useful to find new criteria that guarantee that the interior equilibrium P of (GAML) is globally asymptotically stable. We do this in Section 4. We also provide an example for which the criteria introduced here can be used to show that P is globally asymptotically stable, but both the Computational Theorem in [13] and the Volterra-Liapunov Theorem of [6] are inconclusive.

2 Transforming (1.5) into (1.1)-(1.2)

In this section we perform a series of substitutions and transformations that convert model (1.5) into a model of the form (1.1)-(1.2).

First let

$$\bar{t} = tD; \quad \bar{S}(\bar{t}) = \frac{S(t)}{S^0}; \quad \bar{x}_i(\bar{t}) = \frac{x_i(t)}{\eta_i S^0}, \ i = 1, 2; \quad \bar{x}_3(\bar{t}) = \frac{x_3(t)}{\eta_1 S^0 z};$$

$$\bar{p}_i(\bar{S}(\bar{t})) = \frac{p_i(S(t))}{D}, \ i = 1, 2, 3; \quad \bar{q}(\bar{x}_1(\bar{t})) = \frac{q(x_1(t))}{D}; \quad \gamma = \frac{\eta_3}{\eta_1 z};$$

$$\lambda_i = \frac{D}{m_i}, \ i = 1, 2, 3; \quad \delta = \frac{D}{n}.$$

and assume that $\gamma = 1$.

Then, omitting the bars to simplify notation, the scaled version of model (1.5) can be written as follows:

$$\begin{aligned} S'(t) &= (1 - S(t)) - x_1(t)\frac{S(t)}{\lambda_1} - x_2(t)\frac{S(t)}{\lambda_2} - x_3(t)\frac{S(t)}{\lambda_3}, \\ x_1'(t) &= x_1(t)\left(-1 + \frac{S(t)}{\lambda_1} - \frac{x_3(t)}{\delta}\right), \\ x_2'(t) &= x_2(t)\left(-1 + \frac{S(t)}{\lambda_2}\right), \\ x_3'(t) &= x_3(t)\left(-1 + \frac{S(t)}{\lambda_3} + \frac{x_1(t)}{\delta}\right), \\ & S(0) \geq 0, \ x_1(0) > 0, \ x_2(0) > 0, \ x_3(0) > 0. \end{aligned} \tag{2.1}$$

Note that λ_i, $i = 1, 2, 3$, and δ are called the *break-even concentrations* of nutrient and prey, respectively.

Adding the four equations in (2.1), it follows that

$$\left(S'(t) + \sum_{i=1}^{3} x_i'(t)\right) = 1 - \left(S(t) + \sum_{i=1}^{3} x_i(t)\right).$$

Therefore,

$$\left(S(t) + \sum_{i=1}^{3} x_i(t)\right) = e^{-t}\left(-1 + S(0) + \sum_{i=1}^{3} x_i(0)\right) + 1.$$

It is clear that for model (2.1), the positive cone is positively invariant, and so it follows that the simplex

$$\mathcal{S} \triangleq \{(S, x_1, x_2, x_3) : \ S + \sum_{i=1}^{3} x_i = 1, \ x_i \geq 0, \ i = 1, 2, 3, \}$$

is globally attracting.

Remark 2.1 It is useful to note that the restriction that the initial concentration of the nutrient in model (2.1) must be nonnegative is only imposed for biological realism, and does not affect the asymptotic outcome of the solutions. If $S(0) < 0$ but $x_i(0) > 0,\ i = 1,2,3$, then it still follows that $x_i(t) > 0,\ i = 1,2,3$ for all positive time, and so if $S(0) < 0$, then $S(t)$ increases, $x_i(t),\ i = 1,2$ decrease, and $x_3(t)$ eventually decreases until at some finite time $T > 0,\ \ S(T) = 0$. Then $S(t)$ will be positive for all $t > T$.

Setting $S(t) = 1 - \sum_{i=1}^{3} x_i(t)$, we can eliminate the S' equation in (2.1) to obtain:

$$\begin{aligned}
x_1'(t) &= x_1(t)\left(\frac{1-\lambda_1}{\lambda_1} - \frac{x_1(t)}{\lambda_1} - \frac{x_2(t)}{\lambda_1} - \left(\frac{1}{\lambda_1} + \frac{1}{\delta}\right)x_3(t)\right),\\
x_2'(t) &= x_2(t)\left(\frac{1-\lambda_2}{\lambda_2} - \frac{x_1(t)}{\lambda_2} - \frac{x_2(t)}{\lambda_2} - \frac{x_3(t)}{\lambda_2}\right),\\
x_3'(t) &= x_3(t)\left(\frac{1-\lambda_3}{\lambda_3} - \left(\frac{1}{\lambda_3} - \frac{1}{\delta}\right)x_1(t) - \frac{x_2(t)}{\lambda_3} - \frac{x_3(t)}{\lambda_3}\right),
\end{aligned} \tag{2.2}$$
$$x_1(0) > 0,\ x_2(0) > 0,\ x_3(0) > 0,\ S(t) = 1 - \sum_{i=1}^{3} x_i(t)$$

where, it follows from Remark 2.1 that there is no restriction on the sign of $S(t)$, and hence it is not necessary to assume that $\sum_{i=1}^{3} x_i(0) \le 1$. Even if $\sum_{i=1}^{3} x_i(0) > 1$, there exists $T > 0$ such that $\sum_{i=1}^{3} x_i(T) = 1$ and $\sum_{i=1}^{3} x_i(t) < 1$ for all $t > T$.

In order to obtain the same form as (1.1), we let

$$\hat{x}_i = \frac{x_i}{1-\lambda_i}, \quad i = 1,2,3,$$

Omitting the hats for convenience of notation, and factoring

$$r_i \triangleq \frac{1-\lambda_i}{\lambda_i}, \quad i = 1,2,3, \tag{2.3}$$

from the ith equation, the model can be rewritten:

$$\begin{aligned}
x_1'(t) &= r_1x_1(t)\left(1 - x_1(t) - \frac{1-\lambda_2}{1-\lambda_1}x_2(t) - \frac{(1-\lambda_3)(\lambda_1+\delta)}{\delta(1-\lambda_1)}x_3(t)\right),\\
x_2'(t) &= r_2x_2(t)\left(1 - \frac{1-\lambda_1}{1-\lambda_2}x_1(t) - x_2(t) - \frac{1-\lambda_3}{1-\lambda_2}x_3(t)\right),\\
x_3'(t) &= r_3x_3(t)\left(1 - \frac{(1-\lambda_1)(\delta-\lambda_3)}{\delta(1-\lambda_3)}x_1(t) - \frac{1-\lambda_2}{1-\lambda_3}x_2(t) - x_3(t)\right),
\end{aligned} \tag{2.4}$$
$$x_1(0) > 0,\ x_2(0) > 0,\ x_3(0) > 0,\ S(t) = 1 - \sum_{i=1}^{3}(1-\lambda_i)x_i(t).$$

Again we emphasize that there is no restriction on the sign of $S(0)$. This is a classical Lotka-Volterra model. For our analogy, we require more assumptions on the parameters in order to control the sign and relative magnitudes of the coefficients. Assume that the species are labelled so that

$$0 < \lambda_1 < \lambda_2 < \lambda_3 < 1. \tag{2.5}$$

Under this assumption, Butler and Wolkowicz, [1] proved that if x_3 does not consume x_1 (i.e., $n = 0$ or equivalently $\delta = \infty$), but instead consumes only S, then x_1 would be the sole survivor in a contest against x_2 or against both x_2 and x_3 and in the absence of x_1, x_2 would survive and drive x_3 to extinction. In this sense x_1 is the *strongest* competitor for resource S, and x_3 is the *weakest* competitor for resource S.

It follows from (2.5) that $r_i > 0,\ i = 1, 2, 3$, and so it makes sense to interpret each r_i as the intrinsic growth rate of the ith species.

If we now allow x_3 to consume both S and x_1, and in addition we assume that

$$\delta > \lambda_3, \tag{2.6}$$

so that $\alpha_3 > 0$, then model (2.2) has been transformed into the form of model (1.1), the Gause-Lotka-Volterra model of three species competition, with

$$\alpha_1 = \frac{1-\lambda_2}{1-\lambda_1}, \quad \alpha_2 = \frac{1-\lambda_3}{1-\lambda_2}, \quad \alpha_3 = \frac{(1-\lambda_1)(\delta-\lambda_3)}{\delta(1-\lambda_3)}, \tag{2.7}$$

$$\beta_1 = \frac{(1-\lambda_3)(\lambda_1+\delta)}{\delta(1-\lambda_1)}, \quad \beta_2 = \frac{1-\lambda_1}{1-\lambda_2}, \quad \beta_3 = \frac{1-\lambda_2}{1-\lambda_3}, \tag{2.8}$$

where $\alpha_i > 0$ and $\beta_i > 0,\ i = 1, 2, 3$.

By (2.5), it is clear that $\alpha_i < 1,\ i = 1, 2$, and $\beta_i > 1, i = 2, 3$. However, $\beta_1 > 1$, if, and only if, we also assume that

$$0 < \delta < \frac{\lambda_1(1-\lambda_3)}{\lambda_3-\lambda_1}, \tag{2.9}$$

and $\alpha_3 < 1$, if, and only if, in addition to (2.5), we assume that

$$0 < \delta < \frac{\lambda_3(1-\lambda_1)}{\lambda_3-\lambda_1}. \tag{2.10}$$

Note that if (2.5) holds, then (2.9) implies (2.10).

Therefore, model (2.4) is in the form of model (1.1). If (2.5)-(2.9) hold, then (1.2) also holds and there is a heteroclinic cycle on the boundary, connecting the three singles species equilibria, e_1, e_2, and e_3. Thus we have shown that if $\gamma = \frac{\eta_3}{\eta_1 z} = 1$, then we have transformed model (1.5), a model of three species competition in a chemostat for a single, nonreproducing, growth-limiting nutrient in which one of the competitors, x_3, also predates on one of the other competitors, x_1, into the form of a generalized asymmetric May-Leonard model (1.1)-(1.2) of three species competition. On the other hand, if the inequality in (2.6) is reversed, then model (2.4) is of the same form as model (1.1), but α_3 is negative. The classical interpretation would be that instead of three species competition, x_1 and x_2 compete, but x_3 predates on x_1.

3 Dynamics of the chemostat model (2.1)

Let the equilibria of model (2.1) be denoted:

$$E_0 \triangleq (1,0,0,0); \quad E_{\lambda_1} \triangleq (\lambda_1, 1-\lambda_1, 0, 0); \quad E_{\lambda_2} \triangleq (\lambda_2, 0, 1-\lambda_2, 0);$$

$$E_{\lambda_3} \triangleq (\lambda_3, 0, 0, 1-\lambda_3); \quad E^* \triangleq (S^*, x_1^*, 0, x_3^*); \quad \tilde{E} \triangleq (\lambda_2, \tilde{x}_1, \tilde{x}_2, \tilde{x}_3),$$

where

$$S^* = \frac{\lambda_1\lambda_3}{\lambda_1\lambda_3 + \delta(\lambda_3-\lambda_1)}; \quad x_1^* = \delta(1 - \frac{S^*}{\lambda_3}); \quad x_3^* = \delta(-1 + \frac{S^*}{\lambda_1});$$

$$\tilde{x}_1 = \delta(1 - \frac{\lambda_2}{\lambda_3}); \quad \tilde{x}_2 = 1 - \lambda_2 - \delta\lambda_2(\frac{\lambda_3 - \lambda_1}{\lambda_3\lambda_1}); \quad \tilde{x}_3 = \delta(-1 + \frac{\lambda_2}{\lambda_1}).$$

Criteria for the existence and for the stability of the equilibria of model (2.1) are summarized in Table 3.1.

Table 3.1 Equilibria - Existence and Stability for (2.1) (assuming $\lambda_1 < \lambda_j,\ j = 2, 3$)

	Existence[†]	Globally Asymptotically Stable[‡] (assuming the equilibrium exists)
E_0	always	$\lambda_i \geq 1,\ i = 1, 2, 3$
E_{λ_1}	$\lambda_1 < 1$	$S^* < \lambda_1$
E_{λ_2}	$\lambda_2 < 1$	never
E_{λ_3}	$\lambda_3 < 1$	never
E^*	$\lambda_1 < S^* < \lambda_3$	$\lambda_1 < S^* < \lambda_2$
$\tilde{E}$	$\lambda_1 < \lambda_2 < \lambda_3$ and $S^* > \lambda_2$	whenever it exists

[†] An equilibrium is assumed to exist if, and only if, all of its components are nonnegative.
[‡] Global asymptotical stability is with respect to solutions initiating in the interior of the positive cone.

Note that under the assumption that $\lambda_1 < \lambda_j,\ j = 2, 3$, it follows that $0 < S^* < 1$, and that one of the equilibria, E_0, E_{λ_1}, E^*, or $\tilde{E}$ is globally asymptotically stable. This can be proved using the Liapunov functions summarized in Table 3.2 and the slightly modified version of the LaSalle Invariance Principle in Wolkowicz and Lu [12]. The proofs of the global asymptotic stability of the equilibria were first given in [4].

Table 3.2 Summary of Liapunov functions for (2.1) $V = V(S, x_1, x_2, x_3)$

E_0	$V = S - 1 - \ln(S) + x_1 + x_2 + x_3$ $\dot{V} = -\frac{(S-1)^2}{S} + \sum_{i=1}^{3} x_i(\frac{1-\lambda_i}{\lambda_i})$
E_{λ_1}	$V = S - \lambda_1 - \lambda_1 \ln(\frac{S}{\lambda_1}) + x_1 - (1 - \lambda_1) - (1 - \lambda_1) \ln \frac{x_1}{1-\lambda_1}$ $\dot{V} = -\frac{(S-\lambda_1)^2}{\lambda_1 S} - x_2(\frac{\lambda_2-\lambda_1}{\lambda_2}) + x_3(-1 + \frac{\lambda_1}{\lambda_3} + \frac{1-\lambda_1}{\delta})$
E^*	$V = S - S^* - S^* \ln(\frac{S}{S^*}) + \sum_{i=1,3}(x_i - x_i^* - x_i^* \ln(\frac{x_i}{x_i^*})) + x_2$ $\dot{V} = -\frac{(S-S^*)^2}{SS^*} + x_2(\frac{S^*-\lambda_2}{\lambda_2})$
$\tilde{E}$	$V = S - \lambda_2 - \lambda_2 \ln(\frac{S}{\lambda_2}) + \sum_{i=1}^{3}(x_i - \tilde{x}_i - \tilde{x}_i \ln(\frac{x_i}{\tilde{x}_i}))$ $\dot{V} = -\frac{1}{S\lambda_2}(S - \lambda_2)^2$

Remark 3.1 In fact, one can also prove that if instead, we assume that $\lambda_2 < \lambda_j,\ j = 1, 3$ and $\lambda_2 < 1$, then E_{λ_2} is globally asymptotically stable, or that if $\lambda_3 < \lambda_j,\ j = 1, 2$ and $\lambda_3 < 1$, then E_{λ_3} is globally asymptotically stable. Hence, model (2.1) only admits very simple dynamics. In particular, there is always a single, globally asymptotically stable equilibrium point that attracts all solutions with positive initial conditions.

4 Implications for the dynamics of model (1.1)

Solving (2.7)-(2.8) for the λ_i, $i = 1,2,3$, and δ in terms of $\alpha_i, i = 1,2,3$ and β_1, is equivalent to solving the linear system of equations:

$$\begin{bmatrix} -\alpha_1 & 1 & 0 & 0 \\ 0 & -\alpha_2 & 1 & 0 \\ 0 & 0 & 1 & \alpha_1\alpha_2\alpha_3 - 1 \\ -\alpha_1\alpha_2 & 0 & 0 & \beta_1 - \alpha_1\alpha_2 \end{bmatrix} \begin{bmatrix} \lambda_1 \\ \lambda_2 \\ \lambda_3 \\ \delta \end{bmatrix} = \begin{bmatrix} 1-\alpha_1 \\ 1-\alpha_2 \\ 0 \\ 0 \end{bmatrix}$$

We obtain the unique solution:

$$\lambda_1 = \frac{(\beta_1 - \alpha_1\alpha_2)(1-\alpha_1\alpha_2)}{\alpha_1\alpha_2(1-\beta_1-\alpha_1\alpha_2\alpha_3+\alpha_1\alpha_2)},$$

$$\lambda_2 = \frac{\beta_1 - 2\alpha_1\alpha_2 - \alpha_1\alpha_2^2\alpha_3 + \alpha_2 - \alpha_2\beta_1 + \alpha_1\alpha_2^2 + \alpha_1^2\alpha_2^2\alpha_3}{\alpha_2(1-\beta_1-\alpha_1\alpha_2\alpha_3+\alpha_1\alpha_2)},$$

$$\lambda_3 = \frac{(1-\alpha_1\alpha_2\alpha_3)(1-\alpha_1\alpha_2)}{1-\beta_1-\alpha_1\alpha_2\alpha_3+\alpha_1\alpha_2},$$

$$\delta = \frac{1-\alpha_1\alpha_2}{1-\beta_1-\alpha_1\alpha_2\alpha_3+\alpha_1\alpha_2}.$$

Provided α_i, and β_i, $i = 1,2,3$, are chosen so that $\beta_2 = \frac{1}{\alpha_1}$, $\beta_3 = \frac{1}{\alpha_2}$, $\lambda_i > 0$, $i = 1,2,3$, and $\delta > 0$, the results in the previous section hold, and so model (1.1) has simple dynamics, i.e. there is always a globally asymptotically stable equilibrium that attracts all solutions with positive initial conditions. This is true, even if some or all of the r_i, α_i, β_i are negative.

To determine which equilibrium is globally asymptotically stable, use Table 3.1 and Remark 3.1 at the end of the previous section, and note that E_{λ_i} in (2.1) corresponds to the single species survival equilibria, e_i, $i = 1,2,3$, for system (1.1), E_0 corresponds to the washout equilibrium, $e_0 \triangleq (0,0,0)$, E^* corresponds to the two species survival equilibrium $e^* \triangleq (x_1^*, 0, x_2^*)$, and $\tilde{E}$ corresponds to the equilibrium with all three components positive, $\tilde{e} \triangleq (\tilde{x}_1, \tilde{x}_2, \tilde{x}_3)$.

In order to have a globally attracting equilibrium in the interior of the positive cone with a repelling heteroclinic cycle on the boundary of the positive cone, select $0 < \alpha_i < 1$, $i = 1,2,3$. Then, provided that in addition,

$$\beta_1 < 1 + \alpha_1\alpha_2(1-\alpha_3) \triangleq \beta_M, \tag{4.1}$$

so that the denominators are all positive, it follows that $\lambda_1 > 0$ and $0 < \lambda_3 < \delta$. For $\lambda_2 > 0$, one must also assume that the numerator in the expression for λ_2 above is also positive, i.e.,

$$\beta_1 > \frac{\alpha_2(2\alpha_1 + \alpha_1\alpha_2\alpha_3 - 1 - \alpha_1\alpha_2 - \alpha_1^2\alpha_2\alpha_3)}{1-\alpha_2} \triangleq \beta_m. \tag{4.2}$$

Note that,

$$\beta_M > 1 \text{ and } \beta_M - \beta_m = (1-\alpha_1\alpha_2)(1-\alpha_1\alpha_2\alpha_3) > 0,$$

since $0 < \alpha_i < 1$, $i = 1,2,3$. Therefore, it is always possible to select β_i, $i = 1,2,3$, so that $\beta_2 = \frac{1}{\alpha_1}$, $\beta_3 = \frac{1}{\alpha_2}$, and $\max(1, \beta_m) < \beta_1 < \beta_M$, and hence (1.2) holds.

If

$$\beta_1 = \alpha_1\alpha_2(2 - \alpha_1\alpha_2\alpha_3) \triangleq \beta_{crit},$$

then

$$\lambda_1 = \lambda_2 = \lambda_3 = S^* = 1,$$

and if

$$\beta_1 < \beta_{crit},$$

then

$$\lambda_1 < \lambda_2 < \lambda_3 < S^* < 1.$$

Note that,

$$\beta_{crit} - \beta_m = \alpha_2 \frac{(1-\alpha_1\alpha_2)(1-\alpha_1\alpha_2\alpha_3)}{1-\alpha_2} > 0,$$

Also,

$$\beta_M - \beta_{crit} = (1-\alpha_1\alpha_2)(1-\alpha_2\alpha_2\alpha_3) > 0,$$

so that

$$\beta_m < \beta_{crit} < \beta_M.$$

Note also, that if

$$\frac{1}{\alpha_1\alpha_2} + \alpha_1\alpha_2\alpha_3 < 2,$$

then

$$\beta_{crit} > 1.$$

Therefore, we have just proved,

Theorem 4.1 *In model (1.1)-(1.2), if*

$$0 < \alpha_i < 1, \ i = 1,2,3, \quad \frac{1}{\alpha_1\alpha_2} + \alpha_1\alpha_2\alpha_3 < 2, \quad \textit{and} \quad \max(1, \beta_m) < \beta_1 < \beta_{crit},$$

then there is a a repelling heteroclinic cycle on the boundary of the positive cone and a unique positive equilibrium that is globally asymptotically stable with respect to all orbits initiating in the interior of the positive cone.

EXAMPLE Selecting

$$\alpha_1 = \frac{9}{10}, \ \alpha_2 = \frac{8}{9}, \ \alpha_3 = \frac{5}{12},$$

$$\beta_1 = \frac{6}{5}, \ \beta_2 = \frac{10}{9}, \ \beta_3 = \frac{9}{8},$$

$$r_1 = \frac{5}{3}, \ r_2 = \frac{9}{7}, \ r_3 = 1,$$

in model (1.1)-(1.2) corresponds to taking

$$\lambda_1 = \frac{3}{8}, \ \lambda_2 = \frac{7}{16}, \ \lambda_3 = \frac{1}{2}, \ \delta = \frac{3}{4}, \ S^* = \frac{2}{3}$$

in model (2.1). For these parameters, it follows from Theorem 4.1 that both models (1.1) and (2.1) have a repelling heteroclinic cycle on the boundary and a globally attracting equilibrium in the interior of the positive cone. It is interesting to note that neither the Zeeman Computational Theorem (see [13] for more details and proof), nor the Volterra-Lyapunov Stability Theorem (see ([6], Section 15.3) for more details and a proof) can be used to show the global stability of the interior equilibrium in this example. We provide the statements of both results here for the readers' convenience and demonstrate that a hypothesis fails in each case.

We first introduce notation similar to that used in [13]. Model (1.1) and (2.1) can be written in the form $x' = \chi(b - Ax) = \chi A(p - x)$, where

$$x = \begin{bmatrix} x_1 \\ x_2 \\ x_3 \end{bmatrix}, \quad b = \begin{bmatrix} r_1 \\ r_2 \\ r_3 \end{bmatrix}, \quad p = A^{-1}b,$$

$$\chi = diag(x) = \begin{pmatrix} x_1 & 0 & 0 \\ 0 & x_2 & 0 \\ 0 & 0 & x_3 \end{pmatrix}, \quad A = \begin{pmatrix} r_1 & r_1\alpha_1 & r_1\beta_1 \\ r_2\beta_2 & r_2 & r_2\alpha_2 \\ r_3 & r_3\alpha_3 & r_3 \end{pmatrix}.$$

For an arbitrary $n \times n$ matrix M define the symetric matrix

$$M^S \triangleq \frac{1}{2}(M + M^T),$$

and

$$M^R \triangleq M_{nn} + M_{11} - M_{n1} - M_{1n},$$

where M_{ij} denotes the $(n-1) \times (n-1)$ submtrix of M obtained by deleting row i and column j from M.

Then,

$$M^{SR} \triangleq (M^S)^R.$$

Theorem 4.2 *(Computational Theorem ([13], Theorem 6.7)) Given the competitive system $x' = \chi A(p - x)$, with a unique interior equilibrium point $P = p = A^{-1}b \in int\mathbb{R}^n_+$. Let h^T be a strictly positive left eigenvector of $\mathcal{P}A$ and let $\mathcal{H} = diag(h)$. If all the eigenvalues of $(A\mathcal{H}^{-1})^{SR}$ are negative (positive), then P is a global repellor (attractor).*

Verifying the hypotheses of the Computational Theorem using Maple, we find (to 4 decimal places) that

$$b = \begin{bmatrix} \frac{5}{3} \\ \frac{9}{7} \\ 1 \end{bmatrix}, \quad A = \begin{pmatrix} \frac{5}{3} & \frac{3}{2} & 2 \\ \frac{10}{7} & \frac{9}{7} & \frac{8}{7} \\ \frac{5}{12} & \frac{9}{8} & 1 \end{pmatrix}, \quad p = A^{-1}b = \begin{pmatrix} \frac{3}{20} \\ \frac{11}{18} \\ \frac{1}{4} \end{pmatrix}, \quad h = \begin{pmatrix} 0.9889 \\ 1.0375 \\ 1.0 \end{pmatrix},$$

$$(A\mathcal{H}^{-1})^S = \begin{pmatrix} 1.6667 & 1.429 & 1.1972 \\ 1.4292 & 1.2255 & 1.1012 \\ 1.1972 & 1.1012 & 0.9889 \end{pmatrix}, \quad (A\mathcal{H}^{-1})^{SR} = \begin{pmatrix} 0.0338 & 0.1077 \\ 0.1077 & 0.0119 \end{pmatrix}.$$

$(A\mathcal{H}^{-1})^{SR}$ has a positive and a negative eigenvalue and hence is neither positive nor negative definite, and so the Computational Theorem is inconclusive.

Theorem 4.3 *(Volterra-Liapunov Theorem [6]) If there exists a diagonal matrix D with positive diagonal entries such that $(DA)^S$ is positive definite, then $P = p$ is a globally asymptotically stable equilibrium point of the competitive system $x' = \chi A(p - x)$.*

Recalling that the principal minors of a positive definite matrix must all be strictly positive, since $\det A_{33} = 0$, by a similar argument to that given in Lemma 7.2 and 7.3 of [13], no diagonal matrix D with positive diagonal entries exists such that $(DA)^S$ is positive definite, and so the Volterra-Liapunov Theorem cannot be used to show the global stability of this interior equilibrium point.

References

[1] Butler, G.J. and Wolkowicz, G.S.K. *A mathematical model of the chemostat with a general class of functions describing nutrient uptake*, SIAM J. Appl. Math. **45** (1985), 138–151.

[2] Butler, G.J., Hsu, S.-B., and P. Waltman, *Coexistence of competing predators in a chemostat*, J. Math. Biol. **17** (1983), 133–151.

[3] Chi, C.-W., S-B Hsu, S.-B., and Wu, L.-I. *On the asymmetric May-Leonard model of three competing species*, SIAM J. Appl. Math. **58** (1998), 211–226.

[4] Daoussis, S.P. *Predator-mediated competition: Predator feeding on two trophic levels*, M.Sc. thesis, McMaster University, 1992.

[5] Hofbauer, J. and Sigmund, K. *The Theory of Evolution and Dynamical Systems*, London Mathematical Society Student Texts 7, Cambridge University Press, Cambridge, [1988].

[6] Hofbauer, J. and K. Sigmund, K. *Evolutionary Games and Populaton Dynamics*, Cambridge University Press, Cambridge, 1998.

[7] May, R.M. and Leonard, W.J. *Nonlinear aspects of competition between three species* , SIAM J. Appl. Math. **29** (1975), 243–253.

[8] Schuster, P., Sigmund, K. and Wolf, R. *On ω-limit for competition between three species*, SIAM J. Appl. Math. **37** (1979), 49–54.

[9] Smith H.L. and Waltman, P. *The Theory of the Chemostat: Dynamics of Microbial Competition*, Cambridge University Press, New York, NY, 1995.

[10] Waltman, P. *Competition Models in Populations Biology*, SIAM, Philadelphia, PS, 1983.

[11] Wolkowicz, G.S.K., Ballyk, M.M. and Daoussis, S.P. *Interaction in a chemostat: introduction of a competitor can promote greater diversity*, Rocky Mountain J. Math. **25** (1995), 515–543.

[12] Wolkowicz, G.S.K. and Lu, Z. *Global dynamics of a mathematical model of competition in the chemostat: general response functions and differential death rates*, SIAM J. Appl. Math. **52** (1992), 222–233.

[13] Zeeman E.C. and M. L. Zeeman, M.L. *From local to global behavior in competitive Lotka-Volterra systems*, Trans. Amer. Math. Soc., **355** (2002), 713–734.

Fields Institute Communications
Volume **48**, 2006

On Exact Poisson Structures

Yingfei Yi
School of Mathematics
Georgia Institute of Technology
Atlanta, GA 30332, USA
yi@math.gatech.edu

Xiang Zhang
Department of Mathematics
Shanghai Jiaotong University
Shanghai 200240, P. R. China
xzhang@mail.sjtu.edu.cn

Abstract. By studying the exactness of multi-linear vectors on an orientable smooth manifold $\mathbf{M}$, we give some characterizations to exact Poisson structures defined on $\mathbf{M}$ and study general properties of these structures. Following recent works [12, 13, 15], we will pay particular attention to the classification of some special classes of exact Poisson structures such as Jacobian and quasi-homogeneous Poisson structures. A characterization of exact Poisson structures which are invariant under the flow of a class of completely integrable systems will also be given.

1 Introduction

Let $\mathbf{M}$ be an orientable C^∞ smooth manifold of dimension n and let $C^\infty(\mathbf{M})$ be the space of C^∞ smooth functions defined on $\mathbf{M}$. A *Poisson structure* Λ on $\mathbf{M}$ is an algebra structure on $C^\infty(\mathbf{M})$ satisfying the Leibniz identity, i.e.,

$$\Lambda = \{\cdot,\cdot\} : C^\infty(\mathbf{M}) \times C^\infty(\mathbf{M}) \to C^\infty(\mathbf{M}),$$

is a bilinear map such that for arbitrary $f, g, h \in C^\infty(\mathbf{M})$ the following holds:

(a) (Skew-symmetry) $\{f,g\} = -\{g,f\}$,
(b) (Leibniz rule) $\{f,gh\} = \{f,g\}h + g\{f,h\}$,
(c) (Jacobi identity) $\{\{f,g\},h\} = \{f,\{g,h\}\} + \{\{f,h\},g\}$.

2000 *Mathematics Subject Classification.* Primary 53D17, 37K10; Secondary 70G60, 70G45, 70G65.

The first author was partially supported by NSF grant DMS0204119.
The second author is partially supported by NSFC grant 10231020 and Shuguang plan of Shanghai.

With a Poisson structure $\{,\}$, the algebra $(C^\infty(\mathbf{M}), \{,\})$ becomes a Lie algebra (see e.g., [16, 17]). The pair $(\mathbf{M}, \{,\})$ is called a *Poisson Manifold*. In what follows, smooth manifolds always mean orientable C^∞ smooth manifolds.

With respect to a local coordinate system $\{x_i\}$ on $\mathbf{M}$, such a structure can be explicitly defined so that for arbitrary $f, g \in C^\infty(\mathbf{M})$

$$\Lambda(df, dg) = \{f, g\} = \sum_{i,j=1}^{n} w_{ij} \frac{\partial f}{\partial x_i} \frac{\partial g}{\partial x_j}, \tag{1.1}$$

where $w_{ij} \in C^\infty(\mathbf{M})$, $i, j = 1, \ldots, n$, satisfy the identities

$$w_{ij} + w_{ji} = 0, \qquad \sum_{l=1}^{n} \sum_{\sigma \in A_3} w_{l\sigma(i)} \frac{\partial w_{\sigma(j)\sigma(k)}}{\partial x_l} = 0,$$

here A_3 is the group of cyclic permutations acting on (i, j, k). The matrix $J = (w_{ij})$ is called a *structure matrix* associated to Λ. Since an everywhere non-degenerate Poisson structure is necessarily symplectic, Poisson structures are natural extensions to the standard symplectic ones on a smooth manifold.

Let $\mathcal{X}^k(\mathbf{M})$ be the space of smooth k-linear vectors on $\mathbf{M}$ and $\Omega^k(\mathbf{M})$ the space of differential $k-$forms. For a given volume element ω on $\mathbf{M}$ in standard local expression, we consider its induced isomorphism $\Phi : \mathcal{X}^k(\mathbf{M}) \to \Omega^{n-k}(\mathbf{M})$: $u \to \mathbf{i}_u\omega$, where $\mathbf{i}_u\omega$ is the inner product of u and ω. For instance, if $u = f(\mathbf{x})\partial_{i_1} \wedge \partial_{i_2} \wedge \ldots \wedge \partial_{i_k}$ with $1 \le i_1 < i_2 < \ldots < i_k \le n$, and $\omega = dx_1 \wedge dx_2 \wedge \ldots \wedge dx_n$ under the local coordinate $\{x_i\}$, then $\mathbf{i}_u\omega = (-1)^{i_1-1}(-1)^{i_2-2} \ldots (-1)^{i_k-k} f(\mathbf{x}) dx_1 \wedge \widehat{dx_{i_1}} \wedge \ldots \wedge \widehat{dx_{i_2}} \wedge \ldots \wedge \widehat{dx_{i_k}} \wedge \ldots \wedge dx_n$, hereafter $\widehat{x}$ stands for the omission of x. Let

$$D \equiv \Phi^{-1} \circ d \circ \Phi : \mathcal{X}^k(\mathbf{M}) \to \mathcal{X}^{k-1}(\mathbf{M})$$

be the pull-back operator under the isomorphism Φ, where d is the exterior derivative of differential forms. A k-linear vector $\mathbf{X}_k$ is said to be *exact* if $D(\mathbf{X}_k) = 0$. Specifically, a Poisson structure Λ satisfying $D(\Lambda) = 0$ is called an *exact Poisson structure*. We note that symplectic structures are always exact. It is therefore hopeful that an exact Poisson structure resembles a symplectic one to certain extends.

The present paper is devoted to the study of exact Poisson structures with respect to their characterizations and general properties.

In [10], the author showed that the operator D has some important applications. For instance, the operator can be used to compute the Schouten brackets and to verify the Jacobian identity which a Poisson structure should satisfy. In Section 2, we will further study properties of D. In particular, we will show that under the action of D the direct sum $\oplus_{k=0}^{n}\mathcal{X}^k(\mathbb{R}^n)$ forms a complex. We will also study the homology induced by the complex and its topological structure.

These properties of D and the volume preserving property of exact Poisson structures will be used in Section 3 to study three special classes of Poisson structures: the Lie-Poisson, Jacobian, and the quasi-homogeneous structures. We will also investigate the Hamiltonian flows induced by these Poisson structures, with respect to issues such as normal forms and volume-preservation.

It is known that any Jacobian structure is a Poisson structure ([9, 15]) and any Jacobian structure with constant Jacobian coefficient is exact ([15]). In Section 3, we will give a general sufficient condition for an exact Poisson structure to become

Jacobian. We will also study some general properties of the Jacobian structures. Part of our results in this regard generalizes some of those in [15].

As for the quasi-homogeneous Poisson structures, we will give a necessary and sufficient condition for a decomposition of a quasi-homogeneous Poisson structure with respect to exact Poisson structures. This result is an improvement to the corresponding ones in [12, 15]. Restricting to the classical r-Poisson structures - a class of special quadratic Poisson structures, we will obtain a necessary and sufficient condition for a classical r-Poisson structure in $\mathbb{R}^3$ to be exact. Using this result we further prove that any quadratic Jacobian structure is a classical r-Poisson structure in $\mathbb{R}^3$. It is known that any quadratic Poisson structure in $\mathbb{R}^2$ is a classical r-Poisson structure ([12]) but in $\mathbb{R}^3$ the Poisson structure $\overline{\Lambda} = (x_1^2 + \alpha x_2 x_3)\partial_2 \wedge \partial_3$ for $\alpha \neq 0$ is not a classical r-Poisson structure ([13]). However, we note that for $\alpha = 0$ the structure $\overline{\Lambda}$ is a Jacobian structure. This leads to an open problem to *classify all classical r-Poisson structures within the class of quadratic Poisson structures*. A more concrete question in dimension 3 is that *whether a quadratic, non-Jacobian Poisson structure is necessarily not a classical r-Poisson structure.*

In Section 4, we will give a characterization of exact Poisson structures which are preserved by a $\mathbb{T}^q$-dense, completely integrable flow. We will also show that such a $\mathbb{T}^q$-dense flow preserving an exact Poisson structure can be non-Hamiltonian, in contrast to the closed 2-form case in which any $\mathbb{T}^q$-dense flow preserving a symplectic structure is necessary a Hamiltonian ([4]).

2 General properties of exact Poisson structures

2.1 The operator D. We first recall the following properties of the operator D which will be used later on.

Proposition 2.1 ([10]) *The following holds.*

(i) *For any $\mathbf{X} \in \mathcal{X}(\mathbf{M})$, $D(\mathbf{X}) = \mathrm{div}_\omega \mathbf{X}$, where $\mathrm{div}_\omega \mathbf{X}$ denotes the divergence of the vector field $\mathbf{X}$ with respect to the volume element ω.*

(ii) *For any $\mathbf{X}, \mathbf{Y} \in \mathcal{X}(\mathbf{M})$, $D(\mathbf{X} \wedge \mathbf{Y}) = [\mathbf{Y}, \mathbf{X}] + (\mathrm{div}_\omega \mathbf{Y})\mathbf{X} - (\mathrm{div}_\omega \mathbf{X})\mathbf{Y}$, where $[\cdot,\cdot]$ denotes the usual Lie bracket of two vector fields, and $\wedge$ denotes the wedge product of two vectors.*

(iii) *For any $U \in \mathcal{X}^\mu(\mathbf{M})$ and $V \in \mathcal{X}^\nu(\mathbf{M})$,*

$$[U, V] = (-1)^\mu D(U \wedge V) - D(U) \wedge V - (-1)^\mu U \wedge D(V),$$

where $[U, V]$ denotes the Schouten bracket of the multi-linear vector fields U and V, which is defined as the following: if $U = u_1 \wedge \ldots \wedge u_\mu$ and $V = v_1 \wedge \ldots \wedge v_\nu$, then

$$[U, V] = \sum_{s,t} (-1)^{s+t} u_1 \wedge \ldots \wedge \widehat{u}_s \wedge \ldots \wedge u_\mu \wedge [u_s, v_t] \wedge v_1 \wedge \ldots \wedge \widehat{v}_t \wedge \ldots \wedge v_\nu.$$

(iv) *Let Λ be a Poisson structure and $\mathbf{X}_\Lambda = D(\Lambda)$ be its curl vector field. Then the Lie derivative $L_{\mathbf{X}_\Lambda}\Lambda \equiv [\mathbf{X}_\Lambda, \Lambda] = 0$.*

(v) *A skew-symmetric bilinear vector field Λ is an exact Poisson structure if and only if $D(\Lambda) = 0$ and $D(\Lambda \wedge \Lambda) = 0$.*

Remark 2.2 1) Statement (i) of the above proposition implies that a vector field is exact if and only if it is divergent free. Hence it follows from Gauss' Theorem that the volume is invariant under the flow induced by an exact vector field.

2) For a 3-dimensional vector fields $\mathbf{v} = a\partial_x + b\partial_y + c\partial_z$, the curl of $\mathbf{v}$ is defined in the usual way as $\nabla \times \mathbf{v} = (c_y - b_z)\partial_x + (a_z - c_x)\partial_y + (b_x - a_y)\partial_z$. This is in fact the curl vector field $\mathbf{X}_\Lambda = D(\Lambda)$ with $\Lambda = a\partial_y \wedge \partial_z + b\partial_z \wedge \partial_x + c\partial_x \wedge \partial_y$. Thus, the operator D unifies the computations for the divergence and the curl of a given vector field.

Let $\mathcal{X}^*(\mathbb{R}^n) = \oplus_{k=0}^n \mathcal{X}^k(\mathbb{R}^n)$ be the algebra formed by the direct sum of the space of the k-linear vectors. The following result describes certain homological properties induced by the operator D, which will be used later in the classification of exact Poisson structures.

Proposition 2.3 *The following holds.*

(a) $D^2 = 0$.

(b) *The D-homology of $\mathbb{R}^n$ formed by the vector space*

$$H_k(\mathbb{R}^n) = \left((\text{kernel of } D) \cap \mathcal{X}^k(\mathbb{R}^n)\right) / \left((\text{image of } D) \cap \mathcal{X}^k(\mathbb{R}^n)\right),$$

has the topological structures

$$H_k(\mathbb{R}^n) = \begin{cases} \mathbb{R}, & k = n, \\ 0, & 0 \le k < n. \end{cases}$$

Proof It follows from the definition of D, the de Rham cohomology and Poincaré's Lemma (e.g. [2]) in $\mathbb{R}^n$. □

Remark 2.4 Proposition 2.3 implies that $\mathcal{X}^*(\mathbb{R}^n)$ is a complex induced by D and the sequence of vector spaces

$$0 \longrightarrow \mathcal{X}^n(\mathbb{R}^n) \overset{D}{\longrightarrow} \mathcal{X}^{n-1}(\mathbb{R}^n) \overset{D}{\longrightarrow} \cdots \overset{D}{\longrightarrow} \mathcal{X}^0(\mathbb{R}^n) \longrightarrow 0$$

is exact.

2.2 Characterization of exactness. The following lemma will be used in the general characterization of exact Poisson structures to be given in this section.

Lemma 2.5 ([8]) *Let Λ be a smooth Poisson structure in $\mathbb{R}^n$ defined by* (1.1). *Then Λ is exact if and only if*

$$\sum_{j=1}^n \frac{\partial w_{ij}}{\partial x_j} = 0, \qquad i = 1, \ldots, n. \tag{2.1}$$

Proof It follows from the fact that

$$D(\Lambda) = 2\sum_{i=1}^n \left(\sum_{j=1, j\neq i}^n \frac{\partial w_{ij}}{\partial x_j} \right) \frac{\partial}{\partial x_i}. \tag{2.2}$$

□

Recall that a smooth function H defined on a smooth manifold $\mathbf{M}$ is a *first integral* of a smooth vector field $\mathbf{X}$ if $\mathbf{X}(H) \equiv 0$ on $\mathbf{M}$.

Our next result gives a characterization of exact Hamiltonian vector fields.

Theorem 2.6 *Let Λ be a Poisson structure on a smooth Riemannian manifold* $\mathbf{M}$. *Then a Hamiltonian vector field* $\mathbf{X}_H$ *associated to* Λ *and the Hamilton* H *is exact if and only if the curl vector field* $\mathbf{X}_\Lambda = D(\Lambda)$ *and the gradient vector field* ∇H *are everywhere orthogonal on* $\mathbf{M}$, *i.e.,* $D(\Lambda)$ *belongs to the tangent spaces of the level surfaces of* H; *or equivalently,* H *is a first integral of the curl vector field* $D(\Lambda)$.

Proof On any orientable manifold there is a volume form that locally takes the standard one in $\mathbb{R}^n$, so without loss of generality, we can prove the theorem under a local coordinate system $\{x_i\}_{i=1}^n$ on M by taking the standard volume element $\omega = dx_1 \wedge \ldots \wedge dx_n$, where $n = \dim M$.

Let $J = (w_{ij})$ be the structure matrix of Λ and $\mathbf{X}_H = J\nabla H$ be the associated Hamiltonian vector field. Making use of the skew-symmetry of J, calculations yield that

$$D(\mathbf{X}_H) = -\sum_{i=1}^{n}\left(\sum_{j=1}^{n}\frac{\partial w_{ij}}{\partial x_j}\right)\frac{\partial H}{\partial x_i}. \tag{2.3}$$

By (2.2), we further have

$$D(\mathbf{X}_H) = -\frac{1}{2}D(\Lambda)\cdot\nabla H, \tag{2.4}$$

from which the theorem easily follows. □

Remark 2.7 1) From (2.4) and the fact that

$$D(\mathbf{X}_H) = \Phi^{-1}\circ d\circ\Phi(\mathbf{X}_H) = \Phi^{-1}\circ d\circ \mathbf{i}_{\mathbf{X}_H}\omega = \Phi^{-1}(L_{\mathbf{X}_H}\omega),$$

it follows easily that Λ is exact if and only if any corresponding Hamiltonian vector field is exact, or equivalently, the volume element ω is invariant under any Hamiltonian flow associated to Λ. This result is also a consequence of the theorem given in [18].

2) By the Birkhoff Ergodic Theorem, for any Hamiltonian flow ϕ_t on an orientable manifold $\mathbf{M}$ induced by an exact Poisson structure,

$$P(f)(\mathbf{x}) = \lim_{T\to\infty}\frac{1}{T}\int_0^T f\circ\phi_t(\mathbf{x})dt,$$

is a well defined L^1 function for any $f \in L^1(\mathbf{M})$.

There are two issues involved in checking whether a skew symmetric, bilinear vector field Λ is an exact Poisson structure, i.e., Λ needs to be exact and satisfy the Jacobian identity. For $n \geq 3$, it is known that Λ is exact if and only if the $(n-2)$-form

$$\Omega_{n-2} = \sum_{1\le i<j\le n}(-1)^{i+j-1}w_{ij}dx_1\wedge\ldots\wedge\widehat{dx_i}\wedge\ldots\wedge\widehat{dx_j}\wedge\ldots\wedge dx_n$$

is closed, i.e., $d\Omega_{n-2} = 0$ (see for instance [12] or [8]). Moreover, if Λ is exact, then there exists an $(n-3)$-form Ω_{n-3} such that $d\Omega_{n-3} = \Omega_{n-2}$, or equivalently, there exists a 3-linear vector $\mathbf{X}_3$ such that $D\mathbf{X}_3 = \Lambda$. The following proposition establishes certain connections between exactness and the Jacobian identity. In particular, for an exact Poisson structure, the Jacobian identity can be written in a symmetric form.

Proposition 2.8 *Let Λ be a skew symmetric bilinear vector field in $\mathbb{R}^n$ with the structure matrix $J = \{w_{ij}\}$. Then Λ is an exact Poisson structure if and only if*

$$\sum_{j=1}^{n} \frac{\partial w_{ij}}{\partial x_j} = 0, \qquad i = 1, \ldots, n, \tag{2.5}$$

$$\sum_{s=1, s\neq i,j,k}^{n} \left(A_{ijk}^s \cdot \frac{\partial B_{ijk}^s}{\partial x_s} + B_{ijk}^s \cdot \frac{\partial A_{ijk}^s}{\partial x_s} \right) = 0, \qquad 1 \le i < j < k \le n, \tag{2.6}$$

where $A_{ijk}^s = (w_{si}, w_{sj}, w_{sk})$, $B_{ijk}^s = (w_{jk}, w_{ki}, w_{ij})$.

Proof We only prove the necessity, as the sufficiency follows from a similar argument. The condition (2.5) clearly follows from Lemma 2.5. Now for arbitrary integers $1 \le i < j < k \le n$ we have by (2.5) that

$$\begin{aligned}
&\sum_{s=1}^{n} \sum_{\sigma \in A_3} w_{s\sigma(i)} \frac{\partial w_{\sigma(j)\sigma(k)}}{\partial x_s} \\
&= \sum_{s=1, s\neq i,j,k}^{n} \sum_{\sigma \in A_3} w_{s\sigma(i)} \frac{\partial w_{\sigma(j)\sigma(k)}}{\partial x_s} \\
&\quad + w_{jk} \left(\frac{\partial w_{ij}}{\partial x_j} + \frac{\partial w_{ik}}{\partial x_k} \right) + w_{ki} \left(\frac{\partial w_{ji}}{\partial x_i} + \frac{\partial w_{jk}}{\partial x_k} \right) + w_{ij} \left(\frac{\partial w_{ki}}{\partial x_i} + \frac{\partial w_{kj}}{\partial x_j} \right) \\
&= \sum_{s=1, s\neq i,j,k}^{n} \sum_{\sigma \in A_3} w_{s\sigma(i)} \frac{\partial w_{\sigma(j)\sigma(k)}}{\partial x_s} + \sum_{\sigma \in A_3} w_{\sigma(j)\sigma(k)} \sum_{s=1, s\neq i,j,k}^{n} \frac{\partial w_{s\sigma(i)}}{\partial x_s} \\
&= \sum_{s=1, s\neq i,j,k}^{n} \sum_{\sigma \in A_3} \left(w_{s\sigma(i)} \frac{\partial w_{\sigma(j)\sigma(k)}}{\partial x_s} + w_{\sigma(j)\sigma(k)} \frac{\partial w_{s\sigma(i)}}{\partial x_s} \right).
\end{aligned}$$

This proves the condition (2.6). □

We note that the condition (2.6) cannot be simplified further under the exactness condition. In dimension 3, the condition (2.6) is trivially satisfied. This means that exactness condition implies the Jacobian identity in dimension 3.

3 Lie-Poisson, Jacobian, and Quasi-homogeneous structures

In this section, we will apply the general characterization of exactness from the previous section to obtain more precise information for three special Poisson structures: Lie-Poisson, Jacobian, and Quasi-homogeneous Poisson structures.

3.1 Lie-Poisson structures. A *Lie-Poisson structure* in $\mathbb{R}^n$ or $\mathbb{C}^n$ is defined by

$$L = \sum_{i,j,k=1}^{n} c_{ij}^k x_k \partial_i \wedge \partial_j, \tag{3.1}$$

with ${c_{ij}^k}'s$ being the structure constants of an n-dimensional Lie algebra.

It is well known that a Lie-Poisson structure is necessary a Poisson structure ([14]). We have the following characterization.

Theorem 3.1 *Let L be the Lie-Poisson structure defined in* (3.1). *The following holds.*

(a) *L is exact if and only if* $\sum_{j=1}^{n} c_{ij}^j = 0$, $i = 1, \ldots, n$.

(b) *An associated Hamiltonian vector field* $X_H = L(\cdot, dH)$ *is exact if and only if the Hamiltonian* H *is a first integral of the completely integrable vector fields* $\sum_{i=1}^{n} \left(\sum_{j=1}^{n} c_{ij}^j \right) \partial_i$.

(c) *If the Lie algebra* $\mathfrak{g}$ *formed by homogeneous polynomials of degree 1 under the action of the Lie bracket induced by* (3.1) *is nilpotent, then the structure* (3.1) *is affine equivalent to a structure of the type* (3.1) *with* $c_{ij}^k = 0$ *for* $k \geq \min\{i, j\}$, *and consequently, the resulting structure is exact.*

Proof Statements (a) and (b) follow directly from Theorem 2.6. Using Theorem 3.5.4 of [16] and statement (a) above we obtain the statement (c). □

Lie-Poisson structures play important roles in studying normal forms for a class of Poisson structures. Let Λ be an analytic Poisson structure in a neighborhood of the origin in $\mathbb{R}^n$ or $\mathbb{C}^n$ defined by

$$\Lambda = \sum_{i,j,k=1}^{n} (c_{ij}^k x_k + R_{ij}(x)) \partial_i \wedge \partial_j, \tag{3.2}$$

where $R_{ij} = O(|x|^2)$, $i, j = 1, \cdots, n$. Using Theorem 2.6, Theorem 4.1 of [5], Theorem 2.1 of [17], and the generalized Darboux Theorem, one easily has the following.

Proposition 3.2 *Let* Λ *be an analytic Poisson structure of rank* $2m$ *in a neighborhood of the origin in* $\mathbb{R}^n$ *or* $\mathbb{C}^n$. *If the Lie algebra* $\mathfrak{g}$ *with the structure constants formed by the coefficients* $\{c_{ij}^k\}$ *of the linear truncation for the singular part of* Λ *is semi-simple, then* Λ *is analytically equivalent to*

$$P = \sum_{i=1}^{m} \frac{\partial}{\partial p_i} \wedge \frac{\partial}{\partial q_i} + \sum_{1 \leq i < j \leq n-2m} \left(\sum_{k=1}^{n-2m} c_{ij}^k y_k \right) \frac{\partial}{\partial y_i} \wedge \frac{\partial}{\partial y_j}.$$

If, in addition, $\sum_{j=1}^{n-2m} c_{ij}^j = 0$ *for* $i = 1, \ldots, n-2m$, *i.e.,* P *is exact, then any Hamiltonian flow associated to* Λ *is analytically equivalent to a volume-preserving one.*

Remark 3.3 The analyticity in Proposition 3.2 can be replaced by smoothness if the Lie algebra $\mathfrak{g}$ is of compact type. The existence of smooth equivalency in the smooth case can be shown by using Theorem 4.1 of [6].

3.2 Jacobian structures. Let $n \geq 3$ be an integer. Following [9], a *Jacobian bracket* in $\mathbb{R}^n$ is a bilinear map $\{\cdot, \cdot\} : C^\infty(\mathbf{M}) \times C^\infty(\mathbf{M}) \to C^\infty(\mathbf{M})$, satisfying

$$\{f, g\} = u \det(J(f, g, P_1, \ldots, P_{n-2})),$$

where $u, P_i \in C^\infty(\mathbb{R}^n)$, $i = 1, \ldots, n-2$, J denotes the usual Jacobian matrix of $f, g, P_1, \ldots,$ P_{n-2} with respect to the variables $x_1, \ldots, x_n$, and 'det' stands for the determinant of a matrix. With the above setting, the Jacobian bracket is said to be *generated* by $P_1, \ldots, P_{n-2}$ with a *Jacobian coefficient* u. It is easy to see that if $P_1, \cdots, P_{n-2}$ are functionally dependent, then the Jacobian bracket is trivial. In what follows, we always assume that the generators $P_1, \ldots, P_{n-2}$ of a

Jacobian bracket are functionally independent. It was shown in [9] that a *Jacobian structure*, i.e., an algebra structure defined by a Jacobian bracket, is necessary a Poisson structure and in [15] that a Jacobian structure with a constant Jacobian coefficient is always exact.

We now give some conditions under which a Poisson structure becomes Jacobian.

Theorem 3.4 *For Poisson structures in $\mathbb{R}^n$, the following holds.*

(a) *For $n = 2$, a smooth Poisson structure is exact if and only if it is a constant Poisson structure.*

(b) *For $n = 3$, a smooth Poisson structure is exact if and only if it is a Jacobian structure with a constant Jacobian coefficient.*

(c) *For $n > 3$, a smooth Poisson structure is Jacobian with a non-zero constant Jacobian coefficient if and only if it is exact and has rank ≤ 2, and the Lebesgue measure of the set of points at which the structure has rank 0 is zero.*

Proof (a) A Poisson structure has the form $\Lambda = \omega(x,y)\partial_x \wedge \partial_y$. By part (b) of Proposition 2.3, $H_2(\mathbb{R}^2) = \mathbb{R}$. It follows that if Λ is exact, then it is a constant Poisson structure. The converse is obvious.

(b) Assume that Λ is an exact Poisson structure, i.e., $D(\Lambda) = 0$. By part (b) of Proposition 2.3, we have that $H_2(\mathbb{R}^3) = 0$, i.e., the kernel of D is equal to the image of D. Hence, there exists a 3-linear vector $\mathbf{X}_3 = P(x,y,z)\partial_x \wedge \partial_y \wedge \partial_z \in \mathcal{X}^3(\mathbb{R}^3)$ such that $D(\mathbf{X}_3) = \Lambda$. Since $D(\mathbf{X}_3) = P_x\partial_y \wedge \partial_z + P_y\partial_z \wedge \partial_x + P_z\partial_x \wedge \partial_y$, we see that Λ is the Jacobian structure generated by P. The sufficient part follows from [15].

(c) The necessary part of (c) is obvious. To prove the sufficient part, we let Λ be an exact Poisson structure of rank ≤ 2. Let $\{(U_i, \phi_i, g_i)\}$ be a partition of unity on $\mathbb{R}^n$, where $\{U_i\}$ is a locally finite open cover of $\mathbb{R}^n$, each ϕ_i is an isomorphism from U_i to $U_i' = \phi_i(U_i) \subset \mathbb{R}^n$ under which the Poisson structure Λ is transformed into $\Lambda_{\phi_i} = w_{12}^{(i)}\partial_{u_1} \wedge \partial_{u_2}$, each $g_i \geq 0$ has support in U_i, and moreover $g(x) = \sum_i g_i(x) = 1$ in $\mathbb{R}^n$, summing over a finite number of i's. The existence of such triples follows from pages 21-22 of [2] and Theorem 2.5.3 of [1].

Since Λ is exact, the structure Λ_{ϕ_i} is exact. By statement (a) Λ_{ϕ_i} is a constant Poisson structure corresponding to the variables u_1 and u_2. Hence $w_{12}^{(i)}$ is a function which is independent of the variables u_1 and u_2. Let

$$p_{1i}(u_3, \ldots, u_n) = \int_{u_{30}}^{u_3} w_{12}^{(i)}(s, u_4, \ldots, u_n)ds, \quad p_{ji} = u_{j+2}, \text{ for } j = 2, \ldots, n-2.$$

Then Λ_{ϕ_i} is a Jacobian structure generated by $p_{1i}, \ldots, p_{n-2,i}$ on U_i'. Define

$$P_i(x) = \sum_j P_{ij}(x), \quad x \in \mathbb{R}^n,$$

$i = 1, \ldots, n-2$, where

$$P_{ij}(x) = \begin{cases} g_j(x)(\phi_j^{-1})_* p_{ij}(x), & \text{if } x \in U_j, \\ 0, & \text{if } x \notin U_j. \end{cases}$$

In the above, the subscript '$_*$' denotes the change of function p_{ij} under the coordinate transformation ϕ_j^{-1}. At each point $x \in \mathbb{R}^n$, since $\{(U_i, \phi_i, g_i)\}$ is a partition of unity on $\mathbb{R}^n$, there exists a $j \in \mathbb{N}$ such that $P_i(x) = \left(\phi_j^{-1}\right)_* p_{ij}(x)$. The fact $\Lambda = \left(\phi_j^{-1}\right)_* \left(\Lambda_{\phi_j}\right)$ in U_j implies that Λ is a Jacobian structure generated by $P_1, P_2, \cdots, P_{n-2}$ in some subregion containing x of U_j. Consequently, Λ is a Poisson structure generated by $P_1, \ldots, P_{n-2}$. This proves (c). □

Remark 3.5 1) Part (b) of the above theorem was stated in [12] for quadratic polynomial Poisson structures without a proof, and stated in [15] for general polynomial Poisson structures with a proof given for the quadratic case.

2) In dimension 2, a skew-symmetric, bilinear vector always satisfies the Jacobian identity. If it is exact, then it is either symplectic or trivial. In dimension 3, by part (b) of the above theorem, any Hamiltonian vector field associated to an exact Poisson structure is completely integrable.

3) In dimension 2, the set of Poisson structures forms a vector space, which is isomorphic to $C^\infty(\mathbb{R}^2)$. The set of exact Poisson structures is a 1-dimensional subspace isomorphic to $\mathbb{R}$, which is therefore closed. In higher dimension, the set of Poisson structures cannot form a vector space. But the set of exact Poisson structures in $\mathbb{R}^3$ forms a vector space, which is isomorphic to the quotient space $C^\infty(\mathbb{R}^3)/\mathbb{R}$.

4) Using Proposition 2.8, an alternative proof of statement (b) of Theorem 3.4 can be given as follows. Let $\Omega_1 = w_{23}dx_1 + w_{31}dx_2 + w_{12}dx_3$. Then $d\Omega_1 = 0$, i.e., Ω_1 is closed, hence exact. It follows that there exists a 0-form, i.e., a smooth function P such that $dP = \Omega_1$. This shows that the Poisson structure is Jacobian.

Following [15], we now give another characterization of Jacobian structures depending on their Casimirs. For a Poisson structure Λ on $\mathbf{M}$, a function $h \in C^\infty(\mathbf{M})$ is called a *Casimir* of Λ if $\Lambda(df, dh) = \{f, h\} = 0$ for arbitrary $f \in C^\infty(\mathbf{M})$, or equivalently, the Hamiltonian vector field $\mathbf{X}_h = L_h\Lambda = \Lambda(\cdot, dh)$ is trivial. In other words, a Casimir of a Poisson structure Λ is a first integral of any Hamiltonian vector field $L_H\Lambda = \Lambda(\cdot, dH)$. The set of Casimirs of Λ is called the *center* of Λ.

The necessary part of the first statement in the following theorem was stated in [15] with a proof given for the case $n = 4$. We note that the proof for the case $n = 4$ in [15] does not extend to the general case.

Theorem 3.6 *A Poisson structure Λ in $\mathbb{R}^n$ has $n-2$ functionally independent Casimirs if and only if it is a Jacobian structure. Consequently, if Λ has exactly $n-2$ functionally independent Casimirs, then, excluding a set of zero Lebesgue measure, the common level manifold of the Casimirs is symplectic of dimension* 2.

Proof The sufficient part of the first statement is obvious, because for a Jacobian structure in $\mathbb{R}^n$ its $n-2$ generators are already Casimirs.

We now prove the necessary part of the first statement. Let $\{x_i\}$ be a coordinate system in $\mathbb{R}^n$. Then the Poisson structure Λ has the representation

$$\Lambda(df, dg) \equiv \{f, g\} = \sum_{i,j=1}^{n} w_{ij} \frac{\partial f}{\partial x_i}\frac{\partial g}{\partial x_j}, \qquad \text{for } f, g \in C^\infty(\mathbb{R}^n),$$

where $w_{ij} \in C^\infty(\mathbb{R}^n)$. Let $P_1, \ldots, P_{n-2}$ be a set of functionally independent Casimirs. By definition,

$$\Lambda(dx_i, dP_l) = \sum_{j=1}^{n} w_{ij}\frac{\partial P_l}{\partial x_j} = 0, \qquad i = 1, \ldots, n;\ l = 1, \ldots, n-2. \tag{3.3}$$

For any $n-2$ distinct elements $i_1,\ldots,i_{n-2}$ of $1,\ldots,n$, we denote by $\mathcal{J}(x_{i_1},\ldots,x_{i_{n-2}})$ the Jacobian matrix of $P_1,\ldots,P_{n-2}$ with respect to $x_{i_1},\ldots,x_{i_{n-2}}$, i.e.,

$$\mathcal{J}(x_{i_1},\ldots,x_{i_{n-2}}) = \begin{pmatrix} \frac{\partial P_1}{\partial x_{i_1}} & \cdots & \frac{\partial P_1}{\partial x_{i_{n-2}}} \\ \frac{\partial P_2}{\partial x_{i_1}} & \cdots & \frac{\partial P_2}{\partial x_{i_{n-2}}} \\ \vdots & \ddots & \vdots \\ \frac{\partial P_{n-2}}{\partial x_{i_1}} & \cdots & \frac{\partial P_{n-2}}{\partial x_{i_{n-2}}} \end{pmatrix}.$$

Since $P_1,\ldots,P_{n-2}$ are functionally independent, we can assume without loss of generality that $\det(\mathcal{J}(x_1,\ldots,x_{n-2}))\neq 0$.

Note that equation (3.3) with $i=n$ is equivalent to

$$\mathcal{J}(x_1,\ldots,x_{n-2})\begin{pmatrix} w_{n1} \\ \vdots \\ w_{n,n-2}\end{pmatrix} = -w_{n,n-1}\begin{pmatrix} \frac{\partial P_1}{\partial x_{n-1}} \\ \vdots \\ \frac{\partial P_{n-2}}{\partial x_{n-1}}\end{pmatrix}. \tag{3.4}$$

Since $w_{n,n-1}$ can be arbitrary, if we choose

$$w_{n,n-1} = -u(x_1,\ldots,x_n)\det(\mathcal{J}(x_1,\ldots,x_{n-2})), \tag{3.5}$$

where $u\in C^\infty(\mathbb{R}^n)$ is an arbitrary function, then

$$w_{nj} = u\det(\mathcal{J}(x_1,\ldots,x_{j-1},x_{n-1},x_{j+1},\ldots,x_{n-2})), \quad j=1,\ldots,n-2. \tag{3.6}$$

Similarly, working with equation (3.3) for $i=n-1$ we have

$$w_{n-1,j} = -u\det(\mathcal{J}(x_1,\ldots,x_{j-1},x_n,x_{j+1},\ldots,x_{n-2})), \quad j=1,\ldots,n-2. \tag{3.7}$$

Since equation (3.3) with $i=n-2$ is equivalent to

$$\begin{aligned}\mathcal{J}(x_1,\ldots,x_{n-2})\begin{pmatrix} w_{n-2,1} \\ \vdots \\ w_{n-2,n-2}\end{pmatrix} &= -u\det(\mathcal{J}(x_1,\ldots,x_{n-3},x_n))\mathcal{J}(x_{n-1}) \\ &\quad +u\det(\mathcal{J}(x_1,\ldots,x_{n-3},x_{n-1}))\mathcal{J}(x_n),\end{aligned}$$

we have

$$\begin{aligned} w_{n-2,j} &\\ = -u\Bigg(&\frac{\det(\mathcal{J}(x_1,\ldots,x_{n-3},x_n))\det(\mathcal{J}(x_1,\ldots,x_{j-1},x_{n-1},x_{j+1},\ldots,x_{n-2}))}{\det(\mathcal{J}(x_1,\ldots,x_{n-2}))} \\ &-\frac{\det(\mathcal{J}(x_1,\ldots,x_{n-3},x_{n-1}))\det(\mathcal{J}(x_1,\ldots,x_{j-1},x_n,x_{j+1},\ldots,x_{n-2}))}{\det(\mathcal{J}(x_1,\ldots,x_{n-2}))}\Bigg), \end{aligned} \tag{3.8}$$

for $j=1,\ldots,n-3$.

We claim that, for any $j=1,\cdots,n-2$,

$$\begin{aligned} &\det(\mathcal{J}(x_1,\ldots,x_{n-3},x_n))\det(\mathcal{J}(x_1,\ldots,x_{j-1},x_{n-1},x_{j+1},\ldots,x_{n-2})) \\ &\quad -\det(\mathcal{J}(x_1,\ldots,x_{n-3},x_{n-1}))\det(\mathcal{J}(x_1,\ldots,x_{j-1},x_n,x_{j+1},\ldots,x_{n-2})) \\ &= \det(\mathcal{J}(x_1,\ldots,x_{j-1},x_{n-1},x_{j+1},\ldots,x_{n-3},x_n))\det(\mathcal{J}(x_1,\ldots,x_{n-2})). \end{aligned} \tag{3.9}$$

Indeed, let

$$M = \det \begin{pmatrix} \frac{\partial(P_1,\cdots,P_{n-2})}{\partial(x_1,\cdots,x_{n-3})} & \frac{\partial P}{\partial x_{n-2}} & O & \frac{\partial P}{\partial x_{n-1}} & O & \frac{\partial P}{\partial x_n} \\ O & \frac{\partial P}{\partial x_{n-2}} & \frac{\partial(P_1,\cdots,P_{n-2})}{\partial(x_1,\cdots,x_{j-1})} & \frac{\partial P}{\partial x_{n-1}} & \frac{\partial(P_1,\cdots,P_{n-2})}{\partial(x_{j+1},\cdots,x_{n-3})} & \frac{\partial P}{\partial x_n} \end{pmatrix},$$

where

$$\frac{\partial(P_1,\cdots,P_l)}{\partial(x_1,\cdots,x_k)} = \begin{pmatrix} \frac{\partial P_1}{\partial x_1} & \cdots & \frac{\partial P_1}{\partial x_k} \\ \vdots & \ddots & \vdots \\ \frac{\partial P_l}{\partial x_1} & \cdots & \frac{\partial P_l}{\partial x_k} \end{pmatrix},$$

$$\frac{\partial P}{\partial x_k} = \begin{pmatrix} \frac{\partial P_1}{\partial x_k} \\ \vdots \\ \frac{\partial P_{n-2}}{\partial x_k} \end{pmatrix},$$

$l, k = 1, \cdots, n$, and O denotes a zero matrix with appropriate dimension.

By expanding M according to the first $n-2$ rows, we have

$$\begin{aligned} M \;=\; & \det(\mathcal{J}(x_1,\ldots,x_{n-3},x_{n-2}))\det(\mathcal{J}(x_1,\ldots,x_{j-1},x_{n-1},x_{j+1},\ldots,x_{n-3},x_n)) \\ & + \det(\mathcal{J}(x_1,\ldots,x_{n-3},x_{n-1}))\det(\mathcal{J}(x_1,\ldots,x_{j-1},x_n,x_{j+1},\ldots,x_{n-2})) \\ & - \det(\mathcal{J}(x_1,\ldots,x_{n-3},x_n))\det(\mathcal{J}(x_1,\ldots,x_{j-1},x_{n-1},x_{j+1},\ldots,x_{n-2})). \end{aligned}$$

Since $M = 0$, the claim follows.

From (3.8) and (3.9), we have

$$w_{n-2,j} = -u\det(\mathcal{J}(x_1,\ldots,x_{j-1},x_{n-1},x_{j+1},\ldots,x_{n-3},x_n)), \tag{3.10}$$

$j = 1, \ldots, n-3$. Now, equation (3.3) with $1 < k < n-2$ reads

$$\begin{aligned} \mathcal{J}(x_1,\ldots,x_{n-2}) & \begin{pmatrix} w_{k,1} \\ \vdots \\ w_{k,n-2} \end{pmatrix} \\ = \;& -u\det(\mathcal{J}(x_1,\ldots,x_{k-1},x_n,x_{k+1},\ldots,x_{n-2}))\mathcal{J}(x_{n-1}) \\ & +u\det(\mathcal{J}(x_1,\ldots,x_{k-1},x_{n-1},x_{k+1},\ldots,x_{n-2}))\mathcal{J}(x_n). \end{aligned}$$

Similar to the proof for the case $n-2$ by using a slight modification of (3.9), we have

$$w_{kj} = -u\det(\mathcal{J}(x_1,\ldots,x_{j-1},x_{n-1},x_{j+1},\ldots,x_{k-1},x_n,x_{k+1},\ldots,x_{n-2})), \tag{3.11}$$

for $j = 1, \ldots, k-1$.

Combining the formulas (3.5), (3.6), (3.7), (3.10) and (3.11), direct calculation yields that

$$\Lambda(df, dg) = \sum_{i,j=1}^{n} w_{ij}\frac{\partial f}{\partial x_i}\frac{\partial g}{\partial x_j} = u\det(J(f, g, P_1, \ldots, P_{n-2})).$$

This proves the first statement.

To prove the second statement, we denote by $\mathbf{M}$ the common level manifold of the Casimirs of Λ. Then $\dim \mathbf{M} = 2$ except perhaps at a subset of zero Lebesgue measure.

From the proof of the first statement, it follows that the $n-2$ functionally independent Casimirs are the generators of the Jacobian structure Λ. In terms of suitable local coordinates $\{x_i\}$, it is easily seen that Λ has the canonical form

$\partial_{x_{n-1}} \wedge \partial_{x_n} - \partial_{x_n} \wedge \partial_{x_{n-1}}$ in a neighborhood of each point of the manifold $\mathbf{M}$ except perhaps a subset of zero Lebesgue measure. This means that the structure matrix of Λ has rank 2 in $\mathbb{R}^n$ except perhaps a subset of zero Lebesgue measure, which is equal to the dimension of the manifold. Thus, the second statement follows. □

Part (a) of the following theorem was stated in [15] with a proof given for the case $n = 4$.

Theorem 3.7 *Let Λ be a Jacobian structure in $\mathbb{R}^n$ generated by $P_1, \ldots, P_{n-2} \in C^\infty(\mathbb{R}^n)$. The following holds.*

(a) *If $P_1, \ldots, P_{n-2}$ are functionally independent, then the center of Λ is a sub-algebra of $C^\infty(\mathbb{R}^n)$, generated by $P_1, \ldots, P_{n-2}$.*

(b) *For any $H \in C^\infty(\mathbb{R}^n)$, the Hamiltonian vector field $\mathbf{X}_H = \Lambda(\cdot, dH)$ has the canonical form*

$$\begin{cases} \dot{I} = 0, \\ \dot{\phi} = \omega(I), \end{cases}$$

where $I = (I_1, \ldots, I_{n-1})^\top$ and $\omega(I) = 0$ if H is functionally dependent on $P_1, \ldots, P_{n-2}$.

Proof (a) Since there are n functionally independent functions in the algebra $C^\infty(\mathbb{R}^n)$, we can choose two functions $f, g \in C^\infty(\mathbb{R}^n)$ such that $f, g, P_1, \ldots, P_{n-2}$ are functionally independent. Assume that Λ is the Jacobian structure generated by $P_1, \ldots, P_{n-2}$. For any $h \in C^\infty(\mathbb{R}^n)$, denote by $\mathbf{X}_h$ as the vector field $\Lambda(\cdot, dh)$. Then $f, P_1, \ldots, P_n$ are functionally independent first integrals of the vector field $\mathbf{X}_f$, and $g, P_1, \ldots, P_n$ are functionally independent first integrals of the vector field $\mathbf{X}_g$. Since a vector field in dimension n has at most $n-1$ functionally independent first integrals, if it has the maximal number of first integrals, then any other first integral is a function of these $n-1$ functionally independent first integrals.

Suppose that $H \in C^\infty(\mathbb{R}^n)$ is a Casimir of the Jacobian structure Λ. Then for any $h \in C^\infty(\mathbb{R}^n)$, we have $\Lambda(dh, dH) = 0$. Consequently, $\mathbf{X}_f(H) = \Lambda(dH, df) = 0$ and $\mathbf{X}_g(H) = \Lambda(dH, dg) = 0$. This means that H is a first integral of both vector fields $\mathbf{X}_f$ and $\mathbf{X}_g$. Hence, H is a smooth function of $f, P_1, \ldots, P_{n-2}$, and also a smooth function of $g, P_1, \ldots, P_{n-2}$. It follows from the functional independency of f and g that H should be a smooth function of $P_1, \ldots, P_{n-2}$. Hence, the center of the Jacobian Poisson structure is a sub-algebra of $C^\infty(\mathbb{R}^n)$, generated by $P_1, \ldots, P_{n-2}$.

(b) If H is functionally dependent on $P_1, \ldots, P_{n-2}$, then the vector field $\mathbf{X}_H$ is trivial. Hence, it has the canonical form with $\omega(I) \equiv 0$.

If H is functionally independent of $P_1, \ldots, P_{n-2}$, then the vector field $\mathbf{X}_H$ is completely integrable with the $n-1$ functionally independent first integrals $H, P_1, \ldots, P_{n-2}$. Hence, the vector field $\mathbf{X}_H$ has the desired normal form. □

Combining Theorem 3.6 and statement (a) of Theorem 3.7, we easily have the following.

Corollary 3.8 (a) *The center of a Poisson structure in $\mathbb{R}^n$ having functionally independent Casimirs $P_1, \ldots, P_{n-2}$ is a sub-algebra generated by the $n-2$ given Casimirs.*

(b) *In $\mathbb{R}^3$, a Poisson structure has a Casimir if and only if it is a Jacobian structure. And a Poisson structure is exact if and only if it is a Jacobian structure with a constant Jacobian coefficient.*

Our next result characterizes Poisson structures in $\mathbb{R}^3$ with a Casimir whose level surface is a quadric surface.

Theorem 3.9 *For a Poisson structure Λ in $\mathbb{R}^3$, the following holds.*

(a) *Λ has no Casimir which defines a torus.*

(b) *Λ has a Casimir which defines a quadric surface if and only if it is affine equivalent to one of the following module a Jacobian coefficient u:*

- $\Lambda_1 = -z\partial_x \wedge \partial_y + y\partial_x \wedge \partial_z - x\partial_y \wedge \partial_z$, *with the Casimir* $C = x^2 + y^2 + z^2$;
- $\Lambda_2 = z\partial_x \wedge \partial_y + y\partial_x \wedge \partial_z - x\partial_y \wedge \partial_z$, *with the Casimir* $C = x^2 + y^2 - z^2$;
- $\Lambda_3 = \partial_x \wedge \partial_y + 2y\partial_x \wedge \partial_z - 2x\partial_y \wedge \partial_z$, *with the Casimir* $C = x^2 + y^2 - z$;
- $\Lambda_4 = \partial_x \wedge \partial_y - 2y\partial_x \wedge \partial_z - 2x\partial_y \wedge \partial_z$, *with the Casimir* $C = x^2 - y^2 - z$;
- $\Lambda_5 = 2\partial_x \wedge \partial_z + x\partial_y \wedge \partial_z$, *with the Casimir* $C = x^2 - 4y$;
- $\Lambda_6 = y\partial_x \wedge \partial_z - x\partial_y \wedge \partial_z$, *with the Casimir* $C = x^2 + y^2$;
- $\Lambda_7 = y\partial_x \wedge \partial_z + x\partial_y \wedge \partial_z$, *with the Casimir* $C = x^2 - y^2$.

Proof It follows from direct calculations. □

Remark 3.10 1) In $\mathbb{R}^3$, the restriction of a Poisson structure to each submanifold defined by a level surface of its Casimir becomes symplectic except at singular points.

2) In $\mathbb{R}^3$, normal forms can be obtained for exact Poisson structures around each singular point ([7]).

3) Theorem 3.9 can be used to construct completely integrable Hamiltonian systems. In [3], a method of constructing completely integrable Hamiltonian systems starting from a Poisson coalgebra formed by Lie algebra was developed. As an example, using the Casimir associated to Λ_2 above and its deformation, the authors obtained a large class of completely integrable Hamiltonian systems with an arbitrary number of degrees of freedom. In [11], another method of constructing completely integrable Hamiltonian systems was developed, based on the fact that the symplectic manifold defined by the Casimirs of a Poisson structure cannot form a Lie algebra. Applying this method to the Poisson structure Λ_3 above and its associated Casimir, the author also obtained a class of completely integral Hamiltonian systems having arbitrary number of degrees of freedom, including the Calogero system.

3.3 Quasi-homogeneous Poisson structures. A Poisson structure

$$L = \sum_{i,j=1}^{n} w_{ij} \partial_i \wedge \partial_j$$

in $\mathbb{R}^n$ is called *quasi-homogeneous* of weight degree m with respect to a weight $\mathbf{w} = (w_1, \ldots, w_n)$ if every w_{ij} is a quasi-homogeneous polynomial of weight degree $m - 2 + w_i + w_j$ with the same weight $\mathbf{w}$, where m and w_i are positive integers, and $w_1, \ldots, w_n$ do not have common factors. We recall that a polynomial is quasi-homogeneous of weight degree m with respect to the weight $\mathbf{w}$ if each of its monomials, $x_1^{k_1} \ldots x_n^{k_n}$, satisfies $k_1 w_1 + \ldots + k_n w_n = m$. If $w_1 = \ldots = w_n = 1$, a quasi-homogeneous polynomial becomes homogeneous.

The following is an improvement to the results given in Theorem 3.1 of [12] and Theorem 7 of [15].

Theorem 3.11 *Assume that Λ is a quasi-homogeneous Poisson structure of degree m with respect to a weight $\mathbf{w} = (w_1, \ldots, w_n)$. The following holds.*

(a) *The Poisson structure* Λ *can be decomposed into the form* $\Lambda = \Lambda_0 + c\mathbf{X}_\Lambda \wedge \mathbf{X}_E$ *if and only if the Poisson structure* Λ *is homogeneous and* $c = 1/(m+n-1)$, *where* Λ_0 *is an exact Poisson structure,* $\mathbf{X}_\Lambda = D(\Lambda)$ *and* $\mathbf{X}_E = \sum_{i=1}^n w_i x_i \partial_i$.

(b) *The Poisson structure* Λ *can be decomposed as*

$$\Lambda = \pi + \frac{1}{m-2+\sum w_i}\left(\mathbf{X}_\Lambda \wedge \mathbf{X}_E - \Lambda(\mathbf{X}_E)\right),$$

where π *is an exact bilinear vector field, and* $\mathbf{X}_\Lambda, \mathbf{X}_E$ *are as in* (a).

Proof (a) Let $\Lambda_0 = \Lambda - c\mathbf{X}_\Lambda \wedge \mathbf{X}_E$. Using statement (ii) of Proposition 2.1, we have

$$D(\Lambda - c\mathbf{X}_\Lambda \wedge \mathbf{X}_E) = \mathbf{X}_\Lambda - c([\mathbf{X}_E, \mathbf{X}_\Lambda] + (\mathrm{div}_\omega \mathbf{X}_E)\mathbf{X}_\Lambda - (\mathrm{div}_\omega \mathbf{X}_\Lambda)\mathbf{X}_E).$$

For $\mathbf{X}_\Lambda = D(\Lambda)$, combining (2.2) and statement (i) of Proposition 2.1 we have that $\mathrm{div}_\omega \mathbf{X}_\Lambda = 0$. Moreover, $\mathrm{div}_\omega \mathbf{X}_E = \sum_{i=1}^n w_i$. Direct calculations using the identities $\mathbf{X}_E(w_{ij}) = (m-2+w_i+w_j)w_{ij}$ and $\mathbf{X}_E\left(\frac{\partial w_{ij}}{\partial x_j}\right) = \frac{\partial}{\partial x_j}(\mathbf{X}_E w_{ij})$ yield

$$\mathbf{X}_\Lambda(\mathbf{X}_E) = 2\sum_{i=1}^n w_i \left(\sum_{j=1}^n \frac{\partial w_{ij}}{\partial x_j}\right)\partial_i,$$

$$\begin{aligned}
&\mathbf{X}_E(\mathbf{X}_\Lambda)\\
&= 2\sum_{i=1}^n \left(\sum_{j=1}^n \frac{\partial}{\partial x_j}(\mathbf{X}_E w_{ij})\right)\partial_i = 2\sum_{i=1}^n \left(\sum_{j=1}^n (m-2+w_i+w_j)\frac{\partial w_{ij}}{\partial x_j}\right)\partial_i\\
&= (m-2)\mathbf{X}_\Lambda + 2\sum_{i=1}^n w_i \left(\sum_{j=1}^n \frac{\partial w_{ij}}{\partial x_j}\right)\partial_i + 2\sum_{i=1}^n \left(\sum_{j=1}^n w_j \frac{\partial w_{ij}}{\partial x_j}\right)\partial_i.
\end{aligned}$$

Therefore,

$$\begin{aligned}
[\mathbf{X}_E, \mathbf{X}_\Lambda] &= \mathbf{X}_E(\mathbf{X}_\Lambda) - \mathbf{X}_\Lambda(\mathbf{X}_E)\\
&= (m-2)\mathbf{X}_\Lambda + 2\sum_{i=1}^n \left(\sum_{j=1}^n w_j \frac{\partial w_{ij}}{\partial x_j}\right)\partial_i. \qquad (3.12)
\end{aligned}$$

Moreover,

$$\begin{aligned}
&D(\Lambda - c\mathbf{X}_\Lambda \wedge \mathbf{X}_E)\\
&= \left(1 - c\left(m-2+\sum_{i=1}^n w_i\right)\right)\mathbf{X}_\Lambda - 2c\sum_{i=1}^n \left(\sum_{j=1}^n w_j \frac{\partial w_{ij}}{\partial x_j}\right)\partial_i\\
&= 2\sum_{i=1}^n \left(\sum_{j=1}^n \left(1 - c\left(m-2+\sum_{l=1}^n w_l + w_j\right)\right)\frac{\partial w_{ij}}{\partial x_j}\right)\partial_i.
\end{aligned}$$

In order for $D(\Lambda - c\mathbf{X}_\Lambda \wedge \mathbf{X}_E) \equiv 0$ to hold, we should have $1 - c(m-2+\sum_{l=1}^n w_l + w_j) = 0$, $j = 1, \ldots, n$. This is equivalent to $w_1 = \ldots = w_n = 1$, because $w_1, \ldots, w_n$ are relatively prime. It follows that $c = 1/(m-1+n)$, and the Poisson structure Λ is homogeneous of degree m.

Now we have by (3.12) that $[\mathbf{X}_E, \mathbf{X}_\Lambda] = (m-1)\mathbf{X}_\Lambda$. Therefore, by definition of the Schouten bracket (see (iii) of Proposition 2.1) it is easy to see that

$$[\mathbf{X}_\Lambda \wedge \mathbf{X}_E, \mathbf{X}_\Lambda \wedge \mathbf{X}_E] \quad = \quad -\mathbf{X}_E \wedge [\mathbf{X}_\Lambda, \mathbf{X}_E] \wedge \mathbf{X}_\Lambda - \mathbf{X}_\Lambda \wedge [\mathbf{X}_E, \mathbf{X}_\Lambda] \wedge \mathbf{X}_E = 0.$$

This means that $\mathbf{X}_\Lambda \wedge \mathbf{X}_E$ satisfies the Jacobi identity. Consequently, $\Lambda_0 = \Lambda - (1/(m-1+n))\mathbf{X}_\Lambda \wedge \mathbf{X}_E$ is an exact Poisson structure. This proves statement (a).

(b) We let $\pi = \Lambda - (1/(m-2+\sum w_i))\,(\mathbf{X}_\Lambda \wedge \mathbf{X}_E - \Lambda(\mathbf{X}_E))$. Since $\Lambda(\mathbf{X}_E) = \sum_{i,j=1}^n w_j w_{ij} \partial_i \wedge \partial_j$, the statement follows from formula (2.2) and the proof of statement (a). □

We remark that in [15], a decomposition of a homogeneous Poisson structure similar to that in the statement (a) of the above theorem was already obtained but the coefficient c was not correctly computed.

For homogeneous Poisson structures, an important example is related to the quantization of quadratic Poisson structures. It is known that on a real vector space V, one can define a homogeneous, quadratic Poisson structure on V using classical r-matrix on the Lie algebra of linear operators. The converse is true in dimension 2 (see e.g., [12]) but not in general. In [13], the authors gave a counter-example showing that there exists a homogeneous quadratic Poisson structure in dimension 3 which cannot be realized by any classical r-matrix. We recall that a *classical r-matrix* on a Lie algebra $\mathfrak{g}$ is an element r of $\mathfrak{g} \wedge \mathfrak{g}$ for which the Schouten bracket $[r, r] = 0$. Let $\mathcal{P} : M_n \to \mathcal{X}(\mathbb{R}^n)$ be a real linear isomorphism from the set of $n \times n$ real matrices to the set of vector fields, defined by

$$\mathcal{P}(E_{ij}) = x_i \partial_j,$$

where E_{ij} is the matrix with entries all vanishing except one equals 1 on the ith row and the jth column. Then for any classical r-matrix $M_1 \wedge M_2 \in \mathfrak{g} \wedge \mathfrak{g}$, $\mathcal{P}(r) = \mathcal{P}(M_1) \wedge \mathcal{P}(M_2)$ is a quadratic Poisson structure. A *classical r-Poisson structure* is by definition a quadratic Poisson structure which is the image of a r-matrix under $\mathcal{P}$.

The following results provide a partial answer to the problem we posted toward the end of Section 1.

Theorem 3.12 *Let $\mathfrak{g}$ be the 9-dimensional Lie algebra of linear operators in $\mathbb{R}^3$. Assume that $r = \mathbf{A} \wedge \mathbf{B} \in \mathfrak{g} \wedge \mathfrak{g}$, where $\mathbf{A} = (a_{ij})$ and $\mathbf{B} = (b_{ij})$ are real 3×3 matrices. The following holds.*

(a) *$\mathcal{P}(r)$ is a classical r-Poisson structure and is exact if and only if*

$$\mathbf{AB} + \mathrm{tr}(\mathbf{A})\mathbf{B} = \mathbf{BA} + \mathrm{tr}(\mathbf{B})\mathbf{A}, \tag{3.13}$$

where $\mathrm{tr}(\mathbf{A})$ denotes the trace of the matrix $\mathbf{A}$.

(b) *Any quadratic Jacobian structure in dimension 3 is a classical r-Poisson structure.*

Proof (a) By definition we have

$$\mathcal{P}(r) = w_{12}\partial_1 \wedge \partial_2 + w_{23}\partial_2 \wedge \partial_3 + w_{31}\partial_3 \wedge \partial_1,$$

where

$$
\begin{aligned}
w_{12} &= (a_{11}x_1 + a_{21}x_2 + a_{31}x_3)(b_{12}x_1 + b_{22}x_2 + b_{32}x_3) \\
&\quad -(a_{12}x_1 + a_{22}x_2 + a_{32}x_3)(b_{11}x_1 + b_{21}x_2 + b_{31}x_3), \\
w_{23} &= (a_{12}x_1 + a_{22}x_2 + a_{32}x_3)(b_{13}x_1 + b_{23}x_2 + b_{33}x_3) \\
&\quad -(a_{13}x_1 + a_{23}x_2 + a_{33}x_3)(b_{12}x_1 + b_{22}x_2 + b_{32}x_3), \\
w_{31} &= (a_{13}x_1 + a_{23}x_2 + a_{33}x_3)(b_{11}x_1 + b_{21}x_2 + b_{31}x_3) \\
&\quad -(a_{11}x_1 + a_{21}x_2 + a_{31}x_3)(b_{13}x_1 + b_{23}x_2 + b_{33}x_3).
\end{aligned}
\tag{3.14}
$$

Again it follows from definition that r is a classical r-matrix if and only if $\mathcal{P}(r)$ is a Poisson structure, i.e.,

$$
\sum_{l=1}^{3}\left(w_{l1}\frac{\partial w_{23}}{\partial x_l} + w_{l3}\frac{\partial w_{12}}{\partial x_l} + w_{l2}\frac{\partial w_{31}}{\partial x_l}\right) = 0. \tag{3.15}
$$

For the Poisson structure $\mathcal{P}(r)$ to be exact, we should have

$$
\frac{\partial w_{12}}{\partial x_2} = \frac{\partial w_{31}}{\partial x_3}, \quad \frac{\partial w_{23}}{\partial x_3} = \frac{\partial w_{12}}{\partial x_1}, \quad \frac{\partial w_{31}}{\partial x_1} = \frac{\partial w_{23}}{\partial x_2}. \tag{3.16}
$$

Since (3.16) implies (3.15), $\mathcal{P}(r)$ is an exact classical r-Poisson structure if and only if the condition (3.16) holds.

By comparing coefficients of x_i in (3.16) for $i = 1, 2, 3$, we obtain the condition (3.13).

(b) From the proof of statement (a) and Theorem 3.4, it follows that $\mathcal{P}(r)$ is an exact classical r-Poisson structure if and only if it is a Jacobian structure generated by

$$
\begin{aligned}
P_r &= \frac{1}{3}(a_{12}b_{13} - a_{13}b_{12})x^3 + (a_{13}b_{11} - a_{11}b_{13})x^2y \\
&\quad +\frac{1}{2}(a_{13}b_{21} + a_{23}b_{11} - a_{11}b_{23} - a_{21}b_{13})xy^2 + \frac{1}{3}(a_{23}b_{21} - a_{21}b_{23})y^3 \\
&\quad +(a_{11}b_{12} - a_{12}b_{11})x^2z + (a_{11}b_{22} + a_{21}b_{12} - a_{12}b_{21} - a_{22}b_{11})xyz \\
&\quad +(a_{21}b_{22} - a_{22}b_{21})y^2z + \frac{1}{2}(a_{11}b_{32} + a_{31}b_{12} - a_{12}b_{31} - a_{32}b_{11})xz^2 \\
&\quad +\frac{1}{2}(a_{21}b_{32} + a_{31}b_{22} - a_{22}b_{31} - a_{32}b_{21})yz^2 + \frac{1}{3}(a_{31}b_{32} - a_{32}b_{31})z^3,
\end{aligned}
$$

where (a_{ij}) and (b_{ij}) satisfy the condition (3.13).

Now for any quadratic Jacobian structure generated by $P = Ax^3 + Bx^2y + Cxy^2 + Dy^3 + Ex^2z + Fxyz + Gy^2z + Hxz^2 + Iyz^2 + Jz^3$ to be a classical r-Poisson structure, it is sufficient to find a $r = (a_{ij}) \wedge (b_{ij}) \in \mathfrak{g} \wedge \mathfrak{g}$ for which the condition (3.13) is satisfied and $P_r = P$. By direct computations, these conditions are equivalent to

$$
\begin{aligned}
&3Jb_{33} + Ib_{32} + Hb_{31} = 0,\\
&3Ab_{23}b_{33}M_0M_3 - 2Bb_{23}b_{33}M_2M_3 + Cb_{11}b_{33}M_2M_3 - 3Db_{33}^2M_1M_2\\
&\qquad +2Eb_{23}b_{33}M_1M_3 + 2Gb_{23}b_{33}M_1M_2 - Hb_{13}b_{23}M_1M_3 = Ib_{23}^2M_1M_2,\\
&3Ab_{22}b_{23}(b_{33}M_0M_4 - b_{23}M_2M_3) + 2Bb_{23}M_2(b_{13}b_{22}M_3\\
&\qquad +b_{23}b_{33}(b_{12}b_{21} - b_{11}b_{22})) + C(b_{12}b_{23}b_{33}M_4 - b_{13}^2b_{22}M_3)M_2\\
&\qquad +3Db_{33}(b_{12}b_{23} + b_{13}b_{22})M_1M_2 + 2Eb_{22}b_{23}b_{33}M_1M_4\\
&\qquad +Fb_{23}^2(b_{22} + b_{33})M_1M_2 - 2Gb_{13}b_{22}b_{23}M_1M_2 - Hb_{13}b_{22}b_{23}M_1M_4 = 0,\\
&3AM_0^2 - 2BM_0M_2 + CM_2^2 + 2EM_0M_1 - FM_1M_2 + HM_1^2 = 0,
\end{aligned}
\tag{3.17}
$$

$a_{21} = 0$, and,

$$
\begin{aligned}
b_{21}b_{22}b_{23}M_1a_{11} &= -3Ab_{21}^2b_{22}b_{23} + 2Bb_{12}b_{21}^2b_{23} - Cb_{21}(b_{12}b_{13}b_{21} - b_{11}M_1)r\\
&\quad +3D(b_{12}b_{21} + b_{11}b_{22})M_1 + Fb_{21}b_{23}M_1,\\
b_{21}b_{23}M_1a_{12} &= -3Ab_{21}b_{22}b_{23} + 2Bb_{12}b_{21}b_{23} - Cb_{12}b_{13}b_{21} + 3Db_{12}M_1,\\
b_{21}b_{23}M_1a_{13} &= -3Ab_{21}b_{23}^2 + 2Bb_{13}b_{21}b_{23} - Cb_{13}^2b_{21} + 3Db_{13}M_1,\\
b_{21}b_{23}a_{22} &= Cb_{21} + 3Db_{22},\\
b_{21}a_{23} &= 3D,\\
b_{21}b_{23}^2M_1M_2a_{31} &= -3Ab_{21}^2b_{23}b_{33}M_0 + 2Bb_{21}^2b_{23}b_{33}M_2 - Cb_{13}b_{21}^2b_{33}M_2\\
&\quad +3D(b_{23}b_{31} + b_{21}b_{33})M_1M_2 - 2Eb_{21}^2b_{23}b_{33}M_1\\
&\quad -2Gb_{21}b_{23}M_1M_2 + Hb_{13}b_{21}^2b_{23}M_1,\\
b_{21}b_{23}M_1M_2a_{32} &= -3Ab_{21}b_{23}b_{32}M_0 + 2Bb_{21}b_{23}b_{32}M_2 - Cb_{13}b_{21}b_{32}M_2\\
&\quad +3Db_{32}M_1M_2 - 2Eb_{21}b_{23}b_{32}M_1 + Hb_{12}b_{21}b_{23}M_1,\\
b_{21}b_{23}M_1M_2a_{33} &= -3Ab_{21}b_{23}b_{33}M_0 + 2Bb_{21}b_{23}b_{33}M_2 - Cb_{13}b_{21}b_{33}M_2\\
&\quad +3Db_{33}M_1M_2 - 2Eb_{21}b_{23}b_{33}M_1 + Hb_{13}b_{21}b_{23}M_1,
\end{aligned}
\tag{3.18}
$$

where $M_0 = b_{22}b_{33} - b_{23}b_{32}$, $M_1 = b_{12}b_{23} - b_{13}b_{22}$, $M_2 = b_{12}b_{33} - b_{13}b_{32}$, $M_3 = b_{21}b_{33} - b_{23}b_{31}$, and $M_4 = b_{11}b_{23} - b_{13}b_{21}$.

For any given real numbers $A, B, \ldots, J$, system (3.17) has solutions b_{ij}, $1 \leq i, j \leq 3$ for which system (3.18) also has solutions a_{ij}, $1 \leq i, j \leq 3$. Corresponding to these $\{b_{ij}\}$ and $\{a_{ij}\}$, $\mathcal{P}(r)$ with $r = (a_{ij}) \wedge (b_{ij})$ is the Jacobian structure generated by the prescribed homogeneous polynomial P. $\square$

4 Invariant exact Poisson structures under a completely integrable flow

According to [4], a smooth vector field in an n-dimensional smooth manifold $\mathbf{M}^n$ is said to be *integrable in the broad sense* if it has p functionally independent first integrals with $0 \leq p < n$, and an abelian $(n - p)$ dimensional Lie algebra of symmetries which preserve the p first integrals and are linear independent on the manifold except perhaps at a set of zero Lebesgue measure. This notion of integrability generalizes the classical notion of integrability in the Liouville sense for standard Hamiltonian systems.

For a smooth dynamical system on an n-dimensional smooth manifold which is integrable in the broad sense, we let $\mathbf{M}_l^q$, $q = n - p$, be connected components of the general invariant submanifolds formed by the common level surfaces of the p first integrals, parametrized by $l \in \mathbb{R}^n$. The followings were shown in [4].

- If $\mathbf{M}_l^q$ is compact, then it is a torus $\mathbb{T}^q$.
- If $\mathbf{M}_l^q$ is non-compact, then it is a toroidal cylinder $\mathbb{T}^{q-m} \times \mathbb{R}^m$.
- In a toroidal neighborhood $B_p \times \mathbb{T}^{q-m} \times \mathbb{R}^m$ of the toroidal cylinder, where B_p is a p-dimensional ball, there exist local coordinates $I = (I_1, \ldots, I_p) \in B_p$,

$\phi = (\phi_1, \ldots, \phi_{q-m}) \in \mathbb{T}^{q-m}$, and $\rho = (\rho_1, \ldots, \rho_m) \in \mathbb{R}^m$, such that the dynamical system is equivalent to

$$\begin{cases} \dot{I} = 0, \\ \dot{\phi} = \omega(I), \\ \dot{\rho} = Q(I), \end{cases} \tag{4.1}$$

where $\omega(I) = (\omega_1(I), \cdots, \omega_{q-m}(I))^\top$ and $Q(I) = (q_1(I), \cdots, q_m(I))^\top$ are smooth functions. Moreover, in the case that all $\mathbf{M}_l^q$ are compact, then $m = 0$ and the last equation of (4.1) does not appear.

Consequently, if a smooth dynamical system is integrable in the broad sense, then it is completely integrable in the conventional sense.

We now consider the case that all $\mathbf{M}_l^q$ are compact, i.e., the system (4.1) reduces to

$$\begin{cases} \dot{I} = 0, \\ \dot{\phi} = \omega(I), \end{cases} \tag{4.2}$$

where (I, ϕ) lies in a toroidal domain $B_p \times \mathbb{T}^q \subset \mathbf{M}^n$. We denote by $\mathbf{X}$ the vector field corresponding to (4.2). The dynamical system (4.2) is said to be $\mathbb{T}^q$-*dense at a point* $I_0 \in B_p$ if for any $\phi_0 \in \mathbb{T}^q$, the orbit $\{\phi_0 + \omega(I_0)t\}$ is dense on the torus $\mathbb{T}^q$. Let $\Omega_p \subset B_p$ be the set of points $I \in B_p$ at which the trajectories of system (4.2) are $\mathbb{T}^q$-dense. The dynamical system (4.2) or its induced flow is said to be $\mathbb{T}^q$-*dense* in the toroidal domain if the set Ω_p is everywhere dense in B_p. Clearly, the dynamical system (4.2) is $\mathbb{T}^q$-dense if and only if the frequencies $\{\omega_1(I), \cdots, \omega_q(I)\}$ are rationally independent (i.e., $\omega(I)$ is non-resonant) for almost all $I \in B_p$ and if and only if any first integral is a function of I.

Closed 2-forms which are invariant under the $\mathbb{T}^q$-dense dynamical system (4.2) were characterized in [4]. It was also shown that the system (4.2) preserving a symplectic 2-form must be Hamiltonian. However, if a 2-form is not symplectic, then in general it has no corresponding Poisson structure. We now consider the characterization of the invariant exact Poisson structures under the $\mathbb{T}^q$-dense dynamical system (4.2).

Theorem 4.1 *A Poisson structure Λ defined in a toroidal domain $B_p \times \mathbb{T}^q \subset \mathbf{M}^n$ is exact and invariant under the $\mathbb{T}^q$-dense dynamical system* (4.2) *if and only if*

$$\Lambda = a_{ij}(I)\frac{\partial}{\partial I_i} \wedge \frac{\partial}{\partial I_j} + b_{kl}(I)\frac{\partial}{\partial I_k} \wedge \frac{\partial}{\partial \phi_l} + c_{rs}(I)\frac{\partial}{\partial \phi_r} \wedge \frac{\partial}{\partial \phi_s}, \tag{4.3}$$

with the coefficients satisfying

$$\begin{aligned} \mathbf{a}_i \cdot \nabla\omega_k &= 0, \quad \mathbf{b}_l \cdot \nabla\omega_k - \mathbf{b}_k \cdot \nabla\omega_l = 0, \qquad (4.4) \\ \operatorname{div} \mathbf{a}_i &= 0, \qquad\qquad \operatorname{div} \mathbf{b}_l = 0, \qquad (4.5) \end{aligned}$$

where $\mathbf{a}_i = (a_{i1}, \ldots, a_{ip})$, $i = 1, \cdots, p$, *and* $\mathbf{b}_k = (b_{1k}, \ldots, b_{pk})$, $k, l = 1, \cdots, q$.

Proof Assume that in the toroidal domain, the Poisson structure is of the form

$$\Lambda = a_{ij}(I,\phi)\frac{\partial}{\partial I_i} \wedge \frac{\partial}{\partial I_j} + b_{kl}(I,\phi)\frac{\partial}{\partial I_k} \wedge \frac{\partial}{\partial \phi_l} + c_{rs}(I,\phi)\frac{\partial}{\partial \phi_r} \wedge \frac{\partial}{\partial \phi_s}. \tag{4.6}$$

The Lie derivative of Λ with respect to $\mathbf{X} = \sum_{k=1}^{q} \omega_k(I) \frac{\partial}{\partial \phi_k}$ reads

$$\begin{aligned} L_{\mathbf{X}}\Lambda &= \sum_{1 \le i < j \le p} \mathbf{X}(a_{ij}) \frac{\partial}{\partial I_i} \wedge \frac{\partial}{\partial I_j} \\ &+ \sum_{1 \le i \le p, 1 \le k \le q} \left(\mathbf{X}(b_{ik}) + \sum_{r=1, r \neq i}^{p} a_{ri} \frac{\partial \omega_k}{\partial I_r} \right) \frac{\partial}{\partial I_i} \wedge \frac{\partial}{\partial \phi_k} \\ &+ \sum_{1 \le k < l \le q} \left(\mathbf{X}(c_{kl}) - \sum_{r=1}^{p} b_{rl} \frac{\partial \omega_k}{\partial I_r} + \sum_{r=1}^{p} b_{rk} \frac{\partial \omega_k}{\partial I_r} \right) \frac{\partial}{\partial \phi_k} \wedge \frac{\partial}{\partial \phi_l}. \end{aligned}$$

In order for Λ to be invariant under the $\mathbb{T}^q$-dense dynamical system (4.2), the above Lie derivative must vanish, i.e.,

$$\begin{aligned} \dot{a}_{ij} &= \mathbf{X}(a_{ij}) = 0, \\ \dot{b}_{ik} &= \mathbf{X}(b_{ik}) = -\sum_{r=1}^{p} a_{ri} \frac{\partial \omega_k}{\partial I_r}, \\ \dot{c}_{kl} &= \mathbf{X}(c_{kl}) = \sum_{r=1}^{p} b_{rl} \frac{\partial \omega_k}{\partial I_r} - \sum_{r=1}^{p} b_{rk} \frac{\partial \omega_l}{\partial I_r}, \end{aligned}$$

$i, j = 1, \cdots, p$; $k, l = 1, \cdots, q$. The first equation in the above means that each a_{ij} is a first integral of the $\mathbb{T}^q$-dense dynamical system (4.2), hence each a_{ij} is a function which is independent of ϕ. Since the coefficients of Λ are all bounded on any compact set, it follows from the last two equations in the above that

$$\sum_{r=1}^{p} a_{ri} \frac{\partial \omega_k}{\partial I_r} = 0, \quad \sum_{r=1}^{p} b_{rl} \frac{\partial \omega_k}{\partial I_r} - \sum_{r=1}^{p} b_{rk} \frac{\partial \omega_l}{\partial I_r} = 0,$$

$r, i = 1, \cdots, p$; $k, l = 1, \cdots, q$. This is the condition (4.4). Consequently, the Poisson structure Λ has the form (4.3).

By Lemma 2.5, the Poisson structure (4.3) is exact if and only if the condition (4.5) holds. This proves the necessary part.

The sufficient part follows easily from the proof for the necessary part. □

Remark 4.2 Using (4.5), the condition (4.4) can be written as

$$\nabla(\mathbf{a}_i \omega_k) = 0, \qquad \nabla(\mathbf{b}_l \omega_k - \mathbf{b}_k \omega_l) = 0,$$

$i = 1, \cdots, p$; $k, l = 1, \cdots, q$.

Remark 4.3 A $\mathbb{T}^q$-dense dynamical system (4.2) which preserves an exact Poisson structure (4.3) can be non-Hamiltonian. We note that if $\mathbf{X}$ is a Hamiltonian vector field associated to a Poisson structure Λ, then $\mathbf{X} = \Lambda(\cdot, dH)$ for some Hamiltonian function H, which is equivalent to $\operatorname{div}(B_1 H) = \omega_1$ with the compatible conditions

$$\operatorname{div}(A_i H) = 0, \ \operatorname{div}((B_1 \omega_k - B_k \omega_1) H) = 0,$$

where $A_i = (a_{i1}, \ldots, a_{ip}, b_{i1}, \ldots, b_{iq})$ and $B_k = (-b_{1k}, \ldots, -b_{pk}, c_{k1}, \ldots, c_{kq})$, $i = 1, \cdots, p$, $k = 2, \cdots, q$.

We now consider the $\mathbb{T}^q$-dense dynamical system (4.2) in dimension 3.

1) Let $p = 1$ and $q = 2$. The Poisson structure (4.3) reads

$$\Lambda = b_{11}(I)\frac{\partial}{\partial I} \wedge \frac{\partial}{\partial \phi_1} + b_{12}(I)\frac{\partial}{\partial I} \wedge \frac{\partial}{\partial \phi_2} + c_{12}(I)\frac{\partial}{\partial \phi_1} \wedge \frac{\partial}{\partial \phi_2}.$$

The conditions (4.4) and (4.5) are reduced to

$$b_{11}\frac{\partial \omega_2}{\partial I} = b_{12}\frac{\partial \omega_1}{\partial I}, \tag{4.7}$$

and b_{11}, b_{12}=constants, respectively.

The system (4.2) is a Hamiltonian if and only if there exists a Hamiltonian function $H(I, \phi_1, \phi_2)$ such that

$$b_{11}\frac{\partial H}{\partial \phi_1} + b_{12}\frac{\partial H}{\partial \phi_2} = 0, \quad -b_{11}\frac{\partial H}{\partial I} + c_{12}\frac{\partial H}{\partial \phi_2} = \omega_1, \quad -b_{12}\frac{\partial H}{\partial I} - c_{12}\frac{\partial H}{\partial \phi_1} = \omega_2. \tag{4.8}$$

So, we must have

$$b_{12}\omega_1 = b_{11}\omega_2. \tag{4.9}$$

But the condition (4.7) is not sufficient to assure (4.9) in general.

However, if $b_{12}\omega_1(0) = b_{11}\omega_2(0)$, then the system (4.8) always has a solution provided that (4.7) holds, consequently the system (4.2) is Hamiltonian.

2) Let $p = 2$ and $q = 1$. We claim that the $\mathbb{T}^q$-dense dynamical system (4.2) is always Hamiltonian. Indeed, the Poisson structure (4.3) has the form

$$\Lambda = a_{12}(I)\frac{\partial}{\partial I_1} \wedge \frac{\partial}{\partial I_2} + b_{11}(I)\frac{\partial}{\partial I_1} \wedge \frac{\partial}{\partial \phi} + b_{21}(I)\frac{\partial}{\partial I_2} \wedge \frac{\partial}{\partial \phi}.$$

The conditions (4.4) and (4.5) are reduced to $a_{12}\dfrac{\partial \omega}{\partial I_i} = 0$, $i = 1, 2$, a_{12} is a constant, and

$$\frac{\partial b_{11}}{\partial I_1} + \frac{\partial b_{21}}{\partial I_2} = 0. \tag{4.10}$$

Note that the system (4.2) is Hamiltonian, i.e., $\mathbf{X} = \Lambda(\cdot, dH(I))$ for some Hamiltonian function H, if and only if $a_{12} = 0$, and

$$b_{11}(I)\frac{\partial H}{\partial I_1} + b_{21}(I)\frac{\partial H}{\partial I_2} = -\omega(I). \tag{4.11}$$

Since the condition (4.10) guarantees that the characteristic equation of (4.11) has a smooth solution, we can choose suitable b_{11}, b_{12} for which (4.11) has a global smooth solution $H(I)$. This means that we can choose a Poisson structure Λ such that the dynamical system (4.2) is Hamiltonian induced by Λ.

It is an open problem to *characterize exact Poisson structures invariant under a* $\mathbb{T}^q$*-dense dynamical system such that the dynamical system is Hamiltonian induced by the Poisson structures.*

References

[1] R. Abraham and J.E. Marsden, *Foundations of Mechanics*, 2nd edition, The Benjamin/Cummings publishing Company, Reading, Massachusetts, 1978.

[2] R. Bott and L.W. Tu, *Differential Forms in Algebraic Topology*, Spring-Verlag, New York, 1982.

[3] A. Ballesteros and O. Ragnisco, A systematic construction of completely integrable Hamiltonians from coalgebras, *J. Phys. A* **31** (1998), 3791–3813.

[4] O. I. Bogoyavlenskij, Extended integrability and bi-Hamiltonian systems, *Commun. Math. Phys.* **196** (1998), 19–51.

[5] J. F. Conn, Normal forms for analytic Poisson structures, *Ann. Math.* **119** (1984), 577–601.

[6] J. F. CONN, Normal forms for smooth Poisson structures, *Ann. Math.* **121** (1985), 565–593.
[7] I. CRUZ AND H. MENA-MATOS, Normal forms for locally exact Poisson structures in $\mathbb{R}^3$, *J. Geom. Phys.* **43** (2002), 27–32.
[8] I. CRUZ AND H. MENA-MATOS, Normal forms for two classes of exact Poisson structures in dimension 4, preprint.
[9] J. GRABOWSKI, G. MARMO AND A. M. PERELOMOV, Poisson structures: towards a classification, *Mod. Phys. Lett. A* **8** (1993), 1719–1733.
[10] J. L. KOSZUL, Crochets de Schouten-Nijenhuis et cohomologie, *Astérisque, Soc. Math. de France* hors série (1985), 257–271.
[11] S. P. KASPERCZUK, Poisson structures and integrable systems, *Physica A* **284** (2000), 113–123.
[12] Z. LIU AND P. XU, On quadratic Poisson structures, *Lett. Math. Phys.* **26** (1992), 33–42.
[13] D. MANCHON, M. MASMOUDI AND A. ROUX, On quantization of quadratic Poisson structures, *Commun. Math. Phys.* **225** (2002), 121–130.
[14] P. J. OLVER, *Applications of Lie Groups to Differential Equations*, Spring-Verlag, New York, 1993.
[15] R. PRZYBYSZ, On one class of exact Poisson structures, *J. Math. Phys.* **42** (2001), 1913–1920.
[16] V. A. VARADARAJAN, *Lie Groups, Lie Algebras, and Their Representations*, Spring-Verlag, New York, 1984.
[17] A. WEINSTEIN, The local structure of Poisson manifolds, *J. Diff. Geometry* **18** (1983), 523–557.
[18] A. WEINSTEIN, The modular automorphism group of a Poisson manifold, *J. Geom. Phys.* **23** (1997), 379-394.

Titles in This Series